D1237235

This book applies the technique of stopping times to convergence problems in probability and analysis. There are applications in sequential analysis. Convergence of stochastic processes indexed by directed sets is studied and solutions are given to problems left open in Krickeberg's theory of martingales and submartingales. The rewording of Vitali covering conditions in terms of stopping times establishes connections with the theory of stochastic processes and derivation. A study of martingales yields laws of large numbers for martingale differences, with application to "star-mixing" processes. Convergence of processes taking values in Banach spaces is related to geometric properties of these spaces. There is a self-contained section on operator ergodic theorems: the superadditive, Chacon-Ornstein, and Chacon theorems. A unified treatment of martingale and ergodic theory is in part based on a "three-function inequality."

ENCYCLOPEDIA OF MATHEMATICS AND ITS APPLICATIONS

EDITED BY G.-C. ROTA
Volume 47

Stopping Times and Directed Processes

ENCYCLOPEDIA OF MATHEMATICS AND ITS APPLICATIONS

ENCYCLOPEDIA OF MATHEMATICS AND ITS APPLICATIONS

Stopping Times and Directed Processes

G. A. EDGAR AND LOUIS SUCHESTON

Department of Mathematics
The Ohio State University

CAMBRIDGE
UNIVERSITY PRESS

Published by the Press Syndicate of the University of Cambridge
The Pitt Building, Trumpington Street, Cambridge CB2 1RP
40 West 20th Street, New York, NY 10011-4211, USA
10 Stamford Road, Oakleigh, Victoria 3166, Australia

First published 1992

Printed in the United States of America

Library of Congress Cataloging-in-Publication Data
Edgar, Gerald A., 1949–
Stopping times and directed processes / G. A. Edgar and Louis
Sucheston.
p.   cm. – (Encyclopedia of mathematics and its applications;
v. 47)
Includes bibliographical references and index.
ISBN 0-521-35023-9
1. Convergence.   2. Probabilities.   3. Martingales.
I. Sucheston, Louis.   II. Title.   III. Series.
QA273.43.E34   1992
519.2'6 – dc20   91-44388
                                            CIP

A catalog record for this book is available from the British Library.

ISBN 0-521-35023-9 hardback

Contents

Keywords are shown to provide a fuller description of what is in a section.

Preface

The main themes of this book are: stochastic, almost sure, and essential convergence; stopping times; martingales and amarts; processes indexed by directed sets, multiparameter processes, and Banach-valued processes.

We begin in Chapter 1 with the notion of the stopping time, central to the book. That this notion is important in continuous parameter martingale theory and sequential analysis (briefly touched on in Chapter 3) is well known. This book differs from others in that many of the discrete parameter results are proved via processes (amarts) defined in terms of stopping times—in fact only simple stopping times. The Radon-Nikodým Theorem receives an amart proof. That this theorem follows from the martingale theorem is well known, but here martingales come later, so that the Radon-Nikodým theorem is available to define the conditional expectation and the martingale. In Chapter 4, the rewording of the Vitali covering conditions in terms of stopping times clarifies connections with the theory of stochastic processes.

The main result, the *Amart Convergence Theorem*, is proved by elementary arguments. This—together with a general *Sequential Sufficiency Theorem* (1.1.3), showing how in metric spaces the convergence of increasing sequences implies that of nets—is used to obtain stochastic convergence of L_1-bounded ordered amarts on directed sets. This in turn implies stochastic convergence of quasimartingales—even on directed sets. Quasimartingales include the L_1-bounded submartingales studied by Krickeberg on directed sets. We believe the proofs by this method are shorter than those existing in the literature.

Chapter 2 is a self-contained presentation of Orlicz spaces. The emphasis is on the "heart" of an Orlicz space—the closure of the integrable simple functions. The largest Orlicz space is $L_1 + L_\infty$, and its heart R_0 offers the right setting for much of the ergodic and martingale theory in infinite measure spaces. The union of the L_p-spaces, $p \geq 1$, used for this purpose in some books (e.g., Dellacherie & Meyer [1982], p. 34), is properly contained in R_0. In multiparameter theory, the space R_k, the heart of the Orlicz space $L \log^k L$, plays a major role, as shown in Chapter 9. Except for Fava's multiparameter version of the Dunford-Schwartz theorem in Krengel [1985], this is the first book to treat the theory of spaces R_k.

Chapter 3 deals with inequalities. We claim to offer in this book a partially unified treatment of martingale and ergodic theory. The one-parameter element of the unification is a three-function maximal inequality. Assuming a weak inequality, the passage to the strong maximal inequality relates appropriate Orlicz spaces and their hearts, and provides a unified approach that can be used also in harmonic analysis. We do not offer a unified treatment of weak inequalities, usually the most difficult part of a convergence proof. The discussion of a sharp inequality for martingale transforms due to Burkholder [1986] provides an illuminating probabilistic insight. "Prophet inequalities" have an interesting probabilistic interpretation, and seem to fit in the present context, since both stopping times and transforms of processes are considered.

Chapter 4 deals with stochastic processes indexed by directed sets. Convergence of martingales and submartingales was studied by Krickeberg in the fifties and early sixties. An exposition is given in the memoir of Krickeberg & Pauc [1963] and in the book of Hayes & Pauc [1970]. In that convergence theory, three basic problems remained open: Is the Vitali condition $(V) = (V_\infty)$ necessary for convergence of L_1-bounded martingales? Is the Vitali condition $(V^O) = (V')$ necessary for convergence of L_1-bounded submartingales? Is the Vitali condition (V_p) necessary for convergence of L_q-bounded martingales? (Here $1 < p < \infty$ and $1/p + 1/q = 1$.) In all three cases, the sufficiency was established by Krickeberg. All three questions were answered around 1980: the first two in the negative and the third in the positive. This is the first book that contains an account of this progress, and of subsequent developments. Classes of processes were introduced for whose convergence the conditions (V) and (V^O) are necessary and sufficient—amarts and ordered amarts. Amarts may be applied in situations in derivation theory where martingales and submartingales do not suffice. Finally, a covering condition (C) is exhibited, both sufficient and (in the presence of a countable cofinal subset) necessary for convergence of L_1-bounded martingales.

The analogous progress in the classical derivation theory is presented in Chapter 7. Here the necessity of the analog of the condition (V_p)—and an Orlicz space generalization—was earlier obtained by C. A. Hayes. Chapter 7 also contains a self-contained presentation of the basics of abstract derivation theory. Similarities between derivation theory and martingale theory on directed sets are easily seen.

Chapter 5 deals with Banach spaces. Knowledge of the elements of the theory of Banach spaces is required only for this chapter and parts of Chapter 8. The approach to convergence theorems is geometrical in that we consider convergence theorems that exactly match geometric properties of a Banach space. The best known of these theorems is the characterization of the Radon-Nikodým property due to A. & C. Ionescu Tulcea and S. D. Chatterji. Other more recent geometric characterizations of convergence of martingales are also discussed (5.3.30 and 5.3.34). There are also geometric properties characterized in terms of amarts, such as the Radon-Nikodým property for the dual of the Banach space and reflexivity. Among applications of convergence of martingales we list the Choquet-Edgar integral representation theorem, and the Ryll-Nardzewski fixed point theorem. The amart theory led to the following result that can be stated without amarts:

scalar convergence of stochastic processes implies weak a.s. convergence if and only if the dual Banach space has the Radon-Nikodým property. We regret that this book does not include the important characterization of unconditionality of L_p bases and convergence of martingale transforms in terms of Burkholder's biconvexity—space limitations precluded it.

Chapter 6 deals with martingales: maximal inequalities, L_p laws of large numbers for martingale differences (extended to a class of mixing processes), decompositions, convergence of transforms of martingales, Burkholder's square-function inequalities. The Maharam lifting theorem is derived from the martingale theorem.

In Chapter 8 we derive L_1 ergodic theorems (the general Chacon-Ornstein theorem, the Chacon theorem) from the superadditive ratio ergodic theorem. The identification of the limit by this method is particularly convenient.

Chapter 9 provides a second instance of the unity of ergodic and martingale theory in our treatment. A general principle is proved that shows how multiparameter convergence theorems may follow from one-parameter maximal and convergence theorems, provided that the function space has the property called "order continuous norm"—the L_p spaces have it, and Orlicz spaces have it under (Δ_2). *Hearts* of Orlicz spaces always have order continuous norm. We believe that nearly all the known multiparameter ergodic and martingale theorems with convergence to infinity over "rectangles"—the indices converge to infinity independently—can be obtained by this method. Multiparameter maximal theorems are not needed anymore for convergence.

This is a research book in that most of the content has previously been published only in research articles, and there is some novelty in presentation. On the other hand, we believe that our approach is particularly simple and economical and therefore the book is of some pedagogical interest. In a graduate course taught to students with knowledge of measure theory only, we would cover Chapter 1, some basic results from Chapter 2, most of Chapters 3 and 6, essential convergence of martingales from the beginning of Chapter 4, derivation from the beginning of Chapter 7, and the conservative Chacon-Ornstein theorem from Chapter 8. The book could also be used for a more advanced topics course in probability, ergodic theory (Chapters 8 and 9), derivation (Chapters 4 and 7), or Banach spaces (Chapter 5). Any such course would begin with Chapter 1 and the main results from Chapters 2 and 3.

We are grateful to our many colleagues, collaborators, and students whose work appears here: especially Mustafa Akcoglu, Ken Astbury, Don Austin, S. N. Bagchi, Alexandra Bellow, Antoine Brunel, R. V. Chacon, Bong Dae Choi, Leo Egghe, Hans Föllmer, Jean-Pierre Fouque, Nikos Frangos, Nassif Ghoussoub, Ulrich Krengel, Michael Lin, Annie Millet, László Szabó, Michel Talagrand, and Zhongde Yan. Thanks are due to the National Science Foundation for support of some of the research that now appears in the book. We wish to thank the Department of Mathematics of Ohio State University for bringing us together in the first place, and providing the atmosphere that made the book possible. Thanks are also due to Cambridge University Press, especially editors David Tranah and Lauren Cowles, for their patience and help. Finally, we wish to express our deep gratitude to Annie Millet, who has—as the reader for the Cambridge University Press—corrected many errors and made suggestions for improvements.

Columbus, Ohio   G. A. Edgar
January, 1992   Louis Sucheston

1

Stopping times

We will begin with the material that will be used throughout the book. The idea of the *stopping time*, especially the simple stopping time, is central. The setting in which this naturally occurs involves Moore-Smith convergence, or convergence of nets or generalized sequences. This will be useful even if we are interested only in sequences of real-valued random variables; but will be even more useful when we consider derivation (Chapter 7) and processes indexed by directed sets (Chapter 4).

Given a stochastic process (X_n), a *stopping time* is a random variable τ taking values in $\mathbb{N} \cup \{\infty\}$ such that, for each k, the event $\{\tau = k\}$ is determined by the first k random variables $X_1, X_2, \cdots, X_k$. A process (X_n) is an *amart* iff for every increasing sequence τ_n of bounded stopping times, $\mathbf{E}[X_{\tau_n}]$ converges. (For variants of this definition, see Section 1.2.) The main result of this chapter is the amart convergence theorem for the index set $\mathbb{N}$, proved in Section 1.2. The argument, using stopping times, is elementary, and may be followed by a reader with only a basic knowledge of the measure theory. To make the point, we will sketch the proof of almost sure convergence of an amart (X_n) with integrable supremum. The basic observation is that there is an increasing sequence τ_n of simple stopping times such that X_{τ_n} converges in probability to $X^* = \limsup X_n$. The reason for this is that lim sup—or any other accumulation point—manifests itself infinitely often on the way to infinity; it is like a light shining on the horizon. (This is not true of the supremum, which can all too easily be missed.) Thus, after we obtain X_{τ_1} close in probability to X^*, we can find $\tau_2 > \tau_1$ such that X_{τ_2} is even closer to X^*, etc. See (1.2.4). Similarly, we obtain an increasing sequence σ_n of simple stopping times such that X_{σ_n} converges in probability to $X_* = \liminf X_n$. Integrating X_{τ_n} and X_{σ_n} and using the amart property, we can conclude that $X^* = X_*$. The a.s. convergence of L_1-bounded amarts follows by truncation or stopping: The class of L_1-bounded amarts is closed under both operations.

First we prove a general principle (1.1.3), showing how in metric spaces the behavior of sequences determines that of nets. From the amart convergence theorem, using (1.1.3), we easily obtain stochastic convergence of L_1-bounded ordered amarts indexed by directed sets, which in turn implies Krickeberg's theorem that L_1-bounded submartingales converge stochastically. Quasimartingales are defined on directed sets, and are also shown to be ordered amarts, so that they also converge stochastically, and— under proper assumptions—essentially. To understand the difference between martingales, submartingales, quasimartingales, and amarts, assume the index set in $\mathbb{N}$, and consider a deterministic process (Ω has only one element). Then amarts are exactly convergent sequences of numbers; thus it is not surprising that also in general amarts will be sufficient for convergence, and for the existence of limits in the continuous parameter case (Edgar & Sucheston [1976b]). A deterministic quasimartingale is a sequence (or function) of bounded variation; the analogy of the Lebesgue-Stieltjes integral rightly indicates that the full notion of quasimartingale will be required for stochastic integration.

The sequential sufficiency theorem (1.1.3) is especially adapted to the needs of this book, first because it does not require convergence of *all* increasing sequences, and second because it does not assume that the metric space is complete, permitting in (5.3.37) a sequential proof of convergence in Pettis norm of amarts with values in a Banach space.

Another important idea of the book is *derivation*. As an illustration of a derivation theorem, Section 1.3 contains a proof of the Radon-Nikodým theorem. Other derivation theorems will be seen later in the book.

Section 1.4 discusses the conditional expectation. After it has been defined, the conventional definition for martingale can be seen to be equivalent to the definition in terms of stopping times.

1.1. Definitions

Directed sets

We begin with some of the basic definitions of our subject. The first goal is the discussion of stopping times. As a preliminary to that we consider Moore-Smith convergence.

(1.1.1) Definition. A *directed set* is a nonempty set J, together with a binary relation $\leq$ satisfying:

    (1) if $t_1 \leq t_2$ and $t_2 \leq t_3$, then $t_1 \leq t_3$;
    (2) $t \leq t$ for all $t \in J$;
    (3) if $t_1, t_2 \in J$, then there exists $t_3 \in J$ with $t_1 \leq t_3$ and $t_2 \leq t_3$.

A directed set is sometimes called "filtering to the right" by French writers. When a directed set is specified, we have in mind the "direction" shown by the ordering: when t gets larger according to the ordering, then t is getting closer to the goal we are interested in. The most familiar example of a directed set is the set $\mathbb{N}$ of positive integers with its usual ordering.

A set J with a relation $\leq$ is a *dual directed set* (or "filtering to the left") iff the opposite relation $\leq'$ defined by

$$t_1 \leq' t_2 \quad \Longleftrightarrow \quad t_2 \leq t_1$$

makes J a directed set. When we refer to a dual directed set, we are interested in the "direction" specified so that t gets closer to the goal when it gets smaller according to the ordering $\leq$. For example, the set $\mathbb{N}$ of all positive integers is a directed set. Larger numbers are considered "closer to ∞"; so we are likely to be interested in limits of the form

$$\lim_{n \to \infty} \quad .$$

The set $-\mathbb{N}$ is a dual directed set; we will use it when we are interested in limits of the form

$$\lim_{n \to -\infty} \quad .$$

A *net* (or *generalized sequence*) is a function whose domain is a directed set. Normally we will write $(a_t)_{t \in J}$ for the net f defined by $f(t) = a_t$. Convergence can be defined using any directed set in much the same way that it is defined for sequences indexed by $\mathbb{N}$. Convergence of these nets, or generalized sequences, is often called "Moore-Smith convergence."

(1.1.2) Definition. Let J be a directed set, let S be a topological space, and let $(a_t)_{t \in J} \subseteq S$ be a net in S. We say that (a_t) *converges* to a point $a \in S$ iff, for every neighborhood U of a, there exists $t_0 \in J$ such that for all $t \geq t_0$, we have $a_t \in U$.

If S is a metric space, with metric ρ, then this may be restated like this: the net $(a_t) \subseteq S$ converges to $a \in S$ iff, for every $\varepsilon > 0$, there exists $t_0 \in J$ such that $\rho(a_t, a) < \varepsilon$ for all $t \geq t_0$. Notation:

$$\lim_{t \in J} a_t = a;$$

the "direction" in J is understood.

A few of the elementary properties of nets are stated in the Complements, (1.1.8)–(1.1.11), below. However, a few facts that will be used often are proved here.

A net $(a_t)_{t \in J}$ in a metric space (S, ρ) is a *Cauchy net* iff, for every $\varepsilon > 0$, there exists $t_0 \in J$ such that for all $t \geq t_0$ we have $\rho(a_t, a_{t_0}) < \varepsilon$. Every convergent net is a Cauchy net. In general, the converse is false: in the metric space of rational numbers, even some Cauchy sequences fail to converge. A metric space S is called *complete* iff every Cauchy sequence converges in S. In fact, this implies also that every Cauchy net converges in S. This is part of the next result:

(1.1.3) Sequential sufficiency theorem. *Let S be a metric space with metric ρ and let (a_t) be a net in S. Suppose either:*

(1) *(a_{t_n}) converges in S for every increasing sequence (t_n) in J; or*
(2) *there exist indices $s_n \in J$ such that the sequence $(a_{t_n})_{n \in \mathbb{N}}$ converges in S for every increasing sequence (t_n) in J with $t_n \geq s_n$ for all n; or*
(3) *(a_t) is a Cauchy net and S is complete.*

Then the net (a_t) converges.

Proof. Hypothesis (1) implies hypothesis (2), so case (1) of the theorem follows from case (2). It is therefore enough to prove cases (2) and (3).

We show now that if (a_t) satisfies (2), then it is a Cauchy net. Indeed, if (a_t) is not a Cauchy net, then there exists $\varepsilon > 0$ so that for every $t_0 \in J$, there exists $t \geq t_0$ with $\rho(a_{t_0}, a_t) \geq 2\varepsilon$. We show there are indices $t_0 > s_0$, $t_1 > s_1$ such that $\rho(a_{t_0}, a_{t_1}) > \varepsilon$. Indeed, there is t_1' with $t_1' \geq t_0$ and $t_1' \geq s_1$. If $\rho(a_{t_0}, a_{t_1'}) > \varepsilon$, set $t_1 = t_1'$ and we are done. If not, $\rho(a_{t_0}, a_{t_1'}) \leq \varepsilon$, there is $t_1 > t_1'$ (hence $t_1 > s_1$) such that $\rho(a_{t_1}, a_{t_1'}) > 2\varepsilon$. Then $\rho(a_{t_0}, a_{t_1}) \geq \rho(a_{t_1}, a_{t_1'}) - \rho(a_{t_0}, a_{t_1'}) \geq 2\varepsilon - \varepsilon = \varepsilon$. Since the index set J is directed, there exists an element $t_1 \in J$ such that $t_1 \geq t_0$, $t_1 \geq s_1$, and $\rho(a_{t_0}, a_{t_1}) \geq \varepsilon$. This can be applied recursively to define a sequence $t_1 \leq t_2 \leq \cdots$ such that $t_n \geq s_n$ and $\rho(a_{t_n}, a_{t_{n+1}}) \geq \varepsilon$ for all n, contradicting (2).

Now suppose (a_t) is a net satisfying either (2) or (3). Since (a_t) is a Cauchy net, we may construct recursively a sequence $t_1 \leq t_2 \leq \cdots$ such that for all $t \geq t_n$, we have $\rho(a_{t_n}, a_t) < 2^{-n}$. In case (2), again since J is directed, we may choose t_n so that in addition $t_n \geq s_n$. Now $(a_{t_n})_{n \in \mathbb{N}}$ converges in S; in case (2) this is true by hypothesis, and in case (3) since (a_{t_n}) is a Cauchy sequence and S is complete. Write a for the limit of the sequence (a_{t_n}). We claim that, in fact, the entire net (a_t) converges to a. To see this, note that $\rho(a_{t_m}, a_{t_n}) < 2^{-m}$ for all $n \geq m$; so $\rho(a_{t_m}, a) \leq \rho(a_{t_m}, a_{t_n}) + \rho(a_{t_n}, a) < 2^{-m} + \rho(a_{t_n}, a)$. The last term can be made as small as we like, so $\rho(a_{t_m}, a) < 2^{-m}$. Thus, for all $t \geq t_m$, we have $\rho(a, a_t) \leq \rho(a, a_{t_m}) + \rho(a_{t_m}, a_t) < 2^{-m} + 2^{-m} = 2^{-m+1}$. ∎

Stochastic basis

Let $(\Omega, \mathcal{F}, \mathbf{P})$ be a probability space; that is: Ω is a nonempty set, $\mathcal{F}$ is a σ-algebra of subsets of Ω, and $\mathbf{P}$ is a countably additive, non-negative measure on $\mathcal{F}$ with $\mathbf{P}(\Omega) = 1$. A sequence $(\mathcal{F}_n)_{n \in \mathbb{N}}$ of sub-σ-algebras of $\mathcal{F}$ is a *stochastic basis* iff it satisfies the monotonicity condition: if $m \leq n$, then $\mathcal{F}_m \subseteq \mathcal{F}_n$.

A probabilist refers to Ω as the *sample space*, the elements of $\mathcal{F}$ as *events*, and the number $\mathbf{P}(A)$ as the *probability* of the event A. In this terminology, the index set $\mathbb{N}$ can be considered to be "time," and $\mathcal{F}_n$ is the collection of events "prior to time n" (in the wide sense: $\leq n$). For an event $A \in \mathcal{F}_n$, if we are given everything that is known at time n, then we should be able to determine whether or not event A has occurred.

The data $(\Omega, \mathcal{F}, \mathbf{P})$ and $(\mathcal{F}_n)_{n \in \mathbb{N}}$ will be fixed throughout most of the book.

Suppose a stochastic basis $(\mathcal{F}_n)$ is given. A sequence $(X_n)_{n \in \mathbb{N}}$ of random variables is said to be *adapted* to the stochastic basis iff X_n is $\mathcal{F}_n$-measurable for all $n \in \mathbb{N}$. Such an adapted sequence (X_n) of (real-valued, finite) random variables may be called a *stochastic process* indexed by $\mathbb{N}$.

If a sequence (X_n) of random variables is given, but not a stochastic basis, there is a canonical way to construct a stochastic basis: the σ-algebra $\mathcal{F}_n$ should be the least σ-algebra of subsets of Ω so that the random variables $X_1, X_2, \cdots, X_n$ are measurable. We write $\mathcal{F}_n = \sigma(X_1, X_2, \cdots, X_n)$ for this σ-algebra.

Stopping times

A function $\sigma \colon \Omega \to \mathbb{N} \cup \{\infty\}$ is a *stopping time* [for the stochastic basis $(\mathcal{F}_n)$] iff it satisfies

$$\{\sigma = n\} \in \mathcal{F}_n \qquad \text{for all } n \in \mathbb{N}.$$

An equivalent definition is:

$$\{\sigma \le n\} \in \mathcal{F}_n \qquad \text{for all } n \in \mathbb{N}.$$

To see that this is equivalent, note that

$$\{\sigma \le n\} = \bigcup_{k=1}^{n} \{\sigma = k\}$$

$$\{\sigma = n\} = \{\sigma \le n\} \setminus \{\sigma \le n-1\}.$$

In the probabilistic interpretation, the idea is that it should be known at time n whether $\sigma = n$ is true or not; that is, it should be possible to tell at any given time, without knowledge of the future, whether or not time σ has arrived. We say σ is a *simple* stopping time iff it has finitely many values and does not take the value ∞.

The set of all simple stopping times for the stochastic basis $(\mathcal{F}_n)$ will be denoted $\Sigma\big((\mathcal{F}_n)\big)$, or simply Σ when the stochastic basis is understood.

The set Σ has a natural ordering. If $\sigma, \tau \in \Sigma$, we say that $\sigma \le \tau$ iff $\sigma(\omega) \le \tau(\omega)$ for almost every $\omega \in \Omega$. We may say that σ happens *before* τ. The constants from $\mathbb{N}$ belong to Σ. In fact, the constants are *cofinal* in Σ in the sense that for any $\sigma \in \Sigma$, there is a constant $m \in \mathbb{N}$ with $\sigma \le m$. (Recall that the elements of Σ are *simple* stopping times.) The ordered set Σ is directed. In fact, if $\sigma, \tau \in \Sigma$, then there is a constant m with $\sigma \le m$ and $\tau \le m$. This makes it possible to consider Moore-Smith convergence for nets indexed by Σ. We will do this in Section 1.2 below.

We will write down a few of the elementary properties of the stopping times. They are mainly common sense. For example, if I can tell when σ happens and I can tell when τ happens, then I can tell when the first of the two happens.

(1.1.4) Proposition. *If $\sigma, \tau \in \Sigma$, then the maximum $\sigma \vee \tau$ and the minimum $\sigma \wedge \tau$ also belong to Σ.*

Proof. Let $n \in \mathbb{N}$ be given.

$$\{\sigma \vee \tau \le n\} = \{\sigma \le n\} \cap \{\tau \le n\},$$

$$\{\sigma \wedge \tau \le n\} = \{\sigma \le n\} \cup \{\tau \le n\}. \qquad \blacksquare$$

If $\tau \in \Sigma$, then the σ-algebra $\mathcal{F}_\tau$ of "events prior to time τ" may also be defined. An event A happens prior to time τ iff, when time τ arrives, I know whether A has happened. Technically,

$$\mathcal{F}_\tau = \big\{ A \in \mathcal{F} : \{\tau = n\} \cap A \in \mathcal{F}_n \text{ for all } n \in \mathbb{N} \big\}.$$

It is easy to see that this defines a σ-algebra. It can be easily checked that if τ is the constant m, then $\mathcal{F}_\tau = \mathcal{F}_m$. Also, if $\sigma \le \tau$, then $\mathcal{F}_\sigma \subseteq \mathcal{F}_\tau$.

If two stopping times σ, τ are given, with $\sigma \le \tau$, then a new stopping time may be defined in terms of an event $A \in \mathcal{F}_\sigma$: wait until time σ, and stop then if A has happened, otherwise continue waiting until time τ. Here it is written formally:

(1.1.5) Waiting lemma. *Let $\sigma, \tau \in \Sigma$ with $\sigma \leq \tau$, and let $A \in \mathcal{F}_\sigma$. If ρ is defined by*

$$\rho(\omega) = \begin{cases} \sigma(\omega) & \text{if } \omega \in A \\ \tau(\omega) & \text{if } \omega \notin A, \end{cases}$$

then $\rho \in \Sigma$, $\sigma \leq \rho \leq \tau$, and

(1.1.5a) $$\mathbf{1}_A \left(X_\sigma - X_\tau \right) = X_\rho - X_\tau.$$

Proof. Let $n \in \mathbb{N}$ be given. We must show that $\{\rho = n\} \in \mathcal{F}_n$. Now $A \in \mathcal{F}_\sigma$, so we have $\{\sigma = n\} \cap A \in \mathcal{F}_n$. Since $\sigma \leq \tau$, we have $\mathcal{F}_\sigma \subseteq \mathcal{F}_\tau$, so the complementary set $\Omega \setminus A$ belongs to $\mathcal{F}_\tau$, and $\{\tau = n\} \setminus A \in \mathcal{F}_n$. Therefore

$$\{\rho = n\} = \left(\{\sigma = n\} \cap A \right) \cup \left(\{\tau = n\} \setminus A \right) \in \mathcal{F}_n.$$

For equation (1.1.5a), consider separately what happens on A and on $\Omega \setminus A$. ∎

Optional stopping

Let (X_n) be an adapted stochastic process. If σ is a random variable with values in $\mathbb{N}$, then it makes sense to talk about X_σ, that is, the value that the process has at the time σ. Technically, X_σ is a random variable given by

$$\left(X_\sigma \right)(\omega) = X_{\sigma(\omega)}(\omega).$$

If X_n represents the fortune of a gambler at time n, and he has the option of stopping whenever he wants to (but does not know the future), then the time he chooses to stop must be a "stopping time" as it has been defined. This means that X_σ is his fortune when he stops. Then X_σ not only is a random variable, but is $\mathcal{F}_\sigma$-measurable:

(1.1.6) Proposition. *Let $\sigma \in \Sigma$ and let (X_n) be an adapted process. Then X_σ is $\mathcal{F}_\sigma$-measurable.*

Proof. For all $n \in \mathbb{N}$ and all $\lambda \in \mathbb{R}$,

$$\{\sigma = n\} \cap \{X_\sigma \leq \lambda\} = \{\sigma = n\} \cap \{X_n \leq \lambda\} \in \mathcal{F}_n.$$

Therefore $\{X_\sigma \leq \lambda\} \in \mathcal{F}_\sigma$, so X_σ is $\mathcal{F}_\sigma$-measurable. ∎

A useful consequence of this proposition is the possibility of "optional sampling." Let (X_n) be a process adapted to the stochastic basis $(\mathcal{F}_n)$. Let $\sigma_1 \leq \sigma_2 \leq \cdots$ be an increasing sequence of stopping times. Then the *sampled* process is the sequence $Y_k = X_{\sigma_k}$ of random variables, which is adapted to the stochastic basis $(\mathcal{G}_k)_{k \in \mathbb{N}}$ defined by $\mathcal{G}_k = \mathcal{F}_{\sigma_k}$.

The random variable $\sup_n |X_n|$ is known as a "maximal function." The next result is our first example of an inequality for the distribution of the maximal function; other maximal inequalities will be seen later.

(1.1.7)   Maximal inequality. *Let (X_n) be an adapted sequence of random variables. Then for every $\lambda > 0$,*

$$\mathbf{P}\left\{\sup_{n\in\mathbb{N}}|X_n| > \lambda\right\} \leq \frac{1}{\lambda}\sup_{\sigma\in\Sigma}\mathbf{E}\left[|X_\sigma|\right].$$

Proof. Let $N \in \mathbb{N}$ and define $A_N = \{\max_{n\leq N}|X_n| > \lambda\}$. Then clearly $A_N \in \mathcal{F}_N$. Define

$$\tau(\omega) = \begin{cases} \min\left\{n \in \mathbb{N} : 1 \leq n \leq N, |X_n(\omega)| > \lambda\right\} & \text{if } \omega \in A_N, \\ N & \text{if } \omega \notin A_N. \end{cases}$$

Then $\tau \in \Sigma$. Now

$$(1.1.7a) \qquad \sup_{\sigma\in\Sigma}\mathbf{E}\left[|X_\sigma|\right] \geq \mathbf{E}\left[|X_\tau|\right] \geq \mathbf{E}\left[|X_\tau|\mathbf{1}_{A_N}\right] \geq \lambda\mathbf{P}(A_N).$$

Now as $N \to \infty$, the set A_N increases to $\{\sup_n |X_n| > \lambda\}$, so the result follows by the monotone convergence theorem. ∎

Complements

(1.1.8) (Uniqueness of limits.) Let (a_t) be a net in a metric space S. Then (a_t) converges to at most one point of S.

(1.1.9) (Subnets.) Let J and L be two directed sets. A function $\varphi\colon L \to J$ is called *cofinal* iff, for every $t \in J$ there exists $s_0 \in L$ such that $\varphi(s) \geq t$ for all $s \geq s_0$. If $(a_t)_{t\in J}$ is a net in S, then the net $(b_s)_{s\in L}$ is a *subnet* of (a_t) iff there is a cofinal function $\varphi\colon L \to J$ such that $b_s = a_{\varphi(s)}$ for all $s \in L$. Prove: if (a_t) converges to $a \in S$, then the subnet (b_s) also converges to a.

(1.1.10) (Fréchet property.) Let (a_t) be a net in a topological space S. Let $a \in S$. Suppose that every subnet (b_s) of (a_t) admits a further subnet (c_u) that converges to a. Then (a_t) itself converges to a. (See, for example, Kelley [1955], Chapter 2.)

(1.1.11) (Boundedness.) A convergent real-valued net need not be bounded.

(1.1.12) (Measurable approximation lemma.) Let $(\mathcal{F}_n)$ be a stochastic basis. Write $\mathcal{F}_\infty$ for the σ-algebra generated by the union $\bigcup_{n\in\mathbb{N}}\mathcal{F}_n$. Let Y be an $\mathcal{F}_\infty$-measurable random variable. Then for every $\varepsilon > 0$, there exist an integer $n \in \mathbb{N}$ and a random variable X, measurable with respect to $\mathcal{F}_n$, such that $\mathbf{P}\{|X - Y| > \varepsilon\} < \varepsilon$. To prove this, observe successively that the set of random variables Y with this approximation property: (a) is a linear space; (b) is closed under pointwise a.e. convergence; (c) contains indicator functions $\mathbf{1}_B$ for $B \in \mathcal{F}_n$; (d) contains $\mathbf{1}_B$ for $B \in \mathcal{F}_\infty$; (e) contains all random variables measurable with respect to $\mathcal{F}_\infty$.

(1.1.13) (Optional stopping of the stochastic basis.) If $(\mathcal{F}_n)$ is a stochastic basis, and $\tau \in \Sigma$, then $\mathcal{F}_\tau$ is a σ-algebra.

(1.1.14) (Monotonicity of stopping.) If $\sigma \leq \tau$, then $\mathcal{F}_\sigma \subseteq \mathcal{F}_\tau$.

(1.1.15) (Generalized waiting lemma.) Let $\sigma \in \Sigma$ be given, and for each $m \in \mathbb{N}$ with $\{\sigma = m\} \neq \emptyset$, let $\tau^{(m)} \in \Sigma$ be given with $\tau^{(m)} \geq m$ on $\{\sigma = m\}$. Then τ defined by $\tau(\omega) = \tau^{(m)}(\omega)$ on $\{\sigma = m\}$ belongs to Σ and $\tau \geq \sigma$.

(1.1.16) (Weak L_1.) A random variable Y is said to belong to "weak L_1" iff there is a constant C such that $\mathbf{P}\{|Y| > \lambda\} \leq C/\lambda$ for all $\lambda > 0$. Prove that all L_1 functions belong to weak L_1, but that the converse is false.

(1.1.17) (Maximal weak L_1.) Suppose (X_n) is a stochastic process such that

$$\sup_{\sigma \in \Sigma} \mathbf{E}\left[|X_\sigma|\right] < \infty.$$

Then (by the maximal inequality 1.1.7) the corresponding maximal function $\sup_n |X_n|$ belongs to weak L_1. Show that the maximal function need not belong to L_1.

(1.1.18) (Reversed stochastic basis.) A *reversed* stochastic basis on $(\Omega, \mathcal{F}, \mathbf{P})$ is a family $(\mathcal{F}_n)_{n \in -\mathbb{N}}$ of σ-algebras satisfying the monotonicity condition

$$\mathcal{F} \supseteq \mathcal{F}_{-1} \supseteq \mathcal{F}_{-2} \supseteq \mathcal{F}_{-3} \supseteq \cdots .$$

Simple stopping times are defined as before; we will continue to write Σ for the set of all simple stopping times. (It is now a dual directed set.)

Prove the maximal inequality in this setting: Let $(X_n)_{n \in -\mathbb{N}}$ be a sequence of random variables adapted to the reversed stochastic basis $(\mathcal{F}_n)$. Then for every $\lambda > 0$,

$$\mathbf{P}\left\{\sup_{n \in -\mathbb{N}} |X_n| > \lambda\right\} \leq \frac{1}{\lambda} \sup_{\sigma \in \Sigma} \mathbf{E}\left[|X_\sigma|\right]$$

(Edgar & Sucheston [1976a]).

Remarks

Moore-Smith convergence was proposed by Moore [1915] and Moore & Smith [1922]. Its usefulness in general topology was displayed by Birkhoff [1937], and carried out in detail by Kelley [1950]. A particular case of the sequential sufficiency theorem 1.1.3 was proved by Neveu [1975]: A family indexed by a directed set converges in a complete metric space if there is convergence for all increasing sequences.

The importance of stopping times was emphasized by Doob [1953].

The maximal inequality 1.1.7 is from Chacon & Sucheston [1975].

1.2. The amart convergence theorem

Let a probability space $(\Omega, \mathcal{F}, \mathbf{P})$ and a stochastic basis $(\mathcal{F}_n)$ be fixed. We will write, as before, Σ for the set of all simple stopping times for $(\mathcal{F}_n)$. If (X_n) is a stochastic process, and $\sigma \in \Sigma$, then the stopped random variable X_σ makes sense. If X_n is integrable for all n, then of course X_σ is integrable, since σ has only finitely many values.

An adapted sequence (X_n) of integrable random variables is called an *amart* iff the net $\big(\mathbf{E}[X_\sigma]\big)_{\sigma \in \Sigma}$ of real numbers converges. That is, there is a real number a with the property: for every $\varepsilon > 0$, there exists $\sigma_0 \in \Sigma$ such that, for all $\sigma \in \Sigma$ with $\sigma \geq \sigma_0$, we have

$$\big|\mathbf{E}[X_\sigma] - a\big| < \varepsilon.$$

Since the metric for the real line is complete, this is equivalent to a Cauchy condition: for every $\varepsilon > 0$, there exists $\sigma_0 \in \Sigma$ such that, for all $\sigma, \tau \geq \sigma_0$, we have $\big|\mathbf{E}[X_\sigma] - \mathbf{E}[X_\tau]\big| < \varepsilon$.

The definition may, in turn, be phrased in terms of sequences only: The integrable stochastic process (X_n) is an amart if and only if, for every sequence $\sigma_1 \leq \sigma_2 \leq \cdots$ in Σ with $\sigma_n \to \infty$, the sequence $\big(\mathbf{E}[X_{\sigma_n}]\big)_{n \in \mathbb{N}}$ converges.

If the net $\mathbf{E}[X_\tau]$ is constant, then of course it converges. An adapted sequence X_n with this property is called a *martingale*. Some other interesting examples of amarts are given below; see, for example, (1.4.4).

Observe that in the definition of amarts it is essential to use simple stopping times: otherwise the famous original gambling martingale, in which the player doubles his stakes each time he loses, would not be an amart (see Section 3.2).

The lattice property

The collection of all L_1-bounded amarts satisfies a very useful *lattice property*: if (X_n) and (Y_n) are both L_1-bounded amarts, then so are the pointwise maximum $(X_n \vee Y_n)$ and the pointwise minimum $(X_n \wedge Y_n)$. This will be proved below (1.2.2). A similar result is true for semiamarts. A stochastic process (X_n) is a *semiamart* iff $\sup_{\sigma \in \Sigma} \big|\mathbf{E}[X_\sigma]\big| < \infty$.

(1.2.1) Lemma. *Let (X_n) be an amart. Then (X_n) is a semiamart.*

Proof. The net $\mathbf{E}[X_\sigma]$ converges, so it is a Cauchy net. There is $N \in \mathbb{N}$ such that $\big|\mathbf{E}[X_N] - \mathbf{E}[X_\sigma]\big| < 1$ for all $\sigma \in \Sigma$ with $\sigma \geq N$. If σ is any simple stopping time, then $\sigma \vee N$ is a simple stopping time $\geq N$. But

$$\big|\mathbf{E}[X_{\sigma \wedge N}]\big| \leq \mathbf{E}\left[\max_{1 \leq n \leq N} |X_n|\right]$$

and $\big|\mathbf{E}[X_{\sigma \vee N} - \mathbf{E}[X_N]]\big| < 1$, so

$$\big|\mathbf{E}[X_\sigma]\big| = \big|\mathbf{E}[X_{\sigma \wedge N}] + \mathbf{E}[X_{\sigma \vee N}] - \mathbf{E}[X_N]\big|$$

$$\leq \mathbf{E}\left[\max_{1 \leq n \leq N} |X_n|\right] + 1 < \infty. \qquad \blacksquare$$

(1.2.2) Theorem.

(1) If (X_n) is an L_1-bounded semiamart, then so is (X_n^+).

(2) If (X_n) is an amart, then the limit $\lim_{\sigma \in \Sigma} \mathbf{E}[X_\sigma^+]$ exists in $[0, \infty]$.

(3) If (X_n) is an L_1-bounded amart, then so is (X_n^+).

(4) If (X_n) and (Y_n) are L_1-bounded semiamarts, then so are the processes $(X_n \vee Y_n)$ and $(X_n \wedge Y_n)$.

(5) If (X_n) and (Y_n) are L_1-bounded amarts, then so are $(X_n \vee Y_n)$ and $(X_n \wedge Y_n)$.

Proof. Given σ and τ in Σ with $\sigma \leq \tau$, by the waiting lemma (1.1.5), with $A = \{X_\sigma > 0\}$, the random variable ρ defined by

$$\rho(\omega) = \begin{cases} \sigma(\omega) & \text{if } X_\sigma(\omega) > 0 \\ \tau(\omega) & \text{if } X_\sigma(\omega) \leq 0 \end{cases}$$

is a stopping time. By (1.1.5a), we have

$$X_\rho - X_\tau = 1_A(X_\sigma - X_\tau) \geq X_\sigma^+ - X_\tau^+.$$

Taking expectation, we have

(1.2.2a) $\qquad \mathbf{E}[X_\sigma^+] - \mathbf{E}[X_\tau^+] \leq \mathbf{E}[X_\rho] - \mathbf{E}[X_\tau].$

To prove (1), let $\sigma \in \Sigma$ be arbitrary, and choose $\tau \geq \sigma$ constant, $\tau = m$. Then $\mathbf{E}[X_\sigma^+] \leq \mathbf{E}[X_\rho] + \mathbf{E}[X_\tau^-] \leq \sup_{\pi \in \Sigma} \mathbf{E}[X_\pi] + \sup_{n \in \mathbb{N}} \mathbf{E}[|X_n|]$.

To prove (2), choose $\sigma_n \in \Sigma$, $\sigma_n \to \infty$, such that

$$\mathbf{E}[X_{\sigma_n}^+] \to \limsup_{\sigma} \mathbf{E}[X_\sigma^+]$$

and then choose $\tau_n \geq \sigma_n$ such that

$$\mathbf{E}[X_{\tau_n}^+] \to \liminf_{\sigma} \mathbf{E}[X_\sigma^+].$$

Since (X_n) is an amart, we have $\mathbf{E}[X_\sigma^+] - \mathbf{E}[X_{\tau_n}^+] \to 0$ by inequality (1.2.2a). Therefore $\limsup \mathbf{E}[X_\sigma^+] = \liminf \mathbf{E}[X_\sigma^+]$, possibly both infinite.

For (3), combine (1) and (2). For (4), observe that $x \vee y = (x - y)^+ + y$ and $x \wedge y = x - (x - y)^+$; then apply (1). For (5), apply (3), using the same identities. ∎

(1.2.3) Corollary. Let $(X_n)_{n \in \mathbb{N}}$ be an L_1-bounded amart.

(a) If λ is a positive constant, then the truncation $\big((-\lambda) \vee X_n \wedge \lambda\big)_{n \in \mathbb{N}}$ is also an amart.

(b) $\sup_{\sigma \in \Sigma} \mathbf{E}[|X_\sigma|] < \infty$.

(c) $\sup_{n \in \mathbb{N}} |X_n| < \infty$ a.s.

Proof. For (a), apply "the lattice property," that is, part (5) of the theorem. For (b), observe that $|x| = 2x^+ - x$. For (c), apply the maximal inequality (1.1.7) using (b). ∎

Convergence

We are ready now for the first convergence theorem. We begin with a useful observation on approximation of cluster points by stopping times.

(1.2.4) Cluster point approximation theorem. *Let* (X_n) *be an adapted stochastic process, let* $\mathcal{F}_\infty = \sigma(\bigcup_{n \in \mathbb{N}} \mathcal{F}_n)$, *and let* Y *be an* $\mathcal{F}_\infty$-*measurable random variable. Suppose that* $Y(\omega)$ *is a cluster point of the sequence* $(X_n(\omega))_{n \in \mathbb{N}}$ *for every* $\omega \in \Omega$. *Then there exists a sequence* $\sigma_1 \leq \sigma_2 \leq \cdots$ *in* Σ *such that* $\sigma_n \to \infty$ *and* $\lim_{n \to \infty} X_{\sigma_n} = Y$ *a.s.*

Proof. Given any $N \in \mathbb{N}$ and $\varepsilon > 0$, we will construct a stopping time $\sigma \in \Sigma$ with $\sigma \geq N$ and $\mathbf{P}\{|X_\sigma - Y| \leq \varepsilon\} > 1 - \varepsilon$. This may then be applied recursively to produce an increasing sequence $\sigma_n \in \mathbb{N}$ with $\sigma_n \to \infty$ such that X_{σ_n} converges to Y stochastically (that is, "in probability"). Then there is a subsequence that converges a.s.

First, since Y is $\mathcal{F}_\infty$-measurable, by (1.1.12) there is $N' \geq N$ and an $\mathcal{F}_{N'}$-measurable random variable Y' such that

$$(1.2.4a) \qquad \mathbf{P}\left\{|Y - Y'| < \frac{\varepsilon}{2}\right\} > 1 - \frac{\varepsilon}{2}.$$

But

$$\left\{|Y - Y'| < \frac{\varepsilon}{2}\right\} \subseteq \left\{\text{there exists } n \geq N' \text{ such that } |X_n - Y'| < \frac{\varepsilon}{2}\right\}.$$

Therefore, there is an integer $N'' > N'$ such that $\mathbf{P}(B) > 1 - \varepsilon/2$, where

$$B = \left\{\text{there exists } n \text{ with } N' \leq n \leq N'' \text{ such that } |X_n - Y'| < \frac{\varepsilon}{2}\right\}.$$

Define the simple stopping time σ as follows:

$$\sigma(\omega) = \begin{cases} \inf\left\{n : N' \leq n \leq N'', |X_n(\omega) - Y'(\omega)| < \varepsilon/2\right\} & \text{if } \omega \in B \\ N'' & \text{otherwise.} \end{cases}$$

Then $\sigma \geq N$ and

$$(1.2.4b) \qquad \mathbf{P}\left\{|X_\sigma - Y'| < \frac{\varepsilon}{2}\right\} > 1 - \frac{\varepsilon}{2}.$$

Combining (1.2.4a) and (1.2.4b), we obtain $\mathbf{P}\{|X_\sigma - Y| \leq \varepsilon\} > 1 - \varepsilon.$ ∎

The main theorem is now easy.

(1.2.5) Amart convergence theorem. *Let* $(X_n)_{n \in \mathbb{N}}$ *be an* L_1-*bounded amart. Then* (X_n) *converges a.s.*

Proof. First, consider the special case where $\sup_n |X_n|$ is integrable. Let $X^* = \limsup_n X_n$ and $X_* = \liminf_n X_n$. Both of these random variables satisfy the hypothesis of the cluster point approximation theorem

(1.2.4). So there exist simple stopping times $\sigma_1 \leq \sigma_2 \leq \cdots$ with $\sigma_n \to \infty$ and $X_{\sigma_n} \to X^*$ a.s. and simple stopping times $\tau_1 \leq \tau_2 \leq \cdots$ with $\tau_n \to \infty$ and $X_{\tau_n} \to X_*$ a.s. Now by the defining property of an amart, $\lim_n \mathbf{E}\,[X_{\sigma_n}] = \lim_n \mathbf{E}\,[X_{\tau_n}]$. Finally, by the dominated convergence theorem, $\mathbf{E}\,[X^* - X_*] = \lim_n \mathbf{E}\,[X_{\sigma_n} - X_{\tau_n}] = 0$, so $X^* = X_*$ a.s.

Now consider the general L_1-bounded amart (X_n). If $\lambda > 0$ is fixed, then the truncated process $(-\lambda) \vee X_n \wedge \lambda$ is an amart that has supremum $\leq \lambda$, and therefore integrable. So by the first case, the truncated process converges a.s. Now the truncated process agrees with (X_n) itself on the set $\Omega_\lambda = \{\sup_n |X_n| \leq \lambda\}$. Therefore (X_n) converges a.s. on the set Ω_λ. By Corollary (1.2.3(c)), we have $\sup_{n \in \mathbb{N}} |X_n| < \infty$ a.s., so Ω is (up to a null set) the countable union of sets Ω_λ. Therefore (X_n) converges a.s. ∎

Complements

(1.2.6) (Optional sampling for amarts.) Let $(\mathcal{F}_n)$ be a stochastic basis, and let (X_n) be an amart for that stochastic basis. Let $\sigma_1 \leq \sigma_2 \leq \cdots$ be an increasing sequence of simple stopping times for $(\mathcal{F}_n)$. Then the process (Y_k) defined by $Y_k = X_{\sigma_k}$ is an amart for the stochastic basis $(\mathcal{G}_k)$ defined by $\mathcal{G}_k = \mathcal{F}_{\sigma_k}$ (Edgar & Sucheston [1976a], p. 199).

(1.2.7) (Reversed amarts.) Let $(\mathcal{F}_n)_{n \in -\mathbb{N}}$ be a reversed stochastic basis. The adapted family $(X_n)_{n \in -\mathbb{N}}$ of integrable random variables is an *amart* iff the net $\mathbf{E}\,[X_\sigma]$ converges according to the dual directed set Σ of simple stopping times. Here is the lattice property; note that the hypothesis if L_1-boundedness is not required: If (X_n) and (Y_n) are reversed amarts, then so are $(X_n \vee Y_n)$ and $(X_n \wedge Y_n)$ (Edgar & Sucheston [1976a]).

(1.2.8) (Reversed convergence.) Let $(X_n)_{n \in -\mathbb{N}}$ be a reversed amart. Then (X_n) is uniformly integrable (Edgar & Sucheston [1976a], Theorem 2.9) and converges a.s. and in L_1.

(1.2.9) (Chacon's inequality.) Let (X_n) be an L_1-bounded process. Then

$$\mathbf{E}\left[\limsup_{n \in \mathbb{N}} X_n - \liminf_{n \in \mathbb{N}} X_n\right] \leq \limsup_{\sigma, \tau \in \Sigma} \mathbf{E}\,[X_\sigma - X_\tau].$$

This may be proved using a variant of the argument given above for the amart convergence theorem; see Edgar & Sucheston [1976a]. It is due to Chacon [1974].

(1.2.10) (Approximation by stopping times.) Let (X_n) be an adapted integrable process. Then

$$\mathbf{E}\left[\limsup_{n \in \mathbb{N}} X_n\right] \leq \limsup_{\tau \in \Sigma} \mathbf{E}\,[X_\tau].$$

(1.2.11) (Approximation of sup.) Suppose all X_n are $\mathcal{F}_1$-measurable. Then

$$\mathbf{E}\left[\sup_n |X_n|\right] = \sup_{\sigma \in \Sigma} \mathbf{E}\,[|X_\sigma|].$$

If, in addition, (X_n) is a semiamart, then $\mathbf{E}\left[\sup_n |X_n|\right] < \infty$. (See Edgar & Sucheston [1976a], Proposition 2.4.)

(1.2.12) (Restriction lemma.) Let (X_n) be an amart, and let $A \in \mathcal{F}_m$. Then the process $(X_n \, \mathbf{1}_A)_{n \geq m}$ is an amart.

(1.2.13) (Associated charge.) Let (X_n) be an amart. Then for each $A \in \bigcup_n \mathcal{F}_n$, the limit

$$\mu(A) = \lim_{n \to \infty} \mathbf{E}\left[X_n \, \mathbf{1}_A\right]$$

exists and μ is a finitely additive set function.

Remarks

The term "amart" comes from "asymptotic martingale." Theorem 1.2.4 on approximation of limit points is from Austin, Edgar, Ionescu Tulcea [1974]. The lattice property was given explicitly in Edgar & Sucheston [1976a], but implicitly already in Austin, Edgar, Ionescu Tulcea [1974].

The amart convergence theorem was stated explicitly in Austin, Edgar, Ionescu-Tulcea [1974]. The key element of the theorem is the use of simple stopping times. Their proof used the method of "up-crossings." The proof given here, based on truncation, is from Edgar & Sucheston [1976a]. Earlier versions of this theorem and related theorems are found in Baxter [1974], Lamb [1973], Mertens [1972], Meyer [1966].

1.3. Directed processes and the Radon-Nikodým theorem

We will see many "derivation theorems" in this book. In this section we will prove the Radon-Nikodým theorem as an elementary example. For this purpose (and for its usefulness in the future), we discuss processes with index set more general than $\mathbb{N}$. In Chapter 4 we will discuss more thoroughly the theory of amarts, ordered amarts, and other processes, indexed by a directed set. But some basic results will be proved here.

Processes indexed by directed sets

Let $(\Omega, \mathcal{F}, \mathbf{P})$ be a probability space, and let J be a directed set. The family $(\mathcal{F}_t)_{t \in J}$ of σ-algebras contained in $\mathcal{F}$ is a *stochastic basis* iff it satisfies the monotonicity condition: $\mathcal{F}_s \subseteq \mathcal{F}_t$ whenever $s \leq t$ in J. The family $(X_t)_{t \in J}$ of random variables is *adapted* to $(\mathcal{F}_t)$ iff, for all $t \in J$, the random variable X_t is $\mathcal{F}_t$-measurable. A function $\sigma \colon \Omega \to J$ (a J-valued random variable) is a *simple stopping time* for $(\mathcal{F}_t)$ iff σ has finitely many values and $\{\sigma = t\} \in \mathcal{F}_t$ for all $t \in J$. (Because σ has finitely many values, this measurability condition is equivalent to: $\{\sigma \leq t\} \in \mathcal{F}_t$ for all t.) We will write $\Sigma\left((\mathcal{F}_t)_{t \in J}\right)$ or Σ for the set of all simple stopping times for $(\mathcal{F}_t)$. It is a directed set itself as before. An *amart* for $(\mathcal{F}_t)$ is an adapted family (X_t) of integrable random variables such that the net $\left(\mathbf{E}\left[X_\sigma\right]\right)_{\sigma \in \Sigma}$ converges.

An *ordered stopping time* is a simple stopping time τ such that the elements $t_1, t_2, \cdots, t_m$ in the range of τ are linearly ordered, say $t_1 < t_2 < \cdots < t_m$. We denote by Σ^O the set of ordered stopping times. Then Σ^O

is a directed set under $\leq$. An integrable stochastic process $(X_n)_{n\in\mathbb{N}}$ is an *ordered amart* iff the net $(\mathbf{E}\,[X_\tau])_{\tau\in\Sigma^\circ}$ converges.

Clearly every amart is an ordered amart. When the directed set is $J = \mathbb{N}$, the notions of amart and ordered amart coincide. So each of these notions is a natural extension to directed processes of an amart indexed by $\mathbb{N}$. For general index set, stochastic convergence (convergence in probability) of ordered amarts is proved. Now we will see below (1.4.3) that L_1-bounded submartingales are ordered amarts. So this convergence theorem implies the stochastic convergence of L_1-bounded submartingales. On the other hand, amarts—unlike ordered amarts—converge pointwise a.e. (or essentially) under the covering condition (V) often satisfied in derivation theory (4.2.8 and 4.2.11). Thus each of the two notions (amart and ordered amart) has its applications.

Recall that a net $(X_t)_{t\in J}$ of random variables converges *stochastically* (or *in probability*) to X_∞ iff

$$\lim_{t\in J}\mathbf{P}\{|X_t - X_\infty| > \varepsilon\} = 0$$

for every $\varepsilon > 0$; and converges *in mean* iff $\lim_t \mathbf{E}\,[|X_t - X_\infty|] = 0$. Stochastic convergence is determined by a metric, for example

$$\rho(X,Y) = \mathbf{E}\,[|X - Y| \wedge 1], \quad \text{or} \quad \rho(X,Y) = \mathbf{E}\left[\frac{|X - Y|}{1 + |X - Y|}\right].$$

(These equivalent metrics are complete, but we do not need that fact here.) Recall that a family $(X_t)_{t\in J}$ of random variables is *uniformly integrable* iff

$$\lim_{\lambda\to\infty}\ \sup_{t\in J}\mathbf{E}\,[|X_t|\,\mathbf{1}_{\{|X_t|>\lambda\}}] = 0.$$

(1.3.1) Theorem. *Let J be a directed set, and let $(\mathcal{F}_t)_{t\in J}$ be a stochastic basis. Let $(X_t)_{t\in J}$ be an ordered amart (or amart). If (X_t) is L_1-bounded, then it converges stochastically; if (X_t) is uniformly integrable, then it converges in mean.*

Proof. By the sequential sufficiency theorem (1.1.3), it is enough to show that there exist $s_n \in J$ such that $(X_{t_n})_{n\in\mathbb{N}}$ converges stochastically for each increasing sequence $t_1 \leq t_2 \leq \cdots$ in J with $t_n \geq s_n$. Choose s_n increasing so that $|\mathbf{E}\,[X_{\tau_1} - X_{\tau_2}]| < 2^{-n}$ for $\tau_1, \tau_2 \in \Sigma^\circ$, $\tau_1, \tau_2 \geq s_n$. Suppose t_n increases and $t_n \geq s_n$. We claim that (X_{t_n}) is an amart for the stochastic basis $(\mathcal{G}_n)$ defined by $\mathcal{G}_n = \mathcal{F}_{t_n}$. Given $\varepsilon > 0$, choose N with $2^{-N} < \varepsilon$; for $\sigma \in \Sigma((\mathcal{G}_n))$ with $\sigma \geq N$, we have $t_\sigma \geq t_N \geq s_N$. Also $t_\sigma \in \Sigma((\mathcal{F}_t))$, since $\{\sigma = s\} \in \mathcal{G}_s = \mathcal{F}_{t_s}$ and $\sigma = s$ implies $t_\sigma = t_s$, so the sets $\{t_\sigma = t_s\}$ are countable unions of sets in $\mathcal{F}_{t_s} = \mathcal{G}_s$, and hence are in $\mathcal{G}_s$. Then $|\mathbf{E}\,[X_{t_\sigma} - X_{t_N}]| < 2^{-N} \leq \varepsilon$. Thus $(\mathbf{E}\,[X_{t_\sigma}])_{\sigma\in\Sigma((\mathcal{G}_n))}$ is

Cauchy. Thus (X_{t_n}) is an amart. It therefore converges a.s. by the amart convergence theorem (1.2.5), and thus it converges stochastically.

On a uniformly integrable set, stochastic convergence coincides with convergence in mean. (See Theorem (2.3.4).) ∎

The question of pointwise convergence of (X_t) will be considered below in Chapter 4.

This amart convergence theorem will be used to prove the Radon-Nikodým theorem. Notice that the Radon-Nikodým theorem has not been used up to this point—we have not even mentioned conditional expectation.

Let $(\Omega, \mathcal{F}, \mathbf{P})$ be a probability space. A set function $\mu \colon \mathcal{F} \to \mathbb{R}$ is *countably additive* iff, for any sequence $(A_n) \subseteq \mathcal{F}$ of pairwise disjoint sets,

$$\mu\left(\bigcup_{n=1}^{\infty} A_n\right) = \sum_{n=1}^{\infty} \mu(A_n).$$

We say that μ is *absolutely continuous* (with respect to $\mathbf{P}$) iff $\mu(A) = 0$ whenever $\mathbf{P}(A) = 0$. The *variation* of μ on $A \in \mathcal{F}$ is

$$|\mu|(A) = \sup \sum_{i=1}^{n} |\mu(C_i)|,$$

where the supremum is taken over all finite pairwise disjoint sequences $(C_i)_{i=1}^{n} \subseteq \mathcal{F}$ with $C_i \subseteq A$.

Our proof uses some facts about signed measures. They are proved in the Complements (below) for completeness, and so that it can be verified that they do not use the Radon-Nikodým theorem. The facts are:

(1) Let $\mu \colon \mathcal{F} \to \mathbb{R}$ be a countably additive set function. If $A_n \in \mathcal{F}$ and $A_1 \supseteq A_2 \supseteq \cdots$, then $\mu(\bigcap A_n) = \lim \mu(A_n)$. If $A_1 \subseteq A_2 \subseteq \cdots$, then $\mu(\bigcup A_n) = \lim \mu(A_n)$ (1.3.4).

(2) A real-valued countably additive set function μ has finite variation (1.3.5).

(3) Let μ be a countably additive set function that is absolutely continuous with respect to $\mathbf{P}$. Then for every $\varepsilon > 0$, there exists $\delta > 0$ such that if $A \in \mathcal{F}$ and $\mathbf{P}(A) < \delta$, then $|\mu(A)| < \varepsilon$ (1.3.6).

(1.3.2) Radon-Nikodým theorem. *Let $(\Omega, \mathcal{F}, \mathbf{P})$ be a probability space, and suppose $\mu \colon \mathcal{F} \to \mathbb{R}$ is countably additive and absolutely continuous. Then there exists an integrable random variable Y such that*

$$\mu(A) = \mathbf{E}\left[Y \mathbf{1}_A\right]$$

for all $A \in \mathcal{F}$.

Proof. A *measurable partition* of Ω is a finite subset $t \subseteq \mathcal{F}$, consisting of sets A with $\mathbf{P}(A) > 0$, pairwise disjoint [if $A_1, A_2 \in t$, and $A_1 \neq A_2$, then

$A_1 \cap A_2 = \emptyset]$, and having union Ω. Write J for the set of all measurable partitions. If $s, t \in J$, then we say that t *essentially refines* s, and write $t \geq s$, iff for each $B \in t$ there exists some $A \in s$ with $B \subseteq A$ a.s. With this ordering, J is a directed set.

If $t \in J$, let $\mathcal{F}_t$ be the σ-algebra generated by t; thus $\mathcal{F}_t$ is a finite algebra and t is the set of atoms of $\mathcal{F}_t$. Clearly, if $s \leq t$ in J, then $\mathcal{F}_s \subseteq \mathcal{F}_t$. Next, for $t \in J$, define a random variable

$$X_t = \sum_{A \in t} \frac{\mu(A)}{\mathbf{P}(A)} \mathbf{1}_A.$$

First we claim that (X_t) is an amart. [In fact it is a martingale: that is, $\mathbf{E}[X_\sigma]$ is constant.] Suppose $\sigma \colon \Omega \to J$ is a simple stopping time. Then $\{\sigma = t\} \in \mathcal{F}_t$ for each t, so $\{\sigma = t\}$ is a union of sets $A \in t$. So

$$X_\sigma = \sum_{t \in J} X_t \mathbf{1}_{\{\sigma = t\}} = \sum_t \sum_{A \in t} \frac{\mu(A)}{\mathbf{P}(A)} \mathbf{1}_A \mathbf{1}_{\{\sigma = t\}} = \sum_t \sum_{\substack{A \in t \\ A \subseteq \{\sigma = t\}}} \frac{\mu(A)}{\mathbf{P}(A)} \mathbf{1}_A.$$

Thus

$$\mathbf{E}[X_\sigma] = \sum_t \sum_{\substack{A \in t \\ A \subseteq \{\sigma = t\}}} \frac{\mu(A)}{\mathbf{P}(A)} \cdot \mathbf{P}(A)$$

$$= \sum_t \mu \left(\bigcup_{\substack{A \in t \\ A \subseteq \{\sigma = t\}}} A \right)$$

$$= \sum_t \mu\{\sigma = t\} = \mu(\Omega).$$

This shows that the net $(\mathbf{E}[X_\sigma])$ is constant, and therefore convergent. So (X_t) is an amart.

Next we must show that (X_t) is L_1-bounded. Let $t \in J$. Then

$$\mathbf{E}[|X_t|] = \sum_{A \in t} \frac{|\mu(A)|}{\mathbf{P}(A)} \cdot \mathbf{P}(A) \leq |\mu|(\Omega).$$

But μ has finite variation, by (1.3.5), so (X_t) is L_1-bounded.

We claim next that $(X_t)_{t \in J}$ is uniformly integrable. Let $\varepsilon > 0$. Then, since μ is absolutely continuous, by (1.3.6) there exists $\delta > 0$ such that $|\mu(A)| \leq \varepsilon$ whenever $\mathbf{P}(A) \leq \delta$. Let $\lambda = |\mu|(\Omega)/\delta$. For all $t \in J$,

$$\mathbf{E}[X_t \mathbf{1}_{\{X_t > \lambda\}}] = \sum_{\substack{A \in t \\ \mu(A) > \lambda \mathbf{P}(A)}} \frac{\mu(A)}{\mathbf{P}(A)} \mathbf{P}(A) = \mu \left(\bigcup_{\substack{A \in t \\ \mu(A) > \lambda \mathbf{P}(A)}} A \right) \leq \varepsilon,$$

since

$$\mathbf{P}\left(\bigcup_{\substack{A\in t \\ \mu(A)>\lambda\mathbf{P}(A)}} A\right) \le \frac{1}{\lambda} \sum_{\substack{A\in t \\ \mu(A)>\lambda\mathbf{P}(A)}} \mu(A) \le \delta.$$

Similarly

$$\mathbf{E}\left[|X_t|\,\mathbf{1}_{\{X_t<-\lambda\}}\right] \le \varepsilon.$$

So (X_t) is uniformly integrable.

Now (X_t) is an L_1-bounded amart that is uniformly integrable. There-
fore (X_t) converges in mean, say to $Y \in L_1(\Omega, \mathcal{F}, \mathbf{P})$. If $A_0 \in \mathcal{F}$, let
$t_0 = \{A, \Omega \setminus A_0\} \in J$, and note that if $t \ge t_0$, then

$$\mathbf{E}\left[X_t \mathbf{1}_{A_0}\right] = \sum_{\substack{A\in t \\ A\subseteq A_0}} \mu(A) = \mu(A_0).$$

Thus

$$|\mathbf{E}\left[Y \mathbf{1}_{A_0}\right] - \mu(A_0)|$$
$$\le \lim_t |\mathbf{E}\left[Y \mathbf{1}_{A_0}\right] - \mathbf{E}\left[X_t \mathbf{1}_{A_0}\right]| + \lim_t |\mathbf{E}\left[X_t \mathbf{1}_{A_0}\right] - \mu(A_0)|$$
$$\le \lim_t \mathbf{E}\left[|Y - X_t|\right] + 0 = 0.$$

So $\mathbf{E}\left[Y \mathbf{1}_{A_0}\right] = \mu(A_0)$ for all $A_0 \in \mathcal{F}$. ∎

The random variable Y is called the *Radon-Nikodým derivative* of μ with
respect to $\mathbf{P}$; we write $Y = d\mu/d\mathbf{P}$ or $\mu = Y\,\mathbf{P}$. Once this basic Radon-Nikodým theorem has been proved, many variations can be done as well. For
example, $\mathbf{P}$ can be replaced by an infinite measure, μ can take values in the
complex numbers, or in a finite-dimensional vector space $\mathbb{R}^n$. These will
not be proved here. We will discuss measures μ with values in a Banach
space in Chapter 5.

Complements

(1.3.3) (Reversed amarts.) Let J be a dual directed set. The family
$(\mathcal{F}_t)_{t\in J}$ of sub-σ-algebras of $\mathcal{F}$ is a *reversed stochastic basis* iff $s \le t$ implies
$\mathcal{F}_s \subseteq \mathcal{F}_t$. Let $(X_t)_{t\in J}$ be a *reversed amart*; that is: $\mathbf{E}\left[X_\sigma\right]$ converges for
σ in the dual directed set $\Sigma((\mathcal{F}_t))$. Then (X_t) converges stochastically
and in mean. As usual, L_1-boundedness need not be postulated (Edgar &
Sucheston [1976a]).

(1.3.4) (Countable additivity.) Let $\mu\colon \mathcal{F} \to \mathbb{R}$ be a countably additive
set function. If $A_n \in \mathcal{F}$ and $A_1 \supseteq A_2 \supseteq \cdots$, then $\mu(\bigcap A_n) = \lim \mu(A_n)$. If
$A_1 \subseteq A_2 \subseteq \cdots$, then $\mu(\bigcup A_n) = \lim \mu(A_n)$.

To prove the first one, let $B_1 = A_1$ and $B_n = A_n \setminus A_{n-1}$ for $n \geq 2$. Then the sets B_n are pairwise disjoint, and

$$\mu\left(\bigcup_{n=1}^{\infty} A_n\right) = \mu\left(\bigcup_{n=1}^{\infty} B_n\right) = \sum_{n=1}^{\infty} \mu(B_n)$$

$$= \lim_{n\to\infty} \sum_{k=1}^{n} \mu(B_k) = \lim_{n\to\infty} \mu(A_n).$$

For the second part, apply the first part to the complements $\Omega \setminus A_n$.

(1.3.5) (Variation.) A real-valued, countably additive set function μ has finite variation. First, we claim that $\sup_{A\in\mathcal{F}} \mu(A) < \infty$. If not, then there exist sets $A_n \in \mathcal{F}$ with $\mu(A_{n+1}) \geq \mu(\bigcup_{k=1}^{n} A_k) + 1$. Then

$$\mu\left(\bigcup_{n=1}^{\infty} A_k\right) = \mu(A_1) + \sum_{n=1}^{\infty} \mu\left(A_{n+1} \setminus \bigcup_{k=1}^{n} A_k\right) = \infty,$$

contradicting the fact that μ has real values. Similarly, $\inf_{A\in\mathcal{F}} \mu(A) > -\infty$. Now if $(C_i)_{i=1}^{n}$ are pairwise disjoint sets, write

$$I_+ = \{\, i : 1 \leq i \leq n, \mu(C_i) \geq 0 \,\}$$
$$I_- = \{\, i : 1 \leq i \leq n, \mu(C_i) < 0 \,\}.$$

Then

$$\sum_{i=1}^{n} |\mu(C_i)| = \sum_{i\in I_+} \mu(C_i) - \sum_{i\in I_-} \mu(C_i)$$

$$= \mu\left(\bigcup_{i\in I_+} C_i\right) - \mu\left(\bigcup_{i\in I_-} C_i\right)$$

$$\leq \sup_{A\in\mathcal{F}} \mu(A) - \inf_{A\in\mathcal{F}} \mu(A) < \infty.$$

Thus $|\mu|(\Omega) < \infty$.

(1.3.6) (Absolute continuity.) Let μ be a countably additive set function that is absolutely continuous with respect to $\mathbf{P}$. Then for every $\varepsilon > 0$, there exists $\delta > 0$ such that if $A \in \mathcal{F}$ and $\mathbf{P}(A) < \delta$, then $|\mu(A)| < \varepsilon$. Indeed, suppose the conclusion is false. Then there would exist $\varepsilon > 0$ and sets $A_n \in \mathcal{F}$ with $\mathbf{P}(A) < 2^{-n}$ but $|\mu(A_n)| \geq \varepsilon$. But then the set $A = \bigcap_{n=1}^{\infty} \bigcup_{k=n}^{\infty} A_n$ satisfies

$$\mathbf{P}(A) = \lim_n \mathbf{P}\left(\bigcup_{k=n}^{\infty} A_k\right) \leq \lim_n \sum_{k=n}^{\infty} \mathbf{P}(A_k) = 0,$$

$$|\mu|(A) = \lim_n |\mu|\left(\bigcup_{k=n}^{\infty} A_k\right) \geq \limsup_n |\mu(A_n)| \geq \varepsilon.$$

Then there exists $B \subseteq A$ with $\mathbf{P}(B) = 0$ and $\mu(B) \neq 0$. This contradicts absolute continuity.

(1.3.7) (Hahn decomposition.) We may use the Radon-Nikodým theorem to prove the Hahn decomposition theorem: Let $\mu \colon \mathcal{F} \to \mathbb{R}$ be countably additive. Then there exist measurable sets P and N with $P \cup N = \Omega$ and $P \cap N = \emptyset$ such that $\mu(A) \geq 0$ for all $A \subseteq P$ and $\mu(A) \leq 0$ for all $A \subseteq N$. Indeed, let $\mathbf{P}$ be a probability measure equivalent with $|\mu|$; let X be the Radon-Nikodým derivative $d\mu/d\mathbf{P}$, and set $P = \{X \geq 0\}$, $N = \{X < 0\}$.

Remarks

Stochastic convergence of amarts indexed by directed sets comes from Edgar & Sucheston [1976a]. Ordered amarts are from Millet & Sucheston [1980b]. The process

$$X_t = \sum_{A \in t} \frac{\mu(A)}{\mathbf{P}(A)} \, 1_A$$

is a famous martingale, appearing in Doob [1953], page 343, and earlier in Danish and other papers.

1.4. Conditional expectations

A special case of the Radon-Nikodým derivative, important for martingale theory, is the conditional expectation. We will deal with conditional expectation in a probability space now. Later (Section 2.3) we will discuss conditional expectation in an infinite measure space.

An intuitive way to think of the conditional expectation $\mathbf{E}^{\mathcal{G}}[X]$ is: the expected value of X given the information of the σ-algebra $\mathcal{G}$.

Definition and basic properties

Let $(\Omega, \mathcal{F}, \mathbf{P})$ be a probability space, and let $\mathcal{G} \subseteq \mathcal{F}$ be a σ-algebra. Suppose X is an integrable random variable, that is $X \in L_1(\Omega, \mathcal{F}, \mathbf{P})$. A *conditional expectation* of X given $\mathcal{G}$ is a random variable Y, integrable and measurable with respect to $\mathcal{G}$, $Y \in L_1(\Omega, \mathcal{G}, \mathbf{P})$, such that

$$\mathbf{E}\left[Y \, 1_A\right] = \mathbf{E}\left[X \, 1_A\right]$$

for all $A \in \mathcal{G}$.

The existence of conditional expectations is a simple consequence of the Radon-Nikodým theorem: on the σ-algebra $\mathcal{G}$, the set function $\mu(A) = \mathbf{E}\left[X \, 1_A\right]$ is absolutely continuous with respect to the restriction $\mathbf{P}|\mathcal{G}$ of $\mathbf{P}$. We define $\mathbf{E}^{\mathcal{G}} X$ as the Radon-Nikodým derivative of μ with respect to $\mathbf{P}$ on $\mathcal{G}$. It is also clear that Y is unique up to sets of measure zero, so we often speak of *the* conditional expectation of X given $\mathcal{G}$, and write $Y = \mathbf{E}^{\mathcal{G}} X$ or $\mathbf{E}\left[X \mid \mathcal{G}\right]$.

The properties of the conditional expectation are first those of the expectation (the Lebesgue integral), which is $\mathbf{E}^{\mathcal{N}}$ for the trivial σ-algebra $\mathcal{N} = \{\Omega, \emptyset\}$: linearity, positivity, contraction property with respect to the L_p norm. The proofs are similar (or use the properties of the ordinary expectation to shorten the arguments). Thus Lebesgue's monotone and bounded convergence theorems hold for conditional expectation: $\lim_n X_n = X$ a.s. implies $\lim \mathbf{E}^{\mathcal{G}}[X_n] = \mathbf{E}^{\mathcal{G}}[X]$ a.s. assuming either each X_n integrable and (X_n) monotone, or $\sup |X_n|$ integrable. However, $\lim X_n = X$ a.s. does not necessarily imply $\lim \mathbf{E}^{\mathcal{G}}[X_n] = \mathbf{E}^{\mathcal{G}}[X]$ a.s. if (X_n) is only assumed to be uniformly integrable (1.4.15). Special properties of the conditional expectation are the following:

(1.4.1) Proposition.

   (a) *Factorization: If X is measurable with respect to $\mathcal{G}$, then*

$$\mathbf{E}^{\mathcal{G}}[XY] = X\mathbf{E}^{\mathcal{G}}[Y],$$

   provided both sides exist.

   (b) *The smaller algebra prevails: if $\mathcal{G}_1$ and $\mathcal{G}_2$ are σ-algebras with $\mathcal{G}_1 \subseteq \mathcal{G}_2$ and $X \in L_1$, then*

$$\mathbf{E}^{\mathcal{G}_1}\left[\mathbf{E}^{\mathcal{G}_2}[X]\right] = \mathbf{E}^{\mathcal{G}_2}\left[\mathbf{E}^{\mathcal{G}_1}[X]\right] = \mathbf{E}^{\mathcal{G}_1}[X].$$

Proof. (a) Since both sides are measurable with respect to $\mathcal{G}$, it suffices to show that the right side satisfies the definition of the left side, that is: for each $A \in \mathcal{G}$,

$$\mathbf{E}[XY\mathbf{1}_A] = \mathbf{E}\left[\left(X\mathbf{E}^{\mathcal{G}}[Y]\right)\mathbf{1}_A\right].$$

If $X = \mathbf{1}_B$ for some $B \in \mathcal{G}$, then this is easy: both sides reduce to $\mathbf{E}[Y\mathbf{1}_{A\cap B}]$. For a general $X \in L_1$, use linearity and approximation by simple functions.

   (b) Since $\mathbf{E}^{\mathcal{G}_1}[X]$ is measurable with respect to $\mathcal{G}_2$, the second equality follows. If $A \in \mathcal{G}_1$, then $\mathbf{E}\left[\mathbf{1}_A\mathbf{E}^{\mathcal{G}_2}[X]\right] = \mathbf{E}[\mathbf{1}_A X] = \mathbf{E}\left[\mathbf{1}_A\mathbf{E}^{\mathcal{G}_1}[X]\right]$, so $\mathbf{E}^{\mathcal{G}_1}\left[\mathbf{E}^{\mathcal{G}_2}[X]\right] = \mathbf{E}^{\mathcal{G}_1}[X]$. ∎

One last property deals with the stopped σ-algebra $\mathcal{F}_\sigma$.

(1.4.2) Localization theorem. *Let $(\Omega, \mathcal{F}, \mathbf{P})$ be a probability space, let J be a directed set, let $(\mathcal{F}_t)_{t \in J}$ be a stochastic basis, and let $X \in L_1$. If $Y_t = \mathbf{E}^{\mathcal{F}_t}[X]$ for all $t \in J$, then $Y_\sigma = \mathbf{E}^{\mathcal{F}_\sigma}[X]$ for all $\sigma \in \Sigma$.*

Proof. The assertion may be stated in another way: For each t, we have $\mathbf{E}^{\mathcal{F}_\sigma}[X] = \mathbf{E}^{\mathcal{F}_t}[X]$ on the set $\{\sigma = t\}$.

For the proof, write $\{t_1, \cdots, t_n\}$ for the set of values of σ, and let

$$Z = \sum_i \mathbf{E}^{\mathcal{F}_{t_i}}[X]\,\mathbf{1}_{\{\sigma = t_i\}}.$$

We claim that $Z = \mathbf{E}^{\mathcal{F}_\sigma}[X]$. First we must show that Z is $\mathcal{F}_\sigma$-measurable. Now if $\lambda \in \mathbb{R}$, then for all t_i we have

$$\{Z > \lambda\} \cap \{\sigma = t_i\} = \{\mathbf{E}^{\mathcal{F}_n}[X] > \lambda\} \cap \{\sigma = t_i\} \in \mathcal{F}_{t_i}.$$

Therefore $\{Z > \lambda\} \in \mathcal{F}_\sigma$; this shows that Z is $\mathcal{F}_\sigma$-measurable.
Next, if $A \in \mathcal{F}_\sigma$, we must show $\mathbf{E}[X \, 1_A] = \mathbf{E}[Z \, 1_A]$. But we have

$$A \cap \{\sigma = t_i\} \in \mathcal{F}_{t_i}$$

for all t_i, so

$$\mathbf{E}[X \, 1_A] = \sum_i \mathbf{E}\left[X \, 1_{A \cap \{\sigma = t_i\}}\right] = \sum_i \mathbf{E}\left[\mathbf{E}^{\mathcal{F}_{t_i}}[X] \, 1_{A \cap \{\sigma = t_i\}}\right] = \mathbf{E}[Z \, 1_A].$$

This completes the proof. ∎

Martingales and related processes

There are special classes of amarts that are naturally defined in terms of the conditional expectation. Now that conditional expectations have been discussed, we may consider these classes. The most important is the martingale.

Let $(\Omega, \mathcal{F}, \mathbf{P})$ be a probability space, let J be a directed set, let $(\mathcal{F}_t)_{t \in J}$ be a stochastic basis, and let $(X_t)_{t \in J}$ be adapted and integrable. Then we say (X_t) is a *martingale* iff $\mathbf{E}^{\mathcal{F}_s}[X_t] = X_s$ for $s \leq t$; we say (X_t) is a *submartingale* iff $\mathbf{E}^{\mathcal{F}_s}[X_t] \geq X_s$ for $s \leq t$; we say (X_t) is a *supermartingale* iff $\mathbf{E}^{\mathcal{F}_s}[X_t] \leq X_s$ for $s \leq t$.

Note the use of *ordered* stopping times in the characterizations that follow.

(1.4.3) Theorem. *Let (X_t) be an adapted and integrable process. Then:*

(i) *(X_t) is a submartingale if and only if the net $\left(\mathbf{E}[X_\sigma]\right)_{\sigma \in \Sigma^O}$ is increasing [in the sense that $\mathbf{E}[X_\sigma] \leq \mathbf{E}[X_\tau]$ for $\sigma \leq \tau$] if and only if the process $(X_\tau)_{\tau \in \Sigma^O}$ is a submartingale [that is, $\mathbf{E}^{\mathcal{F}_\sigma}[X_\tau] \geq X_\sigma$ for $\sigma, \tau \in \Sigma^O$, $\sigma \leq \tau$].*

(ii) *(X_t) is a supermartingale if and only if the net $\left(\mathbf{E}[X_\sigma]\right)_{\sigma \in \Sigma^O}$ is decreasing if and only if the process $(X_\tau)_{\tau \in \Sigma^O}$ is a supermartingale.*

(iii) *(X_t) is a martingale if and only if the net $\left(\mathbf{E}[X_\tau]\right)_{\tau \in \Sigma^O}$ is constant if and only if the net $\left(\mathbf{E}[X_\tau]\right)_{\tau \in \Sigma}$ is constant.*

Proof. (i) Let (X_t) be a submartingale. We first prove that $\mathbf{E}^{\mathcal{F}_\sigma}[X_\tau] \geq X_\sigma$ for $\sigma, \tau \in \Sigma^O$, $\sigma \leq \tau$. By the localization theorem 1.4.2, it suffices to show that $\mathbf{E}^{\mathcal{F}_s}[X_\tau] \geq X_s$ on the set $\{\sigma = s\}$. Suppose that τ takes values $t_1 \leq t_2 \leq \cdots \leq t_m$. For a fixed value s of σ, we use induction on the index n between 1 and m defined by $t_n = \max\{\tau(\omega) : \sigma(\omega) = s\}$. Since $\sigma \leq \tau$, certainly $t_n \geq s$. For $t_n = s$, we have $\tau = s$ on $\{\sigma = s\}$, and therefore $\mathbf{E}^{\mathcal{F}_s}[X_\tau] = X_s$ on $\{\sigma = s\}$. For the inductive step, suppose $t_n > s$, and define $\tau' \in \Sigma^O$ by

$$\tau'(\omega) = \begin{cases} \tau(\omega) & \text{if } \sigma(\omega) = s \text{ and } \tau(\omega) < t_n \\ t_{n-1} & \text{if } \sigma(\omega) = s \text{ and } \tau(\omega) = t_n \\ \tau(\omega) & \text{if } \sigma(\omega) \neq s. \end{cases}$$

Then $\tau \geq \tau' \geq \sigma$. By the induction hypothesis, $\mathbf{E}^{\mathcal{F}_s}[X_{\tau'}] \geq X_s$ on $\{\sigma = s\}$. Also, $\{\tau' < \tau\} = \{\tau' \leq t_{n-1}\} \cap \{\sigma = s\}$. Now on $\{\sigma = s\}$,

$$\mathbf{E}^{\mathcal{F}_{t_{n-1}}}[X_\tau] = \mathbf{E}^{\mathcal{F}_{t_{n-1}}}\left[X_{\tau'} + (X_{t_n} - X_{t_{n-1}})\mathbf{1}_{\{\tau' < \tau\}}\right]$$
$$= \mathbf{E}^{\mathcal{F}_{t_{n-1}}}[X_{\tau'}] + \mathbf{1}_{\{\tau' < \tau\}}\mathbf{E}^{\mathcal{F}_{t_{n-1}}}\left[X_{t_n} - X_{t_{n-1}}\right]$$
$$\geq \mathbf{E}^{\mathcal{F}_{t_{n-1}}}[X_{\tau'}] + 0$$
$$= X_{\tau'}.$$

Hence, still supposing $t_n > s$, on $\{\sigma = s\}$, we have

$$\mathbf{E}^{\mathcal{F}_s}[X_\tau] = \mathbf{E}^{\mathcal{F}_s}\left[\mathbf{E}^{\mathcal{F}_{t_{n-1}}}[X_\tau]\right] \geq \mathbf{E}^{\mathcal{F}_s}[X_{\tau'}] \geq X_s.$$

This shows $\mathbf{E}^{\mathcal{F}_\sigma}[X_\tau] \geq X_\sigma$ and completes the proof of $\mathbf{E}^{\mathcal{F}_\sigma}[X_\tau] \geq X_\sigma$ for $\sigma, \tau \in \Sigma^O$, $\sigma \leq \tau$.

On integrating, we obtain $\mathbf{E}[X_\tau] = \mathbf{E}\left[\mathbf{E}^{\mathcal{F}_\sigma}[X_\tau]\right] \geq \mathbf{E}[X_\sigma]$ for $\sigma \leq \tau$.

Next suppose that the net $(\mathbf{E}[X_\tau])_{\tau \in \Sigma^O}$ is increasing. We must show that $\mathbf{E}^{\mathcal{F}_s}[X_t] \geq X_s$ if $s \leq t$. Let $A \in \mathcal{F}_s$. Then

$$\sigma(\omega) = \begin{cases} s & \text{if } \omega \notin A \\ t & \text{if } \omega \in A \end{cases}$$

is an ordered simple stopping time, and $\sigma \geq s$. Therefore $\mathbf{E}[X_\sigma] \geq \mathbf{E}[X_s]$. That is:

$$\mathbf{E}[X_t \mathbf{1}_A] + \mathbf{E}\left[X_s \mathbf{1}_{\Omega \setminus A}\right] \geq \mathbf{E}[X_s \mathbf{1}_A] + \mathbf{E}\left[X_s \mathbf{1}_{\Omega \setminus A}\right].$$

Therefore $\mathbf{E}[X_t \mathbf{1}_A] \geq \mathbf{E}[X_s \mathbf{1}_A]$. This is true for all $A \in \mathcal{F}_s$, so

$$\mathbf{E}^{\mathcal{F}_s}[X_t] \geq X_s,$$

as required.

(ii) Since $(-X_t)$ is a submartingale if and only if (X_t) is a supermartingale, (ii) follows from (i).

(iii) Suppose (X_t) is a martingale. Let $\sigma \in \Sigma$ be given. Choose $t \geq \sigma$. Then

$$\mathbf{E}[X_\sigma] = \sum_{s \leq t} \mathbf{E}\left[X_s \mathbf{1}_{\{\sigma = s\}}\right]$$
$$= \sum_{s \leq t} \mathbf{E}\left[X_t \mathbf{1}_{\{\sigma = s\}}\right]$$
$$= \mathbf{E}[X_t].$$

We can see that this value is independent of t by integrating the equation $\mathbf{E}^{\mathcal{F}_s}[X_t] = X_s$; so in fact $(\mathbf{E}[X_\sigma])_{\sigma \in \Sigma}$ is independent of σ.

If $(\mathbf{E}\,[X_\sigma])_{\sigma\in\Sigma}$ is constant, then certainly $(\mathbf{E}\,[X_\sigma])_{\sigma\in\Sigma^\circ}$ is also constant. Finally, if $(\mathbf{E}\,[X_\sigma])_{\sigma\in\Sigma^\circ}$ is constant, then it is both increasing and decreasing, so by (i) and (ii) the process (X_t) is both a submartingale and a supermartingale, therefore a martingale. ∎

Let $(X_t)_{t\in J}$ be an integrable process. We say that (X_t) is a *quasimartingale* iff there is a constant M such that

$$\sum_{i=1}^{n-1} \mathbf{E}\left[\left|\mathbf{E}\,[X_{t_{i+1}}\,|\,\mathcal{F}_{t_i}] - X_{t_i}\right|\right] \le M$$

for all finite increasing sequences $t_1 < t_2 < \cdots < t_n$ in J.

(1.4.4) Theorem. *Every quasimartingale (X_t) is an ordered amart.*

Proof. Let (X_t) be a quasimartingale:

$$(1.4.4a) \qquad M = \sup \sum_{i=1}^{n-1} \mathbf{E}\left[\left|\mathbf{E}\,[X_{t_{i+1}}\,|\,\mathcal{F}_{t_i}] - X_{t_i}\right|\right] < \infty,$$

where the sup is over all finite sequences $t_1 < \cdots < t_n$. We will show that $(\mathbf{E}\,[X_\tau])_{\tau\in\Sigma^\circ}$ is a Cauchy net. Let $\varepsilon > 0$ be given. Choose $s_1 < s_2 < \cdots < s_m$ so that

$$(1.4.4b) \qquad \sum_{i=1}^{m-1} \mathbf{E}\left[\left|\mathbf{E}\,[X_{s_{i+1}}\,|\,\mathcal{F}_{s_i}] - X_{s_i}\right|\right] > M - \varepsilon.$$

Now let $\tau \in \Sigma^{\mathrm{O}}$ with $\tau > s_m$. Write $t_1 < t_2 < \cdots < t_n$ for the set of values of τ. Now if we apply the inequality (1.4.4a) with the finite sequence $s_1 < \cdots < s_m < t_1 < \cdots < t_n$, then subtract (1.4.4b), we obtain

$$\sum_{j=1}^{n-1} \mathbf{E}\left[\left|X_{t_j} - \mathbf{E}\,[X_{t_{j+1}}\,|\,\mathcal{F}_{t_j}]\right|\right] < \varepsilon.$$

Now for $t_i \le t_j$ we have $\{\tau = t_i\} \in \mathcal{F}_{t_j}$, so we have

$$\left|\mathbf{E}\,[X_\tau] - \mathbf{E}\,[X_{t_n}]\right| = \left|\sum_{i=1}^{n-1} \mathbf{E}\left[(X_{t_i} - X_{t_n})\,\mathbf{1}_{\{\tau=t_i\}}\right]\right|$$

$$= \left|\sum_{i=1}^{n-1}\sum_{j=i}^{n-1} \mathbf{E}\left[(X_{t_j} - X_{t_{j+1}})\,\mathbf{1}_{\{\tau=t_i\}}\right]\right|$$

$$= \left|\sum_{j=1}^{n-1}\sum_{i=1}^{j} \mathbf{E}\left[(X_{t_j} - \mathbf{E}\,[X_{t_{j+1}}\,|\,\mathcal{F}_{t_j}])\,\mathbf{1}_{\{\tau=t_i\}}\right]\right|$$

$$\leq \sum_{j=1}^{n-1} \sum_{i=1}^{j} \mathbf{E}\left[\left|X_{t_j} - \mathbf{E}\left[X_{t_{j+1}} \mid \mathcal{F}_{t_j}\right]\right| 1_{\{\tau=t_i\}}\right]$$

$$\leq \sum_{j=1}^{n-1} \mathbf{E}\left[\left|X_{t_j} - \mathbf{E}\left[X_{t_{j+1}} \mid \mathcal{F}_{t_j}\right]\right|\right] < \varepsilon.$$

So if $\tau_1, \tau_2 \geq s_m$, we have $\left|\mathbf{E}\left[X_{\tau_1}\right] - \mathbf{E}\left[X_{\tau_2}\right]\right| < 2\varepsilon$. Therefore (X_t) is an ordered amart. ∎

Riesz decomposition

The Riesz decomposition (1.4.6) for an amart or an ordered amart shows that the process is "close" to a martingale.

First consider an alternative ordering for the set Σ^O of ordered stopping times. If $\sigma, \tau \in \Sigma^O$, we write $\sigma \lll \tau$ iff there is $t \in J$ with $\sigma \leq t$ and $t \leq \tau$. Note that Σ^O is directed under the order $\lll$, as well as under the usual order $\leq$.

For convergence of nets, the difference does not matter. Indeed, $\sigma \lll \tau$ implies $\sigma \leq \tau$, so if a net (a_τ) converges according to $\leq$, then it trivially converges according to $\lll$. Conversely, suppose (a_τ) converges to a according to $\lll$. Then given $\varepsilon > 0$ there is $\sigma \in \Sigma^O$ such that $|a_\tau - a| < \varepsilon$ for all $\tau \in \Sigma^O$ with $\sigma \lll \tau$. But there is $t \in J$ with $t \geq \sigma$, and therefore $|a_\tau - a| < \varepsilon$ for all $\tau \geq t$. Thus a_τ converges to a according to $\leq$. (See also (1.1.9).)

When we take a limit over pairs of stopping times, it does make a difference whether we restrict to pairs σ, τ with $\sigma \leq \tau$ or to pairs with $\sigma \lll \tau$. The ordering $\lll$ appears in the following useful characterizations of amarts and ordered amarts.

(1.4.5) Difference property. *Let J be a directed set, and let $(\mathcal{F}_t)_{t \in J}$ be a stochastic basis. Let $(X_t)_{t \in J}$ be an adapted integrable process.*

(i) *(X_t) is an ordered amart if and only if*

(1.4.5a) $$\lim_{\substack{\sigma \lll \tau \\ \sigma, \tau \in \Sigma^O}} \left\|\mathbf{E}^{\mathcal{F}_\sigma}\left[X_\tau\right] - X_\sigma\right\|_1 = 0.$$

(ii) *(X_t) is an amart if and only if*

(1.4.5b) $$\lim_{\substack{\sigma \leq \tau \\ \sigma, \tau \in \Sigma}} \left\|\mathbf{E}^{\mathcal{F}_\sigma}\left[X_\tau\right] - X_\sigma\right\|_1 = 0.$$

Proof. (i) Suppose that the difference property (1.4.5a) holds. Then for $\sigma \lll \tau$,

$$\begin{aligned}
\left|\mathbf{E}\left[X_\tau\right] - \mathbf{E}\left[X_\sigma\right]\right| &= \left|\mathbf{E}\left[\mathbf{E}^{\mathcal{F}_\sigma}\left[X_\tau\right] - X_\sigma\right]\right| \\
&\leq \mathbf{E}\left[\left|\mathbf{E}^{\mathcal{F}_\sigma}\left[X_\tau\right] - X_\sigma\right|\right] \\
&= \left\|\mathbf{E}^{\mathcal{F}_\sigma}\left[X_\tau\right] - X_\sigma\right\|_1 \to 0
\end{aligned}$$

so (X_t) is an ordered amart.

Conversely, let $\varepsilon > 0$; choose $s \in J$ such that $\sigma \geq s$, $\tau \geq s$ and $\sigma, \tau \in \Sigma^O$ implies

$$\left| \mathbf{E}\left[X_\sigma\right] - \mathbf{E}\left[X_\tau\right] \right| \leq \varepsilon.$$

Let $s \leq \sigma \lll \tau$; for any $A \in \mathcal{F}_\sigma$ define $\rho = \sigma$ on A, and $\rho = \tau$ on $\Omega \setminus A$. Then $\rho \in \Sigma^O$. Furthermore

$$\mathbf{E}\left[1_A \left(X_\sigma - \mathbf{E}^{\mathcal{F}_\sigma}\left[X_\tau\right]\right)\right] = \mathbf{E}\left[X_\rho\right] - \mathbf{E}\left[X_\tau\right] \leq \varepsilon.$$

Now take $A = \{X_\sigma \geq \mathbf{E}^{\mathcal{F}_\sigma}\left[X_\tau\right]\}$ to see that

$$\mathbf{E}\left[\left(X_\sigma - \mathbf{E}^{\mathcal{F}_\sigma}\left[X_\tau\right]\right)^+\right] \leq \varepsilon.$$

Similarly

$$\mathbf{E}\left[\left(X_\sigma - \mathbf{E}^{\mathcal{F}_\sigma}\left[X_\tau\right]\right)^-\right] \leq \varepsilon,$$

so $\|X_\sigma - \mathbf{E}^{\mathcal{F}_\sigma}\left[X_\tau\right]\|_1 \leq 2\varepsilon$. This proves the difference property (1.4.5a).

(ii) This proof is the same: substitute "amart" for "ordered amart," "Σ" for "Σ^O," and "$\leq$" for "$\lll$." ∎

(1.4.6) Riesz decomposition. *Let J be a directed set, and let $(\mathcal{F}_t)_{t \in J}$ be a stochastic basis. (i) Let (X_t) be an ordered amart. Then X_t can be uniquely written as $X_t = Y_t + Z_t$, where Y_t is a martingale and $(Z_\tau)_{\tau \in \Sigma^O}$ converges to 0 in mean. (ii) Let (X_t) be an amart. Then X_t can be uniquely written as $X_t = Y_t + Z_t$, where Y_t is a martingale and $(Z_\tau)_{\tau \in \Sigma}$ converges to 0 in mean.*

Proof. (i) For any $\sigma \in \Sigma^O$, the net $(\mathbf{E}^{\mathcal{F}_\sigma}\left[X_\tau\right])_{\tau \in \Sigma^O}$ is Cauchy in L_1 norm by the difference property. Write Y_σ for the limit. Conditional expectations are continuous for L_1-norm, so if $\sigma_1 \lll \sigma_2$ we have

$$\mathbf{E}^{\mathcal{F}_{\sigma_1}}\left[Y_{\sigma_2}\right] = \lim_\tau \mathbf{E}^{\mathcal{F}_{\sigma_1}}\left[\mathbf{E}^{\mathcal{F}_{\sigma_2}}\left[X_\tau\right]\right] = \lim_\tau \mathbf{E}^{\mathcal{F}_{\sigma_1}}\left[X_\tau\right] = Y_{\sigma_1}.$$

So (Y_t) is a martingale. The difference $Z_t = X_t - Y_t$ is an ordered amart, and by the difference property $\|Z_\tau\|_1 \to 0$.

For uniqueness, assume there is another decomposition $X_t = \widehat{Y}_t + \widehat{Z}_t$. For $A \in \mathcal{F}_t$, we have

$$\mathbf{E}\left[1_A Y_t\right] = \lim_\tau \mathbf{E}\left[1_A Y_\tau\right] = \lim_\tau \mathbf{E}\left[1_A X_\tau\right] = \lim_\tau \mathbf{E}\left[1_A \widehat{Y}_\tau\right] = \mathbf{E}\left[1_A \widehat{Y}_t\right].$$

Hence $Y_t = \widehat{Y}_t$.

(ii) This proof is the same: substitute "amart" for "ordered amart," "Σ" for "Σ^O," and "$\leq$" for "$\lll$." ∎

The sequential case

The most important setting of the theory is of course the case $J = \mathbb{N}$. Then a submartingale is defined by $X_m \leq \mathbf{E}^{\mathcal{F}_m}[X_n]$ for $m \leq n$, a supermartingale by $X_m \geq \mathbf{E}^{\mathcal{F}_m}[X_n]$ for $m \leq n$, and a martingale by $X_m = \mathbf{E}^{\mathcal{F}_m}[X_n]$ for $m \leq n$. It is easy to see by induction that these properties need to be checked only for $n = m + 1$. A quasimartingale is an integrable process (X_n) for which there is a constant C such that

$$\sum_{i=1}^{n} \|\mathbf{E}^{\mathcal{F}_i}[X_{i+1}] - X_i\|_1 \leq C$$

for all n. Equivalently (1.4.30) there is a constant C such that

$$\sum_{i=1}^{m-1} |\mathbf{E}[X_{\tau_{i+1}} - X_{\tau_i}]| \leq C$$

for any sequence $\tau_1 \leq \tau_2 \leq \cdots \leq \tau_m$ of simple stopping times.

When $J = \mathbb{N}$, all stopping times are ordered. So Theorems (1.4.3) and (1.2.5) lead to:

(1.4.7) Theorem. *Let $(X_n)_{n \in \mathbb{N}}$ be an L_1-bounded supermartingale, submartingale, quasimartingale, or martingale. Then (X_n) is an amart, and converges a.s. If (X_n) is also uniformly integrable, then it converges in mean.*

Complements

(1.4.8) (Atomic example.) An *atom* in a σ-algebra $\mathcal{G}$ is a set $A \in \mathcal{G}$ such that if $B \in \mathcal{G}$ and $B \subseteq A$, then either $\mathbf{P}(B) = 0$ or $\mathbf{P}(A \setminus B) = 0$. If A is an atom of $\mathcal{G}$, then $\mathbf{E}^{\mathcal{G}}[X]$ has the value $(1/\mathbf{P}(A))\mathbf{E}[X \mathbf{1}_A]$ on A.

(1.4.9) (Nonlocalization.) The localization of $\mathbf{E}^{\mathcal{G}}[X]$ (1.4.2) works only in the variable $\mathcal{G}$, not in the variable X. That is, if $Y_n = \mathbf{E}^{\mathcal{G}}[X_n]$ for all n, then it does not follow that $Y_\sigma = \mathbf{E}^{\mathcal{G}}[X_\sigma]$. For example, let $\mathcal{G} = \{\emptyset, \Omega\}$, let $A \in \mathcal{F}_1$ satisfy $\mathbf{P}(A) = 1/2$, let $\sigma = 1$ on A and 2 otherwise, and let $X_1 = 1\,\mathbf{1}_A$, $X_2 = 2\,\mathbf{1}_A$.

(1.4.10) (Reversed martingales.) Let $(\mathcal{F}_n)$ be a reversed stochastic basis. The process $(X_n)_{n \in -\mathbb{N}}$ is a *reversed martingale* iff $\mathbf{E}[X_\sigma]$ is constant $(\sigma \in \Sigma)$, or, equivalently, $\mathbf{E}^{\mathcal{F}_m}[X_n] = X_m$ for $m \leq n$ in $-\mathbb{N}$. Every reversed martingale clearly has the form $X_n = \mathbf{E}^{\mathcal{F}_n}[Y]$ for all n: in fact $Y = X_{-1}$ will do. So a reversed martingale is automatically uniformly integrable.

(1.4.11) (Counterexample for convergence.) It is not true that an amart (X_n) that is not L_1-bounded converges a.s. It is not even true that a martingale converges a.s. on the set $\{\sup_n |X_n| < \infty\}$. Let $Y_1, Y_2, \cdots$ be independent random variables with $\mathbf{P}\{Y_k = -1\} = 1 - 2^{-k}$ and

$\mathbf{P}\{Y_k = 2^k - 1\} = 2^{-k}$. Let $\tau = \inf\{n : Y_n \neq -1\}$. Then τ is a stopping time (not a simple stopping time), and

$$\mathbf{P}\{\tau = \infty\} = \prod_{k=1}^{\infty} \left(1 - 2^{-k}\right) > 0.$$

Define $X_n = \sum_{k=1}^{n}(-1)^k Y_k \mathbf{1}_{\{\tau \geq k\}}$. Then (X_n) is an amart (even a martingale). Write $X^* = \sup_n |X_n|$. Then $\mathbf{P}\{X^* > 2^n\} \leq \sum_{k=n+1}^{\infty} 2^{-k} = 2^{-n}$, so it follows that $2^n \mathbf{P}\{X^* > 2^n\} \leq 1$ for all integers $n \geq 1$. Then $\lambda \mathbf{P}\{X^* > \lambda\} \leq 2$ for all real $\lambda \geq 1$. Thus we have $X^* < \infty$ a.s., but on the set $\{\tau = \infty\}$, the process X_n alternates between 1 and 0, and does not converge. (We owe this example to D. L. Burkholder.)

(1.4.12) (Conditional expectation for nonintegrable random variables.) If $X \geq 0$ is a random variable (possibly with the value ∞), the conditional expectation $\mathbf{E}^{\mathcal{G}}[X]$ may be defined by

$$Y = \lim_{n \to \infty} \mathbf{E}^{\mathcal{G}}[X \wedge n].$$

Even if X is not integrable, the defining relations

$$\mathbf{E}[Y \mathbf{1}_A] = \mathbf{E}[X \mathbf{1}_A] \qquad \text{for all } A \in \mathcal{G}$$

remain true. More generally, if X is a random variable such that $\mathbf{E}[X^+] < \infty$ or $\mathbf{E}[X^-] < \infty$, then we may use the definition

$$\mathbf{E}^{\mathcal{G}}[X] = \mathbf{E}^{\mathcal{G}}[X^+] - \mathbf{E}^{\mathcal{G}}[X^-],$$

since the right side is a.s. not of the form $\infty - \infty$.

(1.4.13) (Contraction property.) The operator $\mathbf{E}^{\mathcal{G}}$ is a contraction on L_1: that is, $\|\mathbf{E}^{\mathcal{G}}[X]\|_1 \leq \|X\|_1$. Indeed, let $A = \{\mathbf{E}^{\mathcal{G}}[X] > 0\}$. Then $A \in \mathcal{G}$, so $\mathbf{E}\left[|\mathbf{E}^{\mathcal{G}}[X]|\right] = \mathbf{E}\left[\mathbf{1}_A \mathbf{E}^{\mathcal{G}}[X]\right] - \mathbf{E}\left[\mathbf{1}_{\Omega \setminus A} \mathbf{E}^{\mathcal{G}}[X]\right] = \mathbf{E}[\mathbf{1}_A X] - \mathbf{E}[\mathbf{1}_{\Omega \setminus A} X] \leq \mathbf{E}\left[|X|\right]$.

(1.4.14) (Markovian operator.) The conditional expectation operator $\mathbf{E}^{\mathcal{G}}$ is a *Markovian* operator, that is, a positive linear operator on L_1 that preserves expectation. Every Markovian operator is a contraction on L_1. Other examples of Markovian operators appear in Chapter 8. More generally, the conditional expectation operator is a contraction on Orlicz spaces (2.3.11), hence L_p spaces $(1 \leq p \leq \infty)$. In fact, $\mathbf{E}^{\mathcal{G}}$ is a contraction on rearrangement invariant Banach function spaces (Lindenstrauss & Tzafriri [1979], p. 122).

(1.4.15) (Non-preservation of convergence.) Convergence a.e. need not be preserved by $\mathbf{E}^{\mathcal{G}}$, even in the presence of uniform integrability. Given a nonnegative sequence (X_n) that converges a.e. to 0 but such that $\sup |X_n|$

is not integrable, there is a σ-algebra $\mathcal{G}$ such that $\limsup \mathbf{E}^{\mathcal{G}}[X_n] = \infty$ (Blackwell & Dubins [1963]).

(1.4.16) (Lebesgue decomposition of signed charges.) Let $\mathcal{G} \subseteq \mathcal{F}$ be an algebra of sets. A *charge* on $\mathcal{G}$ is a finitely additive set function $\mu\colon \mathcal{G} \to \mathbb{R}$. If μ has finite variation, then there is a decomposition $\mu = \mu_a + \mu_s$; where μ_a is absolutely continuous in the sense that $\mu_a(A) = \mathbf{E}[Y 1_A]$ for all A, for some integrable random variable Y; and μ_s is singular in the sense that for every $\varepsilon > 0$ there is $B \in \mathcal{G}$ with $|\mu_s|(B) < \varepsilon$ but $\mathbf{P}(B) > 1-\varepsilon$. Prove this in a manner similar to (1.3.2), above. First suppose $\mu \geq 0$. Let J be the set of finite partitions of Ω by sets of $\mathcal{G}$. Then J is directed by refinement. (Note: changing on a null set is not allowed, since it is possible that $\mu(B) \neq 0$ even when $\mathbf{P}(B) = 0$.) Then for $t \in J$, let

$$X_t = \sum \frac{\mu(A)}{\mathbf{P}(A)} 1_A,$$

where the sum is over all $A \in t$ with $\mathbf{P}(A) \neq 0$. Then (X_t) is an L_1-bounded supermartingale, so it converges stochastically, say to Y. Then $\mu_a(B) = \mathbf{E}[Y 1_B]$, $B \in \mathcal{G}$, and $\mu_s(B) = \mu(B) - \mu_a(B)$ define the required decomposition. For general μ, use the finiteness of the variation of μ to write μ as the difference of two nonnegative charges.

(1.4.17) (A decomposition for martingales.) Let (X_t) be an L_1-bounded martingale for the stochastic basis $(\mathcal{F}_t)$. Then (X_t) can be written $X_t = Y_t + Z_t$, where Y_t is a martingale of the form $Y_t = \mathbf{E}^{\mathcal{F}_t}[Y]$ for some integrable random variable Y, and Z_t is singular in the sense that the charge $\nu(A) = \lim_t \mathbf{E}[Z_t 1_A]$ is singular as in (1.4.16). Prove this by applying the Lebesgue decomposition (1.4.16) to the charge $\mu(A) = \lim_t \mathbf{E}[X_t 1_A]$, for $A \in \mathcal{G} = \bigcup_t \mathcal{F}_t$. Show that Y may be chosen to be the stochastic limit of X_t (Theorem 1.3.1); then clearly $Z_t \to 0$ stochastically.

(1.4.18) (Maximal inequality.) The maximal inequality (1.1.7) is simpler for martingales than for general amarts, since the expression $\sup_{\sigma \in \Sigma} \mathbf{E}[|X_\sigma|]$ may be written in other forms: Let (X_n) be a martingale or a nonnegative submartingale. Then the net $\mathbf{E}[|X_\sigma|]$ is increasing, so

$$\sup_{\sigma \in \Sigma} \mathbf{E}[|X_\sigma|] = \lim_{\sigma \in \Sigma} \mathbf{E}[|X_\sigma|] = \sup_{n \in \mathbb{N}} \mathbf{E}[|X_n|] = \lim_{n \in \mathbb{N}} \mathbf{E}[|X_n|].$$

To see this, observe that if (X_n) is a martingale, then $(|X_n|)$ is a nonnegative submartingale. So it is enough to consider the case of a nonnegative submartingale. By (1.4.3), the net $(\mathbf{E}[X_\sigma])_{\sigma \in \Sigma}$ is increasing, and the sequence $(\mathbf{E}[X_n])_{n \in \mathbb{N}}$ is a subnet of it, therefore has the same limit (finite or infinite). So we have *Doob's maximal inequality*:

$$\mathbf{P}\{\sup |X_n| \geq \lambda\} \leq \frac{1}{\lambda} \sup_n \mathbf{E}[|X_n|].$$

(1.4.19) (Nonsimple stopping times.) Let (X_n) be an L_1-bounded martingale or nonnegative submartingale. Then it converges, write $X_\infty = \lim X_n$. If τ is any stopping time, with values in $\mathbb{N} \cup \{\infty\}$, then

$$\mathbf{E}\left[|X_\tau|\right] \leq \sup_n \mathbf{E}\left[|X_n|\right].$$

To prove this, write $M = \sup_n \mathbf{E}[X_n]$. Since (X_n) is an L_1-bounded martingale (or nonnegative submartingale), so is $(X_{\tau \wedge n})_{n \in \mathbb{N}}$. The stopping times $\tau \wedge n$ belong to Σ, so certainly $\mathbf{E}[X_{\tau \wedge n}] \leq M$. But $\lim_n X_{\tau \wedge n} = X_\tau$ (consider both the sets $A = \{\tau = \infty\}$ and $\Omega \setminus A = \{\tau < \infty\}$). Apply Fatou's Lemma.

(1.4.20) (Approximation by stopping times.) Let (X_n) be an adapted integrable process. Let $\tau \in \Sigma$. Then

$$\mathbf{E}\left[\sup_{n \geq \tau} \left|\mathbf{E}^{\mathcal{F}_\tau}[X_n]\right|\right] \leq \sup_{\sigma \geq \tau} \mathbf{E}\left[\left|\mathbf{E}^{\mathcal{F}_\tau}[X_\sigma]\right|\right].$$

(1.4.21) (Difference property in $\mathbb{N}$.) Let (X_n) be an adapted integrable process. Then the following are equivalent:

(1) (X_n) is an amart;
(2) for every $\varepsilon > 0$, there is $m_0 \in \mathbb{N}$ such that $\sigma, \tau \in \Sigma$ and $m_0 \leq \sigma \leq \tau$ imply

$$\mathbf{E}\left[\left|\mathbf{E}^{\mathcal{F}_\sigma}[X_\tau] - X_\sigma\right|\right] < \varepsilon;$$

(3) for every $\varepsilon > 0$, there is $m_0 \in \mathbb{N}$ such that $m \in \mathbb{N}$, $\tau \in \Sigma$, and $m_0 \leq m \leq \tau$ imply

$$\mathbf{E}\left[\left|\mathbf{E}^{\mathcal{F}_m}[X_\tau] - X_m\right|\right] < \varepsilon.$$

(1.4.22) (Riesz decomposition in $\mathbb{N}$.) Let (X_n) be an amart. Then X_n has a unique decomposition $X_n = Y_n + Z_n$, where (Y_n) is a martingale and (Z_n) is an amart potential. In fact, Y_m is the L_1 norm limit $\lim_n \mathbf{E}^{\mathcal{F}_m}[X_n]$. [By (1.4.20) and (1.2.3(b)), the supremum $\sup_n \mathbf{E}^{\mathcal{F}_m}[X_n]$ is integrable for each m.] Also $\lim_{\sigma \in \Sigma} \mathbf{E}\left[|Z_\sigma|\right] = 0$ and $\lim Z_n = 0$ a.s.

More information on amart potentials is given in (4.1.15). The Riesz decomposition will be generalized below to Banach spaces (5.2.13 and 5.2.27), and semiamarts (1.4.26).

(1.4.23) (Uniform difference property.) Let X_n be an amart. Then for any $\varepsilon > 0$ there is $m \in \mathbb{N}$ so that for all $\sigma, \tau \in \Sigma$ with $m \leq \sigma \leq \tau$, we have

$$\sup_{A \in \mathcal{F}_\sigma} \mathbf{E}\left[\left(X_\sigma - \mathbf{E}^{\mathcal{F}_\sigma}[X_\tau]\right) 1_A\right] < \varepsilon.$$

(1.4.24) (*Doob potentials.*) Amart potentials can be related to the potentials that are used in classical martingale theory. A *Doob potential* is a positive supermartingale S_n with $\lim_n \mathbf{E}\left[|S_n|\right] = 0$. In particular $\lim_n \mathbf{E}\left[S_n \mathbf{1}_A\right] = 0$ for all $A \in \bigcup_{n=1}^{\infty} \mathcal{F}_n$, so it is an amart potential, and $S_n \to 0$. We will see (Theorem 4.1.15) that an adapted process (Z_n) is an amart potential if and only if there is a Doob potential S_n with $|Z_n| \le S_n$ for all n.

(1.4.25) (*A stopped inequality.*) Use Chacon's inequality (1.2.9) to prove: If (X_n) is an L_1-bounded process, then

$$\mathbf{E}\left[\limsup_{n,m} |X_n - X_m|\right] \le 2 \limsup_{\tau \ge \sigma} \mathbf{E}\left[\left|\mathbf{E}^{\mathcal{F}_\sigma}\left[X_\tau\right] - X_\sigma\right|\right].$$

(1.4.26) (*Semiamart Riesz decomposition.*) Suppose $\mathbf{E}\left[X_\tau\right]$ is bounded. Then $X_n = Y_n + Z_n$, where (Y_n) is a martingale and (Z_n) is L_1-bounded and oscillates about 0 in the sense:

$$\liminf_n \frac{1}{n} \sum_{i=1}^{n} \mathbf{E}\left[Z_i \mathbf{1}_A\right] \le 0 \le \limsup_n \frac{1}{n} \sum_{i=1}^{n} \mathbf{E}\left[Z_i \mathbf{1}_A\right]$$

for all $A \in \bigcup \mathcal{F}_m$.

Hint: Fix m. By (1.2.11),

$$\sup_n \mathbf{E}^{\mathcal{F}_m}\left[X_n\right] \in L_1.$$

Let Y_m be a weak truncated limit of a subsequence of

$$\left(\frac{1}{n} \sum_{i=1}^{n} \mathbf{E}^{\mathcal{F}_m}\left[X_n\right]\right)_{n \in \mathbb{N}}$$

(8.1.1).

It can also be shown that $X_n - Y_n$ is dominated by the Snell envelope

$$\operatorname*{ess\,sup}_{\substack{\sigma \in \Sigma \\ \sigma \ge n}} \mathbf{E}^{\mathcal{F}_n}\left[X_\sigma\right]$$

(Krengel & Sucheston [1978], p. 204).

(1.4.27) (*Reversed difference property.*) Let $(\mathcal{F}_n)_{n \in -\mathbb{N}}$ be a reversed stochastic basis. The adapted integrable process $(X_n)_{n \in -\mathbb{N}}$ is an amart if and only if for every $\varepsilon > 0$, there is $m_0 \in -\mathbb{N}$ such that for all $\sigma, \tau \in \Sigma$ with $\sigma \le \tau \le m_0$,

$$\mathbf{E}\left[\left|\mathbf{E}^{\mathcal{F}_\sigma}\left[X_\tau\right] - X_\sigma\right|\right] < \varepsilon.$$

(1.4.28) (Reversed amart Riesz decomposition.) Let $(X_n)_{n \in \mathbb{N}}$ be a reversed amart. The process is uniformly integrable and converges to a random variable $X_{-\infty}$. Then the sequence $Z_n = X_n - X_{-\infty}$ is adapted; set $Y_n = X_{-\infty}$ for all n.

(1.4.29) (Optional sampling.) Let (X_n) be a martingale [submartingale] for $(\mathcal{F}_n)$. If $\sigma_1 \leq \sigma_2 \leq \cdots$ in Σ, then (X_{σ_k}) is a martingale [submartingale] for $(\mathcal{F}_{\sigma_k})$. Optional sampling for quasimartingales follows from the next complement.

(1.4.30) (Alternative definition of quasimartingale.) Let $(X_n)_{n \in \mathbb{N}}$ be a sequence of real valued random variables adapted to an stochastic basis $(\mathcal{F}_n)_{n \in \mathbb{N}}$. Let C be a positive constant. The following are equivalent:

   (i) $\mathbf{E}\left[\sum_{i=1}^{\infty} \left|\mathbf{E}^{\mathcal{F}_i}[X_{i+1}] - X_i\right|\right] \leq C.$
   (ii) $\mathbf{E}\left[\sum_{j=1}^{m-1} \left|\mathbf{E}^{\mathcal{F}_{\tau_i}}[X_{\tau_{i+1}}] - X_{\tau_i}\right|\right] \leq C$ for any sequence $\tau_1 \leq \tau_2 \leq \cdots \leq \tau_m$ of simple stopping times.
   (iii) $\sum_{j=1}^{m-1} \left|\mathbf{E}[X_{\tau_{i+1}}] - \mathbf{E}[X_{\tau_i}]\right| \leq C$ for any sequence $\tau_1 \leq \tau_2 \leq \cdots \leq \tau_m$ of simple stopping times.

The techniques in the proof of this are like those in the proof of (1.4.4) (R. Wittmann, private communication).

(1.4.31) (Continuous parameter amarts.) Let $(\Omega, \mathcal{F}, \mathbf{P})$ be a *complete* probability space (subsets of null sets are in $\mathcal{F}$). Let the index set J be an interval in the line $\mathbb{R}$. Suppose the stochastic basis $(\mathcal{F}_t)_{t \in J}$ is *right continuous*, i.e., for each s,

$$\mathcal{F}_s = \bigcap_{t > s} \mathcal{F}_t,$$

and each $\mathcal{F}_t$ contains all null sets in $\mathcal{F}$. An *optional time* is a function $\tau \colon \Omega \to J \cup \{\infty\}$ such that $\{\tau \leq t\} \in \mathcal{F}_t$ for all $t \in J$. An optional time τ is called *simple* iff it takes finitely many finite values. A *continuous parameter ascending amart* is a process $(X_t)_{t \in J}$ such that $\mathbf{E}[X_{\tau_n}]$ converges for every ascending (= increasing) sequence (τ_n) of simple optional times. The definition of a *descending amart* is the same, except that the sequence τ_n is descending (= decreasing). An *amart* is a process that is both an ascending amart and a descending amart.

The theory of continuous-parameter amarts parallels that of continuous parameter martingales (Doob [1953] or Dellacherie & Meyer [1982]). We recall some basic definitions. A *trajectory* of a process (X_t) at ω is the net $(X_t(\omega))_{t \in J}$. A *modification* of a process (X_t) is a process (Y_t) such that $X_t = Y_t$ a.s.; the exceptional null set may depend on t. The processes (X_t) and (Y_t) are *indistinguishable* iff there is a single null set, outside of which $X_t = Y_t$ for all t.

Under mild boundedness assumptions, a continuous parameter amart has a modification every trajectory of which has right and left limits. If an amart is right continuous in probability, then it has a modification every trajectory of which is right continuous.

A related but stronger notion is the hyperamart. A process (X_t) is a *hyperamart* iff $\mathbf{E}\left[X_{\tau_n}\right]$ converges for all monotone sequences of bounded but not necessarily simple optional times. If (X_t) is a hyperamart, then there is a process Y_t indistinguishable from X_t with the regularity properties (such as right and left limits). Reference: Edgar & Sucheston [1976b].

(1.4.32) (Continuous parameter quasimartingales.) Let J be an interval in $\mathbb{R}$, and let $(\mathcal{F}_t)_{t\in J}$ be a stochastic basis with the properties stated above (1.4.31). A process $(X_t)_{t\in J}$ is a *martingale* iff $X_s = \mathbf{E}^{\mathcal{F}_s}[X_t]$ for $s < t$. The process (X_t) is a *submartingale* iff $X_s \le \mathbf{E}^{\mathcal{F}_s}[X_t]$ for $s < t$ and a *supermartingale* iff $X_s \ge \mathbf{E}^{\mathcal{F}_s}[X_t]$ for $s < t$. The process (X_t) is a *quasimartingale* iff there is a constant M so that

$$\sum_{i=1}^{n} \left\| \mathbf{E}\left[X_{t_i} \mid \mathcal{F}_{t_{i-1}}\right] - X_{t_{i-1}} \right\|_1 \le M$$

for all finite sequences $t_1 < t_1 < \cdots < t_n$ in J. (See Fisk [1965], Orey [1967], Rao [1969].) All quasimartingales (in particular martingales, L_1-bounded submartingales, L_1-bounded supermartingales) are continuous parameter amarts.

Remarks

Roughly speaking, there are two antecedents to the martingale theory. One is the derivation theory of R. de Possel (for example de Possel [1936]). Another, probabilistic, approach originated with Paul Lévy [1937]; he generalized sums of independent variables with 0 expectations by "centering" (assuming 0 conditional expectations given the past). As for sub- and supermartingales (called by Doob upper and lower semimartingales), they seem to have appeared first in the derivation theory. The quest for the best formulation went back and forth between random variables and set-functions, as between Doob and Andersen-Jessen (see Doob [1953], pp. 630 ff). At the end, successors and students of Andersen-Jessen showed that their set-function notions could be just as general or more general; Lamb [1973] gave a set-function formulation equivalent to (but preceding) the random-variable formulation of the amart in Austin, Edgar, Ionescu Tulcea [1974]. But almost no one paid attention any more and random variables prevailed. The reason seems to be that, because of the Radon-Nikodým theorem, set functions bring no greater generality, and the intuitive meaning of stochastic processes and stopping times is lost in the set-function approach.

Krickeberg [1957], [1959], initiating the topic, proved stochastic convergence of L_1-bounded submartingales indexed by directed sets. Millet & Sucheston [1980b] proved stochastic convergence of L_1-bounded ordered amarts. The proof given above is quite different. The difference property is from Edgar & Sucheston [1976c]; the Riesz decomposition is from Edgar & Sucheston [1976a]; we have used a proof due to Astbury [1978].

The theory of continuous parameter amarts is from Edgar & Sucheston [1976b]. Hyperamarts appear earlier in the work of Meyer [1971] and Mertens [1972]; for an intermediate notion, with optional times taking countably many values, see Doob [1975]. A Banach-valued version of continuous parameter amarts appears in Choi & Sucheston [1981].

2

Infinite measure and Orlicz spaces

In this chapter we will prepare some of the tools to be used later.

One important possibility is the use of an infinite measure space, rather than a probability space. For probability in the narrow sense, only finite measure spaces are normally used. Attempts to do probability in infinite measure spaces have had little success. In most of the book, we consider primarily the case of finite measure space. But there are reasons for considering infinite measure spaces. The techniques to be developed for pointwise convergence theorems can tell us something also in infinite measure spaces. The ergodic theorems in Chapter 8 often have their natural setting in infinite measure spaces. J. A. Hartigan [1983] argues that an infinite measure space provides a rigorous foundation for Bayesian statistics. Our material on derivation requires the possibility of infinite measures to cover the most common cases, such as Euclidean space $\mathbb{R}^n$ with Lebesgue measure.

Another important tool to be used is the Orlicz space. The class of Orlicz spaces generalizes the class of function spaces L_p. If the primary concern is finite measure space, then Orlicz spaces are basic. A necessary and sufficient condition for uniform integrability of a set of functions is boundedness in an Orlicz norm. Zygmund's Orlicz space $L \log L$, and, more generally, Orlicz spaces $L \log^k L$, appear naturally in considerations of integrability of supremum and related multiparameter convergence theorems. For consideration of infinite measure spaces, Orlicz spaces will not suffice, unless property (Δ_2) holds. But in principal cases, the "heart" of an Orlicz space (the closure of the integrable simple functions) plays an important role. Measure-theoretic arguments often require approximation of functions by simple functions; therefore much of the theory applies only to hearts of Orlicz spaces. The hearts of the Orlicz spaces $L \log^k L$, the R_k spaces introduced by N. Fava, are especially important. A novelty of our approach consists in identifying useful spaces as hearts of appropriate Orlicz spaces, and applying the general theory. For instance, the Fava space R_0, which is the right setting for the Dunford-Schwartz and the martingale theorems in infinite measure spaces, is identified as the heart of the largest Orlicz space $L_1 + L_\infty$; it follows at once that R_0 is an order-continuous Banach lattice.

We will normally write $(\Omega, \mathcal{F}, \mu)$ for a (possibly infinite) measure space. We may say "measurable function" rather than "random variable" for a function $f \colon \Omega \to \mathbb{R}$. Normally we will write $\int f \, d\mu$ for the integral, and reserve $\mathbf{E}[f]$ for the case of a probability space. Conditional expectations $\mathbf{E}_\mu[f \mid \mathcal{G}]$ will be considered later, but $\int f \, d\mu$ is *not* the special case of $\mathbf{E}_\mu[f \mid \mathcal{G}]$ for trivial σ-algebra $\mathcal{G}$. We may also use the phrase "almost everywhere" rather than "almost surely" in this context.

2.1. Orlicz spaces

Much of modern probability theory, especially the part concerned with convergence theorems, deals with appropriate *spaces* of measurable functions. The most useful are the spaces L_p, but these spaces are not enough for everything we want to do here. It is therefore necessary to consider the generalization known as the Orlicz spaces.

We have attempted to make the exposition self-contained. It requires only a knowledge of the spaces L_p and their basic properties, such as contained in Cohn [1980] or Royden [1968].

Orlicz functions and their conjugates

The Orlicz function generalizes the function t^p used for the definition of the space L_p. The generality in the definition determines the generality of the Orlicz spaces that will result. We allow the values 0 and ∞ for the function so that spaces like $L \log L$ and L_∞ will be included. Arithmetic with ∞ follows the usual conventions, including $0 \cdot \infty = 0$.

(2.1.1) Definition. An *Orlicz function* is a function

$$\Phi \colon [0, \infty) \to [0, \infty]$$

satisfying:

(1) $\Phi(0) = 0$.
(2) Φ is left-continuous: $\lim_{u \uparrow x} \Phi(u) = \Phi(x)$.
(3) Φ is increasing: if $u_1 \leq u_2$, then $\Phi(u_1) \leq \Phi(u_2)$.
(4) Φ is convex: $\Phi\big(au_1 + (1-a)u_2\big) \leq a\Phi(u_1) + (1-a)\Phi(u_2)$, for $0 \leq a \leq 1$.
(5) Φ is nontrivial: $\Phi(u) > 0$ for some $u > 0$ and $\Phi(u) < \infty$ for some $u > 0$.

The convexity (4) implies that Φ is continuous except possibly at a single point, where it jumps to ∞, so condition (2) is needed only at that one point.

An Orlicz function is increasing, so (see 7.1.4) it is differentiable a.e. and (see 7.1.11) its derivative $\varphi = \Phi'$ satisfies

$$(2.1.1a) \qquad \Phi(u) = \int_0^u \varphi(x)\, dx, \qquad 0 \leq u < \infty.$$

If $\Phi(u) = \infty$ for some values of u, then we take by convention $\varphi(u) = \infty$ also, so that (2.1.1a) remains correct. Since Φ is increasing, we have $\varphi \geq 0$. Since Φ is convex, the derivative φ is increasing. Then φ is continuous except at countably many points, so φ can be made left-continuous without destroying (2.1.1a).

(2.1.2) **Definition.** An *Orlicz derivative* is a function

$$\varphi \colon (0, \infty) \to [0, \infty]$$

satisfying:

(1) φ is increasing.
(2) φ is left-continuous.
(3) φ is nontrivial: $\varphi(x) > 0$ for some x and $\varphi(x) < \infty$ for some x.

It is not hard to verify that if φ is an Orlicz derivative, then (2.1.1a) defines an Orlicz function Φ with derivative φ.

If φ is strictly increasing, then it has an inverse function ψ. In any case, we may define a "generalized left-continuous inverse" ψ of φ as follows:

$$\psi(y) = \inf \left\{ x \in (0, \infty) : \varphi(x) \geq y \right\}.$$

An example is shown in Figure (2.1.2). The new function ψ is also an Orlicz derivative, so

$$\Psi(v) = \int_0^v \psi(y)\, dy$$

defines an Orlicz function. We say that Ψ is the *conjugate* Orlicz function to Φ. Notice that the construction works in reverse: The derivative of Ψ is ψ; the generalized left-continuous inverse of ψ is φ; and $\Phi(u) = \int_0^u \varphi(x)\, dx$.

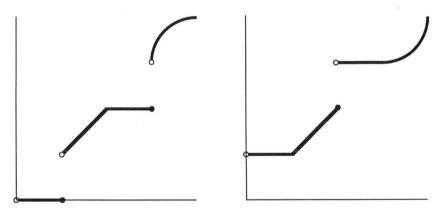

Figure (2.1.2). An Orlicz derivative and its
generalized left-continuous inverse.

Let us consider more carefully the relation between φ and ψ. If φ is discontinuous at $x = a$, then $\psi(y) = a$ for all y with $\varphi(a-) < y \leq \varphi(a+)$. Since Φ is convex, the function $\Phi(u)/u$ is increasing. We have $\Phi(u)/u \to \infty$ as $u \to \infty$ if and only if $\varphi(x) \to \infty$ as $x \to \infty$; this is equivalent to $\psi(y) < \infty$ for all y. (In this case, we say that φ is *unbounded*, ψ is *finite*, or Ψ is *finite*.) On the other hand, if $\lim_{x \to \infty} \varphi(x) = d$ is finite, then $\psi(v) = \Psi(v) = \infty$ for all $v > d$. (Then we say that φ is *bounded*, ψ is *infinite*, or Ψ is *infinite*.)

(2.1.3) Definition. A *Young partition* is a pair (E_1, E_2) of open subsets of $(0, \infty) \times (0, \infty)$ such that

   (1) $E_1 \cap E_2 = \emptyset, E_1 \neq \emptyset, \; E_2 \neq \emptyset$.

   (2) E_1 is south-east hereditary: if $(x, y) \in E_1$, $x' \geq x$, and $y' \leq y$, then $(x', y') \in E_1$.

   (3) E_2 is north-west hereditary.

   (4) $\overline{E_1} \cup E_2 = E_1 \cup \overline{E_2} = (0, \infty) \times (0, \infty)$.

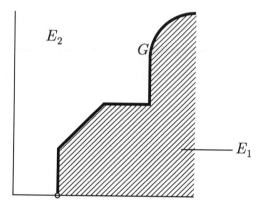

Figure (2.1.3). Young partition.

If (E_1, E_2) is a Young partition, then

$$\varphi(x) = \sup \{ y : (x, y) \in E_1 \}$$

is an Orlicz derivative with generalized left-continuous inverse

$$\psi(y) = \sup \{ x : (x, y) \in E_2 \} .$$

Conversely, if φ is an Orlicz derivative with inverse ψ, then

$$E_1 = \{ (x, y) \in (0, \infty) \times (0, \infty) : y < \varphi(x) \}$$
$$E_2 = \{ (x, y) \in (0, \infty) \times (0, \infty) : x < \psi(y) \}$$

is a Young partition. The set

$$G = \{ (x, \varphi(x)) : x \in (0, \infty) \} \cup \{ (\psi(y), y) : y \in (0, \infty) \}$$

is the common boundary of E_1 and E_2. See Figure (2.1.3). The set G has two-dimensional Lebesgue measure 0, since it meets each line of slope -1 in only one point.

The Young partition leads to a useful inequality.

(2.1.4) Theorem (Young's inequality). *Let Φ and Ψ be conjugate Orlicz functions with derivatives φ and ψ. Then*

$$uv \leq \Phi(u) + \Psi(v)$$

for all $u, v > 0$. Equality holds if and only if $v = \varphi(u)$ or $u = \psi(v)$.

Proof. Figure (2.1.4). ∎

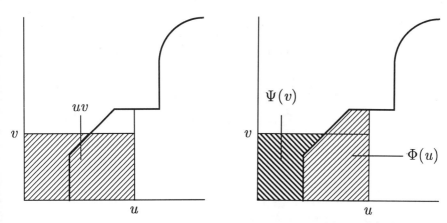

Figure (2.1.4). Young's inequality: $uv \leq \Phi(u) + \Psi(v)$.

(2.1.5) Corollary. $\Phi(u) = \sup\{uv - \Psi(v) : v \geq 0\}$ *for $u \geq 0$, and the sup is achieved if $\varphi(u) < \infty$.*

Proof. Fix $u > 0$. Then $\Phi(u) + \Psi(v) \geq uv$ for all v and uv is finite, so (even if one of $\Phi(u), \Psi(v)$ is infinite) we have

$$\Phi(u) \geq \sup\{uv - \Psi(v) : v \geq 0\}.$$

Now if $\varphi(u) < \infty$, then for $v = \varphi(u)$ we have $\Phi(u) + \Psi(v) = uv$, so both $\Phi(u)$ and $\Psi(v)$ are finite and

$$\Phi(u) = \max\{uv - \Psi(v) : v \geq 0\}.$$

On the other hand, if $\varphi(u) = \infty$, then Ψ is bounded, so

$$\sup\{uv - \Psi(v) : v \geq 0\} = \infty.$$

∎

Orlicz spaces

We will now consider a measure space $(\Omega, \mathcal{F}, \mu)$ where μ is σ-finite (and nonnegative). Certain properties of the measure space will be important in the description of the Orlicz space.

(2.1.6) Definition. Let $(\Omega, \mathcal{F}, \mu)$ be a measure space. Then we say μ is *infinite* iff $\mu(\Omega) = \infty$; otherwise we say μ is *finite*. We say that μ *has arbitrarily small sets* iff

$$\inf \{ \mu(A) : A \in \mathcal{F}, \ \mu(A) > 0 \} = 0.$$

The canonical examples are the following. The unit interval $[0, 1]$ with Lebesgue measure: This is finite and has arbitrarily small sets. The interval $[0, \infty)$ with Lebesgue measure: This is infinite and has arbitrarily small sets. The positive integers $\mathbb{N}$ with counting measure: This is infinite and has no arbitrarily small sets. The fourth possibility (finite measure with no arbitrarily small sets) is typified by a finite set $\{1, 2, \cdots, n\}$ with counting measure; this is less interesting, since all of the Orlicz spaces consist of the same functions (it is, however, important in the isometric theory of Banach spaces).

Let Φ and Ψ be conjugate Orlicz functions with derivatives φ and ψ. They will be fixed throughout most of Section 2.1.

(2.1.7) Definition. Let $f \colon \Omega \to \mathbb{R}$ be a measurable function. The *Orlicz modular* of f for Φ is

$$M_\Phi (f) = \int \Phi(|f|) \, d\mu.$$

The *Young class* of Φ is the set $Y_\Phi = Y_\Phi(\Omega, \mathcal{F}, \mu)$ consisting of all measurable functions $f \colon \Omega \to \mathbb{R}$ with $M_\Phi (f) < \infty$.

Note that Φ is convex, so M_Φ is convex in this sense:

(2.1.8) Proposition. *Let f, g be measurable functions, and let a satisfy $0 < a < 1$. Then $M_\Phi (af + (1 - a)g) \leq a M_\Phi (f) + (1 - a)M_\Phi (g)$.*

Proof.

$$M_\Phi (af + (1 - a)g) = \int \Phi(|af + (1 - a)g|) \, d\mu$$

$$\leq \int \Phi(a|f| + (1 - a)|g|) \, d\mu$$

$$\leq \int \left(a\Phi(|f|) + (1 - a)\Phi(|g|) \right) d\mu$$

$$= a M_\Phi (f) + (1 - a)M_\Phi (g) . \qquad \blacksquare$$

This implies that the Young class Y_Φ is a convex set, and in fact that the *Orlicz ball*

$$B_\Phi = \{ f : M_\Phi (f) \leq 1 \}$$

is a convex set.

In general, the Young class is not a linear space. It is possible that $f \in Y_\Phi$ but $2f \notin Y_\Phi$ (see 2.1.25). We must therefore consider more complex definitions.

(2.1.9) Definition. The *Orlicz space* for Φ is the set

$$L_\Phi = L_\Phi(\Omega, \mathcal{F}, \mu)$$

of all measurable functions $f \colon \Omega \to \mathbb{R}$ such that $M_\Phi\left(f/a\right) < \infty$ for some $a > 0$. The *heart* of the Orlicz space for Φ is the set $H_\Phi = H_\Phi(\Omega, \mathcal{F}, \mu)$ of all measurable functions $f \colon \Omega \to \mathbb{R}$ such that $M_\Phi\left(f/a\right) < \infty$ for all $a > 0$. The *Luxemburg norm* of a measurable function f is

$$\|f\|_\Phi = \inf\left\{\, a > 0 : M_\Phi\left(f/a\right) \le 1 \right\},$$

where by convention $\inf \emptyset = \infty$. If $(\Omega, \mathcal{F}, \mu)$ is the discrete space $\mathbb{N}$ with counting measure, then we write $l_\Phi = L_\Phi(\mathbb{N})$ and $h_\Phi = H_\Phi(\mathbb{N})$. The spaces l_Φ are known as *Orlicz sequence spaces*.

Note that if $f \in L_\Phi$, then $M_\Phi\left(f/n\right) < \infty$ for some integer n. But $|f|/n \to 0$ a.e., so $\Phi(|f|/n) \to 0$ a.e. (Here we use the assumption that $\Phi(u) < \infty$ for some $u > 0$.) Thus, by the dominated convergence theorem, $M_\Phi\left(f/n\right) \le 1$ for some n, and $\|f\|_\Phi < \infty$. Conversely, if $\|f\|_\Phi < \infty$, then clearly $f \in L_\Phi$. Another description of L_Φ is thus the set of all f with $\|f\|_\Phi < \infty$.

We insert here a few basic calculations.

(2.1.10) Proposition.

(1) If $0 < \|f\|_\Phi < \infty$, then $M_\Phi\left(f/\|f\|_\Phi\right) \le 1$.
(2) If $\|f\|_\Phi \le 1$, then $M_\Phi\left(f\right) \le \|f\|_\Phi \le 1$.
(3) If $\|f\|_\Phi > 1$, then $M_\Phi\left(f\right) \ge \|f\|_\Phi > 1$.
(4) If $f \in H_\Phi$, then $\|f\|_\Phi = 1$ is equivalent to $M_\Phi\left(f\right) = 1$.
(5) $\|f_n - f\|_\Phi \to 0$ if and only if $M_\Phi\left(k(f_n - f)\right) \to 0$ for all $k > 0$.
(6) If $\|f_n - f\|_\Phi \to 0$, then f_n converges in measure to f, that is: for every $\varepsilon > 0$, we have $\lim_{n\to\infty} \mu\{|f_n - f| > \varepsilon\} = 0$.

Proof. (1) For $a > \|f\|_\Phi$, we have $M_\Phi\left(f/a\right) \le 1$. Now as a decreases to $\|f\|_\Phi$, the quotient $|f|/a$ increases to $|f|/\|f\|_\Phi$. By the left-continuity of Φ, we have $\Phi(|f|/a) \uparrow \Phi(|f|/\|f\|_\Phi)$, so by the monotone convergence theorem, $M_\Phi\left(f/a\right) \to M_\Phi\left(f/\|f\|_\Phi\right)$. Therefore $M_\Phi\left(f/\|f\|_\Phi\right) \le 1$.

(2) Let $a = \|f\|_\Phi$. Then by convexity, $M_\Phi\left(f\right)/a \le M_\Phi\left(f/a\right) \le 1$, so $M_\Phi\left(f\right) \le a$.

(3) If $1 < a < \|f\|_\Phi$, then by convexity, $M_\Phi\left(f\right)/a \ge M_\Phi\left(f/a\right) > 1$, so $M_\Phi\left(f\right) > a$. Thus $M_\Phi\left(f\right) \ge \|f\|_\Phi$.

(4) By (2) and (3), if $M_\Phi\left(f\right) = 1$, then $\|f\|_\Phi = 1$. Conversely, suppose $\|f\|_\Phi = 1$. By (2), $M_\Phi\left(f\right) \le 1$. For $0 < a < 1$, we have $\|f/a\|_\Phi > 1$, so $M_\Phi\left(f/a\right) > 1$. As $a \to 1$, we have $\Phi(|f|/a) \to \Phi(|f|)$. Now Φ is continuous at all points except possibly where it jumps to ∞. Since $f \in H_\Phi$, we see that $M_\Phi\left(f/a\right) < \infty$, so Φ is finite at the point $|f(\omega)|/a$ for almost all ω. Therefore $M_\Phi\left(f\right) = \lim_{a \uparrow 1} M_\Phi\left(f/a\right) \ge 1$. Thus $M_\Phi\left(f\right) = 1$.

(5) If $\|f - f_n\|_\Phi \to 0$, then $\|k(f - f_n)\|_\Phi \to 0$, so $\|k(f - f_n)\|_\Phi \leq 1$ for large n, so by (2), we have $M_\Phi(k(f - f_n)) \to 0$. Conversely, suppose $M_\Phi(k(f - f_n)) \to 0$ for all k. Then for each k we have $M_\Phi(k(f - f_n)) \leq 1$ for large n, so that $\|f - f_n\|_\Phi \leq 1/k$ for large n. Thus $\|f - f_n\|_\Phi \to 0$.

(6) Let $\varepsilon > 0$. There is u_0 so that $\Phi(u_0) > 0$. If $k = u_0/\varepsilon$, then for any $\delta > 0$, for large n we have $\int \Phi(k|f_n - f|)\, d\mu \leq \delta$, so

$$\delta \geq \int \Phi(k|f_n - f|)\, d\mu \geq \Phi(u_0)\, \mu\{|f_n - f| > \varepsilon\}.$$

Thus $\mu\{|f_n - f| > \varepsilon\} \to 0$.   ∎

One might remark that the hypothesis $f \in H_\Phi$ in part (4) cannot be omitted. The space L_∞ provides a counterexample.

If functions in L_Φ are identified when they agree a.e., then L_Φ is a Banach space:

(2.1.11) Theorem. (a) The set L_Φ is a Banach space. (b) The set H_Φ is a closed linear subspace of L_Φ, and $H_\Phi \subseteq Y_\Phi \subseteq L_\Phi$. (c) (The Fatou property) If $f_n \in L_\Phi$ is an increasing, nonnegative sequence with $\|f_n\|_\Phi \leq 1$ for all n, then the pointwise limit $f = \lim f_n$ belongs to L_Φ and $\|f\|_\Phi \leq 1$.

Proof. We first prove that L_Φ is a normed linear space. Homogeneity $\|\alpha f\|_\Phi = |\alpha|\, \|f\|_\Phi$ is easy from the definition of the norm. Thus if $f \in L_\Phi$, it follows that $\alpha f \in L_\Phi$. For the triangle inequality, let $f, g \in L_\Phi$, and let $a = \|f\|_\Phi$, $b = \|g\|_\Phi$. Then f/a and g/b belong to the Orlicz ball B_Φ. But B_Φ is convex, so

$$\frac{a}{a+b}\frac{f}{a} + \frac{b}{a+b}\frac{g}{b} = \frac{f+g}{a+b}$$

belongs to B_Φ. Thus $\|f + g\|_\Phi \leq a + b$. This shows that $f + g \in L_\Phi$ and $\|f + g\|_\Phi \leq \|f\|_\Phi + \|g\|_\Phi$.

In order for $\|f\|_\Phi = 0$, we must have $M_\Phi(f/a) \leq 1$ for all $a > 0$. Fix $\varepsilon > 0$. Then

$$M_\Phi\left(\frac{f}{a}\right) = \int \Phi\left(\frac{|f|}{a}\right) d\mu \geq \Phi\left(\frac{\varepsilon}{a}\right) \mu\{|f| > \varepsilon\}.$$

Now when $a \to 0$, we have $\Phi(\varepsilon/a) \to \infty$, so $\mu\{|f| > \varepsilon\} = 0$. This is true for all $\varepsilon > 0$, so $f = 0$ a.e.

Now consider a sequence $f_n \in L_\Phi$ with $0 \leq f_1 \leq f_2 \leq \cdots$ and $\|f_n\|_\Phi \leq 1$ for all n. That is, $\int \Phi(|f_n|)\, d\mu \leq 1$. If $f = \lim f_n$, then $\Phi(|f_n|) \to \Phi(|f|)$ by the left-continuity of Φ. Therefore, by the monotone convergence theorem, $\int \Phi(|f|)\, d\mu \leq 1$. Thus, $\|f\|_\Phi \leq 1$.

From this we may deduce completeness: Suppose (f_n) is a Cauchy sequence in L_Φ. There is a subsequence f_{n_k} with $\|f_{n_k} - f_{n_{k-1}}\|_\Phi \leq 2^{-k}$. Now let

$$g = \sum_{k=1}^{\infty} |f_{n_k} - f_{n_{k-1}}|.$$

By convexity and left-continuity of Φ,

$$
\begin{aligned}
\Phi(g) &= \Phi\left(\sum_{k=1}^{\infty} |f_{n_k} - f_{n_{k-1}}|\right) \\
&= \Phi\left(\sum 2^{-k} 2^k |f_{n_k} - f_{n_{k-1}}|\right) \\
&\leq \sum 2^{-k} \Phi\left(2^k |f_{n_k} - f_{n_{k-1}}|\right).
\end{aligned}
$$

Thus $\|g\|_\Phi \leq 1$, so g belongs to L_Φ, so the series converges a.e. to a finite limit (since $\Phi(u) \to \infty$ as $u \to \infty$). Now

$$
f = \sum_{k=1}^{\infty} \left(f_{n_k} - f_{n_{k-1}}\right) + f_{n_0}
$$

also converges a.e., and $|f - f_{n_0}| \leq g$, so $f \in L_\Phi$. But (f_n) is Cauchy, so for any $\varepsilon > 0$, there exists m so that for all $n \geq m$, we have $\|f_n - f_m\|_\Phi \leq \varepsilon$. Then for $n_k \geq m$, we get $\|f_m - f_{n_k}\|_\Phi \leq \varepsilon$. Thus $\|f_m - f\|_\Phi \to 0$ as $m \to \infty$.

Next, consider the subset H_Φ. Clearly it is a linear space. We claim it is closed in L_Φ. Suppose $f_n \in H_\Phi$ and $\|f - f_n\|_\Phi \to 0$. Let $a > 0$. Then $\|f_n/a - f/a\|_\Phi \to 0$, so there is n with $M_\Phi\left(f_n/a - f/a\right) < 1$. Then $(f - f_n)/a \in Y_\Phi$. But also $f_n/a \in Y_\Phi$. By the convexity of Y_Φ, we have $f/2a \in Y_\Phi$. Since a is arbitrary, this shows that $f \in H_\Phi$. ∎

The sequence f_n converges in Φ-*norm* to f iff $\|f_n - f\|_\Phi \to 0$ as usual. Also, f_n converges in Φ-modular iff $M_\Phi(f_n - f) \to 0$. These two modes of convergence are related (2.1.10(5)) but not identical (2.1.15 and 2.1.18).

The Banach space L_Φ is an example of a *Banach lattice*. This includes the following observations. The relation

$$
f \leq g \text{ a.e.}
$$

is a partial order on L_Φ. If $f \leq g$ a.e., then $f + h \leq g + h$ a.e. If $f \in L_\Phi$, $f \geq 0$ a.e., and $a \in \mathbb{R}$, $a \geq 0$, then $af \geq 0$ a.e. If $f, g \in L_\Phi$, then the pointwise maximum $f \vee g$ and pointwise minimum $f \wedge g$ also belong to L_Φ. In particular, the pointwise absolute value $|f|$ belongs to L_Φ and $\||f|\|_\Phi = \|f\|_\Phi$. Observe that these closure properties are true as well for the heart H_Φ. It is also a Banach lattice. [The Fatou property (2.1.11(c)) may fail for H_Φ, however. See the example below (2.1.25(5)).]

The space L_Φ is a *rearrangement invariant* function space. This means that if $f \in L_\Phi$ and g is a rearrangement of f, then $g \in L_\Phi$ and $\|g\|_\Phi = \|f\|_\Phi$. A good way to make this precise is to consider the *distribution functions* of f and g:

(2.1.12) Proposition. *Let f and g be measurable functions. If*

$$\mu\{|f| > t\} = \mu\{|g| > t\}$$

for all $t > 0$, then $M_\Phi(f) = M_\Phi(g)$ and $\|f\|_\Phi = \|g\|_\Phi$.

Proof. Since Φ is strictly increasing (except possibly where it is 0), we have $u_1 < u_2$ if and only if $\Phi(u_1) < \Phi(u_2)$ (provided $\Phi(u_2) > 0$), so we make the substitution $t = \Phi(u)$ in the integral:

$$
\begin{aligned}
M_\Phi(f) &= \int \Phi(|f|)\, d\mu \\
&= \int_0^\infty \mu\{\Phi(|f|) > t\}\, dt \\
&= \int_0^\infty \mu\{|f| > u\}\, \varphi(u)\, du.
\end{aligned}
$$

This depends only on the distribution function $\mu\{|f| > t\}$ of f.   ∎

There are some desirable properties that the spaces H_Φ have that are not (in general) shared by L_Φ.

(2.1.13) Definition. A Banach lattice E has *order continuous norm* iff any downward directed net $(f_t)_{t \in J}$ in E with greatest lower bound 0 satisfies $\lim \|f_t\| = 0$.

(2.1.14) Theorem. *Let Φ be a finite Orlicz function.*

   (a) *H_Φ has order continuous norm.*
   (b) *H_Φ is the closure in L_Φ of the integrable simple functions.*

Proof. (a) We first prove that if $(f_n)_{n \in \mathbb{N}}$ is a decreasing sequence in H_Φ with greatest lower bound 0, then $\lim \|f_n\|_\Phi = 0$. Now $f_n \to 0$ a.e. For any $a > 0$, the sequence $\Phi(f_n/a)$ also converges to 0 a.e. But $f_1 \in H_\Phi$, so $\int \Phi(f_1/a)\, d\mu < \infty$. Thus we have $\lim \int \Phi(f_n/a)\, d\mu = 0$. In particular, then, for large n we have $\int \Phi(f_n/a)\, d\mu \leq 1$, or $\|f_n\|_\Phi \leq a$. This is true for all $a > 0$, so $\lim \|f_n\|_\Phi = 0$.

Notice that a consequence of this property is: If f_n is an increasing sequence in H_Φ that is bounded above by an element of H_Φ, then f_n converges. Indeed, let f be the pointwise limit of f_n. It is dominated by an element of H_Φ, so it also belongs to H_Φ. Then the sequence $(f - f_n)$ is decreasing and has greatest lower bound 0. Thus $\|f - f_n\|_\Phi \to 0$.

Now we may deduce the full order completeness using the sequential sufficiency theorem (1.1.3). If (f_t) is a net directed downward in H_Φ, then any sequence (f_{t_k}) with $t_1 \leq t_2 \leq \cdots$ is a decreasing sequence, which converges in the norm of H_Φ. Therefore the net (f_t) converges in norm. Since it has greatest lower bound 0, the limit must be 0 a.e.

(b) Since Φ is finite, the integrable simple functions are in H_Φ, so their closure also lies in H_Φ. Let $f \in H_\Phi$, $f \geq 0$. There is a sequence (f_n) of integrable simple functions that increases to f a.e. (since μ is assumed to be σ-finite). By order continuity, $\|f_n - f\|_\Phi \to 0$. For general $f \in H_\Phi$, note that $f = f^+ - f^-$, where $f^+ = f \vee 0$ and $f^- = (-f) \vee 0$ also belong to H_Φ. ∎

The space L_Φ may fail the two properties given in Theorem (2.1.14) (see 2.1.25). Since many results in probability theory (and even in integration theory) rely on approximation by simple functions, the density of the integrable simple functions is often indispensable. The (Δ_2) condition, treated below, insures that L_Φ shares this property, but in the absence of the (Δ_2) condition, we will see that the Orlicz heart H_Φ is often the proper space to use, rather than the full Orlicz space L_Φ.

The (Δ_2) condition

We have seen that the Orlicz space L_Φ has certain desirable properties that the Orlicz heart H_Φ may lack [the Fatou property, Theorem (2.1.11(c))]. On the other hand, H_Φ has some desirable properties that L_Φ may lack (order continuous norm, density of simple functions; see (2.1.14)). Therefore, the cases when $L_\Phi = H_\Phi$ are particularly useful. This is the situation we will consider here.

(2.1.15) Proposition. *Let $(\Omega, \mathcal{F}, \mu)$ be a σ-finite measure space, and let Φ be an Orlicz function.*

(1) *The following are equivalent:* (1a) $H_\Phi(\Omega, \mathcal{F}, \mu) = L_\Phi(\Omega, \mathcal{F}, \mu)$. (1b) $Y_\Phi(\Omega, \mathcal{F}, \mu) = 2Y_\Phi(\Omega, \mathcal{F}, \mu)$.

(2) *The following are equivalent:* (2a) *If $M_\Phi (f_n) \to 0$, then we have also $M_\Phi (2f_n) \to 0$.* (2b) $\|f - f_n\|_\Phi \to 0$ *if and only if $M_\Phi (f - f_n) \to 0$ (norm convergence is equivalent to modular convergence).*

(3) *Suppose μ is infinite or Φ is strictly positive. Then* (1) $\Longrightarrow$ (2).

(4) *Suppose μ has arbitrarily small sets or Φ is finite. Then* (2) $\Longrightarrow$ (1).

Proof. (1) If $H_\Phi = L_\Phi$, then $Y_\Phi \subseteq 2Y_\Phi \subseteq L_\Phi = H_\Phi \subseteq Y_\Phi$, so $Y_\Phi = 2Y_\Phi$. Conversely, suppose $Y_\Phi = 2Y_\Phi$. If $f \in Y_\Phi$, then $2^n f \in Y_\Phi$ for all n, so if $a \in \mathbb{R}$, we have $|af| \leq |a||f| \leq 2^n |f|$ for some n, and $af \in Y_\Phi$. Thus $Y_\Phi = H_\Phi$. If $f \in L_\Phi$, then $f/a \in Y_\Phi = H_\Phi$ for some a, so $f \in H_\Phi$.

(2) If (2a) holds, then $M_\Phi (f_n) \to 0$ implies $\lim_n M_\Phi (2^m f_n) = 0$ for any m, and thus $\lim_n M_\Phi (kf_n) = 0$ for any $k \in \mathbb{R}$. By (2.1.10(5)), this is equivalent to $\|f_n\|_\Phi \to 0$.

(3) Suppose (1) holds. If μ is infinite, then Φ is strictly positive. Indeed, if $\Phi(u) = 0$ for some $u > 0$, then there is u_0 with $\Phi(u_0) = 0$ but $\Phi(2u_0) > 0$. So the constant u_0 belongs to Y_Φ, but the constant $2u_0$ does not. Thus we may suppose Φ is strictly positive.

Assume (for purposes of contradiction) that there exists a sequence $f_n \in Y_\Phi$ with $M_\Phi (f_n) \to 0$ but $M_\Phi (2f_n) \not\to 0$. Now $M_\Phi (2f_n) = \infty$ for some n

would contradict (1b). We may assume $\infty > M_\Phi(2f_n) \geq c > 0$. Taking a subsequence, we may assume $M_\Phi(f_n) \leq 2^{-n}$. Consider $f = \sup |f_n|$. We have

$$
\begin{aligned}
\int \Phi(|f|)\,d\mu &= \int \Phi(\sup |f_n|)\,d\mu \\
&= \int \sup \Phi(|f_n|)\,d\mu \qquad \text{(by left-continuity)} \\
&\leq \int \sum \Phi(|f_n|)\,d\mu \\
&\leq \sum \int \Phi(|f_n|)\,d\mu \qquad \text{(by Fatou's Lemma)} \\
&\leq \sum 2^{-n} < \infty.
\end{aligned}
$$

Thus $f \in Y_\Phi$. Therefore $2f \in Y_\Phi$. Now $\int \Phi(|f_n|)\,d\mu \leq 2^{-n}$, so $\Phi(|f_n|) \to 0$ a.e. Since Φ is strictly positive, this means $|f_n| \to 0$ a.e. Thus $\Phi(2|f_n|) \to 0$ a.e. But $\Phi(2|f_n|) \leq \Phi(2f)$, which is integrable, so by the dominated convergence theorem, we have $M_\Phi(2f_n) \to 0$.

(4) Suppose (2a) holds. If μ has arbitrarily small sets, then Φ is finite. Indeed, if $\Phi(u) = \infty$ for some u, then there is u_0 with $\Phi(u_0) < \infty$ but $\Phi(2u_0) = \infty$. Now if A_n is a sequence of sets with $0 < \mu(A_n)$ and $\mu(A_n) \to 0$, then $f_n = u_0 \mathbf{1}_{A_n}$ satisfies $M_\Phi(f_n) \to 0$ but $M_\Phi(2f_n) \nrightarrow 0$, so we may suppose Φ is finite.

Suppose $Y_\Phi \neq 2Y_\Phi$. Then there is $f \in Y_\Phi$ with $2f \notin Y_\Phi$. Let f_n be integrable simple functions with $f_n \uparrow |f|$. Then $g_n = |f| - f_n$ decreases to 0. Also, $\Phi(g_n) \leq \Phi(|f|)$, which is integrable, so by the dominated convergence theorem, $M_\Phi(g_n) \to 0$. Now $M_\Phi(2|f|) = \infty$, and

$$
M_\Phi(2|f|) \leq \frac{1}{2}\left[\int \Phi(4f_n)\,d\mu + \int \Phi(4g_n)\,d\mu \right].
$$

The first term on the right is finite (since Φ is finite), so $M_\Phi(4g_n) = \infty$. We thus have a sequence (g_n) with $M_\Phi(g_n) \to 0$ but $M_\Phi(4g_n) = \infty$. [Using functions of the form $f_n - f_m$, we may obtain a sequence (h_n) with $M_\Phi(h_n) \to 0$, $M_\Phi(2h_n) < \infty$, but $M_\Phi(2h_n) \nrightarrow 0$.] ∎

(2.1.16) Definition. The Orlicz function Φ satisfies *condition (Δ_2)* at ∞ iff there exist $u_0 \in (0, \infty)$ and $k \in \mathbb{R}$ such that

(2.1.16a) $\qquad\qquad \Phi(2u) < k\Phi(u) \qquad$ for all $u \geq u_0$.

The Orlicz function Φ satisfies *condition (Δ_2)* at 0 iff there exist $u_0 \in (0, \infty)$ and $k \in \mathbb{R}$ such that

(2.1.16b) $\qquad\qquad \Phi(2u) < k\Phi(u) \qquad$ for $0 < u \leq u_0$.

The strict inequality in (2.1.16a) requires that $\Phi(u) < \infty$ for all u. The strict inequality in (2.1.16b) requires that $\Phi(u) > 0$ for all $u > 0$.

(2.1.17) Theorem. *Let $(\Omega, \mathcal{F}, \mu)$ be a σ-finite measure space, and let Φ be an Orlicz function.*

1. *If Φ satisfies (Δ_2) at 0 and ∞, then $H_\Phi(\Omega, \mathcal{F}, \mu) = L_\Phi(\Omega, \mathcal{F}, \mu)$. If $H_\Phi\big([0,\infty)\big) = L_\Phi\big([0,\infty)\big)$, then Φ satisfies (Δ_2) at 0 and ∞.*
2. *Suppose μ is finite. If Φ satisfies (Δ_2) at ∞, then $H_\Phi(\Omega, \mathcal{F}, \mu) = L_\Phi(\Omega, \mathcal{F}, \mu)$. If $H_\Phi\big([0,1]\big) = L_\Phi\big([0,1]\big)$, then Φ satisfies (Δ_2) at ∞.*
3. *Suppose μ has no arbitrarily small sets. If Φ is finite and satisfies (Δ_2) at 0, then $H_\Phi(\Omega, \mathcal{F}, \mu) = L_\Phi(\Omega, \mathcal{F}, \mu)$. If $h_\Phi = l_\Phi$, then Φ satisfies (Δ_2) at 0.*

Proof. (1) Suppose Φ satisfies (Δ_2) at 0 and ∞. Then $0 < \Phi(u) < \infty$ for all $u \in (0, \infty)$. The quotient $\Phi(2u)/\Phi(u)$ is continuous on $(0, \infty)$, so we have

$$(2.1.17a) \qquad \sup_{u \in (0,\infty)} \frac{\Phi(2u)}{\Phi(u)} < \infty.$$

Thus $\Phi(2u) < k\Phi(u)$ for all u, so clearly $M_\Phi(2f) \leq kM_\Phi(f)$.

Conversely, suppose $H_\Phi\big([0,\infty)\big) = L_\Phi\big([0,\infty)\big)$, but the (Δ_2) condition fails (either at 0 or ∞). If $\Phi(u) = \infty$ for some u, then there is u_0 with $\Phi(u_0) < \infty$ but $\Phi(2u_0) > 0$, so that $f = u_0 \mathbf{1}_{[0,1]}$ shows that $Y_\Phi \neq 2Y_\Phi$. If $\Phi(u) = 0$ for some $u > 0$, then there is u_0 with $\Phi(u_0) = 0$ but $\Phi(2u_0) = \infty$, so that $f = u_0 \mathbf{1}_{[0,\infty)}$ shows that $Y_\Phi \neq 2Y_\Phi$. Thus we may suppose that Φ is positive and finite.

Since the (Δ_2) condition fails, there exist $u_n \in (0, \infty)$ with $\Phi(2u_n) > 2^n \Phi(u_n)$. Let A_n be disjoint intervals in $[0, \infty)$ with

$$\mu(A_n) = 2^{-n}/\Phi(u_n).$$

The function

$$f = \sum_{n=1}^{\infty} u_n \mathbf{1}_{A_n}$$

satisfies

$$M_\Phi(f) = \int \Phi(|f|)\, d\mu = \sum \frac{\Phi(u_n)\, 2^{-n}}{\Phi(u_n)} = \sum 2^{-n} < \infty$$

but

$$M_\Phi(2f) = \int \Phi(2|f|)\, d\mu = \sum \frac{\Phi(2u_n)\, 2^{-n}}{\Phi(u_n)} \geq \sum 1 = \infty.$$

(2) Now Φ satisfies (Δ_2) possibly at ∞ only, say

$$\Phi(2u) < k\Phi(u) \qquad \text{for } u \geq u_0.$$

By the strict inequality, we have $\Phi(u) < \infty$ for all $u \in (0, \infty)$. Let $f \in Y_\Phi$. Then

$$M_\Phi\,(2f) = \int \Phi\big(2|f|\big)\,d\mu$$

$$\leq \int_{\{|f| \geq u_0\}} k\Phi\big(|f|\big) + \int_{\{|f| < u_0\}} \Phi\big(2u_0\big)\,d\mu$$

$$\leq k M_\Phi\,(f) + \Phi(2u_0)\mu(\Omega) < \infty.$$

Thus $2f \in Y_\Phi$.

Conversely, suppose $H_\Phi\big([0,1]\big) = L_\Phi\big([0,1]\big)$. Then Φ is finite as in the previous case. If the (Δ_2) condition fails at ∞, then u_n may be chosen as in case (1), but with $u_n \to \infty$, so also $\Phi(u_n) \to \infty$. We may suppose $\Phi(u_n) \geq 1$ for all n. Now the intervals A_n, chosen as before, have total length at most 1, so they may be chosen in $[0,1]$. The rest is the same as case (1).

(3) Now Φ satisfies (Δ_2) at 0, so

$$\Phi(2u) < k\Phi(u) \qquad \text{for } 0 < u \leq u_0.$$

By strict inequality, we have $\Phi(u) > 0$ for all u. Now μ has no arbitrarily small sets, say $\mu(A) \geq c > 0$ whenever $\mu(A) > 0$. Let $f \in Y_\Phi$. We have

$$\int \Phi(|f|)\,d\mu \geq \Phi(u_0)\,\mu\{|f| \geq u_0\},$$

so $\mu\{|f| \geq u_0\} < \infty$. Since μ is bounded below by $c > 0$, the set $A = \{|f| \geq u_0\}$ consists of a finite number of atoms, so $\Phi(2|f|)$ is bounded on the set. Finally,

$$M_\Phi\,(2f) = \int \Phi\big(2|f|\big)\,d\mu$$

$$\leq \int_{\{|f| \leq u_0\}} k\Phi\big(|f|\big)\,d\mu + \int_{\{|f| > u_0\}} \Phi\big(2|f|\big)\,d\mu$$

$$\leq k M_\Phi\,(f) + \max_A \Phi\big(2|f|\big)\,\mu\{|f| > u_0\}$$

$$< \infty.$$

Conversely, suppose $l_\Phi = h_\Phi$. Then Φ is finite and positive as in case (1). Suppose that the (Δ_2) condition at 0 is false. Then we may choose $u_n \to 0$ with $\Phi(2u_n) > 2^n \Phi(u_n)$. We may assume $\Phi(u_n) < 2^{-n}$. Now the disjoint sets $A_n \subseteq \mathbb{N}$ may be taken with $\mu(A_n) = 2^k$ where $2^{-n} < \Phi(u_n)\,2^k \leq 2^{-n+1}$. Then, for $f = \sum_{n=1}^\infty u_n\,\mathbf{1}_{A_n}$, we have

$$M_\Phi\,(f) = \sum \Phi(u_n)\,\mu(A_n) \leq \sum 2^{-n+1} < \infty$$

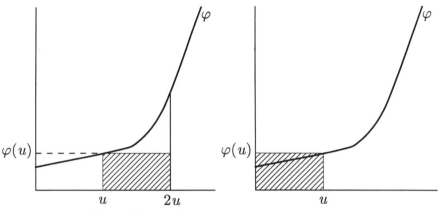

Figure (2.1.19a).   Figure (2.1.19b).

but

$$M_\Phi(2f) = \sum \Phi(2u_n)\,\mu(A_n) \geq \sum_{n=1}^{\infty} 1 = \infty.$$

Thus $l_\Phi \neq h_\Phi$.   ∎

We will normally say simply: condition (Δ_2) is satisfied, and understand the particular one of the three possibilities appropriate for the measure space under discussion. In particular, if $(\Omega, \mathcal{F}, \mu)$ is a probability space, then we will mean that (Δ_2) is satisfied at ∞.

(2.1.18) Corollary. *Let Φ be a strictly positive Orlicz function satisfying the (Δ_2) condition. Then:*

(A) *Integrable simple functions are dense in L_Φ.*

(B) *If f_n is an increasing sequence in L_Φ and $\sup \|f_n\|_\Phi < \infty$, then f_n converges in norm.*

(C) *Modular convergence is equivalent to norm convergence in L_Φ; that is: $\|f - f_n\|_\Phi \to 0$ if and only if $M_\Phi(f - f_n) \to 0$.*

Proof. By the (Δ_2) condition, $L_\Phi = H_\Phi$. Part (A) is from (2.1.14(b)). Part (C) is from (2.1.15(3)). For (B), let f be the pointwise supremum of f_n. Then by (2.1.11), we have $f \in L_\Phi$. Now $(f - f_n)$ is a decreasing sequence with greatest lower bound 0. By (2.1.14(a)), L_Φ has order continuous norm, so $\|f - f_n\|_\Phi \to 0$.   ∎

Complements

(2.1.19) (Inequalities.) Let Φ be an Orlicz function and φ its derivative. Use Figures (2.1.19a)–(2.1.19c) to prove inequalities:

(a) $\Phi(u) \leq u\varphi(u)$;

(b) $\Phi(2u) \geq \Phi(u) + u\varphi(u)$;

(c) $\Phi(n) \leq \Phi(n-1) + \varphi(n)$, if $n \geq 1$.

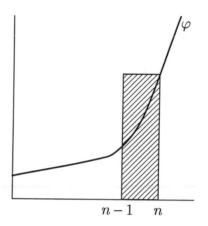

Figure (2.1.19c).

(2.1.20) (Norm of indicator.) Let $A \in \mathcal{F}$, $0 < \mu(A) < \infty$. Then $\|\mathbf{1}_A\|_\Phi = 1/a$, where

$$a = \sup \left\{ u > 0 : \Phi(u) \leq \frac{1}{\mu(A)} \right\}.$$

Thus either a satisfies $\Phi(a) = 1/\mu(A)$ or else Φ jumps to ∞ at a. Corollary: if Φ is finite, then $\mu(A) \to 0$ if and only if $\|\mathbf{1}_A\|_\Phi \to 0$.

(2.1.21) (Comparison of L_Φ and L_1.) If $\mu(\Omega) < \infty$, then there is a constant $C > 0$ with $\|f\|_1 \leq C\|f\|_\Phi$. The constant C depends only on the function Φ and the value $\mu(\Omega)$. To see this, choose $a > 0$ with $\Phi(a) > 0$, then choose b with $0 < b \leq \Phi(a)$. Let $C = (a/b) + a\mu(\Omega)$. By convexity of Φ, we have $\Phi(u) \geq (b/a)u$ for $u \geq a$. Now if $f \in L_\Phi$, let $r = \|f\|_\Phi$ and compute

$$\frac{1}{r} \int |f|\, d\mu = \int_{\{|f| \geq ar\}} \frac{|f|}{r}\, d\mu + \int_{\{|f| < ar\}} \frac{|f|}{r}\, d\mu$$

$$\leq \frac{a}{b} \int \Phi\left(\frac{|f|}{r}\right) d\mu + a\mu(\Omega) \leq \frac{a}{b} + a\mu(\Omega) = C.$$

That is, $\|f\|_1 \leq Cr = C\|f\|_\Phi$.

(2.1.22) Suppose the functions g_n have only integer values. If $\|g_n\|_\Phi \to 0$, then $\|g_n\|_1 \to 0$. To see this, choose $a > 0$ and $0 < b \leq \Phi(a)$. We claim: if $\|g_n\|_\Phi < 1/a$, then $\|g_n\|_1 \leq (2a/b)\|g_n\|_\Phi$, so certainly $\|g_n\|_1 \to 0$. Let $A = \{g_n \neq 0\}$. Then $\mathbf{1}_A \leq |g_n|$, so $\|\mathbf{1}_A\|_\Phi \leq \|g_n\|_\Phi \leq 1/a$. Thus $1 \geq \int \Phi(\mathbf{1}_A/(1/a))\, d\mu = \mu(A)\Phi(a) \geq \mu(A) \cdot b$. Thus $\mu(A) \leq 1/b$. Now for $C = (a/b) + a\mu(A) \leq (2a/b)$, we may apply the previous result (2.1.21): $\|g_n\|_1 \leq C\|g_n\|_\Phi = (2a/b)\|g\|_\Phi$.

(2.1.23) Suppose Φ is finite and $|g_n| \leq 1$ for all n. If $\|g_n\|_1 \to 0$, then $\|g_n\|_\Phi \to 0$. (Prove as in (2.1.20).)

(2.1.24) Property (B) of (2.1.18) is equivalent to a Banach space condition known as *weak sequential completeness* (see Lindenstrauss & Tzafriri [1979], p. 34).

(2.1.25) (Failure of the (Δ_2) condition.) The Orlicz function

$$\Phi(u) = e^u - u - 1$$

is finite and satisfies the (Δ_2) condition at 0, but fails the (Δ_2) condition at ∞. It can be used to provide explicit examples that illustrate the problems that arise when that happens. Let $(\Omega, \mathcal{F}, \mu)$ be $[0,1]$ with Lebesgue measure.

(1) $L_\Phi \neq H_\Phi$ and $Y_\Phi \neq 2Y_\Phi$: For the function $f(t) = \log(1/t)$, we have $f \in L_\Phi$ and $f \in 2Y_\Phi$ since $M_\Phi(f/2) < \infty$, but $f \notin H_\Phi$ and $f \notin Y_\Phi$ since $M_\Phi(f) = \infty$.

(2) L_Φ does not have order continuous norm: The sequence

$$f_n(t) = \log\frac{1}{t}\,\mathbf{1}_{(0,1/n)}(t)$$

decreases, and has greatest lower bound 0, but does not converge to 0 in norm, since $\|f_n\|_\Phi \geq 1$ for all n.

(3) Modular convergence is not equivalent to norm convergence in L_Φ: The functions

$$f_n(t) = \frac{1}{2}\log\frac{1}{t}\,\mathbf{1}_{(0,1/n)}(t)$$

satisfy $M_\Phi(f_n) \to 0$, but $\|f_n\|_\Phi \geq 1/2$.

(4) Weak sequential completeness fails in L_Φ: The sequence

$$f_n(t) = \log\frac{1}{t}\,\mathbf{1}_{(1/n,1)}(t)$$

is increasing and bounded in norm, but not norm convergent.

(5) H_Φ fails the Fatou property: The previous sequence (f_n) belongs to H_Φ, is increasing, bounded in norm, but its pointwise limit is not in H_Φ.

(2.1.26) (Separability.) If Φ is finite, then the integrable simple functions are dense in H_Φ. So $H_\Phi([0,1])$ is separable. In fact, if $\Phi(u)/u \to \infty$ as $u \to \infty$, then the Haar functions constitute a basis for $H_\Phi([0,1])$ (Krasnosel'skii & Rutickii [1961], p. 106).

In (2.1.17) we saw that if the (Δ_2) condition fails at ∞, then $H_\Phi([0,1]) \neq L_\Phi([0,1])$. In fact in this case more is true:

(2.1.27) Theorem. *If (Δ_2) fails at ∞, then $L_\Phi([0,1])$ is nonseparable.*

Proof. We will prove that L_Φ is nonseparable by constructing an uncountable family of functions in L_Φ, each of distance at least 1 from the others. Since (Δ_2) fails, there exist u_n increasing to ∞ with $\Phi(2u_n) > 2^n\Phi(u_n)$. We may assume $\Phi(u_1) \geq 1$. Choose intervals A_n in $[0,1]$ with $\mu(A_n) = 2^{-n}/\Phi(u_n)$. This may be done so that the right endpoint of A_n is the left endpoint of A_{n+1} for each n, and the endpoints cluster at 1. Define a function f by

$$f(x) = \sum_{n=1}^{\infty} 3u_n \, 1_{A_n}.$$

Now $f \in L_\Phi$, since $f/3 \in Y_\Phi$. For each α with $0 < \alpha < 1$, define

$$f_\alpha(x) = \begin{cases} f(x+1-\alpha) & 0 \leq x \leq \alpha \\ f(x-\alpha) & \alpha < x \leq 1. \end{cases}$$

Then for $\alpha \neq \beta$ we have $\int \Phi(f_\alpha - f_\beta)\,d\mu = \infty$, so $\|f_\alpha - f_\beta\|_\Phi > 1$. ∎

(2.1.28) (Equivalents to (Δ_2).) Let $(\Omega, \mathcal{F}, \mu)$ be a nonatomic σ-finite measure space, let Φ be an Orlicz function, and let $L = L_\Phi(\Omega, \mathcal{F}, \mu)$ with norm denoted $\|\ \|$. The following contitions are equivalent.

   (i) The function Φ satisfies (Δ_2) (both at ∞ and at 0).

   (ii) (C_1) If $f \in L^+$, $g \in L^+$, $\|f\| > 0$, then $\|f + g\| > \|g\|$.

   (iii) (C) For each $f' \in L^+$ and each number $\alpha > 0$, there is a number $\beta = \beta(f', \alpha)$ such that if $f \in L^+$, $f \leq f'$, $\|f\| \geq \alpha$, $g \in L^+$, $\|g\| \leq 1$, then $\|f + g\| \geq \|g\| + \beta$.

   (iv) G. Birkhoff's uniform monotonicity condition (UMB): Given an $\varepsilon > 0$, there is a $\delta > 0$ such that if $f \in L^+$, $g \in L^+$, $\|g\| = 1$, and $\|f + g\| \leq 1 + \delta$, then $\|f\| \leq \varepsilon$

(Akcoglu & Sucheston [1985a]).

Remarks

Most of this material on Orlicz spaces was adapted from the following sources: Zaanen [1983], Chapter 19, for the basic material (the most complete account we found); Lindenstrauss & Tzafriri [1977], Chapter 4, for the sequence spaces l_Φ and h_Φ; Lindenstrauss & Tzafriri [1979], Chapter 2, for the function spaces on $[0,1]$ and $[0,\infty)$; Akcoglu & Sucheston [1985a] for much of the discussion of H_Φ; Krasnosel'skii & Rutickii [1961] also discusses the (Δ_2) condition and the spaces H_Φ (called there E_M). There is some overlap in these references, and all of them have additional material not mentioned here.

The Young classes Y_Φ were introduced by W. H. Young [1912]. In general they are not linear spaces, and even when they are, Young did not norm them. W. Orlicz [1932], [1936] introduced the Banach spaces L_Φ, defining the norm in terms of the conjugate. The definition of the norm given above, which uses only Φ, is due to W. A. J. Luxemburg [1955]. The (Δ_2) condition appears already in Orlicz [1932]. The second paper, Orlicz [1936], showed how to avoid requiring the (Δ_2) condition. A. C. Zaanen [1949] extended the definition of the Orlicz function to allow the value ∞, so that L_∞ became an Orlicz space.

Spaces H_Φ originate in Morse & Transue [1950].

2.2. More on Orlicz spaces

We consider now the classical examples. Let p be given, $1 < p < \infty$. Then

$$\Phi_p(u) = \frac{u^p}{p}$$

is an Orlicz function with conjugate Φ_q, where

$$\frac{1}{p} + \frac{1}{q} = 1.$$

Clearly, the Young class Y_{Φ_p} is the usual space L_p. The Orlicz space L_{Φ_p} and its heart H_{Φ_p} are both equal to L_p, but the Luxemburg norm is

$$\|f\|_{\Phi_p} = \frac{1}{p^{1/p}} \|f\|_p.$$

We may also include

$$\Phi_1(u) = u$$

and its conjugate

$$\Phi_\infty(u) = \begin{cases} 0 & 0 \le u \le 1 \\ \infty & 1 < u < \infty. \end{cases}$$

Then $L_{\Phi_1} = H_{\Phi_1} = L_1$ and $\|f\|_{\Phi_1} = \|f\|_1$. Also, $L_{\Phi_\infty} = L_\infty$ and $\|f\|_{\Phi_\infty} = \|f\|_\infty$. This is a good example where the derivative $\varphi_1 = \Phi_1{}'$ is bounded, so the conjugate Φ_∞ is infinite. Note that Φ_∞ fails the (Δ_2) condition both at 0 and at ∞. This example also illustrates a case where $H_{\Phi_\infty} = 0$, and is not equal to the closure of the integrable simple functions.

Comparison of Orlicz spaces

It is possible for different Orlicz functions to yield the same set of functions for the Orlicz space, with equivalent norms. A related possibility is that one Orlicz space is a subset of another. These possibilities are related to the ways that Orlicz functions can be compared.

(2.2.1) Proposition. *Let $(\Omega, \mathcal{F}, \mu)$ be a σ-finite measure space, and let Φ_1, Φ_2 be Orlicz functions. The following are equivalent.*

(1) *$L_{\Phi_1}(\Omega, \mathcal{F}, \mu) \subseteq L_{\Phi_2}(\Omega, \mathcal{F}, \mu)$ (a subset, possibly not closed).*

(2) *There is a constant k such that $k\|f\|_{\Phi_1} \ge \|f\|_{\Phi_2}$ for all measurable functions f.*

Proof. If $k\|f\|_{\Phi_1} \ge \|f\|_{\Phi_2}$, then clearly $L_{\Phi_1}(\Omega, \mathcal{F}, \mu) \subseteq L_{\Phi_2}(\Omega, \mathcal{F}, \mu)$. Conversely, if $L_{\Phi_1}(\Omega, \mathcal{F}, \mu) \subseteq L_{\Phi_2}(\Omega, \mathcal{F}, \mu)$, then the linear transformation $T \colon L_{\Phi_1} \to L_{\Phi_2}$ defined by $T(f) = f$ has closed graph. [Indeed, if we have $\|f_n - f\|_{\Phi_1} \to 0$, then by (2.1.10(6)) there is a subsequence with $f_{n_k} \to f$ a.e. If also $\|f_n - h\|_{\Phi_2} \to 0$, then there is a further subsequence with $f_{n_{k_j}} \to h$ a.e. Thus $f = h$.] By the closed graph theorem (Rudin [1973], Theorem 2.15), T is bounded, and its norm $\|T\|$ will do for k.   ∎

(2.2.2) Definition. Let Φ_1 and Φ_2 be Orlicz functions. Then Φ_1 *dominates* Φ_2 *at* ∞ iff there exist $a, b, u_0 > 0$ such that

$$b\Phi_1(au) \ge \Phi_2(u) \qquad \text{for all } u \ge u_0.$$

We write $\Phi_1 \succ_\infty \Phi_2$. Similarly, Φ_1 *dominates* Φ_2 *at* 0 iff there exist $a, b, u_0 > 0$ such that

$$b\Phi_1(au) \ge \Phi_2(u) \qquad \text{for all } u \le u_0.$$

We write $\Phi_1 \succ_0 \Phi_2$.

Of course, if the appropriate (Δ_2) condition is satisfied, we may equivalently take $a = 1$ in these definitions. (But in general we may not take $a = 1$; see (2.2.20).)

The comparison of Orlicz spaces can be treated in much the same way as the (Δ_2) condition was treated above.

(2.2.3) Theorem. *Let* $(\Omega, \mathcal{F}, \mu)$ *be a σ-finite measure space, and let* Φ_1, Φ_2 *be Orlicz functions.*

(1) *If* $\Phi_1 \succ_\infty \Phi_2$ *and* $\Phi_1 \succ_0 \Phi_2$, *then* $L_{\Phi_1}(\Omega, \mathcal{F}, \mu) \subseteq L_{\Phi_2}(\Omega, \mathcal{F}, \mu)$. *If* $L_{\Phi_1}([0, \infty)) \subseteq L_{\Phi_2}([0, \infty))$, *then* $\Phi_1 \succ_\infty \Phi_2$ *and* $\Phi_1 \succ_0 \Phi_2$.

(2) *Suppose μ is finite. If* $\Phi_1 \succ_\infty \Phi_2$, *then*

$$L_{\Phi_1}(\Omega, \mathcal{F}, \mu) \subseteq L_{\Phi_2}(\Omega, \mathcal{F}, \mu).$$

If $L_{\Phi_1}([0, 1]) \subseteq L_{\Phi_2}([0, 1])$, *then* $\Phi_1 \succ_\infty \Phi_2$.

(3) *Suppose μ has no arbitrarily small sets and Φ_2 is finite. If* $\Phi_1 \succ_0 \Phi_2$, *then* $L_{\Phi_1}(\Omega, \mathcal{F}, \mu) \subseteq L_{\Phi_2}(\Omega, \mathcal{F}, \mu)$. *If* $l_{\Phi_1} \subseteq l_{\Phi_2}$, *then* $\Phi_1 \succ_0 \Phi_2$.

Proof. (1) Suppose $\Phi_1 \succ_0 \Phi_2$ and $\Phi_1 \succ_\infty \Phi_2$. Thus

$$b'\Phi_1(a'u) \ge \Phi_2(u) \qquad \text{for } u \ge u'$$
$$b''\Phi_1(a''u) \ge \Phi_2(u) \qquad \text{for } u \le u''.$$

We may assume $\Phi_2(u') < \infty$. If not, choose $\tilde{u} < u'$ with $\Phi(\tilde{u}) < \infty$, and let $\tilde{a} = a'u'/\tilde{u}$, so that for $u \ge \tilde{u}$,

$$\Phi_2(u) \le \Phi_2(u'u/\tilde{u}) \le b'\Phi_1(a'u'u/\tilde{u}) = b'\Phi_1(\tilde{a}u).$$

Similarly, we may assume $\Phi_1(a''u'') > 0$. If we take $a = \max\{a', a''\}$,

$$b''' = \max\left\{ \frac{\Phi_2(u)}{\Phi_1(au)} : u'' \le u \le u' \right\},$$

and $b = \max\{b', b'', b''', 1\}$, then we have

$$b\Phi_1(au) \geq \Phi_2(u) \qquad \text{for all } u.$$

Let $f \in L_{\Phi_1}$. Then $M_{\Phi_1}(f/k) \leq 1$ for some k. Now

$$
\begin{aligned}
M_{\Phi_2}\left(\frac{f}{ak}\right) &= \int \Phi_2\left(\frac{|f|}{ak}\right) d\mu \\
&\leq \int b\Phi_1\left(\frac{|f|}{k}\right) d\mu \leq b.
\end{aligned}
$$

Since $b \geq 1$, by convexity of M_{Φ_2} we have $M_{\Phi_2}(f/abk) \leq 1$. Thus $f \in L_{\Phi_2}$.

Conversely, suppose $L_{\Phi_1}([0,\infty)) \subseteq L_{\Phi_2}([0,\infty))$. If $\Phi_1 \succ \Phi_2$ fails, either at 0 or ∞, then there exist $u_n \in (0,\infty)$ with $0 < 2^n\Phi_1(2^n u_n) < \Phi_2(u_n)$. The strict inequality shows $\Phi_1(2^n u_n) < \infty$ and $\Phi_2(u_n) > 0$. Let A_n be disjoint intervals in $[0,\infty)$ with $\mu(A_n) = 2^{-n}/\Phi_1(2^n u_n)$. Let

$$f = \sum_{n=1}^{\infty} 2^n u_n \mathbf{1}_{A_n}.$$

Now

$$M_{\Phi_1}(f) = \sum \Phi_1(2^n u_n) \frac{2^{-n}}{\Phi_1(2^n u_n)} = \sum 2^{-n} < \infty.$$

Thus $f \in L_{\Phi_1}$. If $a > 0$, then $2^n u_n/a > u_n$ for n larger than some n_0, so

$$
\begin{aligned}
M_{\Phi_2}\left(\frac{f}{a}\right) &= \sum \Phi_2\left(\frac{2^n u_n}{a}\right) \frac{2^{-n}}{\Phi_1(2^n u_n)} \\
&\geq \sum_{n=n_0}^{\infty} \Phi_2(u_n) \frac{2^{-n}}{\Phi_1(2^n u_n)} \\
&\geq \sum_{n=n_0}^{\infty} 1 = \infty.
\end{aligned}
$$

Thus $f \notin L_{\Phi_2}$.

(2) Now $\Phi_1 \succ_\infty \Phi_2$; say $b\Phi_1(au) \geq \Phi_2(u)$ for $u \geq u_0$. As before, we may assume $\Phi_2(u_0) < \infty$. If $f \in L_{\Phi_1}$, then $M_{\Phi_1}(f/k) < \infty$ for some k. Then

$$
\begin{aligned}
M_{\Phi_2}\left(\frac{f}{ka}\right) &= \int \Phi_2\left(\frac{|f|}{ka}\right) d\mu \\
&\leq \int_{\{|f| < kau_0\}} \Phi_2(u_0)\, d\mu + \int_{\{|f| \geq kau_0\}} b\Phi_1\left(\frac{|f|}{k}\right) d\mu \\
&\leq \Phi_2(u_0)\, \mu(\Omega) + bM_{\Phi_1}\left(\frac{f}{k}\right) < \infty.
\end{aligned}
$$

Thus $f \in L_{\Phi_2}$.

For the converse, if $\Phi_1 \succ_\infty \Phi_2$ fails, then the points u_n as in case (1) may be chosen with $u_n \to \infty$, so also $\Phi_1(2^n u_n) \to \infty$, and we may assume $\Phi_1(2^n u_n) \geq 1$. Then the intervals A_n, chosen as before, have total length at most 1, so they may be chosen in $[0,1]$.

(3) Now $\Phi_1 \succ_0 \Phi_2$; say $b\Phi_1(au) \geq \Phi_2(u)$ for $u \leq u_0$. As in case (1), we may assume $\Phi_1(au_0) > 0$. If $f \in L_{\Phi_1}$, then $M_{\Phi_1}(f/k) < \infty$ for some k. Now

$$\int \Phi_1\left(\frac{|f|}{k}\right) d\mu \geq \Phi_1(au_0)\mu\{|f| \geq aku_0\},$$

so $\mu\{|f| \geq aku_0\} < \infty$. Since μ has no arbitrarily small sets, we see that the set $\{|f| \geq aku_0\}$ consists of a finite number of atoms, so $\Phi_2(|f|/ka)$ is bounded on the set. Now

$$M_{\Phi_2}\left(\frac{f}{ka}\right) = \int \Phi_2\left(\frac{|f|}{ka}\right) d\mu$$

$$\leq \int_{\{|f| < aku_0\}} b\Phi_1\left(\frac{|f|}{k}\right) d\mu + \int_{\{|f| \geq aku_0\}} \Phi_2\left(\frac{|f|}{ka}\right) d\mu$$

$$< \infty.$$

Thus $f \in L_{\Phi_2}$.

For the converse, if $\Phi_1 \succ_0 \Phi_2$ fails, then the points u_n as in case (1) may be chosen with $u_n \to 0$. We may suppose $\Phi(u_n) \leq 1$, and hence $\Phi_1(2^n u_n) \leq 2^{-n}$. Then disjoint sets $A_n \subseteq \mathbb{N}$ should be chosen, with $\mu(A_n) = 2^k$, where $2^{-n} < \Phi_1(2^n u_n) 2^k \leq 2^{-n+1}$. The function f defined in case (1) satisfies $f \in l_{\Phi_1}$ but $f \notin l_{\Phi_2}$. ∎

We will normally write simply $\Phi_1 \succ \Phi_2$ for the condition of domination appropriate for the measure space being considered. If both $\Phi_1 \succ \Phi_2$ and $\Phi_2 \succ \Phi_1$, then we will say that Φ_1 and Φ_2 are *equivalent*, and write $\Phi_1 \approx \Phi_2$.

Largest and smallest Orlicz functions

Consider the following conjugate pair of Orlicz functions (see Figures (2.2.4a) and (2.2.4b)).

$$\Phi_{\min}(u) = \begin{cases} 0 & 0 \leq u \leq 1 \\ u - 1 & 1 < u < \infty \end{cases}$$

$$\Phi_{\max}(u) = \begin{cases} u & 0 \leq u \leq 1 \\ \infty & 1 < u < \infty. \end{cases}$$

(2.2.4) Proposition. *If Φ is any Orlicz function, then $\Phi_{\max} \succ \Phi \succ \Phi_{\min}$ (at both 0 and ∞). If f is a measurable function, $f \in L_{\Phi_{\max}}$ if and only if $f \in L_1$ and $f \in L_\infty$, and then $\|f\|_{\Phi_{\max}} = \|f\|_1 \vee \|f\|_\infty$. Also, $f \in L_{\Phi_{\min}}$*

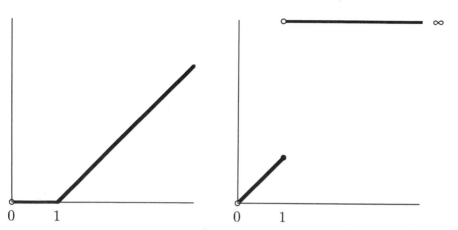

Figure (2.2.4a). The smallest and largest Orlicz functions.

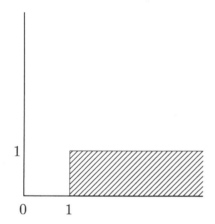

Figure (2.2.4b). The Young partition for $\Phi_{\min}$ and $\Phi_{\max}$.

if and only if f can be written as a sum $f = f_1 + f_\infty$ with $f_1 \in L_1$ and $f_\infty \in L_\infty$, and then

$$\|f\|_{\Phi_{\min}} = \inf\big\{\, \|f_1\|_1 \vee \|f_\infty\|_\infty : f = f_1 + f_\infty \,\big\}.$$

Proof. Let Φ be an Orlicz function. Then there is $u_0 > 0$ with $\Phi(u_0) > 0$. If $\Phi(u_0) = \infty$, let $b = 1$, and if $\Phi(u_0) < \infty$, let $b = 1/\Phi(u_0)$. Let $a = u_0$. Then, if $0 \le u \le 1$, we have $\Phi_{\min}(u) = 0 \le b\Phi(au)$. And if $1 < u$, then by convexity of Φ, we have $b\Phi(au) \ge bu\Phi(a) = u \ge \Phi_{\min}(u)$. Thus $\Phi \succ \Phi_{\min}$.

Next, there is $u_1 > 0$ with $\Phi(u_1) < \infty$. Let $b = \Phi(u_1)$ and $a = 1/u_1$. If $u > u_1$, we have $b\Phi_{\max}(au) = \infty \ge \Phi(u)$, and if $0 \le u \le u_1$, then by convexity of Φ, we have $\Phi(u) \le (u/u_1)\Phi(u_1) = b\Phi_{\max}(au)$. Thus $\Phi_{\max} \succ \Phi$.

Suppose $f \in L_1 \cap L_\infty$. Write $a = \|f\|_1 \vee \|f\|_\infty$. Then

$$M_{\Phi_{\max}}\left(\frac{f}{a}\right) = \int \Phi_{\max}\left(\frac{|f|}{a}\right) d\mu = \int \frac{|f|}{a} d\mu \leq 1,$$

so $f \in L_{\Phi_{\max}}$ and

$$\|f\|_{\Phi_{\max}} \leq a = \|f\|_1 \vee \|f\|_\infty.$$

Conversely, let $f \in L_{\Phi_{\max}}$. Write $a = \|f\|_{\Phi_{\max}}$. Then

$$\int |f| \, d\mu = a \int \frac{|f|}{a} d\mu \leq a \int \Phi_{\max}\left(\frac{|f|}{a}\right) \leq a.$$

Hence $f \in L_1$ and $\|f\|_1 \leq \|f\|_{\Phi_{\max}}$. Also, if $\mu\{|f| > a\} > 0$, then we would have $1 \geq \int \Phi_{\max}(|f|/a) \, d\mu \geq \int_{\{|f|>a\}} \infty \, d\mu = \infty$, which is false. So $f \in L_\infty$ and $\|f\|_\infty \leq \|f\|_{\Phi_{\max}}$.

Now suppose $f \in L_{\Phi_{\min}}$. Write $a = \|f\|_{\Phi_{\min}}$. Define

$$f_\infty = \begin{cases} f & \text{if } |f| \leq a \\ a\,\mathrm{sgn}(f) & \text{if } |f| > a \end{cases}$$

$$f_1 = f - f_\infty = \begin{cases} 0 & \text{if } |f| \leq a \\ (|f| - a)\,\mathrm{sgn}(f) & \text{if } |f| > a. \end{cases}$$

Then $\|f_\infty\|_\infty \leq a$, and

$$\|f_1\|_1 = \int |f_1| \, d\mu$$

$$= \int_{\{|f|>a\}} (|f| - a) \, d\mu$$

$$= a \int_{\{|f|/a>1\}} \left(\frac{|f|}{a} - 1\right) d\mu$$

$$= a \int \Phi_{\min}\left(\frac{|f|}{a}\right) d\mu \leq a.$$

Thus $\|f_1\|_1 \vee \|f_\infty\|_\infty \leq \|f\|_{\Phi_{\min}}$.

Conversely, suppose f can be written $f = f_1 + f_\infty$ with $f_1 \in L_1$ and $f_\infty \in L_\infty$. Write $a = \|f_1\|_1 \vee \|f_\infty\|_\infty$. Then $|f_\infty|/a \leq 1$ a.e., so

$$\int \Phi_{\min}\left(\frac{|f|}{a}\right) d\mu \leq \int \Phi_{\min}\left(\frac{|f_\infty|}{a} + \frac{|f_1|}{a}\right) d\mu$$

$$\leq \int \Phi_{\min}\left(\frac{|f_1|}{a} + 1\right) d\mu$$

$$= \int \frac{|f_1|}{a} d\mu = \frac{1}{a}\|f_1\|_1 \leq 1.$$

Thus $f \in L_{\Phi_{\min}}$ and $\|f\|_{\Phi_{\min}} \leq \|f_1\|_1 \vee \|f_\infty\|_\infty$. ∎

The space $L_{\Phi_{\max}}$ is the smallest Orlicz space. We will sometimes write $L_{\min} = L_{\Phi_{\max}}$. As suggested by the above description, it is called $L_1 \cap L_\infty$. If μ is finite, then $L_{\Phi_{\max}}(\Omega, \mathcal{F}, \mu)$ is just $L_\infty(\Omega, \mathcal{F}, \mu)$ with an equivalent norm. Similarly, if μ has no arbitrarily small sets, then $L_{\Phi_{\max}}(\Omega, \mathcal{F}, \mu)$ is $L_1(\Omega, \mathcal{F}, \mu)$ with an equivalent norm.

The space $L_{\Phi_{\min}}$ is the largest Orlicz space. We will sometimes write $L_{\max} = L_{\Phi_{\min}}$ It is often called $L_1 + L_\infty$. The Luxemburg norm (described in the theorem) differs from the usual norm

$$\|f\|_{L_1+L_\infty} = \inf \{ \|f_1\|_1 + \|f_\infty\|_\infty : f = f_1 + f_\infty \},$$

but

$$\|f\|_{\Phi_{\min}} \le \|f\|_{L_1+L_\infty} \le 2\|f\|_{\Phi_{\min}}.$$

If μ is finite, then $L_{\Phi_{\min}}(\Omega, \mathcal{F}, \mu) = L_1(\Omega, \mathcal{F}, \mu)$ with an equivalent norm. If μ has no arbitrarily small sets, then $L_{\Phi_{\min}}(\Omega, \mathcal{F}, \mu) = L_\infty(\Omega, \mathcal{F}, \mu)$ with an equivalent norm.

We will see later that the largest Orlicz heart $H_{\Phi_{\min}}(\Omega, \mathcal{F}, \mu)$ is useful in probability and ergodic theory. Now $\Phi_{\min}$ satisfies the (Δ_2) condition at ∞, so if μ is finite, then we have $H_{\Phi_{\min}} = L_{\Phi_{\min}} = L_1$. But $\Phi_{\min}$ fails the (Δ_2) condition at 0, so $L_1 + L_\infty = L_{\Phi_{\min}}([0,\infty)) \supsetneqq H_{\Phi_{\min}}([0,\infty))$. This is the space R_0 of Fava, discussed below. If $\Omega = \mathbb{N}$ and μ is counting measure, then $L_{\Phi_{\min}} = l_\infty$ and $H_{\Phi_{\min}} = c_0$.

We next briefly consider the spaces L_1 and L_∞.

(2.2.5) Proposition. *Let Φ be an Orlicz function. (1) If $\Phi(u) = \infty$ for some u, then $L_\Phi(\Omega, \mathcal{F}, \mu) \subseteq L_\infty(\Omega, \mathcal{F}, \mu)$ and $H_\Phi(\Omega, \mathcal{F}, \mu) = \{0\}$. If $H_\Phi([0,1]) \subseteq L_\infty([0,1])$, then $\Phi(u) = \infty$ for some u. (2) If $\Phi(u) = 0$ for some $u > 0$, then $L_\Phi(\Omega, \mathcal{F}, \mu) \supseteq L_\infty(\Omega, \mathcal{F}, \mu)$. If $l_\Phi \supseteq c_0$, then $\Phi(u) = 0$ for some $u > 0$. (3) $L_\Phi([0,\infty)) = L_\infty([0,\infty))$ if and only if $\Phi(u) = \infty$ for some u and $\Phi(u) = 0$ for some $u > 0$.*

Proof. (1) Suppose $\Phi(u_0) = \infty$. If $f \in L_\Phi$, then $M_\Phi(f/a) < \infty$ for some $a > 0$. Then, however, $|f|/a < u_0$ a.e., so $f \in L_\infty$.

Conversely, suppose Φ is finite. Choose positive numbers c_k so that $c_k \Phi(k^2) \le 2^{-k}$ and $c_k \le 2^{-k}$ for $k = 1, 2, \cdots$. Let A_k be disjoint intervals in $[0,1]$ with $\mu(A_k) = c_k$. Let $f = \sum_{k=1}^{\infty} k \, \mathbf{1}_{A_k}$. Then f is not in L_∞. If $a > 0$, then

$$M_\Phi(af) = \sum_{k \le a} \Phi(ak)c_k + \sum_{k > a} \Phi(ak)c_k$$

$$\le \sum_{k \le a} \Phi(ak)c_k + \sum_{k=1}^{\infty} \Phi(k^2)c_k < \infty,$$

so $f \in H_\Phi$.

(2) Suppose $\Phi(u_0) = 0$. Then if $f \in L_\infty$, take $a = \|f\|_\infty$, and we have $M_\Phi(f/au_0) = 0$, so $f \in L_\Phi$.

Conversely, suppose Φ is positive. Choose disjoint finite sets $A_n \subseteq \mathbb{N}$ with

$$\mu(A_k) \geq \frac{1}{\Phi(1/k^2)}.$$

Then

$$f = \sum_{k=1}^{\infty} \frac{1}{k} \mathbf{1}_{A_k}$$

is in c_0 but for any $a > 0$, we have

$$M_\Phi\left(\frac{f}{a}\right) \geq \sum_{k>a} \Phi\left(\frac{1}{ak}\right) \mu(A_k)$$

$$\geq \sum_{k>a} \Phi\left(\frac{1}{k^2}\right) \mu(A_k) = \infty.$$

Thus $f \notin l_\Phi$.

(3) Combine the arguments of (1) and (2).   ∎

(2.2.6) Proposition. *Let Φ be an Orlicz function, and let φ be its Orlicz derivative. (1) If φ is bounded above, then $H_\Phi(\Omega, \mathcal{F}, \mu) \supseteq L_1(\Omega, \mathcal{F}, \mu)$. If $L_\Phi([0,1]) \supseteq L_1([0,1])$, then φ is bounded above. (2) If φ is bounded away from 0, then $L_\Phi(\Omega, \mathcal{F}, \mu) \subseteq L_1(\Omega, \mathcal{F}, \mu)$. If $h_\Phi \subseteq l_1$, then φ is bounded away from 0. (3) $L_\Phi([0,\infty)) = L_1([0,\infty))$ if and only if φ is bounded away from 0 and ∞.*

Proof. (1) Suppose $\varphi(x) \leq C$ for all x. Then $\Phi(u) \leq Cu$ for all u. Let $f \in L_1$. For $a > 0$ we have

$$M_\Phi(af) = \int \Phi(a|f|) \, d\mu \leq aC \int |f| \, d\mu < \infty.$$

Thus $f \in H_\Phi$.

Conversely, suppose φ is unbounded, so that $\Phi(u)/u \to \infty$ as $u \to \infty$. Let a_n be such that $\Phi(a_n) \geq 2^n a_n$ and $a_n \geq 1$. Choose disjoint intervals A_n in $[0,1]$ with $\mu(A_n) = 2^{-n}/a_n$. Then for $f = \sum n a_n \mathbf{1}_{A_n}$ we have

$$\int |f| \, d\mu = \sum n a_n \mu(A_n) = \sum n 2^{-n} < \infty,$$

but for any $k > 0$,

$$\int \Phi\left(\frac{|f|}{k}\right) d\mu = \sum \Phi\left(\frac{n a_n}{k}\right) \mu(A_n) \geq \sum_{n=k}^{\infty} 1 = \infty.$$

Thus $L_1([0,1]) \nsubseteq L_\Phi([0,1])$.

(2) Suppose $\varphi(x) \geq r > 0$ for all x. Then $\Phi(u) \geq ru$ for all u. Let $f \in L_\Phi$. Then there is $a > 0$ with $M_\Phi(f/a) < \infty$. Then

$$\int |f| \, d\mu = a \int \frac{|f|}{a} \, d\mu \leq \frac{a}{r} \int \Phi\left(\frac{|f|}{a}\right) d\mu < \infty.$$

Thus $f \in L_1$.

Conversely, suppose φ is not bounded away from 0. For $u > 0$, we have $\Phi(u)/u \leq \varphi(u)$, so $\Phi(u)/u \to 0$ as $u \to 0$. Let a_n be such that $\Phi(a_n)/a_n \leq 2^{-n}/n$ and $a_n \leq 1$. Choose disjoint sets $A_n \subseteq \mathbb{N}$ with

$$\frac{n}{a_n} \leq \mu(A_n) < \frac{2n}{a_n}.$$

Let $f = \sum (a_n/n) \mathbf{1}_{A_n}$. Then:

$$\int |f| \, d\mu = \sum \frac{a_n}{n} \mu(A_n) \geq \sum 1 = \infty.$$

For any $c > 0$,

$$\int \Phi(c|f|) \, d\mu = \sum \Phi\left(\frac{ca_n}{n}\right) \mu(A_n)$$

$$\leq \sum_{n \leq c} \Phi\left(\frac{ca_n}{n}\right) \mu(A_n) + \sum_{n > c} \Phi(a_n) \mu(A_n)$$

$$\leq \sum_{n \leq c} \Phi\left(\frac{ca_n}{n}\right) \mu(A_n) + \sum_{n > c} 2^{-n+1} < \infty.$$

Thus $f \in h_\Phi$ but $f \notin l_1$.

(3) For this part, combine the arguments of (1) and (2). $\qquad\blacksquare$

Duality for Orlicz spaces

If E is a Banach space, then the *dual* of E (or *conjugate* of E) is the set E^* of all bounded linear functionals $x^* \colon E \to \mathbb{R}$. The norm

$$\|x^*\| = \sup\{|x^*(x)| : x \in E, \ \|x\| \leq 1\}$$

makes E^* a Banach space. We discuss here briefly a few duality theorems for Orlicz spaces. The key is Young's inequality (2.1.4). The following will sometimes also be referred to as Young's inequality.

(2.2.7) Proposition. *Let Φ and Ψ be conjugate Orlicz functions. If $f \in L_\Phi$ and $g \in L_\Psi$, then the product fg is integrable, and*

$$\int |fg| \, d\mu \leq 2\|f\|_\Phi \|g\|_\Psi.$$

Proof. First, by Young's inequality,

$$\int |fg|\, d\mu \leq \int \big(\Phi(|f|) + \Psi(|g|)\big)\, d\mu = M_\Phi\,(f) + M_\Psi(g).$$

Now if $a = \|f\|_\Phi$ and $b = \|g\|_\Psi$, then, applying Young's inequality to f/a and g/b, we obtain

$$\int |fg|\, d\mu \leq ab\big(M_\Phi\,(f/a) + M_\Psi(g/b)\big) \leq 2ab.\qquad \blacksquare$$

An important consequence of the preceding result is this: if $g \in L_\Psi$, then a bounded linear functional θ_g on L_Φ may be defined by

$$\theta_g(f) = \int fg\, d\mu.$$

The *Orlicz norm* of g is defined as the norm of this functional, and written $\|g\|_\Phi^*$:

$$\|g\|_\Phi^* = \sup\left\{\left|\int fg\, d\mu\right| : f \in L_\Phi, \|f\|_\Phi \leq 1\right\}$$

$$= \sup\left\{\left|\int fg\, d\mu\right| : M_\Phi\,(f) \leq 1\right\}.$$

The last expression is the norm originally used by Orlicz for the space L_Ψ. We will show below that this norm is equivalent to the one we are using, called the Luxemburg norm. Here are some simple variants of the Orlicz definition.

(2.2.8) Proposition. *Let g be a measurable function.*

(1) $$\|g\|_\Phi^* = \sup\left\{\int |fg|\, d\mu : f \in L_\Phi, \|f\|_\Phi \leq 1\right\}.$$

(2) $$\|g\|_\Phi^* = \sup\left\{\int |fg|\, d\mu : f \text{ integrable simple, } \|f\|_\Phi \leq 1\right\}.$$

(3) *If Φ is finite, then*

$$\|g\|_\Phi^* = \sup\left\{\int |fg|\, d\mu : f \in H_\Phi, \|f\|_\Phi \leq 1\right\}.$$

Proof. (1) Replace f by $\operatorname{sgn}(fg)$.

(2) Suppose $\int |fg|\, d\mu \leq a$ for all integrable simple functions f with $\|f\|_\Phi \leq 1$. We must show that $\int |fg|\, d\mu \leq a$ for any $f \in L_\Phi$ with $\|f\|_\Phi \leq 1$. There is a sequence (f_n) of integrable simple functions with $f_n \uparrow |f|$ a.e. Clearly $\|f_n\|_\Phi \leq 1$, so $\int |f_n g|\, d\mu \leq a$ for all n. But $|f_n g| \uparrow |fg|$ a.e., so by the monotone convergence theorem, $\int |fg|\, d\mu \leq a$.

(3) If Φ is finite, then the integrable simple functions are in H_Φ. $\blacksquare$

Now we can consider the connection of the Orlicz norm to the Luxemburg norm and the Young modular:

(2.2.9) Theorem. *Let g be a measurable function.*

(1) $$\|g\|_\Psi \le \|g\|_\Phi^* \le 2\|g\|_\Psi.$$

(2) $$\|g\|_\Phi^* \le 1 + M_\Psi(g).$$

Proof. The second inequality in (1) follows from (2.2.7).

For (2), let $\|f\|_\Phi = 1$. Then

$$\left| \int fg \, d\mu \right| \le \int \left(\Phi(|f|) + \Psi(|g|) \right) d\mu$$
$$= M_\Phi(f) + M_\Psi(g) \le 1 + M_\Psi(g).$$

So $\|g\|_\Phi^* \le 1 + M_\Psi(g)$.

Next we turn to the first inequality in (1). Let $g \in L_\Psi$, and write $a = \|g\|_\Phi^*$. We must show $\|g\|_\Psi \le a$; that is: $\int \Psi(|g|/a) \, d\mu \le 1$.

First consider the case where g is an integrable simple function, and $\Psi(|g|/a)$ is finite a.e. Let $f = \psi(|g|/a)$. By the case of equality in Young's inequality (2.1.4), we have

$$\int \frac{|fg|}{a} \, d\mu = \int \left[\Phi(|f|) \, d\mu + \Psi\left(\frac{|g|}{a} \right) \right] d\mu = M_\Phi(f) + M_\Psi\left(\frac{g}{a} \right).$$

Now if $M_\Phi(f) \le 1$, then $\int |fg| \, d\mu \le \|g\|_\Phi^* = a$, so we have

$$M_\Psi\left(\frac{g}{a} \right) \le M_\Phi(f) + M_\Psi\left(\frac{g}{a} \right) = \int \frac{|fg|}{a} \, d\mu \le 1.$$

On the other hand, if $M_\Phi(f) = b > 1$, then $M_\Phi(f/b) \le M_\Phi(f)/b$, so (by the definition of $\|g\|_\Phi^*$) we have $\int |fg| \, d\mu \le b\|g\|_\Phi^* = ab$, and thus

$$M_\Phi(f) + M_\Psi\left(\frac{g}{a} \right) = \frac{1}{a} \int |fg| \, d\mu \le \frac{ab}{a} = M_\Phi(f),$$

so that $M_\Psi(g/a) = 0$. Thus in both cases $M_\Psi(g/a) \le 1$, and we have $\|g\|_\Psi \le a = \|g\|_\Phi^*$.

Next consider the case where g is an integrable simple function, but $\Psi(|g|/a)$ is not a.e. finite. This means that Ψ is infinite. There is d so that $\Psi(v) = \infty$ for $v > d$ and $\Psi(v) < \infty$ for $v < d$. Then $\varphi(x) \to d$ as $x \to \infty$, so $\Phi(u) \le du$ for all u. We claim that $|g| \le ad$ a.e. Indeed, if $\mu\{|g| > ad\} > 0$, there is a set $A \subseteq \{|g| > ad\}$ with $0 < \mu(A) < \infty$. For $f = (1/d\mu(A)) \mathbf{1}_A$, we have $M_\Phi(f) = \Phi(1/d\mu(A)) \mu(A) \le 1$, so $\|f\|_\Phi \le 1$. Then

$$a = \|g\|_\Psi^* \ge \int |fg| \, d\mu > \frac{ad}{d\mu(A)} \mu(A) = a,$$

a contradiction, so $|g| \leq ad$ a.e. If $0 < \alpha < 1$, we have $\alpha|g|/a \leq \alpha d < d$, so $\Psi(\alpha|g|/a)$ is finite. We may then proceed as in the previous case to conclude $M_\Psi(\alpha g/a) \leq \alpha < 1$. But then when $\alpha \uparrow 1$, we have $M_\Psi(g/a) \leq 1$, since Ψ is left-continuous.

Finally, consider general $g \in L_\Psi$. Since μ is σ-finite, there is a sequence (g_n) of integrable simple functions with $g_n \uparrow |g|$ a.e. Now $M_\Psi(g_n/\|g_n\|_\Phi^*) \leq 1$ and $\|g_n\|_\Phi^* \leq \|g\|_\Phi^* = a$, so $M_\Psi(g_n/a) \leq 1$. Again, Ψ is left-continuous, so $M_\Psi(g/a) \leq 1$. Thus $\|g\|_\Psi \leq a$. ∎

(2.2.10) Corollary. *Suppose g is a measurable function. If $fg \in L_1$ for all $f \in L_\Phi$, then $g \in L_\Psi$. Suppose that Φ is finite. If $fg \in L_1$ for all $f \in H_\Phi$, then $g \in L_\Psi$.*

Proof. Suppose $fg \in L_1$ for all $f \in L_\Phi$. Observe that the linear transformation $T: L_\Phi \to L_1$ defined by $T(f) = fg$ has closed graph. [Indeed, if we have $\|f_n - f\|_\Phi \to 0$, then (by 2.1.10(6)) there is a subsequence with $f_{n_k} \to f$ a.e., so $f_{n_k}g \to fg$ a.e. If $\|f_n g - h\|_1 \to 0$, then there is a further subsequence with $f_{n_{k_j}}g \to h$ a.e. Thus $fg = h$.] So $\|g\|_\Phi^* < \infty$ by the closed graph theorem. Therefore $\|g\|_\Psi < \infty$ and $g \in L_\Psi$.

In the second case, where Φ is finite, we know that H_Φ is enough to compute $\|g\|_\Phi^*$. So again $\|g\|_\Phi^* < \infty$ by the closed graph theorem and $g \in L_\Psi$. ∎

The preceding result is close to showing that L_Ψ is the dual of L_Φ. But we know that L_1 is not the dual of L_∞. Here is a good illustration of the difference between the spaces L_Φ and H_Φ.

(2.2.11) Theorem. *Suppose Ψ is finite. Then the dual of H_Ψ (with the Luxemburg norm) is L_Φ (with the Orlicz norm).*

Proof. Since Ψ is finite, the space H_Ψ is the closure of the integrable simple functions. Let x^* be a bounded linear functional on H_Ψ. For each $A \in \mathcal{F}$ with $\mu(A) < \infty$, define $\nu(A) = x^*(1_A)$. Now if $A_n \downarrow \emptyset$, then [since H_Ψ has order continuous norm (2.1.14(a))] we have $\nu(A_n) \to 0$. So ν is a finite signed measure on each A where μ is finite. If $\mu(A) = 0$ then $\nu(A) = 0$. By the Radon-Nikodým theorem, there is a measurable function f such that $\int_A f \, d\mu = x^*(1_A)$ for all A with $\mu(A) < \infty$. Then by linearity, for all integrable simple functions we have $\int fg \, d\mu = x^*(g)$, so $\int |fg| \, d\mu \leq \|x^*\| \|g\|_\Psi$. Thus $\|f\|_\Psi^* \leq \|x^*\| < \infty$, so $f \in L_\Phi$. The integrable simple functions are dense in H_Ψ, so $\int fg \, d\mu = x^*(g)$ for all $g \in H_\Psi$. By (2.2.8), $\|f\|_\Psi^* = \|x^*\|$. ∎

Note that Φ and Ψ are finite in the following (2.2.25).

(2.2.12) Corollary. *Suppose Φ and Ψ are conjugate finite Orlicz functions. If Ψ satisfies the (Δ_2) condition, then the dual of L_Ψ is L_Φ. If Φ satisfies the (Δ_2) condition, then the bidual of H_Ψ is L_Ψ. If both Ψ and Φ satisfy condition (Δ_2), then L_Φ and L_Ψ are reflexive.*

The heart of a sum of Orlicz spaces

We consider an Orlicz function Φ that satisfies condition (Δ_2) at ∞; that is, there exist u_0 and M so that

$$\Phi(2u) < M\Phi(u) \qquad \text{for all } u \geq u_0.$$

If $(\Omega, \mathcal{F}, \mu)$ is finite, then $L_\Phi = H_\Phi$ and $L_\Phi \supseteq L_\infty$, so there is no point to discussion of the heart of $L_\Phi + L_\infty$. Therefore, let us assume that $(\Omega, \mathcal{F}, \mu)$ is infinite.

The space $L_\Phi + L_\infty$ is again a Banach lattice. The usual norm for this space is:

$$\|f\|_{L_\Phi + L_\infty} = \inf \{ \|f_1\|_\Phi + \|f_2\|_\infty : f = f_1 + f_2 \}.$$

If $\Phi(u) = 0$ for some $u > 0$, then $L_\Phi \supseteq L_\infty$, so it is of little interest to consider $L_\Phi + L_\infty$. Therefore it will sometimes be assumed below that this does not happen.

In the next result, subscript 's' for 'shift' was chosen because the graph of Φ_s is the graph of Φ shifted to the right by one unit.

(2.2.13) Proposition. *Let Φ be an Orlicz function. Let Φ_s be defined by*

$$(2.2.13a) \qquad \Phi_s(u) = \begin{cases} 0 & \text{if } 0 \leq u \leq 1, \\ \Phi(u-1) & \text{if } u > 1. \end{cases}$$

Then L_{Φ_s} is the space $L_\Phi + L_\infty$, with Luxemburg norm

$$\|f\|_{\Phi_s} = \inf \{ \|f_1\|_\Phi \vee \|f_2\|_\infty : f = f_1 + f_2 \}.$$

so that

$$\|f\|_{\Phi_s} \leq \|f\|_{L_\Phi + L_\infty} \leq 2\|f\|_{\Phi_s}.$$

Proof. Note that $M_{\Phi_s}(f) = \int_{\{|f|>1\}} \Phi(|f| - 1) \, d\mu$.

Suppose $f = f_1 + f_2$, where $f_1 \in L_\Phi$ and $f_2 \in L_\infty$. If $a = \|f_1\|_\Phi \vee \|f_2\|_\infty$, then $|f| - a \leq |f_1|$, so

$$M_{\Phi_s}\left(\frac{f}{a}\right) = \int_{\{|f|>a\}} \Phi\left(\frac{|f| - a}{a}\right) d\mu \leq \int \Phi\left(\frac{|f_1|}{a}\right) d\mu \leq 1.$$

Thus $\|f\|_{\Phi_s} \leq a$. This shows

$$\|f\|_{\Phi_s} \leq \inf \{ \|f_1\|_\Phi \vee \|f_2\|_\infty : f = f_1 + f_2 \}.$$

On the other hand, let $f \in L_{\Phi_s}$. Write $a = \|f\|_{\Phi_s}$. Then we have, first, that $M_{\Phi_s}(f/a) \leq 1$, or

$$\int_{\{|f|>a\}} \Phi\left(\frac{|f| - a}{a}\right) d\mu \leq 1.$$

Let $f_1 = \operatorname{sgn} f(|f| - a) \mathbf{1}_{\{|f|>a\}}$ and $f_2 = \operatorname{sgn} f(|f| \wedge a)$, so $f = f_1 + f_2$. Then $\|f_2\|_\infty \leq a$ and $\int \Phi(f_1/a) \, d\mu \leq 1$, so $\|f_1\|_\Phi \leq a$. Thus

$$\|f\|_{\Phi_s} \geq \inf \{ \|f_1\|_\Phi \vee \|f_2\|_\infty : f = f_1 + f_2 \}. \qquad \blacksquare$$

The heart of such a space $L_\Phi + L_\infty$ can be characterized in various ways.

(2.2.14) Proposition. *Let Φ be an Orlicz function satisfying (Δ_2) at ∞ and $\Phi(u) > 0$ for all $u > 0$. Let Φ_s be defined by (2.2.13a). Let f be a measurable function. The following are equivalent.*

(a) $f \in H_{\Phi_s}$.
(b) $f \in L_{\Phi_s}$ *and* $\mu\{|f| > t\} < \infty$ *for all* $t > 0$.
(c) $\int_{\{|f|>t\}} \Phi(|f|)\, d\mu < \infty$ *for all* $t > 0$.

Proof. (a) $\Longrightarrow$ (c). Suppose $f \in H_{\Phi_s}$. Fix $t > 0$. Write $u = t/(1+t)$. Then $x \geq t$ implies $x \leq (x/u) - 1$, so on the set $\{|f| \geq t\}$ we have $|f| \leq |f|/u - 1$. Also, $u < t$, so $\{|f| > t\} \subseteq \{|f| > u\}$. Thus

$$
\int_{\{|f|>t\}} \Phi(|f|)\, d\mu \leq \int_{\{|f|>t\}} \Phi\left(\frac{|f|}{u} - 1\right) d\mu
$$
$$
\leq \int_{\{|f|>u\}} \Phi\left(\frac{|f|}{u} - 1\right) d\mu
$$
$$
= \int \Phi_s\left(\frac{|f|}{u}\right) d\mu < \infty.
$$

(c) $\Longrightarrow$ (b). Since $\Phi(t) > 0$, we have

$$
\mu\{|f| > t\} \leq \frac{1}{\Phi(t)} \int_{\{|f|>t\}} \Phi(|f|)\, d\mu.
$$

(b) $\Longrightarrow$ (a). Suppose $f \in L_{\Phi_s}$ and $\mu\{|f| > t\} < \infty$ for all $t > 0$. Write $a = \|f\|_{\Phi_s}$. Given $\varepsilon > 0$, write $f_1 = f\,\mathbf{1}_{\{|f|>\varepsilon\}}$ and $f_2 = f\,\mathbf{1}_{\{|f|\leq\varepsilon\}}$. We claim that $f_1 \in H_{\Phi_s}$. Indeed, let $b > 0$ be given. Since Φ satisfies (Δ_2) at ∞, so does Φ_s, and we have

$$
M = \sup_{u \geq \varepsilon/a} \frac{\Phi_s(au/b)}{\Phi_s(u)} < \infty.
$$

Then

$$
\int_{\{|f|>\varepsilon\}} \Phi_s\left(\frac{|f_1|}{b}\right) d\mu = \int_{\{|f|>\varepsilon\}} \Phi_s\left(\frac{|f|}{b}\right) d\mu
$$
$$
\leq M \int \Phi_s\left(\frac{|f|}{a}\right) d\mu \leq M.
$$

Hence $f_1 \in H_{\Phi_s}$. Also,

$$
M_{\Phi_s}\left(\frac{f_2}{\varepsilon}\right) = \int_{\{|f|\leq\varepsilon\}} \Phi_s\left(\frac{|f|}{\varepsilon}\right) d\mu \leq \int \Phi_s(0)\, d\mu = 0.
$$

Therefore $\|f_2\|_{\Phi_s} \leq \varepsilon$. The distance of f from H_{Φ_s} is at most ε. Since H_{Φ_s} is closed, we have $f \in H_{\Phi_s}$. ∎

(2.2.15) Let us consider again the largest Orlicz space $L_1 + L_\infty$. The Orlicz function that may be used is

$$(2.2.15a) \qquad \Phi_{\min}(u) = \begin{cases} 0 & \text{if } 0 \leq u \leq 1 \\ u - 1 & \text{if } 1 \leq u. \end{cases}$$

As we have seen (2.2.4), this produces the Luxemburg norm

$$\|f\|_{\Phi_{\min}} = \inf\left\{\, \|f_1\|_1 \vee \|f_\infty\|_\infty : f = f_1 + f_\infty, f_1 \in L_1, f_\infty \in L_\infty \,\right\}.$$

The heart $H_{\Phi_{\min}}$ of this Orlicz space is also known as Fava's space R_0. Thus $f \in R_0$ if and only if $f \in L_1 + L_\infty$ and $\mu\{|f| > t\} < \infty$ for all $t > 0$. Also, $f \in R_0$ if and only if $\int_{\{|f|>t\}} |f|\, d\mu < \infty$ for all $t > 0$. (See (2.3.2) for the uniform version of this condition.)

(2.2.16) Proposition. *A set $\mathcal{K}$ is bounded in $L_{\Phi_{\min}} = L_1 + L_\infty$ if and only if there exist α, M such that*

$$\int_{\{|f|\geq\alpha\}} |f|\, d\mu \leq M$$

for all $f \in \mathcal{K}$.

Proof. Suppose $\|f\|_{\Phi_{\min}} \leq A$ for all $f \in \mathcal{K}$. Let $\alpha > 2A, M > 2A$. If $f \in \mathcal{K}$, there is a decomposition $f = f_1 + f_\infty$ with $\|f_1\|_1 < M/2$ and $\|f_\infty\|_\infty < \alpha/2$. Now on the set $\{|f| \geq \alpha\}$ we have $|f_\infty| < \alpha/2$ so $|f_1| > \alpha/2$, and

$$\frac{\alpha}{2}\mu\{|f| \geq \alpha\} \leq \frac{\alpha}{2}\mu\left\{|f_1| > \frac{\alpha}{2}\right\}$$
$$\leq \int |f_1|\, d\mu \leq \frac{M}{2}.$$

Thus

$$\int_{\{|f|\geq\alpha\}} |f|\, d\mu \leq \int_{\{|f|\geq\alpha\}} \left(\frac{\alpha}{2} + |f_1|\right) d\mu \leq \frac{M}{2} + \frac{M}{2} = M. \qquad \blacksquare$$

We consider next Fava's spaces R_k. They are (in one sense) generalizations of the space R_0.

Let $k \geq 0$ be real. (In most applications, k will be an integer; but real values certainly make sense.) Consider the function Φ_k defined by

$$(2.2.16a) \qquad \Phi_k(u) = \begin{cases} 0 & \text{if } u \leq 1 \\ u(\log u)^k & \text{if } u > 1. \end{cases}$$

[We may abbreviate this as $\Phi_k(u) = u(\log^+ u)^k$. See Figure (2.2.16).]
For $k \geq 1$, this is an Orlicz function. The space L_{Φ_k} is traditionally called
$L \log^k L$. The heart H_{Φ_k} is Fava's space R_k. Note that Φ_k satisfies (Δ_2)
at ∞, so if μ is a finite measure, then we have $R_k = L \log^k L$. But Φ_k fails
(Δ_2) at 0, so if μ is infinite, we may have $R_k \neq L \log^k L$. In any case, R_k
is the closure of the integrable simple functions in $L \log^k L$.

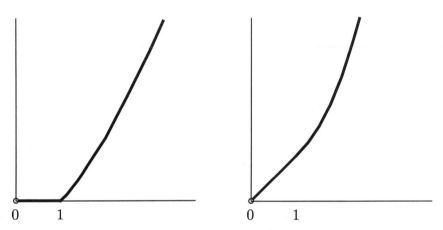

Figure (2.2.16). Orlicz function $u \log^+ u$ and its conjugate.

When $k = 0$, the function Φ_0 is not an Orlicz function. But it is "equiva-
lent" to the Orlicz function $\Phi_{\min}$ in (2.2.4a), $\Phi_0 \approx \Phi_{\min}$, so $L_{\Phi_0} = L_{\Phi_{\min}} = L_1 + L_\infty$.

Since R_k $(k \geq 0)$ is an Orlicz heart, it is also a Banach lattice with
order continuous norm (2.1.14(a)). We may also note that if $k < m$, then
$R_k \supseteq R_m$. Clearly, also, if $k \geq 0$ and $p > 1$, then $R_k \supseteq L_p$. We will see
below that certain classical results known to hold for all spaces L_p with
$p > 1$ but known to fail for L_1 in fact hold on an appropriate space R_k.

As in the case of R_0, we may describe R_k in many equivalent ways. We
have $f \in R_k$ if and only if

$$\int_{\{|f|>a\}} |f| \left(\log |f| - \log a \right)^k d\mu < \infty$$

for all $a > 0$. Or: $f \in R_k$ if and only if $f \in L \log^k L$ and $\mu\{|f| \geq a\} < \infty$
for all $a > 0$. Or: $f \in R_k$ if and only if, for all $\varepsilon > 0$, there exist g and h
such that $f = g + h$, $\int |g|(\log^+ |g|)^k d\mu < \infty$, $\mu\{g \neq 0\} < \infty$, $\|h\|_\infty < \varepsilon$.

Complements

(2.2.17) There is, in fact, a formula for the Orlicz norm not involving the conjugate Orlicz function.

$$\|f\|_{\Psi}^* = \inf_{k>0} \frac{1}{k}\left(1 + \int \Phi(k|f|)\, d\mu\right)$$

(Krasnosel'skii & Rutickii [1961], p. 92).

(2.2.18) (Inequality.) Let Φ be an Orlicz function with derivative φ and conjugate Ψ. Suppose $\Phi(2u) \le C\Phi(u)$ for all u. Then $\Psi(\varphi(u)) \le (C-2)\Phi(u)$ for all u. To see this, use 2.1.19b and the case of equality in Young's inequality 2.1.4: If $\Phi(u) < \infty$, then $\Psi(\varphi(u)) = u\varphi(u) - \Phi(u) \le \Phi(2u) - 2\Phi(u) \le (C-2)\Phi(u)$.

(2.2.19) (Definition of '$\succ$'.) Let Φ_1 and Φ_2 be Orlicz functions. The definition of $\Phi_1 \succ_\infty \Phi_2$ is: There exist $a, b, u_0 > 0$ such that

$$b\Phi_1(au) \ge \Phi_2(u) \qquad \text{for all } u \ge u_0.$$

The following are equivalent: (a) $\Phi_1 \succ_\infty \Phi_2$; (b) there exist $a, u_0 > 0$ such that $\Phi_1(au) \ge \Phi_2(u)$ for all $u \ge u_0$. A similar equivalence holds at 0.

(2.2.20) (Definition of '$\succ$'.) Find Orlicz functions showing that the following are not equivalent: (a) $\Phi_1 \succ_\infty \Phi_2$; (b) there exist $b, u_0 > 0$ such that $b\Phi_1(u) \ge \Phi_2(u)$ for all $u \ge u_0$.

(2.2.21) (Convolution on $\mathbb{R}^n$.) Let Ω be n-dimensional Euclidean space $\mathbb{R}^n$ and let μ be n-dimensional Lebesgue measure. Let Φ and Ψ be conjugate Orlicz functions. Suppose Φ satisfies the (Δ_2) condition. If $f \in L_\Phi(\mathbb{R}^n)$ and $g \in L_\Psi(\mathbb{R}^n)$, then the convolution

$$k(t) = \int_{\mathbb{R}^n} f(t-x)g(x)\, dx = \int_{\mathbb{R}^n} f(x)g(t-x)\, dx$$

exists and is continuous on $\mathbb{R}^n$. If Ψ also satisfies the (Δ_2) condition, then $\lim_{\|t\|\to\infty} k(t) = 0$ (Zaanen [1983], p. 600).

(2.2.22) Let Ω be n-dimensional Euclidean space $\mathbb{R}^n$ and let μ be n-dimensional Lebesgue measure. Suppose Φ satisfies the (Δ_2) condition. If $f \in L_\Phi(\mathbb{R}^n)$, then, for any $\varepsilon > 0$, there exists a continuous function $g\colon \mathbb{R}^n \to \mathbb{R}$ with compact support such that $\|f - g\|_\Phi < \varepsilon$.

For $h \in \mathbb{R}^n$, let the translate f_h be defined by $f_h(x) = f(x-h)$. Then $\|f_h - f\|_\Phi \to 0$ as $\|h\| \to 0$ (Zaanen [1983], p. 599).

(2.2.23) Let A be a subset of $H_\Phi(\mathbb{R}^n)$. Then A is relatively compact if and only if

(1) $\sup_{f \in A} \|f\|_\Phi < \infty$; and

(2) for every $\varepsilon > 0$ there is $\delta > 0$ such that for all $h \in \mathbb{R}^n$ with $\|h\| < \delta$ and all $f \in A$, we have $\|f_h - f\|_\Phi < \varepsilon$

(Krasnosel'skii & Rutickii [1961], p. 100).

(2.2.24) (Linear functionals.) Let Φ be a finite Orlicz function with conjugate Ψ. Let α be a (possibly discontinuous) linear functional on L_Φ. Suppose

(a) $\alpha(f) \leq 1$ for all $f \leq 0$;

(b) there is $\varepsilon > 0$ such that $\alpha(f) \leq 1$ for all f with $\|f\|_\Phi \leq \varepsilon$.

Then there is $g \in L_\Psi$, $g \geq 0$ such that

$$\alpha(f) = \int fg\,d\mu$$

for all $f \in L_\Phi$. To see this, observe that by (a) α is a positive linear functional, and by (b) α is continuous (with norm $\leq 1/\varepsilon$). Define a set-function ν on the sets $A \in \mathcal{F}$ with $\mu(A) < \infty$ by

$$\nu(A) = \alpha(\mathbf{1}_A).$$

By (2.1.20) it follows that ν is countably additive (on the ring of sets of finite μ-measure), so by the Radon-Nikodým theorem (and the σ-finiteness of μ) there is g with $\alpha(A) = \int_A g\,d\mu$. The rest follows as in (2.2.11), using monotone convergence.

(2.2.25) (Finiteness.) It is possible for an Orlicz function Ψ to be infinite even if its conjugate Φ satisfies (Δ_2). For example

$$\varphi(x) = \begin{cases} \dfrac{\pi}{4}x & \text{if } 0 \leq x \leq 1 \\ \arctan x & \text{if } x > 1. \end{cases}$$

(Thanks to A. Millet for providing this example.)

Remarks

This section comes from the same sources as the last one: Krasnosel'skii & Rutickii [1961] discuss the comparison of Orlicz spaces in Chapter II, Section 13; and duality of Orlicz spaces in Section 14. Zaanen [1983], Chapter 19, also has a lot on duality.

Fava [1972] introduced the spaces R_k, $k = 0, 1, \cdots$ in the context of ergodic theory. Frangos & Sucheston [1986] identified the spaces R_k, $k \geq 1$, as hearts of Orlicz spaces $L\log^k L$, and showed that R_0 is a Banach lattice with order continuous norm. The identification of R_0 as the heart of $L_1 + L_\infty$, and the other material on the heart of the sum of Orlicz spaces, is from Edgar & Sucheston [1989].

2.3. Uniform integrability and conditional expectation

Uniform integrability is a useful criterion in probability theory. We define it here even for infinite measure spaces, so that the usual role played by L_1 is taken by the largest Orlicz space $L_1 + L_\infty$. Only the basic properties of uniform integrability are proved here, but it will reappear frequently throughout the book. An interesting characterization (2.3.5) of uniform integrability (which goes back to de La Vallée Poussin) can be formulated in terms of Orlicz functions. Then we will discuss conditional expectation on infinite measure spaces.

(2.3.1) Definition. Let $\mathcal{K}$ be a family of real-valued measurable functions defined on $(\Omega, \mathcal{F}, \mu)$. We say that $\mathcal{K}$ is *uniformly integrable* iff, for every $\varepsilon > 0$, there is $\lambda > 0$ such that for all $f \in \mathcal{K}$, we have $\int_{\{|f| \geq \lambda\}} |f| \, d\mu < \varepsilon$.

It would seem logical that a "uniformly integrable" set should consist of integrable functions. This is true if μ is finite, but not in general. A singleton $\{g\}$ satisfies this definition if and only if $g \in L_1 + L_\infty$, that is, g is in some Orlicz space L_Φ. Uniform integrability of a set $\mathcal{K}$ might be considered as the property that the elements of $\mathcal{K}$ belong to $L_1 + L_\infty$ in a uniform way. The main uses of the term "uniformly integrable" will occur when μ is finite.

Some books use a slightly different definition. We will say that $\mathcal{K}$ is *uniformly absolutely continuous* iff, for every $\varepsilon > 0$, there is $\delta > 0$ such that if $f \in \mathcal{K}$ and $A \in \mathcal{F}$ with $\mu(A) < \delta$, then $\int_A |f| \, d\mu < \varepsilon$. The next result shows that this concept is (for most purposes) equivalent to uniform integrability. [The vector-valued versions are different, however; see (5.2.15).]

(2.3.2) Proposition. *Let $\mathcal{K}$ be a family of real-valued measurable functions defined on $(\Omega, \mathcal{F}, \mu)$.*

(1) *If $\mathcal{K}$ is uniformly integrable, then $\mathcal{K}$ is uniformly absolutely continuous.*

(2) *Suppose $\mathcal{K}$ is $L_1 + L_\infty$-bounded. Then $\mathcal{K}$ is uniformly integrable if and only if $\mathcal{K}$ is uniformly absolutely continuous.*

(3) *Suppose μ is atomless. Then $\mathcal{K}$ is uniformly integrable if and only if $\mathcal{K}$ is uniformly absolutely continuous.*

Proof. (1) Suppose $\mathcal{K}$ is uniformly integrable. Let $\varepsilon > 0$ be given. Then there is λ so that for all $f \in \mathcal{K}$, we have $\int_{\{|f| \geq \lambda\}} |f| \, d\mu \leq \varepsilon/2$. Let $\delta < \varepsilon/2\lambda$. Now if $f \in \mathcal{K}$ and $\mu(A) < \delta$, we have

$$\int_A |f| \, d\mu \leq \int_{A \cap \{|f| \geq \lambda\}} |f| \, d\mu + \int_{A \cap \{|f| < \lambda\}} |f| \, d\mu$$
$$\leq \frac{\varepsilon}{2} + \delta\lambda < \varepsilon.$$

(2) Suppose $\mathcal{K}$ is uniformly absolutely continuous and $L_1 + L_\infty$-bounded. Then by (2.2.16) there exist α and M such that

$$\int_{\{|f| \geq \alpha\}} |f| \, d\mu \leq M$$

for all $f \in \mathcal{K}$. Let $\varepsilon > 0$ be given. Now there is $\delta > 0$ so that for $\mu(A) < \delta$ and $f \in \mathcal{K}$ we have $\int_A |f| \, d\mu < \varepsilon$. Let $\lambda > \max\{\alpha, M/\delta\}$. Then

$$\mu\{|f| \geq \lambda\} \leq \frac{1}{\lambda} \int_{\{|f| \geq \lambda\}} |f| \, d\mu$$
$$\leq \frac{1}{\lambda} \int_{\{|f| \geq \alpha\}} |f| \, d\mu \leq \frac{M}{\lambda} < \delta,$$

so

$$\int_{\{|f|\geq\lambda\}} |f|\,d\mu < \varepsilon.$$

(3) Suppose μ is atomless and $\mathcal{K}$ is uniformly absolutely continuous. Let $\varepsilon > 0$ be given. Then there is $\delta > 0$ so that $\int_A |f|\,d\mu < \varepsilon$ whenever $f \in \mathcal{K}$ and $\mu(A) < \delta$. Let $\lambda = 2\varepsilon/\delta$. We claim that $\mu\{|f| \geq \lambda\} < \delta$ for all $f \in \mathcal{K}$, which will prove that $\mathcal{K}$ is uniformly integrable. Suppose not: suppose that, for some $f \in \mathcal{K}$, we have $\mu\{|f| \geq \lambda\} \geq \delta$. Since μ is atomless, there is a set $A \subseteq \{|f| \geq \lambda\}$ with $\mu(A) = \delta/2$. Then

$$\varepsilon > \int_A |f|\,d\mu \geq \frac{\lambda\delta}{2} = \varepsilon,$$

a contradiction.   ∎

There are some obvious sufficient conditions for uniform integrability.

(2.3.3) Proposition. *Let $\mathcal{K}$ be a family of measurable functions.*

(a) *Suppose there is $f \in L_1 + L_\infty$ with $|g| \leq |f|$ a.e. for all $g \in \mathcal{K}$ (we say $\mathcal{K}$ is* dominated *by f). Then $\mathcal{K}$ is uniformly integrable.*

(b) *Suppose there is $f \in L_1 + L_\infty$ with $\mu\{|g| \geq \lambda\} \leq \mu\{|f| \geq \lambda\}$ for all $\lambda > 0$, $g \in \mathcal{K}$ (we say $\mathcal{K}$ is* dominated in distribution *by f). Then $\mathcal{K}$ is uniformly integrable.*

Proof. Part (a) is an easy consequence of (b); so we prove (b). Suppose $f = f_1 + f_\infty$, where $f_1 \in L_1$ and $f_\infty \in L_\infty$. Write $M = \|f_\infty\|_\infty$. For $\lambda > M$, we have for all $g \in \mathcal{K}$,

$$\int_{\{|g|\geq\lambda\}} |g|\,d\mu = \lambda\mu\{|g| \geq \lambda\} + \int_\lambda^\infty \mu\{|g| \geq t\}\,dt$$

$$\leq \lambda\mu\{|f| \geq \lambda\} + \int_\lambda^\infty \mu\{|f| \geq t\}\,dt$$

$$\leq \lambda\mu\{|f_1| + M \geq \lambda\} + \int_\lambda^\infty \mu\{|f_1| + M \geq t\}\,dt$$

$$= \int_{\{|f_1|+M\geq\lambda\}} (|f_1| + M)\,d\mu.$$

This is independent of g and tends to 0 as $\lambda \to \infty$.   ∎

One common use of uniform integrability is to connect convergence in mean with other modes of convergence. Recall that a sequence (f_n) converges to f *in measure* iff, for every $\varepsilon > 0$, we have

$$\lim_{n\to\infty} \mu\{|f_n - f| > \varepsilon\} = 0.$$

(2.3.4) Proposition. *Let $(\Omega, \mathcal{F}, \mu)$ be a finite measure space. Suppose $f_n \in L_1$ for $n \in \mathbb{N}$ and $f \in L_1$. Then $\|f_n - f\|_1 \to 0$ if and only if $\{f_n : n \in \mathbb{N}\}$ is uniformly integrable and $f_n \to f$ in measure.*

Proof. Suppose $\|f_n - f\|_1 \to 0$. Now first, if $\varepsilon > 0$, then

$$\mu\{|f_n - f| > \varepsilon\} \le \frac{1}{\varepsilon}\int |f_n - f| \, d\mu = \frac{1}{\varepsilon}\|f_n - f\|_1.$$

Thus $\mu\{|f_n - f| > \varepsilon\} \to 0$ as $n \to \infty$, so $f_n \to f$ in measure.

Second, we must prove uniform integrability. But $\{f_n : n \in \mathbb{N}\}$ is L_1-bounded, so by (2.3.2(2)) we may prove uniform absolute continuity. For $\varepsilon > 0$, there is N with $\|f_n - f\|_1 < \varepsilon$ for $n \ge N$. Now

$$g = \max_{1 \le n < N} |f_n| \vee |f|$$

is integrable, so there is $\delta > 0$ with

$$\int_A |g| \, d\mu < \varepsilon \qquad \text{whenever } \mu(A) < \delta.$$

Now for $n < N$, if $\mu(A) < \delta$ we have

$$\int_A |f_n| \, d\mu \le \int_A |g| \, d\mu < \varepsilon,$$

and for $n \ge N$, if $\mu(A) < \delta$ we have

$$\int_A |f_n| \, d\mu \le \int_A |f| \, d\mu + \int_A |f_n - f| \, d\mu < 2\varepsilon.$$

Thus $\{f_n\}$ is uniformly absolutely continuous.

Conversely, suppose $\{f_n\}$ is uniformly integrable and $f_n \to f$ in measure. Of course, $\mathcal{K} = \{f\} \cup \{f_n : n \in \mathbb{N}\}$ is also uniformly integrable. Fix $\varepsilon > 0$. There is $\delta > 0$ so that $\int_A |g| \, d\mu < \varepsilon$ whenever $g \in \mathcal{K}$ and $\mu(A) < \delta$. Also, there is N so that if $n \ge N$ we have $\mu\{|f_n - f| > \varepsilon\} < \delta$. Then for $n \ge N$,

$$\int |f_n - f| \, d\mu \le \int_{\{|f_n - f| > \varepsilon\}} (|f_n| + |f|) \, d\mu + \int_{\{|f_n - f| \le \varepsilon\}} |f_n - f| \, d\mu$$
$$= 2\varepsilon + \varepsilon\mu(\Omega).$$

Thus $\|f_n - f\|_1 \to 0$. ∎

(2.3.5) Theorem (criterion of de La Vallée Poussin). *Let $\mathcal{K}$ be a set of functions. The following are equivalent:*

(1) $\mathcal{K}$ *is uniformly integrable.*

(2) $\mathcal{K}$ *is L_Φ-bounded for some finite Orlicz function Φ with*

$$\lim_{u \to \infty} \frac{\Phi(u)}{u} = \infty.$$

(3) $\sup \{ M_\Phi(f) : f \in \mathcal{K} \} < \infty$ *for some finite Orlicz function Φ with*

$$\lim_{u \to \infty} \frac{\Phi(u)}{u} = \infty.$$

Proof. $(2) \Longrightarrow (1)$. Suppose $\mathcal{K}$ is L_Φ-bounded. Say $\|f\|_\Phi \leq M$ for all $f \in \mathcal{K}$. Given $\varepsilon > 0$, let λ be such that $\Phi(u)/u > M/\varepsilon$ for $u \geq \lambda/M$. Then

$$\int_{\{|f| \geq \lambda\}} |f| \, d\mu = M \int_{\{|f| \geq \lambda\}} \frac{|f|}{M} \, d\mu$$

$$< M \frac{\varepsilon}{M} \int_{\{|f| \geq \lambda\}} \Phi\left(\frac{|f|}{M}\right) d\mu \leq \varepsilon.$$

So $\mathcal{K}$ is uniformly integrable.

$(1) \Longrightarrow (3)$. Suppose $\mathcal{K}$ is uniformly integrable. Write

$$C(\lambda) = \sup \left\{ \int_{\{|f| \geq \lambda\}} |f| \, d\mu : f \in \mathcal{K} \right\}.$$

Then C is a decreasing function; $\lim_{\lambda \to \infty} C(\lambda) = 0$ since $\mathcal{K}$ is uniformly integrable. Choose λ_0 with $C(\lambda_0) < 1$. Then $-\log C(\lambda)$ is a positive increasing function of λ for $\lambda \geq \lambda_0$, and $-\log C(\lambda) \to \infty$ as $\lambda \to \infty$. There is a continuous function $\varphi \colon (0, \infty) \to (0, \infty)$ with $\varphi(u) = 0$ for $u \leq \lambda_0$, $\varphi(u) < -\log C(u)$ for $u \geq \lambda_0$, and $\lim_{u \to \infty} \varphi(u) = \infty$. We may assume φ is strictly increasing on (λ_0, ∞). Thus φ is an Orlicz derivative. Let Φ be the corresponding Orlicz function and let ψ be the inverse of φ. Now $\Phi(u)/u \to \infty$ since $\varphi(u) \to \infty$. Since φ is strictly increasing on (λ_0, ∞), we have $\varphi(\psi(y)) = y$ for all $y > 0$. Thus $y < -\log C(\psi(y))$, or $C(\psi(y)) < e^{-y}$. Now for $f \in \mathcal{K}$, we have

$$M_\Phi(f) \leq \int \varphi(|f|) \, |f| \, d\mu$$

$$= \int_\Omega \int_0^{\varphi(|f(\omega)|)} dy \, |f(\omega)| \, d\mu(\omega)$$

$$= \int_0^\infty \int_{\{|f(\omega)| \geq \psi(y)\}} |f(\omega)| \, d\mu(\omega) \, dy$$

$$\leq \int_0^\infty C(\psi(y)) \, dy$$

$$\leq \int_0^\infty e^{-y} \, dy = 1.$$

$(3) \Longrightarrow (2)$. By $(2.1.10(3))$, if we have $\sup \{ M_\Phi(f) : f \in \mathcal{K} \} = N$, then it follows that $\sup \{ \|f\|_\Phi : f \in \mathcal{K} \} \leq N \vee 1$. ∎

(2.3.6) Corollary. *(a) If $\mathcal{K}$ is uniformly integrable, then the convex hull $\operatorname{conv} \mathcal{K}$ is also uniformly integrable. (b) If $\mathcal{K}$ is uniformly integrable, then the closure $\overline{\mathcal{K}}$ in an Orlicz norm $\| \ \|_\Phi$ is also uniformly integrable if $\lim_{u \to \infty} \Phi(u)/u = \infty$. (c) If $\mathcal{K}_1$ and $\mathcal{K}_2$ are uniformly integrable, then*

$$\mathcal{K}_1 + \mathcal{K}_2 = \{ f + g : f \in \mathcal{K}_1, g \in \mathcal{K}_2 \}$$

is also uniformly integrable. (d) If $\mathcal{K}$ is a finite subset of $L_1 + L_\infty$, then $\mathcal{K}$ is uniformly integrable.

Proof. (a) The norm $\| \cdot \|_\Phi$ is a convex function, so $\operatorname{conv} \mathcal{K}$ has the same L_Φ bound as $\mathcal{K}$.

(b) The norm $\| \cdot \|_\Phi$ is a continuous function on the Orlicz space, so $\overline{\mathcal{K}}$ has the same L_Φ bound as $\mathcal{K}$.

(c) Suppose Φ_i is an Orlicz function such that $\mathcal{K}_i$ is bounded in L_{Φ_i} $(i = 1, 2)$. Write $\varphi_i = \Phi_i'$. Then the pointwise minimum $\varphi = \varphi_1 \wedge \varphi_2$ is also an Orlicz derivative. Let Φ be the corresponding Orlicz function. Now $\Phi_i(u)/u \to \infty$, so $\varphi_i(u) \to \infty$, and therefore $\varphi(u) \to \infty$, so $\Phi(u)/u \to \infty$. Also $\Phi \le \Phi_i$ $(i = 1, 2)$. Thus $\mathcal{K}_1$ and $\mathcal{K}_2$ are both bounded in L_Φ. Clearly $\mathcal{K} = \mathcal{K}_1 + \mathcal{K}_2$ is also bounded in L_Φ.

(d) A singleton $\{f\}$ is uniformly integrable if $f \in L_1 + L_\infty$. Then the result follows from (c). ∎

Conditional expectation in infinite measure spaces

(2.3.7) Suppose $(\Omega, \mathcal{F}, \mu)$ is a measure space, $f \in L_1(\Omega, \mathcal{F}, \mu)$ is a measurable function, and $\mathcal{G} \subseteq \mathcal{F}$ is a σ-algebra. The conditional expectation of f given $\mathcal{G}$, written $\mathbf{E}_\mu [f \,|\, \mathcal{G}]$ or $\mathbf{E}_\mu^{\mathcal{G}} f$, should be a function $g \in L_1(\Omega, \mathcal{G}, \mu)$ satisfying

$$(2.3.7a) \qquad \int_A g \, d\mu = \int_A f \, d\mu$$

for all $A \in \mathcal{G}$. If $\mu(\Omega) < \infty$, this can easily be arranged: the Radon-Nikodým theorem may be applied as in the case of a probability measure. But if $\mu(\Omega) = \infty$ it is not possible in general to find such a function g (2.3.19).

(2.3.8) Proposition. *Let $(\Omega, \mathcal{F}, \mu)$ be a σ-finite measure space, and let $\mathcal{G} \subseteq \mathcal{F}$ be a σ-algebra. There is a set $B \in \mathcal{G}$, unique up to sets of measure zero, such that*

(a) *B is σ-finite in $(\Omega, \mathcal{G}, \mu)$: there exist sets $B_n \in \mathcal{G}$ $(n = 1, 2, \cdots)$ with $\mu(B_n) < \infty$ and $B = \bigcup_{n=1}^\infty B_n$.*
(b) *For $C \in \mathcal{G}$, with $C \cap B = \emptyset$, we have either $\mu(C) = 0$ or $\mu(C) = \infty$.*

Proof. B is the *essential supremum* of all sets in $\mathcal{G}$ of finite measure. (This concept is discussed in detail in Section 4.1, but here we will include a self-contained proof.)

Pass to a finite measure ν equivalent to μ on $\mathcal{F}$. Choose a sequence G_n of sets in $\mathcal{G}$ with $\mu(G_n) < \infty$ so that $0 < \nu(G_n)$ and $\nu(G_n)$ converges to

$$\sup \left\{ \nu(G) : \mu(G) < \infty, G \in \mathcal{G} \right\}.$$

By disjoining the sets G_n, we obtain a countable disjoint family $\{H_n\}$ of sets maximal for the condition that $0 < \mu(A) < \infty$. Let

$$B = \bigcup H_n.$$

Then clearly $B \in \mathcal{G}$ and B is σ-finite in $(\Omega, \mathcal{G}, \mu)$. If $C \in \mathcal{G}$ with $C \cap B = \emptyset$, then $C \cap H_n = \emptyset$ for all n. By maximality, we see that either $\mu(C) = 0$ or $\mu(C) = \infty$. ∎

The set B of Proposition (2.3.8) is called the *set of σ-finiteness* of μ with respect to $\mathcal{G}$.

It is easy to define a "conditional expectation" on the class of functions $L_1 + L_\infty$. Suppose $f \in L_1 + L_\infty$ and $\mathcal{G}$ are given. Let B be the set of σ-finiteness of μ with respect to $\mathcal{G}$. Then

$$\nu(A) = \int_{A \cap B} f \, d\mu$$

exists for all $A \in \mathcal{G}$ with $\mu(A) < \infty$. Now B is σ-finite, say $B = \bigcup B_n$, $\mu(B_n) < \infty$, $B_n \cap B_m = \emptyset$ for $n \neq m$. The Radon-Nikodým theorem yields $g_n \in L_1(B_n, \mathcal{G}, \mu)$ with $\nu(A) = \int_A g_n \, d\mu$ for all $A \in \mathcal{G}$, $A \subseteq B_n$. The functions may be pieced together,

$$g = \sum_{n=1}^{\infty} g_n \mathbf{1}_{B_n},$$

to obtain $g \in (L_1 + L_\infty)(\Omega, \mathcal{G}, \mu)$ satisfying

$$\int_A g \, d\mu = \int_A f \, d\mu$$

for all $A \in \mathcal{G}$ with $\mu(A) < \infty$.

(2.3.9) Definition. Let $f \in (L_1 + L_\infty)(\Omega, \mathcal{F}, \mu)$ and let $\mathcal{G} \subseteq \mathcal{F}$ be a σ-algebra. Then g is the *conditional expectation* of f given $\mathcal{G}$, written

$$g = \mathbf{E}_\mu \left[f \mid \mathcal{G} \right] \quad \text{or} \quad g = \mathbf{E}_\mu^{\mathcal{G}} f$$

iff g is $\mathcal{G}$-measurable, and

$$\int_A g \, d\mu = \int_A f \, d\mu$$

for all $A \in \mathcal{G}$ with $\mu(A) < \infty$. Note that (even if f is integrable), there need not exist a $\mathcal{G}$-measurable function g with

$$\int_A g \, d\mu = \int_A f \, d\mu$$

for all $A \in \mathcal{G}$; see (2.3.19).

It can easily be verified that if $g = \mathbf{E}_\mu[f \,|\, \mathcal{G}]$, then we also have

$$\int hg \, d\mu = \int hf \, d\mu$$

for all $\mathcal{G}$-measurable $h \in L_1 \cap L_\infty$.

Occasionally we will write $\mathbf{E}_\mu[f \,|\, \mathcal{G}]$ even when f is not in $L_1 + L_\infty$. This will make sense, for example, if $f \geq 0$, provided we allow $\mathbf{E}_\mu[f \,|\, \mathcal{G}]$ to take the value ∞. (Compare (1.4.8).)

Several of the usual properties may be checked easily: $\mathbf{E}_\mu[f_1 + f_2 \,|\, \mathcal{G}] = \mathbf{E}_\mu[f_1 \,|\, \mathcal{G}] + \mathbf{E}_\mu[f_2 \,|\, \mathcal{G}]$ (conditional expectation is a linear operator). If $f \geq 0$, then $\mathbf{E}_\mu[f \,|\, \mathcal{G}] \geq 0$ (conditional expectation is a positive operator). If g is $\mathcal{G}$-measurable, then $\mathbf{E}_\mu[gf \,|\, \mathcal{G}] = g\mathbf{E}_\mu[f \,|\, \mathcal{G}]$, provided both conditional expectations exist.

Jensen's inequality is an important and useful property, which will be used to prove that conditional expectation is a contraction on all Orlicz spaces.

(2.3.10) Jensen's inequality. *Let* $f \in (L_1 + L_\infty)(\Omega, \mathcal{F}, \mu)$, *let* $\mathcal{G} \subseteq \mathcal{F}$ *be a σ-algebra, and let* $\Phi \colon \mathbb{R} \to \mathbb{R}$ *be a convex function. Then*

$$\Phi\big(\mathbf{E}_\mu[f \,|\, \mathcal{G}]\big) \leq \mathbf{E}_\mu\big[\Phi(f) \,\big|\, \mathcal{G}\big]$$

a.e. on the set of σ-finiteness of $\mathcal{G}$.

Proof. First recall that a convex function on $\mathbb{R}$ is necessarily continuous, so that $\Phi(f)$ is a measurable function. Consider rational numbers m and b such that the graph of the line $y = mx + b$ is below the graph $y = \Phi(x)$; that is, $\Phi(x) \geq mx + b$ for all $x \in \mathbb{R}$. Then for any $A \in \mathcal{G}$ with $\mu(A) < \infty$, we have

$$\int_A \Phi(f) \, d\mu \geq m \int_A f \, d\mu + b\mu(A).$$

Thus

$$\int_A \mathbf{E}_\mu\big[\Phi(f) \,\big|\, \mathcal{G}\big] \, d\mu \geq m \int_A \mathbf{E}_\mu[f \,|\, \mathcal{G}] \, d\mu + b\mu(A)$$

or

$$\int_A \left(\mathbf{E}_\mu\big[\Phi(f) \,\big|\, \mathcal{G}\big] - m\mathbf{E}_\mu[f \,|\, \mathcal{G}] - b \right) d\mu \geq 0.$$

The integrand is $\mathcal{G}$-measurable, and the inequality holds for all $A \in \mathcal{G}$ with $\mu(A) < \infty$, so

$$\mathbf{E}_\mu\big[\Phi(f) \,\big|\, \mathcal{G}\big] - m\mathbf{E}_\mu[f \,|\, \mathcal{G}] - b \geq 0$$

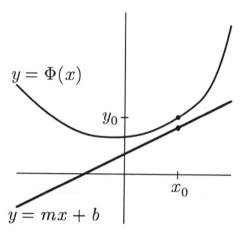

Figure (2.3.10). Diagram for Jensen's inequality.

a.e. on the set B of σ-finiteness of μ with respect to $\mathcal{G}$. There are countably many pairs (m, b) of rationals such that $y = mx + b$ is below $y = \Phi(x)$. Thus there is a single set $N \in \mathcal{G}$ with $\mu(N) = 0$ so that, for all $\omega \in B \setminus N$,

$$\mathbf{E}_\mu \left[\Phi(f) \,|\, \mathcal{G} \right] (\omega) \geq m \mathbf{E}_\mu \left[f \,|\, \mathcal{G} \right] (\omega) + b$$

for all such pairs (m, b).

Now let $\omega \in B \setminus N$, and $\varepsilon > 0$. Write $x_0 = \mathbf{E}_\mu \left[f \,|\, \mathcal{G} \right] (\omega)$ and $y_0 = \Phi(x_0)$. For each $\varepsilon > 0$, there is a pair (m, b) of rationals with $y = mx + b$ below $y = \Phi(x)$, but $y_0 \leq mx_0 + b + \varepsilon$. See Figure (2.3.10). (This is true since the convex function $\Phi(x)$ is left-differentiable at x_0.) Thus we have

$$\mathbf{E}_\mu \left[\Phi(f) \,|\, \mathcal{G} \right] (\omega) \geq \Phi\big(\mathbf{E}_\mu \left[f \,|\, \mathcal{G} \right] (\omega)\big) - \varepsilon.$$

This is true for all $\varepsilon > 0$, so we have

$$\mathbf{E}_\mu \left[\Phi(f) \,|\, \mathcal{G} \right] \geq \Phi\big(\mathbf{E}_\mu \left[f \,|\, \mathcal{G} \right]\big) \qquad \text{a.e. on } B. \qquad \blacksquare$$

Note that if $\Phi(0) = 0$, then we even have $\mathbf{E}_\mu \left[\Phi(f) \,|\, \mathcal{G} \right] \geq \Phi\big(\mathbf{E}_\mu \left[f \,|\, \mathcal{G} \right]\big)$ a.e. on Ω.

(2.3.11) Corollary. *Let Φ be an Orlicz function. If $f \in L_\Phi$, then $\mathbf{E}_\mu \left[f \,|\, \mathcal{G} \right] \in L_\Phi$ and*

$$\left\| \mathbf{E}_\mu \left[f \,|\, \mathcal{G} \right] \right\|_\Phi \leq \| f \|_\Phi .$$

If $f \in H_\Phi$, then $\mathbf{E}_\mu \left[f \,|\, \mathcal{G} \right] \in H_\Phi$.

Proof. Both parts follow from: $M_\Phi\big(\mathbf{E}_\mu \left[f \,|\, \mathcal{G} \right]\big) \leq M_\Phi(f)$. This is a consequence of Theorem (2.3.10), using the convex function $\Phi(|x|)$. $\qquad \blacksquare$

(2.3.12) Corollary. *Let $\mathcal{K}$ be a uniformly integrable set of $\mathcal{F}$-measurable functions, and let $\{\mathcal{G}_i : i \in I\}$ be a family of sub-σ-algebras of $\mathcal{F}$. Then*

$$\tilde{\mathcal{K}} = \{\mathbf{E}_\mu[f \mid \mathcal{G}_i] : f \in \mathcal{K}, i \in I\}$$

is also uniformly integrable.

Proof. By Theorem (2.3.5), there is an Orlicz function with $\Phi(u)/u \to \infty$ such that $\mathcal{K}$ is L_Φ-bounded. By Corollary (2.3.11), the set $\tilde{\mathcal{K}}$ has the same bound in L_Φ. ∎

Complements

(2.3.13) (Uniform integrability and uniform absolute continuity.) Suppose $(\Omega, \mathcal{F}, \mu)$ has an atom: $A \in \mathcal{F}$, $0 < \mu(A) < \infty$, and if $B \subseteq A$, then either $\mu(B) = \mu(A)$ or $\mu(B) = 0$. Let $f_n = n\,1_A$. The set $\{f_n : n \in \mathbb{N}\}$ is uniformly absolutely continuous, but not uniformly integrable.

(2.3.14) (The converse is false in (2.3.3(b)).) Let $(\Omega, \mathcal{F}, \mu)$ be $[0,1]$ with Lebesgue measure. For $k \geq 3$, let

$$f_k = k\,1_{(0,1/(k \log k))}.$$

Then the set $\mathcal{K} = \{f_k : k = 3, 4, \cdots\}$ is uniformly integrable. But if $\mathcal{K}$ is dominated in distribution by f, that is

$$\mu\{|f_k| \geq \lambda\} \leq \mu\{|f| \geq \lambda\}$$

for all k, then we have $\int |f|\,dt = \infty$, so $f \notin L_1 + L_\infty = L_1$ (Clarke [1979]).

(2.3.15) (Mean convergence of nets.) Let J be a directed set, and suppose $f_t \in L_1$ for each $t \in J$. The net (f_t) is said to be *uniformly integrable at infinity* iff, for every $\varepsilon > 0$, there exists $s \in J$ and $\lambda > 0$ such that for all $t \geq s$

$$\int_{\{|f_t| \geq \lambda\}} |f_t|\,d\mu < \varepsilon.$$

Suppose $\mu(\Omega) < \infty$. Then: $\|f_t - f\|_1 \to 0$ if and only if $f_t \to f$ in measure and (f_t) is uniformly integrable at infinity. (Neveu [1965a], p. 54.)

Uniform integrability is connected to weak compactness. The classical theorem along these lines deals with finite measure spaces: *Let μ be a finite measure, and let $\mathcal{K} \subseteq L_1(\mu)$. Then $\mathcal{K}$ is relatively sequentially compact in the weak topology of L_1 if and only if $\mathcal{K}$ is uniformly integrable.* (For example, Dunford & Schwartz [1958], (IV.8.11).) Here is the more general version.

(2.3.16) Proposition. *Let $\mathcal{K} \subseteq L_1 + L_\infty$. Then $\mathcal{K}$ is relatively sequentially compact in the weak topology $\sigma(L_1 + L_\infty, L_1 \cap L_\infty)$ if and only if $\mathcal{K}$ is uniformly integrable and $L_1 + L_\infty$-bounded.*

Proof. Suppose $\mathcal{K}$ is relatively sequentially compact. Then, for every measurable function $h \in L_1 \cap L_\infty$, the set $\left\{ \int fh \, d\mu : f \in \mathcal{K} \right\}$ is a bounded set of scalars. Thus, by the uniform boundedness principle, $\left\{ \|f\|_{L_1 + L_\infty} : f \in \mathcal{K} \right\}$ is bounded.

We claim that $\mathcal{K}$ is uniformly integrable. Suppose not. Then there exist $\alpha > 0$, $A_n \in \mathcal{F}$, $f_n \in \mathcal{K}$ with $\mu(A_n) \to 0$ but $\int_{A_n} |f_n| \, d\mu \geq \alpha$. Taking a subsequence, we may assume that $\mu(A_n) \leq 2^{-n}$ and f_n converges for the weak topology $\sigma(L_1 + L_\infty, L_1 \cap L_\infty)$, say $f_n \to f$. Now $A = \bigcup_{n=1}^\infty A_n$ has finite measure, and $f_n \to f$ weakly in $L_1(A)$. By the classical theorem, $\int_{A_n} |f_n| \, d\mu \to 0$ since $\mu(A_n) \to 0$, a contradiction.

Conversely, suppose $\mathcal{K}$ is uniformly integrable and $L_1 + L_\infty$-bounded. Now Ω is σ-finite, say $A_k \uparrow \Omega$, $\mu(A_k) < \infty$. Let $f_n \in \mathcal{K}$. For each k, the set $\{ f_n \mathbf{1}_{A_k} : n \in \mathbb{N} \}$ is uniformly integrable and bounded in L_1, so by the classical theorem, it is relatively weakly sequentially compact. We may piece together the limits, so there exists a measurable function g with $f_n \mathbf{1}_{A_k} \to g \mathbf{1}_{A_k}$ weakly as $n \to \infty$, for all k. Now we claim that $f_n \to g$ in $\sigma(L_1 + L_\infty, L_1 \cap L_\infty)$. Let $\varepsilon > 0$. Let $h \in L_1 \cap L_\infty$. Choose λ so that

$$\int_{\{|f_n - g| \geq \lambda\}} |f_n - g| \, d\mu < \frac{\varepsilon}{3\|h\|_\infty}$$

for all n. Then choose k so that

$$\int_{\Omega \setminus A_k} |h| \, d\mu < \frac{\varepsilon}{3\lambda}.$$

Then choose N so that for all $n \geq N$, we have

$$\left| \int_{A_k} (f_n - g) h \, d\mu \right| < \frac{\varepsilon}{3}.$$

Then we have, for any $n \geq N$,

$$\left| \int (f_n - g) h \, d\mu \right| \leq \left| \int_{A_k} (f_n - g) h \, d\mu \right| + \int_{\{|f_n - g| < \lambda\} \setminus A_k} |f_n - g| \, |h| \, d\mu$$

$$+ \int_{\{|f_n - g| \geq \lambda\} \setminus A_k} |f_n - g| \, |h| \, d\mu$$

$$\leq \frac{\varepsilon}{3} + \lambda \frac{\varepsilon}{3\lambda} + \|h\|_\infty \frac{\varepsilon}{3\|h\|_\infty} = \varepsilon.$$

This shows that $f_n \to g$ in $\sigma(L_1 + L_\infty, L_1 \cap L_\infty)$. ∎

Another variant uses this definition: $\mathcal{K}$ is *uniformly R_0* iff for every $\gamma > 0$ there exists $M > 0$ such that if $A \in \mathcal{F}$ and $\mu(A) \leq \gamma$, then $\int_A |f| \, d\mu \leq M$ for all $f \in \mathcal{K}$. Obviously a singleton $\{f\}$ is uniformly R_0 if and only if $f \in R_0$.

(2.3.17) Let $\mathcal{K} \subseteq R_0$. Then $\mathcal{K}$ is weakly relatively sequentially compact in the weak topology $\sigma(R_0, L_1 \cap L_\infty)$ if and only if $\mathcal{K}$ is bounded in $L_1 + L_\infty$ norm, $\mathcal{K}$ is uniformly integrable, and $\mathcal{K}$ is uniformly R_0.

(2.3.18) (Weak compactness in c_0.) Describe weak compactness in c_0 by taking $\Omega = \mathbb{N}$ in the preceding.

(2.3.19) (Conditional expectation counterexample.) Let $(\Omega, \mathcal{F}, \mu)$ be $\mathbb{R}$ with Lebesgue measure, let $f = 1_{[0,1]}$, and let $\mathcal{G} = \{\Omega, \emptyset\}$. Then there is no $g \in (L_1 + L_\infty)(\Omega, \mathcal{G}, \mu)$ with

$$\int_A g \, d\mu = \int_A f \, d\mu$$

for all $A \in \mathcal{G}$.

(2.3.20) (Alternative construction of the conditional expectation.) Let $(\Omega, \mathcal{F}, \mu)$ be a σ-finite measure space, and let $\mathcal{G}$ be a sub-σ-algebra of $\mathcal{F}$. Let B be the set of σ-finiteness of μ with respect to $\mathcal{G}$. Then there is a probability measure $\mathbf{P}$ on $\mathcal{F}$ and a Radon-Nikodým derivative $\rho = d\mathbf{P}/d\mu$, where $\{\rho > 0\} = B$. We have

$$\mathbf{E}_\mu[f \mid \mathcal{G}] = \frac{\mathbf{E}_\mathbf{P}[g/\rho \mid \mathcal{G}]}{\mathbf{E}_\mathbf{P}[1/\rho \mid \mathcal{G}]} 1_B.$$

This could be used as the definition of the conditional expectation in an infinite measure space.

(2.3.21) (Uniformly integrable martingales.) Let $(\Omega, \mathcal{F}, \mathbf{P})$ be a probability space. Let $(\mathcal{F}_t)_{t \in J}$ be a stochastic basis indexed by a directed set J. (i) If (X_t) is a uniformly integrable martingale, then X_t converges in L_1 to a random variable X and $X_t = \mathbf{E}^{\mathcal{F}_t}[X]$. (ii) If (X_t) is a uniformly integrable submartingale, then X_t converges in L_1 to a random variable X and $X_t \leq \mathbf{E}^{\mathcal{F}_t}[X]$.

Proof. (i) X_t converges stochastically to a random variable X; and since it is uniformly integrable, it converges to X in L_1 (1.1.3, 1.3.1, and 1.4.7). By Fatou's lemma, $X \in L_1$. Now if $s \leq t$ and $A \in \mathcal{F}_s$, then

$$\mathbf{E}[X_s 1_A] = \mathbf{E}[X_t 1_A]$$

by the martingale property. Taking the limit as $t \to \infty$, we get

$$\mathbf{E}[X_s 1_A] = \mathbf{E}[X 1_A]$$

for all $A \in \mathcal{F}_s$. That is, $X_s = \mathbf{E}^{\mathcal{F}_s}[X]$.

(ii) Proofs for submartingales are similar.   ■

(2.3.22) (Orlicz bounded martingales.) Suppose (X_t) is a uniformly bounded martingale. By the criterion of de La Vallée Poussin, there is an Orlicz function Φ with $\lim \Phi(u)/u \to \infty$ such that $\mathbf{E}\left[\Phi(|X_t|)\right]$ is bounded. (Modular boundedness.) On the other hand, if (X_t) is a martingale, and there is an Orlicz function Φ with $\Phi(u)/u \to \infty$ such that $\mathbf{E}\left[\Phi(|X_t|)\right]$ is bounded, then (X_t) is uniformly integrable. What is less well known is that $\Phi(|X_t|)$ is also uniformly integrable. To see this, observe that since $\Phi(|X_t|)$ converges stochastically to $\Phi(|X|)$, by Fatou's lemma we have $\mathbf{E}\left[\Phi(|X|)\right] < \infty$. But by Jensen's inequality,

$$\Phi(|X_t|) = \Phi\left(\left|\mathbf{E}^{\mathcal{F}_t}[X]\right|\right) \leq \mathbf{E}^{\mathcal{F}_t}\left[\Phi(|X|)\right].$$

The last term is uniformly integrable, so $\Phi(|X_t|)$ is also uniformly integrable.

(2.3.23) (Orlicz norm and Orlicz modular convergence of martingales.) Let Φ be an Orlicz function with $\Phi(u)/u \to \infty$. Let (X_t) be a martingale or a positive submartingale. (i) If (X_t) is bounded in Φ-modular:

$$\sup_t M_\Phi(X_t) < \infty,$$

then (X_t) converges in Φ-modular.

(ii) Suppose also that Φ satisfies condition (Δ_2). If (X_t) is L_Φ-bounded, then (X_t) converges in L_Φ norm.

Proof. The martingale case follows from the positive submartingale case by decomposition into positive and negative parts. So assume (X_t) is a positive submartingale. The limit X of X_t exists in L_1. Then $\Phi(X_t)$ converges stochastically to $\Phi(X)$, and by Fatou's lemma, $\Phi(X) \in L_1$. By (2.3.21) we have

$$X_t \leq \mathbf{E}^{\mathcal{F}_t}[X],$$

so by Jensen's inequality

$$\Phi(X_t) \leq \mathbf{E}^{\mathcal{F}_t}\left[\Phi(X)\right].$$

As in (2.3.22), $\Phi(X_t)$ is uniformly integrable. Thus $\Phi(X_t) \to \Phi(X)$ in L_1. But Φ is convex, so

$$\Phi(x - y) \leq \Phi(x) - \Phi(y)$$

for $x \geq y \geq 0$. Therefore

$$\Phi(|X_t - X|) \leq |\Phi(X_t) - \Phi(X)|.$$

Thus $\Phi(|X_t - X|)$ converges to zero in L_1; that is, X_t converges to X in Φ-modular.

(ii) If (Δ_2) holds, modular convergence is equivalent to L_Φ convergence, and modular boundedness is equivalent to L_Φ boundedness (2.1.18). ∎

Condition (Δ_2) cannot be omitted in (ii). See Mogyoródi [1978] and Bui [1987].

Remarks

Conditional expectations in infinite measure spaces can be found in Chow [1960b] and Dellacherie & Meyer [1978].

3

Inequalities

In this chapter we will prove several different kinds of inequalities. We begin with the "three-function inequality," which relates weak inequalities, such as

$$\mu\{|g| \geq \lambda\} \leq \frac{c}{\lambda} \int_{\{|g| \geq \lambda\}} |f| \, d\mu,$$

to strong inequalities, such as

$$\|g\|_p \leq C \, \|f\|_p.$$

We will replace the space L_p with various Orlicz spaces. The main result (Theorem (3.1.2)) is called the "three-function inequality" since the most general version deals with inequalities

$$(W) \qquad\qquad \mu\{|g| \geq \lambda\} \leq \frac{c}{\lambda} \int_{\{|h| \geq \lambda\}} |f| \, d\mu$$

or

$$(S) \qquad\qquad b \, M_\xi \left(\frac{g}{b}\right) \leq a \, M_\Phi \left(\frac{f}{a}\right) + a \, M_\xi \left(\frac{h}{b}\right).$$

involving three functions f, g, h.

Section 3.2 studies martingale transforms, and proves the basic maximal inequality due to Burkholder.

Section 3.3 deals with some elementary "prophet inequalities." These compare the gain that can be made by a gambler (who knows only the present and past) to the gain that can be made by a prophet (who knows also the future).

3.1. The three-function inequality

The proofs of ergodic and martingale theorems are similar, and to unify them is an old problem. A. & C. Ionescu Tulcea [1963] gave a common generalization of two vector-valued maximal theorems: that of Chacon, and the martingale theorem. In other places the theories converge more obviously. By Fubini, there exists a passage from weak L_1 inequalities to strong maximal inequalities involving either the L_p norm ($1 < p < \infty$), or the $L \log^k L$ norm. This is used in ergodic and martingale theory, and in harmonic analysis where it originated. In this section, this argument is extended to general Orlicz spaces. There is also a reverse inequality.

(3.1.1) Let $(\Omega, \mathcal{F}, \mu)$ be a σ-finite measure space. Let Φ be an Orlicz function, with derivative φ, and conjugate Ψ. We will be interested in the function ξ defined by

$$(3.1.1a) \qquad \xi(u) = \Psi(\varphi(u)).$$

By the case of equality in Young's inequality (2.1.4), for values of u with $\Phi(u) < \infty$ the function ξ may also be written

$$\xi(u) = u\varphi(u) - \Phi(u).$$

For simplicity we will normally assume in this section that $\Phi(u) < \infty$ for all u. In many cases, ξ will again be an Orlicz function, but even when it is not, we will use the notation

$$M_\xi(f) = \int \xi(|f|)\, d\mu.$$

The expression $\|\ \|_\xi$ will be defined by the Luxemburg formula $\|f\|_\xi = \inf\{a : M_\xi(f/a) \le 1\}$. If ξ is not an Orlicz function, then $\|\ \|_\xi$ need not satisfy the triangle inequality.

(3.1.2) Theorem. *Let Φ be a finite Orlicz function with $\Phi(u)/u \to \infty$, and let ξ be defined by (3.1.1a). Let f, g, h be measurable functions. Suppose that*

$$(W) \qquad \mu\{|g| \ge \lambda\} \le \frac{1}{\lambda} \int_{\{|h| \ge \lambda\}} |f|\, d\mu$$

for all $\lambda > 0$. Then for $a, b > 0$, we have

$$(S) \qquad b\, M_\xi\left(\frac{g}{b}\right) \le a\, M_\Phi\left(\frac{f}{a}\right) + a\, M_\xi\left(\frac{h}{b}\right).$$

Proof. Only the absolute values of the functions f, g, h enter into the theorem, so we may assume that these functions are all nonnegative. We may assume that $M_\Phi(f/a)$ and $M_\xi(h/b)$ are both finite. Note that φ is continuous except at countably many points of $(0, \infty)$, since it is nondecreasing. Let C_φ be the set of points of continuity of φ, and let C_ψ be the set of points of continuity of ψ.

First, suppose that g and h are integrable simple functions with values in $T = bC_\varphi \cup \{0\}$. (Note that T is a dense set in $\mathbb{R}^+$.) Then, for each $\omega \in \Omega$, we have $g(\omega)/b \in C_\varphi \cup \{0\}$. If $y \in C_\psi$, then we have

$$y \le \varphi\left(\frac{g(\omega)}{b}\right) \quad \Longleftrightarrow \quad \psi(y) \le \frac{g(\omega)}{b}.$$

Therefore, for almost all $y \in (0, \infty)$, we have

$$\left\{ \omega \in \Omega : y \le \varphi\left(\frac{g(\omega)}{b}\right) \right\} = \left\{ \omega \in \Omega : \psi(y) \le \frac{g(\omega)}{b} \right\}.$$

By Fubini's Theorem, the sets

$$\left\{ (\omega, y) \in \Omega \times (0, \infty) : y \le \varphi\left(\frac{g(\omega)}{b}\right) \right\},$$

$$\left\{ (\omega, y) \in \Omega \times (0, \infty) : \psi(y) \le \frac{g(\omega)}{b} \right\}$$

agree except for a set of measure zero in the product $\Omega \times (0, \infty)$. Similarly, the sets

$$\left\{ (\omega, y) \in \Omega \times (0, \infty) : y \le \varphi\left(\frac{h(\omega)}{b}\right) \right\},$$

$$\left\{ (\omega, y) \in \Omega \times (0, \infty) : \psi(y) \le \frac{h(\omega)}{b} \right\}$$

agree except for a set of measure zero.

Now we have:

$$bM_\xi\left(\frac{g}{b}\right) = b \int \xi\left(\frac{g}{b}\right) d\mu$$

$$= b \int \Psi\left(\varphi\left(\frac{g}{b}\right)\right) d\mu$$

$$= b \int_\Omega \int_0^{\varphi(g/b)} \psi(y) \, dy \, d\mu$$

apply Fubini's theorem to interchange the order

$$= b \int_0^\infty \int_{\{g \ge b\psi(y)\}} d\mu \, \psi(y) \, dy$$

$$= \int_0^\infty \mu\{g \ge b\psi(y)\} \, b\psi(y) \, dy$$

use the hypothesis

$$\le \int_0^\infty \int_{\{h \ge b\psi(y)\}} f \, d\mu \, dy$$

apply Fubini's theorem to exchange back

$$= \int_\Omega \int_0^{\varphi(h/b)} dy \, f \, d\mu$$

$$= a \int \varphi \left(\frac{h}{b} \right) \frac{f}{a} \, d\mu$$

apply Young's inequality (2.1.4)

$$\leq a \int \Phi \left(\frac{f}{a} \right) d\mu + a \int \Psi \left(\varphi \left(\frac{h}{b} \right) \right) d\mu$$

$$= a M_\Phi \left(\frac{f}{a} \right) + a M_\xi \left(\frac{h}{b} \right).$$

For the general case we will approximate g and h by integrable simple functions with values in T in the usual way. Since it is used below, we spell out the approximation. Given n, choose $t_0 = 0 < t_1 < \cdots < t_m$ with $t_i \in bC_\varphi$ and $t_i - t_{i-1} < 2^{-n}$ and $t_m > 2^n$. Define g_n by:

$$\{ g_n = t_i \} = \{ t_i \leq g < t_{i+1} \},$$
$$\{ g_n = t_m \} = \{ t_m \leq g \},$$

and similarly for h_n approximating h. Thus we will have $g_n \uparrow g$ and $h_n \uparrow h$.
Now f, g, h satisfy the hypothesis

$$\mu \{ g \geq \lambda \} \leq \frac{1}{\lambda} \int_{\{ h \geq \lambda \}} f \, d\mu$$

for all $\lambda > 0$. We claim that f, g_n, h_n satisfy the same hypothesis. Consider $\lambda > 0$. If $t_{i-1} < \lambda \leq t_i$, then

$$\{ g_n \geq \lambda \} = \{ g_n \geq t_i \} = \{ g \geq t_i \},$$
$$\{ h_n \geq \lambda \} = \{ h_n \geq t_i \} = \{ h \geq t_i \}.$$

Therefore

$$\mu \{ g_n \geq \lambda \} = \mu \{ g \geq t_i \}$$
$$\leq \frac{1}{t_i} \int_{\{ h \geq t_i \}} f \, d\mu$$
$$= \frac{1}{t_i} \int_{\{ h_n \geq \lambda \}} f \, d\mu$$
$$\leq \frac{1}{\lambda} \int_{\{ h_n \geq \lambda \}} f \, d\mu.$$

This verifies that f, g_n, h_n also satisfy the hypothesis. Therefore the previous case yields

$$b M_\xi \left(\frac{g_n}{b} \right) \leq a M_\Phi \left(\frac{f}{a} \right) + a M_\xi \left(\frac{h_n}{b} \right).$$

Now ξ is left-continuous, so we may pass to the limit to obtain:

$$b\,M_\xi\left(\frac{g}{b}\right) \le a\,M_\Phi\left(\frac{f}{a}\right) + a\,M_\xi\left(\frac{h}{b}\right).$$

∎

To illustrate the result, we display a number of consequences. First take $g = h$: If

$$(W_g) \qquad\qquad \mu\{|g| \ge \lambda\} \le \frac{1}{\lambda}\int_{\{|g|\ge\lambda\}} |f|\,d\mu$$

for all $\lambda > 0$, we have

$$(S'_g) \qquad\qquad b\,M_\xi\left(\frac{g}{b}\right) \le a\,M_\Phi\left(\frac{f}{a}\right) + a\,M_\xi\left(\frac{g}{b}\right).$$

With an assumption, we get more:

(3.1.3) Theorem. *Let Φ be an Orlicz function, and let ξ be defined as in (3.1.1a). Let f and g be measurable functions related by (W_g). If*

$$(F_g) \qquad\qquad \mu\{|g| \ge \lambda\} < \infty \qquad \text{for all } \lambda > 0,$$

then for any $a < b$,

$$(S_g) \qquad\qquad M_\xi\left(\frac{g}{b}\right) \le \frac{a}{b-a}\,M_\Phi\left(\frac{f}{a}\right).$$

Proof. We may assume that f and g are nonnegative functions, since all quantities involved depend only on the absolute values of the functions.

First, consider the special case where $\mu\{g > 0\} < \infty$ (for example, $\mu(\Omega) < \infty$) and g is bounded, say $g \le C$. Then of course

$$M_\xi\left(\frac{g}{b}\right) \le \int_{\{g>0\}} \xi\left(\frac{C}{b}\right) d\mu < \infty.$$

Thus we may solve (S'_g) to get the required conclusion.

Next, suppose only that g is bounded, but possibly $\mu\{g > 0\} = \infty$. For $\varepsilon > 0$, consider $g_\varepsilon = (g - \varepsilon)^+$. It, too, is bounded; and $\mu\{g_\varepsilon > 0\} = \mu\{g > \varepsilon\} < \infty$. Also, g_ε satisfies a weak inequality of the form (W_g):

$$\mu\{g_\varepsilon \ge \lambda\} = \mu\{g \ge \lambda + \varepsilon\} \le \frac{1}{\lambda+\varepsilon}\int_{\{g\ge\lambda+\varepsilon\}} f\,d\mu \le \frac{1}{\lambda}\int_{\{g_\varepsilon\ge\lambda\}} f\,d\mu$$

as required. By the previous case, we have

$$M_\xi\left(\frac{g_\varepsilon}{b}\right) \le \frac{a}{b-a}\,M_\Phi\left(\frac{f}{a}\right).$$

Now as $\varepsilon \downarrow 0$, we have $g_\varepsilon \uparrow g$, so by the monotone convergence theorem (and left-continuity of ξ) we get the conclusion (S_g).

Finally, consider general g. For a constant $C > 0$, let $g^C = g \wedge C$, so if $\lambda > C$,

$$\mu\{g^C \geq \lambda\} = 0 \leq \frac{1}{\lambda} \int_{\{g^C \geq \lambda\}} f \, d\mu,$$

and if $\lambda \leq C$,

$$\mu\{g^C \geq \lambda\} = \mu\{g \geq \lambda\} \leq \frac{1}{\lambda} \int_{\{g \geq \lambda\}} f \, d\mu = \frac{1}{\lambda} \int_{\{g^C \geq \lambda\}} f \, d\mu.$$

Thus by the previous case, we have

$$M_\xi \left(\frac{g^C}{b} \right) \leq \frac{a}{b-a} M_\Phi \left(\frac{f}{a} \right).$$

Now as $C \uparrow \infty$ we have $g^C \uparrow g$, so by the monotone convergence theorem we get (S_g). ∎

(3.1.4) Replace f by cf: If $\mu\{|g| \geq \lambda\} < \infty$ and

$$\mu\{|g| \geq \lambda\} \leq \frac{c}{\lambda} \int_{\{|g| \geq \lambda\}} |f| \, d\mu$$

for all $\lambda > 0$, then for $0 < a < b$, we have

$$M_\xi \left(\frac{g}{b} \right) \leq \frac{a}{b-a} M_\Phi \left(\frac{cf}{a} \right).$$

Or, replace a by ac to obtain for $0 < a < b/c$,

$$M_\xi \left(\frac{g}{b} \right) \leq \frac{ac}{b-ac} M_\Phi \left(\frac{f}{a} \right).$$

(3.1.5) Take $f = g = h$. Assume $\mu\{|f| \geq \lambda\} < \infty$ for all $\lambda > 0$. The inequality

$$\mu\{|f| \geq \lambda\} \leq \frac{1}{\lambda} \int_{\{|f| \geq \lambda\}} |f| \, d\mu$$

is clearly true for all $\lambda > 0$, so by (3.1.3), if $a < b$, we have

$$M_\xi \left(\frac{f}{b} \right) \leq \frac{a}{b-a} M_\Phi \left(\frac{f}{a} \right).$$

Next we obtain a version by taking $f = h$:

(3.1.6) Proposition. *Let f and g be measurable functions related by*

$$(W_f) \qquad\qquad \mu\{|f| \geq \lambda\} \leq \frac{1}{\lambda} \int_{\{|f| \geq \lambda\}} |f| \, d\mu.$$

Suppose either (a)

$$(F_f) \qquad\qquad \mu\{|f| \geq \lambda\} < \infty$$

for all $\lambda > 0$; or (b) Φ satisfies condition Δ_2. Then (S_g) holds for any $a < b$.

Proof. Clearly

$$\mu\{|f| \geq \lambda\} \leq \frac{1}{\lambda} \int_{\{|f| \geq \lambda\}} |f| \, d\mu,$$

so by the three-function inequality (3.1.2),

$$(S_f') \qquad\qquad b\, M_\xi\left(\frac{f}{b}\right) \leq a\, M_\Phi\left(\frac{f}{a}\right) + a\, M_\xi\left(\frac{f}{b}\right).$$

Assume first that $\mu\{|f| \geq \lambda\} < \infty$ for all $\lambda > 0$. By Theorem (3.1.3), we have

$$(S_f) \qquad\qquad M_\xi\left(\frac{f}{b}\right) \leq \frac{a}{b-a} M_\Phi\left(\frac{f}{a}\right).$$

Thus by the three-function inequality (3.1.2),

$$b M_\xi\left(\frac{g}{b}\right) \leq a M_\Phi\left(\frac{f}{a}\right) + a M_\xi\left(\frac{f}{b}\right) \leq a M_\Phi\left(\frac{f}{a}\right) + \frac{a^2}{b-a} M_\Phi\left(\frac{f}{a}\right).$$

This implies (S_g).

Now assume instead condition Δ_2, say $\Phi(2u)/\Phi(u) \leq C$. Then

$$C\Phi(u) \geq \Phi(2u) \geq u\varphi(u),$$

so $\xi(u) = u\varphi(u) - \Phi(u) \leq C\Phi(u)$. Thus $M_\xi(f/a) \leq CM_\Phi(f/a)$. Now if $M_\xi(f/a) = \infty$, then also $M_\Phi(f/a) = \infty$, so (S_g) holds. On the other hand, if $M_\xi(f/a) < \infty$, then $M_\xi(f/b) < \infty$ and we may solve (S_f') to obtain (S_f). The remainder of the proof is the same as the previous case. ∎

(3.1.7) Replace g by g/c: If $\{|f| \geq \lambda\} < \infty$ and

$$\mu\{|g| \geq c\lambda\} \leq \frac{1}{\lambda} \int_{\{|f| \geq \lambda\}} |f| \, d\mu$$

for all $\lambda > 0$, then for $0 < ac < b$, we have

$$M_\xi \left(\frac{g}{b} \right) \leq \frac{ac}{b - ac} M_\Phi \left(\frac{f}{a} \right).$$

(3.1.8) Take $f = g$: If $\{|f| \geq \lambda\} < \infty$ and

$$\mu\{|f| \geq \lambda\} \leq \frac{1}{\lambda} \int_{\{|h| \geq \lambda\}} |f| \, d\mu$$

for all $\lambda > 0$, then for $0 < a < b$, we have

$$M_\xi \left(\frac{f}{b} \right) \leq \frac{a}{b} M_\Phi \left(\frac{f}{a} \right) + \frac{a}{b} M_\xi \left(\frac{h}{b} \right).$$

Note that we obtain the same conclusion (S_g) in both (3.1.3) and (3.1.6). Which approach is better? In the typical application, we will have $g \geq f$, so that, assuming (F_g), (3.1.3) is a stronger result than (3.1.6). For example, we will apply these results below where g is the maximal function constructed from $f \geq 0$ by

$$g = \sup_n T_n f,$$

where T_n are positive operators on function spaces, and $T_0 = I$, so that $f \leq g$. Now if $f \leq g$, then (W_g) is easier to verify than (W_f). On the other hand, in infinite measure spaces, integrability of f implies (F_f) but not (F_g). It is therefore of interest to know that in many cases, (W_g) implies (W_f) with different constants. (See, for example, (8.2.5).)

The inequalities have all been stated using the nonstrict inequalities such as $\mu\{|g| \geq \lambda\}$. They are equivalent to the same formulas with strict inequalities such as $\mu\{|g| > \lambda\}$. This is because of the relations

$$\mu\{|g| > \lambda\} = \lim_n \mu\{|g| \geq \lambda + 1/n\},$$
$$\mu\{|g| \geq \lambda\} = \lim_n \mu\{|g| > \lambda - 1/n\}.$$

For example, by the monotone and dominated convergence theorems

$$\mu\{|g| \geq \lambda\} \leq \frac{1}{\lambda} \int_{\{|h| \geq \lambda\}} |f| \, d\mu \qquad \text{for all } \lambda > 0$$

is equivalent to

$$\mu\{|g| > \lambda\} \leq \frac{1}{\lambda} \int_{\{|h| > \lambda\}} |f| \, d\mu \qquad \text{for all } \lambda > 0.$$

In the case when Φ and the related function ξ are Orlicz functions, our theorem tells us about Orlicz spaces, and Orlicz hearts.

(3.1.9) Corollary. *Let Φ be an Orlicz function, and suppose that the function ξ defined by (3.1.1a) is also an Orlicz function. Then:*

(a) *$L_\Phi \subseteq L_\xi$; $H_\Phi \subseteq H_\xi$.*

Suppose f and g are measurable functions with $\mu\{|g| \geq \lambda\} < \infty$ and

$$(W_g) \qquad\qquad \mu\{|g| \geq \lambda\} \leq \frac{1}{\lambda} \int_{\{|g| \geq \lambda\}} |f| \, d\mu$$

for all $\lambda > 0$. Then:

(b) *$f \in L_\Phi \Longrightarrow g \in L_\xi$.*
(c) *$f \in H_\Phi \Longrightarrow g \in H_\xi$.*

Proof. (a) If $f \in L_\Phi$, then there is $a > 0$ so that $M_\Phi(f/a) < \infty$. Then by (2.1.19b), $\xi(u) \leq \Phi(2u)$ so for $b > 2a$ we have $M_\xi(f/b) < \infty$, and so $f \in L_\xi$. Similarly, if $f \in H_\Phi$, then for every $a > 0$ we have $M_\Phi(f/a) < \infty$. Thus for any $b > 0$, there is a with $0 < 2a < b$, and we have $M_\xi(f/b) < \infty$, so $f \in H_\xi$.

For parts (b) and (c), proceed as in part (a), using (3.1.3). ∎

(3.1.10) Corollary. *Suppose f, g are measurable functions such that $\mu\{|g| \geq \lambda\} < \infty$, and*

$$\mu\{|g| \geq \lambda\} \leq \frac{1}{\lambda} \int_{\{|g| \geq \lambda\}} |f| \, d\mu$$

for all $\lambda > 0$.

(a) *Let $1 < p < \infty$. If $f \in L_p$, then $g \in L_p$ and*

$$\|g\|_p \leq \frac{p}{p-1} \|f\|_p.$$

(b) *Let $k > 1$. If $f \in R_k$, then $g \in R_{k-1}$. If $f \in L\log^k L$, then $g \in L\log^{k-1} L$ and*

$$\|g\|_{L\log^{k-1} L} \leq \frac{k+1}{k} \|f\|_{L\log^k L}.$$

(c) *If $f \in R_1$, then $g \in R_0$ and.*

$$\|g\|_{\Phi_0} \leq 2\|f\|_{L\log L},$$

where Φ_0 is as in (2.2.16a). If $f \in L\log L$ then $g \in L_1 + L_\infty$, and

$$\|g\|_{L_1+L_\infty} \leq 2 + 2 \int |f| \log^+ |f| \, d\mu.$$

Proof. (a) Let $1 < p < \infty$. Consider the Orlicz function $\Phi(u) = u^p/p$. Then $\varphi(u) = u^{p-1}$ and $\xi(u) = ((p-1)/p)u^p$. Thus we have

$$M_\Phi(f) = \frac{1}{p}\|f\|_p^p,$$

$$M_\xi(f) = \frac{p-1}{p}\|f\|_p^p.$$

Now if $f \in L_p$, we may use (3.1.3) with $a = 1$ and $b = p/(p-1)$ to obtain the inequality stated.

(b) Let $k > 1$. Consider the Orlicz function defined in (2.2.16a): $\Phi(u) = \Phi_k(u) = u(\log^+ u)^k$. Then

$$\varphi(u) = (k + \log u)(\log^+ u)^{k-1}, \quad \text{and}$$
$$\xi(u) = ku(\log^+ u)^{k-1} = k\Phi_{k-1}(u).$$

Now if $f \in L\log^k L$, take

$$a = \|f\|_{L\log^k L},$$
$$b = \frac{k+1}{k} a.$$

Then $M_\Phi(f/a) = 1$, so by (3.1.3) we have $M_\xi(g/b) \le a/(b-a) = k$, so that $M_{\Phi_{k-1}}(g/b) = (1/k)M_\xi(g/b) \le 1$. Thus

$$\|g\|_{L\log^{k-1} L} \le b = \frac{k+1}{k}\|f\|_{L\log^k L}.$$

(c) The first inequality is proved in the same way as (b): if $\Phi = \Phi_1$, then $\xi = \Phi_0$. Recall that (since Φ_0 is not an Orlicz function) $\|\ \|_{\Phi_0}$ is not a norm.

For the second inequality, begin with (3.1.3):

$$M_{\Phi_0}\left(\frac{g}{b}\right) \le \frac{a}{b-a}M_{\Phi_1}\left(\frac{f}{a}\right).$$

Then observe that

$$\|g\|_{L_1 + L_\infty} \le b + \int_{\{|g| \ge b\}} |g|\, d\mu$$
$$= b + bM_{\Phi_0}\left(\frac{g}{b}\right)$$
$$\le b + \frac{ab}{b-a}M_{\Phi_1}\left(\frac{f}{a}\right).$$

Finally, set $a = 1$ and $b = 2$. ∎

The constants $p/(p-1)$ in part (a), $(k+1)/k$ in part (b), and 2 in the first inequality of part (c) are the best possible in this result. (See (3.1.14) and (3.1.15), below.)

Reverse inequality

Theorem (3.1.2) has a companion "reverse" theorem. We take $a = b$ and $f = h$. If there is c so that

$$\mu\{|g| \geq \lambda\} \geq \frac{c}{\lambda} \int_{\{|f| \geq \lambda\}} |f| \, d\mu$$

for all $\lambda > 0$, then

$$M_\xi \left(\frac{g}{a}\right) \geq cM_\Phi \left(\frac{f}{a}\right) + cM_\xi \left(\frac{f}{a}\right).$$

It is enough to prove it in the case $a = 1$, since we may replace λ by $a\lambda$. In fact, we will prove a more general result with two terms on the right. (It is not hard to see that it could be done with three or more terms on the right, but our applications use at most two terms.) Note that the constants c_1, c_2 are not required to be positive.

(3.1.11) Theorem. *Let Φ be a finite Orlicz function with $\Phi(u)/u \to \infty$, and let ξ be defined by (3.1.1a). Let f_1, f_2, g be measurable functions. Suppose that there exist constants c_1, c_2 so that*

$$\lambda\mu\{|g| \geq \lambda\} \geq c_1 \int_{\{|f_1| \geq \lambda\}} |f_1| \, d\mu + c_2 \int_{\{|f_2| \geq \lambda\}} |f_2| \, d\mu$$

for all $\lambda > 0$. Then we have

$$M_\xi (g) \geq c_1 M_\Phi (f_1) + c_1 M_\xi (f_1) + c_2 M_\Phi (f_2) + c_2 M_\xi (f_2).$$

Proof. This proof is similar to the proof of Theorem (3.1.2). (We take $a = 1$.) Again, we may assume that f_1, f_2, and g are nonnegative.

Suppose first that f_1, f_2, and g are functions having countably many values, all in the set $T = C_\varphi \cup \{0\}$, as before. Then

$$M_\xi (g) = \int \xi (g) \, d\mu$$

$$= \int \Psi (\varphi (g)) \, d\mu$$

$$= \int_\Omega \int_0^{\varphi(g)} \psi(y) \, dy \, d\mu$$

apply Fubini's theorem to exchange the order

$$= \int_0^\infty \int_{\{g \geq \psi(y)\}} d\mu \, \psi(y) \, dy$$

$$= \int_0^\infty \mu\{g \geq \psi(y)\} \, \psi(y) \, dy$$

use the hypothesis

$$\geq c_1 \int_0^\infty \int_{\{f_1 \geq \psi(y)\}} f_1 \, d\mu \, dy + c_2 \int_0^\infty \int_{\{f_2 \geq \psi(y)\}} f_2 \, d\mu \, dy$$

apply Fubini's theorem (twice) to exchange back

$$= c_1 \int_\Omega \int_0^{\varphi(f_1)} dy \, f_1 \, d\mu + c_2 \int_\Omega \int_0^{\varphi(f_2)} dy \, f_2 \, d\mu$$

$$= c_1 \int \varphi(f_1) \, f_1 \, d\mu + c_2 \int \varphi(f_2) \, f_2 \, d\mu$$

apply the case of equality in Young's inequality

$$= c_1 \int \Phi(f_1) \, d\mu + c_1 \int \Psi(\varphi(f_1)) \, d\mu$$

$$+ c_2 \int \Phi(f_2) \, d\mu + c_2 \int \Psi(\varphi(f_2)) \, d\mu$$

$$= c_1 M_\Phi(f_1) + c_1 M_\xi(f_1) + c_2 M_\Phi(f_2) + c_2 M_\xi(f_2).$$

For the case of general f_1, f_2, and g, approximate as follows. Given n, let countably many values $0 < \cdots < t_{-2} < t_{-1} < t_0$ be chosen so that $t_i \in C_\varphi$, $t_0 \geq 2^n$, $t_i - t_{i-1} \leq 2^{-n}$, $\lim_{i \to -\infty} t_i = 0$, and $t_i/t_{i-1} < 1 + 2^{-n}$. Let

$$\{g_n = t_i\} = \{t_i \leq g < t_{i+1}\},$$
$$\{g_n = t_0\} = \{t_0 \leq g\},$$
$$\{g_n = 0\} = \{g = 0\};$$

so that $g_n \leq g$ with $g_n \to g$. Similarly, f_{1n} should be constructed to approximate f_1 and f_{2n} should be constructed to approximate f_2. Then for $t_{i-1} < \lambda \leq t_i$ we have $\{g_n \geq \lambda\} = \{g \geq t_i\}$, and similarly for f_1 and f_2. If $c_1 \geq 0$, then

$$\frac{c_1}{t_i} \geq \frac{c_1}{t_{i-1}(1 + 2^{-n})} \geq \frac{c_1}{\lambda(1 + 2^{-n})}.$$

If $c_1 < 0$, then

$$\frac{c_1}{t_i} \geq \frac{c_1}{\lambda}.$$

Similarly for c_2. Thus in the case where $c_1, c_2 \geq 0$, we have

$$\mu\{g_n \geq \lambda\} = \mu\{g \geq t_i\} \geq \frac{c_1}{t_i} \int_{\{f_1 \geq t_i\}} f_1 \, d\mu + \frac{c_2}{t_i} \int_{\{f_2 \geq t_i\}} f_2 \, d\mu$$

$$= \frac{c_1}{t_i} \int_{\{f_{1n} \geq \lambda\}} f_{1n} \, d\mu + \frac{c_2}{t_i} \int_{\{f_{2n} \geq \lambda\}} f_{2n} \, d\mu$$

$$\geq \frac{c_1}{\lambda(1 + 2^{-n})} \int_{\{f_{1n} \geq \lambda\}} f_{1n} \, d\mu + \frac{c_2}{\lambda(1 + 2^{-n})} \int_{\{f_{2n} \geq \lambda\}} f_{2n} \, d\mu.$$

Therefore the previous case yields

$$M_\xi(g_n) \geq \frac{c_1}{1 + 2^{-n}} M_\Phi(f_{1n}) + \frac{c_1}{1 + 2^{-n}} M_\xi(f_{1n})$$
$$+ \frac{c_2}{1 + 2^{-n}} M_\Phi(f_{2n}) + \frac{c_2}{1 + 2^{-n}} M_\xi(f_{2n}).$$

Take the limit as $n \to \infty$ to obtain the conclusion.

If $c_1 \geq 0$ and $c_2 < 0$, we have similarly

$$M_\xi(g_n) \geq \frac{c_1}{1 + 2^{-n}} M_\Phi(f_{1n}) + \frac{c_1}{1 + 2^{-n}} M_\xi(f_{1n})$$
$$+ c_2 M_\Phi(f_{2n}) + c_2 M_\xi(f_{2n}).$$

Again, take the limit as $n \to \infty$ to obtain the conclusion. Cases with $c_1 < 0$ are similar. ∎

Theorem (3.1.11) has many simple consequences. The following will be included here. Others will be seen below.

(3.1.12) Corollary. Let Φ be a finite Orlicz function with $\Phi(u)/u \to \infty$, and suppose that the function ξ defined by (3.1.1a) is also an Orlicz function. Suppose f and g are measurable functions satisfying

$$\mu\{|g| \geq \lambda\} \geq \frac{c}{\lambda} \int_{\{|f| \geq \lambda\}} |f| \, d\mu$$

for all $\lambda > 0$. Assume either (i) Φ satisfies (Δ_2) or (ii) $\mu\{|f| \geq \lambda\} < \infty$ for all $\lambda > 0$. Then:

(a) $g \in L_\xi \implies f \in L_\Phi$.
(b) $g \in H_\xi \implies f \in H_\Phi$.

Proof. We prove part (a). The other part is left to the reader. If $g \in L_\xi$, then there is $a > 0$ such that $M_\xi(g/a) < \infty$. But then we have $M_\Phi(f/a) \leq M_\Phi(f/a) + M_\xi(f/a) \leq (1/c)M_\xi(g/a) < \infty$, so $f \in L_\Phi$. ∎

Maximal inequalities for stopped processes

The three-function inequalities may be used to prove refinements of the basic maximal inequality (1.1.7).

If (X_n) is an adapted process, let

$$X_N^* = \max_{1 \leq n \leq N} |X_n|, \qquad X^* = \sup_N X_N^*.$$

Let $\lambda > 0$. If $A_N = \{X_N^* > \lambda\}$, let the stopping time σ be defined by

$$\sigma(\omega) = \begin{cases} \inf\{n : 1 \le n \le N, |X_n(\omega)| > \lambda\} & \text{if } \omega \in A_N, \\ N & \text{otherwise.} \end{cases}$$

The inequality

$$\mu(A_N) \le \frac{1}{\lambda} \int_{A_N} |X_\sigma|\, d\mu$$

was seen above (1.1.7a). By (3.1.3), if $(X_\sigma)_{\sigma \in \Sigma}$ is bounded in L_Φ, and $\mu(A_N) < \infty$ for all $\lambda > 0$, then $(X_N^*)_{N \in \mathbb{N}}$ is bounded in L_ξ, so we have:

(3.1.13) Theorem. *Let Φ and ξ be as before. (i) If $(X_\sigma)_{\sigma \in \Sigma}$ is an integrable net bounded in L_Φ, then $X^* \in L_\xi$. (ii) If (X_n) is a positive submartingale, bounded in L_Φ, then $X^* \in L_\xi$. The estimates in (3.1.10) apply.*

Proof. For (ii), observe that if X_n is a positive submartingale, then $M_\Phi(X_\sigma)$ is an increasing function of $\sigma \in \Sigma$, since (by Jensen's inequality) $\Phi(X_n/a)$ is a submartingale. Thus we have: If $(X_n)_{n \in \mathbb{N}}$ is bounded in L_Φ, then $(X_\sigma)_{\sigma \in \Sigma}$ is also bounded in L_Φ, so $X^* \in L_\xi$. ∎

Complements

(3.1.14) (Best constant.) The constant $p/(p-1)$ in Corollary (3.1.10(a)) cannot be improved. Let $(\Omega, \mathcal{F}, \mu)$ be $(0,1)$ with Lebesgue measure. Let α satisfy $-1/p < \alpha < 0$. Define $g(t) = t^\alpha$ and $f(t) = (\alpha + 1)t^\alpha$. Then $\mu\{|g| \ge \lambda\} = \mu\{t^\alpha > \lambda\} = \mu\{t < \lambda^{1/\alpha}\} = \mu\big((0, \lambda^{1/\alpha}]\big) = \lambda^{1/\alpha}$ and

$$\int_{\{|g| \ge \lambda\}} |f|\, d\mu = (\alpha + 1)\int_0^{\lambda^{1/\alpha}} t^\alpha\, dt$$
$$= \frac{\alpha + 1}{\alpha + 1} \lambda^{(\alpha+1)/\alpha}$$
$$= \lambda \mu\{|g| \ge \lambda\}.$$

But also

$$\|g\|_p^p = \int_0^1 t^{\alpha p}\, dt < \infty,$$

since $\alpha p > -1$. Thus $\|g\|_p/\|f\|_p = 1/(\alpha + 1)$. Finally

$$\sup\left\{\frac{1}{\alpha + 1} : \frac{-1}{p} < \alpha < 0\right\} = \frac{p}{p - 1}.$$

(3.1.15) (Best constant.) The constant $(k+1)/k$ in Corollary (3.1.10(b)) cannot be improved. Let $(\Omega, \mathcal{F}, \mu)$ be $[0, \infty)$ with Lebesgue measure. Let

$\alpha = -1/(k+1)$, so $-1 < \alpha < 0$. Then let $g(t) = t^\alpha$ and $f(t) = (1+\alpha)t^\alpha$. Check that

$$\mu\{|g| \geq \lambda\} = \frac{1}{\lambda} \int_{\{|g| \geq \lambda\}} |f| \, d\mu \qquad \text{for all } \lambda > 0.$$

Write $\Phi_k(u) = u(\log^+ u)^k$. The following are calculus exercises(!):

$$M_{\Phi_k}\left(\frac{f}{a}\right) = \frac{a^{1/\alpha}(-\alpha)^k k!}{(1+\alpha)^{1/\alpha}(\alpha+1)^{k+1}},$$

$$M_{\Phi_{k-1}}\left(\frac{g}{b}\right) = \frac{b^{1/\alpha}(-\alpha)^{k-1}(k-1)!}{(\alpha+1)^k}.$$

Then take $a = \|f\|_{\Phi_k}$ and $b = \|g\|_{\Phi_{k-1}}$, and conclude [since $\alpha = -1/(k+1)$]

$$\frac{a}{b} = \frac{(1+\alpha)^{\alpha+1}}{(-\alpha)^\alpha k^\alpha} = \frac{k}{k+1}.$$

So $\|g\|_{\Phi_{k-1}} = ((k+1)/k)\,\|f\|_{\Phi_k}$.

(3.1.16) (Improved constants.) Prove the elementary inequality:

$$a \log^+ b \leq a \log^+ a + \frac{b}{e}.$$

This is used to prove: If $\mu(\Omega) = 1$, then the second inequality of (3.1.10(c)) may be improved to

$$\int |g| \, d\mu \leq \frac{e}{e-1} + \frac{e}{e-1} \int |f| \log^+ |f| \, d\mu$$

(or see Doob [1953], p. 317, or Neveu [1975], p. 71).

This inequality is true with $e/(e-1)$ replaced by the unique positive solution of the equation $e^{-c} = (c-1)^2$. Note $e/(e-1) \approx 1.582$ and $c \approx 1.478$. This is the best constant for the inequality (D. Gilat [1986]).

(3.1.17) (Variant L_p inequality, Burkholder [1973].) Let X and T be nonnegative measurable functions, and let α, β be positive constants. Suppose, for all $\lambda > 0$,

$$\lambda\mu\{T > \beta\lambda\} \leq \alpha \int_{\{T>\lambda\}} X \, d\mu.$$

Then, for $1 < p < \infty$, we have

$$\|T\|_p \leq \alpha\beta^p \left(\frac{p}{p-1}\right) \|X\|_p.$$

To see this, apply the three-function inequality (3.1.2) with: $\Phi(u) = u^p/p$, $\varphi(u) = u^{p-1}$, $\xi(u) = u^p/q$, where $q = p/(p-1)$ is the conjugate index.

Then with $f = \alpha X$, $g = T/\beta$, $h = T$, the hypothesis of Theorem (3.1.2) is satisfied. The conclusion is

$$\frac{1}{qb^{p-1}}\left(\frac{1}{\beta^p} - \frac{a}{b}\right)\|T\|_p^p \le \frac{\alpha^p}{pa^{p-1}}\|X\|_p^p.$$

Substitute $a = 1$ and $b = q\beta^p$ to obtain the result.

(3.1.18) (Best constant.) Show that $\alpha\beta^p q$ is the best constant in the preceding. It is clearly enough to consider the case $\alpha = 1$. For $\beta \le 1$, use the measure space $(0, 1]$, and $T(\omega) = \omega^s \beta^{-1/s}$, $X(\omega) = (s+1)\omega^s$, where $-1/p < s < 0$; then let $s \to -1/p$ to show that the constant is best possible. What should be done for $\beta > 1$?

(3.1.19) (Hardy-Littlewood.) The best known use of maximal functions is due to Hardy and Littlewood. Let $\Omega = (0, 1)$ and let μ be Lebesgue measure. Suppose $f\colon (0, 1) \to \mathbb{R}$ is an integrable function. Define

$$f^*(x) = \sup\left\{\frac{1}{v - u}\int_u^v |f(t)|\,dt : 0 < u < x < v < 1\right\}.$$

Let $\bar{f}$ be the decreasing rearrangement of $|f|$ on $(0, 1)$. Thus,

$$\mu\{|f| \ge \lambda\} = \mu\{\bar{f} \ge \lambda\},$$

so f and $\bar{f}$ belong simultaneously to any L_Φ or H_Φ, and have the same norm. Let $g(x) = (1/x)\int_0^x \bar{f}(t)\,dt$. Thus

$$\mu\{f^* \ge \lambda\} \le \mu\{g \ge \lambda\},$$

so if g belongs to some L_Φ or H_Φ, then so does f^*, with no larger norm. Now, it is easy to see that

$$\mu\{g > \lambda\} \le \int_{\{g>\lambda\}} \bar{f}(t)\,dt.$$

This yields, then, many corollaries. If $1 < p < \infty$, and $f \in L_p$, then $f^* \in L_p$, and $\|f^*\|_p \le (p/(p-1))\,\|f\|_p$ (see 3.1.10(a)). If $f \in L\log^k L$, then $f^* \in L\log^{k-1} L$. If $f \in L\log L$, then $f^* \in L_1$.

(3.1.20) Let Φ be an Orlicz function with derivative φ and conjugate Ψ. If $\|f\|_\Phi \le 1/2$, and $\{|f| \ge \lambda\} < \infty$ for all $\lambda > 0$, then $\|\varphi(|f|)\|_\Psi \le 1$. To see this, apply (3.1.5) with $b = 1$ and $a = \|f\|_\Phi \le 1/2$. Then

$$\int \Psi(\varphi(|f|))\,d\mu \le \frac{a}{b-a}\int \Phi\left(\frac{|f|}{a}\right)d\mu = \frac{a}{1-a} \le 1.$$

(3.1.21) (Given ξ, find Φ.) Suppose $\xi\colon (0, \infty) \to (0, \infty)$ is nondecreasing, left-continuous, and

$$\int_0^1 \frac{\xi(x)}{x^2}\,dx < \infty.$$

Then

$$\Phi(u) = u \int_0^u \frac{\xi(x)}{x^2}\, dx, \quad u \geq 0,$$

is an Orlicz function, and $\xi(u) = \Psi(\varphi(u))$ a.e., where Ψ is the conjugate and φ is the derivative of Φ.

(3.1.22) (Converse maximal inequality.) Let Φ be an Orlicz function and $\xi(u) = \Psi(\varphi(u))$ as usual. We saw in (3.1.13(ii)) that for a positive martingale (X_n), if $\sup_n \|X_n\|_\Phi < \infty$, then $\|\sup X_n\|_\xi < \infty$. The converse is false: for example take $X_n = X_1$ for all n, where $X_1 \in L\log^{k-1} L$ but $X_1 \notin L\log^k L$. The converse can be proved under additional assumptions:

Theorem. *Let (X_n) be a positive supermartingale such that for a constant $C > 0$ and all n we have $X_{n+1} \leq CX_n$. If $M_\xi(\sup_n X_n) < \infty$ and $M_\Phi(X_1) < \infty$, then $\sup_n M_\Phi(X_n) < \infty$.*

Proof. Fix $\lambda > 0$. Let σ be the stopping time equal to the first n such that $X_n > \lambda$. The set $\{\sigma = k\} \in \mathcal{F}_k$, hence

$$\mathbf{E}\left[1_{\{1 < \sigma \leq n\}} X_n\right] \leq \sum_{k=2}^n \mathbf{E}\left[1_{\{\sigma = k\}} X_k\right] = \mathbf{E}\left[1_{\{1 < \sigma \leq n\}} X_\sigma\right].$$

Also $X_\sigma \leq CX_{\sigma-1} \leq C\lambda$ on the set $\{1 < \sigma \leq n\}$. Therefore

$$\begin{aligned}
\mathbf{E}\left[1_{\{X_n > \lambda\}} X_n\right] &\leq \mathbf{E}\left[1_{\{\sup_{1 \leq i \leq n} X_i > \lambda\}} X_n\right] \\
&= \mathbf{E}\left[1_{\{X_1 > \lambda\}} X_n\right] + \mathbf{E}\left[1_{\{1 < \sigma \leq n\}} X_n\right] \\
&\leq \mathbf{E}\left[1_{\{X_1 > \lambda\}} X_1\right] + \mathbf{E}\left[1_{\{1 < \sigma \leq n\}} X_\sigma\right] \\
&\leq \mathbf{E}\left[1_{\{X_1 > \lambda\}} X_1\right] + C\lambda \mathbf{P}\left\{\sup_{1 \leq i \leq n} X_i > \lambda\right\}.
\end{aligned}$$

Now apply Theorem (3.1.11) with $g = \sup_{1 \leq i \leq n} X_i$, $f_1 = X_n$, $f_2 = X_1$, $c_1 = 1/C$, $c_2 = -1/C$. ∎

(3.1.23) Let (X_n) be a positive supermartingale such that for a constant $C > 0$ and all n we have $X_{n+1} \leq CX_n$. Let $k \geq 1$. If

$$\mathbf{E}\left[\left(\sup_n X_n\right)\left(\log^+\left(\sup_n X_n\right)\right)^{k-1}\right] < \infty \text{ and } \mathbf{E}\left[X_1(\log^+ X_1)^k\right] < \infty,$$

then

$$\sup_n \mathbf{E}\left[X_n(\log^+ X_n)^k\right] < \infty.$$

To see this, choose $\Phi(u) = u(\log^+ u)^k$. Then $\xi(u) = ku(\log^+ u)^{k-1}$.

(3.1.24) (Atomic σ-algebras.) Suppose each $\mathcal{F}_n$ is atomic, and there is a constant C such that $\mathbf{P}(A_n) \leq C\mathbf{P}(A_{n+1})$ for atoms $A_n \in \mathcal{F}_n$ and $A_{n+1} \in \mathcal{F}_{n+1}$ with $A_n \supseteq A_{n+1}$. Show that every positive supermartingale satisfies $X_{n+1} \leq CX_n$, the hypothesis of (3.1.22). The most familiar example is the dyadic stochastic basis $(\mathcal{F}_n)$, where $\mathbf{P}(A_n) = 2\mathbf{P}(A_{n+1})$.

Remarks

The three-function inequality is from Edgar & Sucheston [1989] and [1991]. Inequality (3.1.3) is proved but not stated in Neveu [1975], pages 217–219, for the case $g \in L_\infty$. There is a version of Theorem 3.1.11 in which the weak inequalities are assumed to hold only for sufficiently large λ; this is useful to obtain the converse of the dominated ratio ergodic theorem; see Szabó [1991]. The proof (3.1.14) that $(p-1)/p$ is the best constant is taken from Hardy, Littlewood, & Polya [1952]. For $k = 1$, the martingale case of (3.1.23), and the important stopping time argument are due to Gundy [1969].

3.2. Sharp maximal inequality for martingale transforms

In this section $(\Omega, \mathcal{F}, \mathbf{P})$ will be a fixed probability space. Normally we will be concerned with a sequence (X_n) of random variables and the corresponding stochastic basis defined by $\mathcal{F}_n = \sigma(X_1, X_2, \cdots, X_n)$.

We will use the following notation in this section. If $\mathbf{X} = (X_n)$ is a sequence of random variables, then $\mathbf{X}^* = \sup_n |X_n|$. The L_p-norm of the sequence is $\|\mathbf{X}\|_p = \sup_n \|X_n\|_p$. The notation $\mathbf{X} > 0$ means $X_n > 0$ for all n.

Recall that the sequence $\mathbf{X} = (X_n)$ is a *martingale* iff $X_m = \mathbf{E}^{\mathcal{F}_m}[X_n]$ for $m < n$. Of course, it is enough to have $X_n = \mathbf{E}^{\mathcal{F}_n}[X_{n+1}]$ for all n. Let $\mathbf{Y} = (Y_n)$ be the *difference sequence* of $\mathbf{X}$; that is, $Y_1 = X_1$ and $Y_n = X_n - X_{n-1}$ for $n \geq 2$. Thus, $X_n = \sum_{i=1}^n Y_i$. An equivalent definition of the martingale is: $\mathbf{E}^{\mathcal{F}_n}[Y_{n+1}] = 0$ for $n \geq 1$. We will say that $\mathbf{Y}$ is a *martingale difference sequence*.

A sequence $\mathbf{V} = (V_n)$ of random variables is *predictable* iff V_{n+1} is measurable with respect to $\mathcal{F}_n$. The *transform* of $\mathbf{X}$ by $\mathbf{V}$ is the process $\mathbf{Z} = (Z_n)$ defined by $Z_n = \sum_{i=1}^n V_i Y_i$. We will write $\mathbf{Z} = \mathbf{V} * \mathbf{X}$. If X_n is the fortune of a gambler at time n, then Z_n may be viewed as the result of controlling $\mathbf{X}$ by $\mathbf{V}$. Since $\mathbf{V}$ is predictable, multiplication of Y_i by V_i is equivalent to changing the stakes for the ith game on the basis of information available before the ith game. The transform $\mathbf{Z}$ of a martingale $\mathbf{X}$ by a predictable process $\mathbf{V}$ is again a martingale, provided it is integrable, since

$$\mathbf{E}^{\mathcal{F}_n}[Z_{n+1} - Z_n] = \mathbf{E}^{\mathcal{F}_n}[V_{n+1}Y_{n+1}] = V_{n+1}\mathbf{E}^{\mathcal{F}_n}[Y_{n+1}] = 0.$$

If τ is a stopping time, then the stopped process $(X_{\tau \wedge n})_{n \in \mathbb{N}}$ is the transform of (X_n) by the predictable sequence $V_n = \mathbf{1}_{\{\tau \geq n\}} = 1 - \mathbf{1}_{\{\tau \leq n-1\}}$.

It is easy to see that transforms satisfy identities like:

$$\mathbf{W} * (\mathbf{V} * \mathbf{X}) = (\mathbf{W} * \mathbf{V}) * \mathbf{X},$$
$$\mathbf{W} * (\mathbf{V} * \mathbf{X}) = \mathbf{V} * (\mathbf{W} * \mathbf{X}).$$

The theorem below gives the weak maximal inequality for martingale transforms by a process $\mathbf{V}$ between 0 and 1.

(3.2.1) Theorem. *Let* $\mathbf{X}$ *be a martingale, let* $\mathbf{V}$ *be a predictable process with values in* $[0, 1]$, *and let* $\mathbf{Z} = \mathbf{V} * \mathbf{X}$. *Then for all* $\lambda > 0$,

$$\mathbf{P}\{\mathbf{Z}^* \geq \lambda\} \leq \frac{1}{\lambda}\|\mathbf{X}\|_1.$$

The proof is omitted, since it is a special case of Theorem (3.2.2), below.

Theorem (3.2.1) may appear surprising, since $\mathbf{Z}$ may be badly behaved in some other ways; for example $\mathbf{Z}$ need not be L_1-bounded, even if $\mathbf{X}$ is. (See (6.2.8).) To give an intuitive interpretation of the theorem, assume that $\mathbf{X}$ is nonnegative, and $X_1 = \|\mathbf{X}\|_1 = c < \lambda$. In a fair game, the player is allowed to change the stakes by multiplying them by a number between 0 and 1, using information before each play. In particular, the player can skip a game by letting $V_i = 0$, but he cannot reverse the roles of his opponent and himself, since $\mathbf{V} \geq 0$. The probability that the player's initial fortune c is ever increased to λ is at most $c/\lambda < 1$. In particular, the probability that the player's initial fortune will ever double is less than $1/2$. In fact, controlling a fair game (the martingale $\mathbf{X}$) by a predictable process $\mathbf{V}$ taking values in $[0, 1]$ does not visibly improve this probability, since the inequality in Theorem (3.2.1) is the same as Doob's maximal inequality (1.4.18), corresponding to the case $\mathbf{V} = 1$ and $\mathbf{Z} = \mathbf{X}$.

A more effective way of controlling the game is to allow $\mathbf{V}$ to take values in the interval $[-1, 1]$. Now the gambler may choose, just before game i, to reverse roles with the opponent for game i by choosing $V_i = -1$. In the seventeenth century, the word "martingale" meant the double-or-nothing martingale which can be mathematically described as follows. Let $\mathbf{P}$ be Lebesgue measure on $[0, 1]$, and let $X_n = c2^{n-1} \mathbf{1}_{[0, 2^{-n+1}]}$. Then X_n is a nonnegative martingale starting at c with $\lim_n X_n = 0$. Define $V_1 = 1$ and $V_n = -1$ for $n > 1$. Then

$$Z_n = X_1 - \sum_{i=2}^{n} Y_i = c - (X_n - c) = 2c - X_n.$$

Thus $\sup_n Z_n = 2c - \inf X_n = 2c - 0 = 2c$.

In an infinite game, the gambler can double his initial fortune. Even if an infinite game is impractical, it remains that in a game of sufficient duration, a gambler endowed with very large reserves or credit can double his initial fortune with large probability.

The following maximal inequality, due to Burkholder [1986], allows for $\mathbf{V}$ with both signs. In particular, it gives the bound 2 in the case of the double-or-nothing martingale. The case $a = 0, b = 1$ is Theorem (3.2.1).

(3.2.2) Theorem. *Let* a, b *be real numbers with* $a \leq 0 \leq b$. *Let* $\mathbf{X}$ *be a martingale, let* $\mathbf{V}$ *be a predictable process with values in the interval* $[a, b]$, *and let* $\mathbf{Z} = \mathbf{V} * \mathbf{X}$. *Then for all* $\lambda > 0$,

$$\mathbf{P}\{\mathbf{Z}^* \geq \lambda\} \leq \frac{b-a}{\lambda}\|\mathbf{X}\|_1.$$

The constant $b - a$ *is the best possible.*

Proof. Let $\mathbf{Y}$ be the difference process of $\mathbf{X}$. Set

$$(3.2.2a) \qquad A_n = \sum_{i=1}^{n} (V_i - a) Y_i,$$

$$(3.2.2b) \qquad B_n = \sum_{i=1}^{n} (b - V_i) Y_i.$$

Then

$$(3.2.2c) \qquad A_n + B_n = (b - a) X_n,$$

$$(3.2.2d) \qquad b A_n + a B_n = (b - a) Z_n, \text{ and}$$

$$(3.2.2e) \qquad (A_n - A_{n-1})(B_n - B_{n-1}) \geq 0$$

(with the convention that $A_0 = B_0 = 0$).

Let $u \colon \mathbb{R}^2 \to \mathbb{R}$ be defined by

$$u(x, y) = \begin{cases} 1 + xy & \text{if } |x| < 1 \text{ and } |y| < 1 \\ |x + y| & \text{otherwise.} \end{cases}$$

Note that u is continuous, since the two formulas agree on the boundary square. Next, $0 \leq u(x, y) - |x + y| \leq 1$: the extrema may be easily computed separately inside and outside the square $|x| \vee |y| \leq 1$.

The function u may be estimated by its first-degree Taylor polynomial in two variables as follows. The partial derivative with respect to x exists and is equal to

$$(3.2.2f) \qquad u_x(x, y) = \begin{cases} y & \text{if } |x| < 1 \text{ and } |y| < 1 \\ \operatorname{sgn}(x + y) & \text{otherwise,} \end{cases}$$

except possibly at the values $x = 1$ and $x = -1$. For each fixed y, the function $u_x(x, y)$ is nondecreasing in x and $u(x, y)$ is continuous in x (so for each fixed y the function $u(x, y)$ is absolutely continuous); for fixed x, the function $u_x(x, y)$ is nondecreasing in y. Similarly, the partial derivative with respect to y is

$$u_y(x, y) = \begin{cases} x & \text{if } |x| < 1 \text{ and } |y| < 1 \\ \operatorname{sgn}(x + y) & \text{otherwise.} \end{cases}$$

For fixed x, the function u_y is increasing in y, and u is continuous; for fixed y, the function u_y is increasing in x. Now if x, y, h, k are real, and $hk \geq 0$, then we have

(3.2.2g) $u(x+h, y+k) \geq u(x,y) + u_x(x,y)\, h + u_y(x,y)\, k.$

We prove the case $h > 0$ and $k > 0$:

$$u(x+h, y+k) = u(x, y+k) + \int_0^h u_x(x+t, y+k)\, dt$$

$$= u(x,y) + \int_0^k u_y(x, y+s)\, ds + \int_0^h u_x(x+t, y+k)\, dt$$

$$\geq u(x,y) + \int_0^k u_y(x, y)\, ds + \int_0^h u_x(x, y)\, dt$$

$$= u(x,y) + u_x(x,y)\, h + u_y(x,y)\, k.$$

If we apply (3.2.2g) to the $\mathbb{R}^2$-valued stochastic process $C_n = (A_n, B_n)$ with $x = A_{n-1}$, $y = B_{n-1}$, $h = A_n - A_{n-1}$, $k = B_n - B_{n-1}$, we obtain

(3.2.2h) $u(C_n) \geq u(C_{n-1}) + u_x(C_{n-1})(A_n - A_{n-1}) + u_y(C_{n-1})(B_n - B_{n-1}).$

If $n = 1$, this is the inequality $u(C_1) \geq u(C_0)$. If $n > 1$, then $A_n - A_{n-1} = (V_n - a)Y_n$, so $u_x(C_{n-1})(A_n - A_{n-1})$ can be written as a product of Y_n and a random variable Q which is a bounded measurable function of $V_1, V_2, \cdots, V_n, Y_1, Y_2, \cdots, Y_{n-1}$. Thus Q is $\mathcal{F}_{n-1}$-measurable, and thus

$$\mathbf{E}\left[u_x(C_{n-1})(A_n - A_{n-1})\right] = \mathbf{E}\left[Y_n Q\right] = \mathbf{E}\left[\mathbf{E}^{\mathcal{F}_{n-1}}\left[Y_n\right] Q\right] = 0.$$

Similarly,

$$\mathbf{E}\left[u_y(C_{n-1})(B_n - B_{n-1})\right] = 0.$$

Therefore, by (3.2.2h),

$$\mathbf{E}\left[u(C_n)\right] \geq \mathbf{E}\left[u(C_{n-1})\right] \geq \cdots \geq \mathbf{E}\left[u(C_1)\right] \geq \mathbf{E}\left[u(C_0)\right] = u(0,0) = 1.$$

Using (3.2.2d), we have

$$\mathbf{P}\{|Z_n| \geq 1\} = \mathbf{P}\{|bA_n + aB_n| \geq b - a\}$$

(3.2.2i)
$$= 1 - \mathbf{P}\{|bA_n + aB_n| < b - a\}$$
$$\leq 1 - \mathbf{P}\{|A_n| < 1 \text{ and } |B_n| < 1\}$$
$$\leq \mathbf{E}\left[u(C_n) - I(C_n)\right],$$

where I is the indicator function of the set $\{(x,y) : |x| < 1 \text{ and } |y| < 1\}$.

Now by (3.2.2c) and the inequalities for u, we have

$$u(C_n) - I(C_n) \leq |A_n + B_n| \leq (b - a)|X_n|.$$

Combining this with (3.2.2i), we obtain

(3.2.2j) $$\mathbf{P}\{|Z_n| \geq 1\} \leq (b - a)\|X_n\|_1,$$

which is quite close to the announced inequality for $\mathbf{Z}^*$. We will use a stopping time argument similar to the one often used in proofs of Doob's inequality. Let $\tau = \inf\{n : |Z_n| \geq 1\}$. Now the stopped process $Z_{\tau \wedge n}$ is the transform by $\mathbf{V}$ of the stopped martingale $X_{\tau \wedge n}$. Thus we have from (3.2.2j)

$$\mathbf{P}\{|Z_{\tau \wedge n}| \geq 1\} \leq (b - a)\|X_{\tau \wedge n}\|_1.$$

But $|X_n|$ is a submartingale, so by (1.4.18), we have $\|X_{\tau \wedge n}\|_1 \leq \|\mathbf{X}\|_1$, so if $n \to \infty$ we obtain

$$\mathbf{P}\{\mathbf{Z}^* \geq 1\} \leq (b - a)\|\mathbf{X}\|_1.$$

Applying this to the martingale $\mathbf{X}/\lambda$ yields the maximal inequality stated.

In the proof that the constant $b - a$ is best possible, we may assume that $\lambda = 1$. If $a = 0$, we may assume $b = 1$. Then the deterministic example $X_n = V_n = 1$ shows that the constant 1 is the best possible (also in Doob's inequality). If $a < 0$, consider a martingale $\mathbf{X}$ such that X_1 is the constant $1/(b - a)$ and $X_2 = X_3 = \cdots$ is such that

$$\mathbf{P}\left\{X_2 = \frac{-2}{a}\right\} = \frac{-a}{2(b - a)},$$
$$\mathbf{P}\{X_2 = 0\} = \frac{2b - a}{2(b - a)}.$$

Let $V_1 = b$ and $V_n = a$ for $n \geq 2$. Then $\mathbf{P}\{\mathbf{Z}^* = 1\} = 1$ and $\|\mathbf{X}\|_1 = 1/(b - a)$.   ∎

In Section 6.2, below, we give the proof that the transform of an L_1-bounded martingale converges a.s.

Remarks

References related to this section are: Burkholder [1966] and Burkholder [1986].

3.3. Prophet compared to gambler

Let $n \in \mathbb{N}$, and let the random variable X_i be the fortune of a player at time i, for $1 \leq i \leq n$. We may sometimes use a boldface letter as an abbreviation: $\mathbf{X} = (X_1, X_2, \cdots, X_n)$. The player is to stop playing at a certain time i; his goal is to choose i so that his fortune X_i is as large as possible (on the average). We will consider two types of players: the *prophet* and the *gambler*.

A *prophet* is a player with complete foresight. Since he knows all the values X_i, he may simply choose the largest one, and stop at the appropriate time. Thus he achieves the maximum $X_1 \vee X_2 \vee \cdots \vee X_n$.

A *gambler* is a player without knowledge of the future; but he knows the past and the present, and has knowledge of the odds (that is, the joint distribution of $\mathbf{X}$). Let $\mathcal{F}_i$ be the σ-algebra generated by $X_1, X_2, \cdots, X_i$. Then the time the gambler chooses to stop must be a stopping time for this stochastic basis $(\mathcal{F}_i)_{i=1}^n$. Write Σ_n for this set of stopping times; their values are in the set $\{1, 2, \cdots, n\}$. When the gambler uses the stopping time $\tau \in \Sigma_n$, his fortune when he stops is X_τ. So his expected fortune is $\mathbf{E}[X_\tau]$. Thus the best expected fortune he can achieve is:

$$\sup\{\mathbf{E}[X_\tau] : \tau \in \Sigma_n\}.$$

This is known as the *value* of the process $\mathbf{X} = (X_1, X_2, \cdots, X_n)$. Notation:

$$V = V(\mathbf{X}) = V(X_1, X_2, \cdots, X_n).$$

We will be interested in *prophet inequalities*. They compare the expected gain P of a prophet to the expected gain $G = V$ of a gambler. Of course the prophet has the advantage, so $P \geq G$. A surprising result is that often there are moderate universal constants C (independent of n and of the distributions of the X_i) such that $P \leq CG$. In the independent positive case, $C = 2$ (Theorem 3.3.2) and the constant 2 is optimal. Thus the advantage of knowing the future is not as large as might have been expected.

To be sure, the prophet uses his foresight only to stop the game, not to change the stakes. The gambler uses his nonanticipating skills for the same purpose. A different situation arises when the players are allowed to change the stakes. A prophet inequality (with optimal constant 3) also exists in that case. This will be treated in Theorems (3.3.4) and (3.3.5).

Stopped processes

We prove first a basic lemma; the technique used to define the stopping times σ_i is known as "backward induction," since we proceed from σ_{i+1} to σ_i.

(3.3.1) Lemma. *Let $X_1, X_2, \cdots, X_n$ be independent nonnegative random variables. Define $\sigma_n = n$, and inductively for $i = n-1, n-2, \cdots, 2, 1$:*

$$(3.3.1a) \qquad \sigma_i = \begin{cases} i & \text{on } \{X_i \geq \mathbf{E}[X_{\sigma_{i+1}}]\} \\ \sigma_{i+1} & \text{on } \{X_i < \mathbf{E}[X_{\sigma_{i+1}}]\}. \end{cases}$$

Then we have

$$\mathbf{E}[X_n] + \sum_{i=1}^{n-1} \mathbf{E}\left[(X_i - \mathbf{E}[X_{\sigma_{i+1}}])^+\right] = \mathbf{E}[X_{\sigma_1}].$$

Proof. Let i be given, $1 \leq i < n$. Then we have

$$(3.3.1b) \qquad (X_i - \mathbf{E}\left[X_{\sigma_{i+1}}\right])^+ = 1_{\{X_i \leq \mathbf{E}[X_{\sigma_{i+1}}]\}} \left(X_{\sigma_i} - \mathbf{E}\left[X_{\sigma_{i+1}}\right]\right).$$

Since $1_{\{X_i < \mathbf{E}[X_{\sigma_{i+1}}]\}}$ and $X_{\sigma_{i+1}} - \mathbf{E}\left[X_{\sigma_{i+1}}\right]$ are independent, we obtain

$$\mathbf{E}\left[1_{\{X_i < \mathbf{E}[X_{\sigma_{i+1}}]\}} \left(X_{\sigma_i} - \mathbf{E}\left[X_{\sigma_{i+1}}\right]\right)\right]$$

$$= \mathbf{E}\left[1_{\{X_i < \mathbf{E}[X_{\sigma_{i+1}}]\}} \left(X_{\sigma_{i+1}} - \mathbf{E}\left[X_{\sigma_{i+1}}\right]\right)\right]$$

$$(3.3.1c)$$

$$= \mathbf{P}\left\{X_i < \mathbf{E}\left[X_{\sigma_{i+1}}\right]\right\} \mathbf{E}\left[X_{\sigma_{i+1}} - \mathbf{E}\left[X_{\sigma_{i+1}}\right]\right] = 0.$$

Integrating (3.3.1b) and adding (3.3.1c) to it, we obtain

$$(3.3.1d) \qquad \mathbf{E}\left[\left(X_i - \mathbf{E}\left[X_{\sigma_{i+1}}\right]\right)^+\right] = \mathbf{E}\left[X_{\sigma_i}\right] - \mathbf{E}\left[X_{\sigma_{i+1}}\right].$$

Summing these equations from $i = 1$ to $i = n - 1$ and adding $\mathbf{E}\left[X_n\right] = \mathbf{E}\left[X_{\sigma_n}\right]$, the assertion follows. ∎

Now assume $n \geq 2$. Write $V = V(X_1, X_2, \cdots, X_n)$, $e_i = \mathbf{E}\left[X_i\right]$, and $_1V = V(X_2, \cdots, X_n)$. Clearly $_1V \leq V$, so the theorem below proves slightly more than $P \leq 2V$.

(3.3.2) Theorem. *Let $X_1, X_2, \cdots, X_n$ be independent positive random variables with $0 < \mathbf{E}\left[X_i\right] < \infty$. Then*

$$(3.3.2a) \qquad \mathbf{E}\left[(X_1 \vee X_2 \vee \cdots \vee X_n - {}_1V)^+\right] \leq V,$$

hence

$$(3.3.2b) \qquad \mathbf{E}\left[X_1 \vee X_2 \vee \cdots \vee X_n\right] \leq {}_1V + V \leq 2V.$$

Proof. Apply the lemma to the random variables X_i. By (3.3.1d) we have $\mathbf{E}\left[X_{\sigma_2}\right] \geq \mathbf{E}\left[X_{\sigma_{i+1}}\right]$ and therefore

$$\left(\sup_{1 \leq i \leq n} X_i - \mathbf{E}\left[X_{\sigma_2}\right]\right)^+ \leq X_n + \sum_{i=1}^{n-1} \left(X_i - \mathbf{E}\left[X_{\sigma_{i+1}}\right]\right)^+.$$

Integrating this, and using the lemma, we obtain:

$$\mathbf{E}\left[\left(\sup_{1 \leq i \leq n} X_i - \mathbf{E}\left[X_{\sigma_2}\right]\right)^+\right] \leq \mathbf{E}\left[X_{\sigma_1}\right].$$

Since $\mathbf{E}\left[X_{\sigma_2}\right] \leq {}_1V$ and $\mathbf{E}\left[X_{\sigma_1}\right] \leq V$, we have (3.3.2a). ∎

Note that, in fact, $\mathbf{E}\left[X_{\sigma_1}\right] = V$ and $\mathbf{E}\left[X_{\sigma_2}\right] = {}_1V$. (This is not used below.) See (3.3.8).

Transformed processes

To study transformed (rather than stopped) processes, it will be useful to consider the random variables $X_0, X_1, \cdots X_n$, and sub-σ-algebras $\mathcal{G}_i$ of $\mathcal{F}$ such that each X_i is measurable with respect to $\mathcal{G}_i$. However, we do not assume, in general, that $\mathcal{G}_i \subseteq \mathcal{G}_{i+1}$.

The gambler we considered previously adds at time $i+1$ to his fortune X_i the amount $X_{i+1} - X_i$ if he chooses to continue, and 0 if he does not. We will now allow the gambler to change the stakes: At time $i+1$ the gambler gains the amount $U_{i+1}(X_{i+1} - X_i)$, where the factor U_{i+1} is chosen by the gambler on the basis of the information provided by $\mathcal{G}_i$; that is, U_{i+1} is $\mathcal{G}_i$ predictable. Usually when we consider transforms we have $\mathcal{G}_i = \sigma(X_0, X_1, \cdots, X_i)$; in this case we simply say that $\mathbf{U} = (U_1, U_2, \cdots, U_n)$ is predictable. In our next result, we assume only $\mathcal{G}_i = \sigma(X_i)$; in this case $\mathbf{U}$ is called presently predictable: the player multiplies his stake by the random variable U_{i+1} which depends only on the present. In any event, the gain of the player up to a time m is

$$\sum_{i=0}^{m-1} U_{i+1}(X_{i+1} - X_i) = Z_m.$$

Of course, the sequence $\mathbf{Z} = (Z_1, Z_2, \cdots, Z_n)$ is the transform $\mathbf{U} * \mathbf{X}$ of the process $\mathbf{X}$ by $\mathbf{U}$.

In the following theorem, the gambler transforms the process $\mathbf{X}$ by a presently predictable process $\mathbf{U}$ bounded by a constant c. Because we are only interested in the ratio P/G, we may assume without loss of generality $c = 1$. Let Π_s be the set of all presently predictable processes bounded by 1. On the other hand, the prophet transforms $\mathbf{X}$ by any measurable process $\mathbf{U}$ bounded by 1. Let Δ_s be the set of all measurable processes bounded by 1, so the expected gains of the prophet and gambler are, respectively,

$$P_s = \sup_{\mathbf{U} \in \Delta_s} \mathbf{E}\left[\mathbf{U} * \mathbf{X}\right], \qquad G_s = \sup_{\mathbf{U} \in \Pi_s} \mathbf{E}\left[\mathbf{U} * \mathbf{X}\right].$$

Let Π be the set of nonnegative processes in Π_s, and let Δ be the set of nonnegative processes in Δ_s. Our main concern here will be the expected gains P of the prophet and G of the gambler when they are restricted to these classes.

In the nonnegative case Π, it is obvious that the best choice for the prophet is: bet the maximum if there is gain, and bet 0 if there is loss. Hence

$$P = \sum_{i=1}^{n} \mathbf{E}\left[(X_i - X_{i-1})^+\right].$$

In the signed case Π_s, the best choice for the prophet is: bet the maximum on the winning side. Hence

$$P = \sum_{i=1}^{n} \mathbf{E}\left[|X_i - X_{i-1}|\right].$$

It will be useful to introduce the functional μ defined on L_1 by

$$\mu(X) = \mathbf{E}[X] - \mathbf{E}\left[(X - \mathbf{E}[X])^+\right].$$

Note that $\mu(X) = \mathbf{E}[X] - (1/2)\mathbf{E}\left[|X - \mathbf{E}[X]|\right]$. We will sometimes assume below that the first and last random variables satisfy $\mu(X_0) \leq \mu(X_n)$. We observe that this is not a loss of generality if all the random variables are positive: This may be seen as follows. Since we compare gains, we suppose that both players have the same initial fortune. If $X_1, X_2, \cdots, X_n$ are positive and the players receive X_1, then addition of the random variable $X_0 = 0$ at the beginning does not change anything. Hence in this case we may assume $X_n \geq \mathbf{E}[X_0]$. Now $X_n \geq \mathbf{E}[X_0]$ implies $\mu(X_n) \geq \mu(X_0)$ whether or not X_n is positive. Indeed, subtracting the same constant from X_0 and X_n does not change $\mu(X_n) - \mu(X_0)$; so we may assume $\mathbf{E}[X_0] = 0$ and $X_n \geq 0$. Then $\mu(X_0) \leq 0$ while $\mu(X_n) = \mathbf{E}[X_n] - (1/2)\mathbf{E}\left[|X_n - \mathbf{E}[X_n]|\right] \geq \mathbf{E}[X_n] - (1/2)\mathbf{E}\left[|X_n|\right] - (1/2)\mathbf{E}[X_n] \geq 0$.

Here we will not require the independence of the X_i's; only that the centered random variables $X_i - \mathbf{E}[X_i] = X_i - e_i$ be differences of a weak martingale. A sequence Y_i is a *weak martingale* iff $\mathbf{E}[Y_i \mid Y_{i-1}] = Y_{i-1}$ for all i. This notion is of interest in convergence theory, since L_1-bounded weak martingales converge in probability, while the issue of a.e. convergence remains open (Nelson [1970]).

Before we prove the theorem, we insert two lemmas.

(3.3.3) Lemma. *Let X be an integrable random variable, and let $a < b$ be constants. Then*

$$(b-a)\mathbf{P}\{X \geq b\} \leq \mathbf{E}\left[(X-a)^+\right] - \mathbf{E}\left[(X-b)^+\right] \leq (b-a)\mathbf{P}\{X > a\}.$$

Proof. Note that

$$\mathbf{E}\left[(X-a)^+\right] - \mathbf{E}\left[(X-b)^+\right]$$
$$= \mathbf{E}\left[(X - a - X + b)\mathbf{1}_{\{X \geq b\}}\right] + \mathbf{E}\left[(X-a)\mathbf{1}_{\{a < X < b\}}\right].$$

But on the set $\{a < X < b\}$, we have

$$0 < X - a < b - a,$$

so that

$$\mathbf{E}\left[(b-a)\mathbf{1}_{\{X \geq b\}}\right] \leq \mathbf{E}\left[(X-a)^+\right] - \mathbf{E}\left[(X-b)^+\right]$$
$$\leq \mathbf{E}\left[(b-a)\mathbf{1}_{\{X \geq b\}}\right] + \mathbf{E}\left[(b-a)\mathbf{1}_{\{a < X < b\}}\right],$$

which proves the lemma. ∎

(3.3.4) Lemma. Let X be an integrable random variable, let d be a constant, and $\mathbf{E}[X] = e$. Then

$$\mathbf{E}\left[(X - e)^+\right] \leq \mathbf{E}\left[(X - d)^+\right]\left(1 + \mathbf{P}\{X < e\}\right) + d - e.$$

If $d > e$, the term $\mathbf{P}\{X < e\}$ may be omitted.

Proof. If $d > e$, then by the second inequality in (3.3.3),

$$\mathbf{E}\left[(X - e)^+\right] \leq \mathbf{E}\left[(X - d)^+\right] + d - e.$$

Next, if $d \leq e$, then by the first half of (3.3.3),

$$\begin{aligned}
\mathbf{E}\left[(X - e)^+\right] &\leq \mathbf{E}\left[(X - d)^+\right] + (d - e)\mathbf{P}\{X \geq e\} \\
&= \mathbf{E}\left[(X - d)^+\right] + (d - e) + (d - e)\mathbf{P}\{X < e\} \\
&\leq \mathbf{E}\left[(X - d)^+\right]\left(1 + \mathbf{P}\{X < e\}\right) + d - e
\end{aligned}$$

because $\mathbf{E}\left[(X - d)^+\right] \geq \mathbf{E}[X - d] = e - d$. ∎

(3.3.5) Theorem. Let $\mathbf{X} = (X_0, X_1, \cdots, X_n)$ be an integrable stochastic process. Let $\mathcal{G}_i$ be σ-algebras such that X_i is $\mathcal{G}_i$-measurable. Suppose that

(3.3.5a) $$\mathbf{E}[X_i \mid X_{i-1}] = e_i$$

is constant. Assume $\mu(X_0) \leq \mu(X_n)$ (or assume that all the X_i's are nonnegative). Then the expected gain P of the prophet and the expected gain G of the gambler are related by:

$$P \leq 3G.$$

The inequality is strict unless all the random variables are identically equal to the same constant.

Proof. Since decreasing the σ-algebras $\mathcal{G}_i$ decreases the expected gain G but leaves P unchanged, we may assume $\mathcal{G}_i = \sigma(X_i)$. Let $\mathbf{U} = (U_1, U_2, \cdots, U_n)$ be presently predictable and $0 \leq U_i \leq 1$. For the gambler, using (3.3.5a), we have

$$\mathbf{E}\left[U_i(X_i - X_{i-1})\right] = \mathbf{E}\left[\mathbf{E}^{\mathcal{G}_{i-1}}\left[U_i(X_i - X_{i-1})\right]\right] = \mathbf{E}\left[U_i(e_i - X_{i-1})\right].$$

This is maximal when

$$U_i = \mathbf{1}_{\{X_{i-1} < e_i\}}.$$

Hence $G = \sum_{i=1}^n \mathbf{E}\left[(e_i - X_{i-1})^+\right]$.

On the other hand,

$$P = \sum_{i=1}^{n} \mathbf{E}\left[(X_1 - X_{i-1})^+\right]$$

$$\leq \sum_{i=1}^{n} \left(\mathbf{E}\left[(X_i - e_i)^+\right] + \mathbf{E}\left[(e_i - X_{i-1})^+\right]\right)$$

$$= \sum_{i=1}^{n} \mathbf{E}\left[(X_i - e_i)^+\right] + G.$$

It is therefore sufficient to show

(3.3.5b)
$$\sum_{i=1}^{n} \mathbf{E}\left[(X_i - e_i)^+\right] \leq 2G.$$

The identity $\mathbf{E}\left[(e_i - X_{i-1})^+\right] = \mathbf{E}\left[(X_{i-1} - e_i)^+\right] + e_i - e_{i-1}$ yields

$$G = \sum_{i=1}^{n} \mathbf{E}\left[(X_{i-1} - e_i)^+\right] + e_n - e_0.$$

Applying (3.3.4) with $X = X_i$, $e = e_i$, $d = e_{i+1}$, summing from $i = 0$ to $n - 1$, and observing that $1 + \mathbf{P}\{X_i < e_i\} \leq 2$, we obtain

(3.3.5c)
$$\sum_{i=0}^{n-1} \mathbf{E}\left[(X_i - e_i)^+\right] \leq 2\sum_{i=1}^{n} \mathbf{E}\left[(X_{i-1} - e_i)^+\right] + e_n - e_0.$$

Thus we have

$$\sum_{i=1}^{n} \mathbf{E}\left[(X_i - e_i)^+\right] = \mathbf{E}\left[(X_n - e_n)^+\right] - \mathbf{E}\left[(X_0 - e_0)^+\right]$$

$$+ 2\left(\sum_{i=1}^{n} \mathbf{E}\left[(X_{i-1} - e_i)^+\right] + e_n - e_0\right) - (e_n - e_0)$$

$$= 2G - \mu(X_n) + \mu(X_0).$$

This proves (3.3.5b) because we assume $\mu(X_n) \geq \mu(X_0)$.

Finally, we show that $P < 3G$ unless all the X_i's are equal to e_0. So assume $P = 3G$. The computation just completed shows: (1) $\mu(X_n) = \mu(X_0)$; (2) equality in (3.3.5c); and (3) since the estimate $1 + \mathbf{P}\{X_i < e_i\} \leq 2$ is strict, we must have $\mathbf{E}\left[(X_{i-1} - e_i)^+\right] = 0$, or $X_{i-1} \leq e_i$. Thus $e_{i+1} = \mathbf{E}\left[X_{i+1}\right] \geq e_i$, and $\mathbf{E}\left[(X_i - e_i)^+\right] \leq \mathbf{E}\left[(e_{i+1} - e_i)^+\right] = e_{i+1} - e_i$, with strict inequality unless $e_{i+1} = e_i$ and $X_i = e_i$. But by equality in (3.3.5c), we must have equality in every term $\mathbf{E}\left[(X_i - e_i)^+\right] \leq e_{i+1} - e_i$, so $X_i = e_i = e_{i-1} = \cdots = e_1 = e_0$ for all i. ∎

There is a version depending on n, obtained by applying the stopping time result to transforms. Note that if the random variables $X_1, \cdots, X_n$ are nonnegative, then we can always add $X_0 = 0$ to the beginning without changing P or G.

(3.3.6) Proposition. *Let* $\mathbf{X} = (X_0, X_1, \cdots, X_n)$ *be nonnegative random variables such that* $\mathbf{E}\left[X_i \,|\, X_{i+1}\right] = e_i$ *is constant, and* $X_0 = 0$. *Then*

$$P \leq G + \sum_{i=1}^{n} \mathbf{E}\left[X_i\right].$$

Proof. Note $V(X_i) = e_i$ and $V(X_{i-1}, X_i) = \mathbf{E}\left[X_{i-1} \vee e_i\right]$, so by Theorem (3.3.2) we have

$$\mathbf{E}\left[X_{i-1} \vee X_i\right] \leq e_i + \mathbf{E}\left[X_{i-1} \vee e_i\right] = e_i + \mathbf{E}\left[(e_i - X_{i-1})^+\right] + e_{i-1}.$$

Also $\mathbf{E}\left[X_{i-1} \vee X_i\right] = \mathbf{E}\left[(X_i - X_{i-1})^+\right] + e_{i-1}$, so

$$\mathbf{E}\left[(X_i - X_{i-1})^+\right] + e_{i-1} \leq e_i + \mathbf{E}\left[(e_i - X_{i-1})^+\right] + e_{i-1},$$

$$\sum_{i=1}^{n} \mathbf{E}\left[(X_i - X_{i-1})^+\right] \leq \sum_{i=1}^{n} \mathbf{E}\left[(e_i - X_{i-1})^+\right] + \sum_{i=1}^{n} e_i.$$

Therefore $P \leq G + \sum_{i=1}^{n} e_i$.   ■

The case of signed U

We now allow U_i with $-c \leq U_i \leq c$. For consideration of the ratio P/G, again we may assume $c = 1$. Our notation is the same as before: The expected gain of the prophet is P_s and the expected gain of the gambler is G_s. For the computation, we transform $\mathbf{X}$ by processes $\mathbf{U}$ presently predictable with respect to σ-algebras $(\mathcal{G}_i)$ such that X_i is $\mathcal{G}_i$-measurable.

(3.3.7) Theorem. *Suppose* $\mathbf{E}\left[X_i \,|\, X_{i-1}\right] = e_i$ *is constant. Assume that* $\mu(X_0) \leq \mu(X_n)$ *and* $e_0 \geq e_n$ *(both conditions hold if* $X_n = e_0$*). Then*

$$P_s \leq 3G_s.$$

If $e_n = e_0$, *then* $P_s = 2P$ *and* $G_s = 2G$.

Proof. At stage i, the optimal gambler receives $X_i - X_{i-1}$ on the set $\{X_{i-1} < e_i\}$ and $X_{i-1} - X_i$ on the set $\{X_{i-1} \geq e_i\}$. Hence

$$G_s = \sum_{i=1}^{n} \left(\mathbf{E}\left[(e_i - X_{i-1})^+\right] + \mathbf{E}\left[(e_i - X_{i-1})^-\right] \right).$$

On the other hand, the difference of the summands is

$$\mathbf{E}\left[(e_i - X_{i-1})^+\right] - \mathbf{E}\left[(e_i - X_{i-1})^-\right] = e_i - e_{i-1},$$

hence

$$G_s = 2\sum_{i=1}^{n} \mathbf{E}\left[(e_i - X_{i-1})^+\right] - \sum_{i=1}^{n} \mathbf{E}\left[(e_i - e_{i-1})^-\right] = 2G - e_n + e_0.$$

Similarly, $P_s = 2P - e_n + e_0$.

Now $P \leq 3G$ and $e_0 \geq e_n$ imply $P_s \leq 3G_s$. If $e_n = e_0$, then $P_s = 2P$ and $G_s = 2G$.   ∎

Complements

(3.3.8) (Optimality of backward induction.) In Lemma (3.3.1), we have $\mathbf{E}[X_{\sigma_1}] = V$ and $\mathbf{E}[X_{\sigma_2}] = {}_1V$. It follows inductively (backward on i) that $\mathbf{E}[X_{\sigma_i}]$ is the value of the process $(X_i, \cdots, X_n)$. (See Chow, Robbins & Siegmund [1971], p. 50.)

(3.3.9) (Constant 2 is best in (3.3.2).) Let $n = 2$, $X_1 = 1$ and $X_2 = M > 1$ with probability $1/M$ and $X_2 = 0$ with probability $1 - 1/M$. The expected fortune of the prophet is

$$M \cdot \frac{1}{M} + 1 \cdot \left(1 - \frac{1}{M}\right) = 2 - \frac{1}{M}.$$

The gambler expects to receive 1 regardless of when he stops. Then $P/V = 2 - 1/M$, so 2 is the best constant since M is arbitrary.

(3.3.10) (Bounded case.) In Theorem (3.3.2), if the random variables take values in $[0, 1]$, then the prophet inequality $P \leq 2V$ can be improved to $P \leq 2V - V^2$ (Hill [1983]).

(3.3.11) (Constant 3 is best in (3.3.4).) Given $\varepsilon > 0$, there exist random variables $X_1', X_2', \cdots, X_n'$ with $P' \geq (3 - \varepsilon)G'$, where P' and G' are the expected gains of the prophet and of the gambler for the primed process. The sequence (X_i') is obtained from a sequence (X_i) by inserting many copies of the constant random variable $\mathbf{E}[X_i]$ between X_i and X_{i+1}. The independent positive sequence X_i itself is chosen so that $\sum_{i=1}^n \mathbf{E}[(X_i - e_i)^+] \geq (2 - \varepsilon)G$ and $X_0 = X_n$ (Krengel & Sucheston [1987], p. 1598).

(3.3.12) (An economic interpretation of (3.3.4).) Assume that (Y_n) and (Z_n) are two integrable processes with arbitrary distributions, but independent of each other. Players observe the two processes alternately: say $X_n = Y_n$ for n even and $X_n = Z_n$ for n odd. Then the adjacent X_n's are independent.

To fix the idea, suppose that a conglomerate is sufficiently diversified so that the stock prices of two of its firms, F_1 and F_2, are independent. Let Y_n be the value of c_1 shares of F_1; let Z_n be the value of c_2 shares of F_2. Two of the conglomerate's executives are allowed to trade each year kc_1 shares of F_1 for kc_2 shares of F_2 or vice versa. Each executive chooses each year his own value of k, but the k's are bounded by a fixed constant C. Assume that the junior executive bases his decisions on the present (equivalently, present and past): he is a gambler. The senior executive knows the future: he is a prophet. In practice, there is a device equivalent to the gift of prophecy put at the disposal of senior executives: the right to exercise options, i.e., trading in stock some time in the future at prices prevailing earlier. The theorem implies that the expected gain of the senior executive is less than 3 times that of his junior colleague (Krengel & Sucheston, unpublished, 1987).

Inequalities

Remarks

Prophet inequalities were introduced by Krengel & Sucheston [1978], where it was first shown that there are constants C such that $P/V \leq C$. For stopped processes, Krengel and Sucheston obtained $C = 4$; the same paper contains a proof by D. Garling that $C = 2$ can be taken. The paper also studies the analogous but more difficult problem when the X_i's are averages of independent random variables, obtaining $C = 5.46$. The problem whether the best constant in this case was 2 had remained open until Hill [1986] resolved it in the affirmative.

The proof of Theorem (3.3.2) given here is based on arguments communicated to us by D. Gilat and R. Wittmann. Theorems (3.3.4) and (3.3.5) are from Krengel & Sucheston [1987]. Proposition (3.3.6) was observed by Sucheston & Yan [in press].

4

Directed index set

In this Chapter we present the theory of martingales and amarts indexed by directed sets. After Dieudonné showed that martingales indexed by directed sets in general need not converge essentially, Krickeberg—in a series of papers—proved essential convergence under covering conditions called "Vitali conditions." This theory is presented in an expository article by Krickeberg & Pauc [1963] and in a book by Hayes & Pauc [1970].

Here we offer a new approach and describe the subsequent progress. The condition (V), introduced by Krickeberg to prove the essential convergence of L_1-bounded martingales, was shown not to be necessary. Similarly the condition (V^O), introduced to prove convergence of L_1-bounded submartingales, is now also known not to be necessary. The condition (V$_\Phi$), which Krickeberg showed to be sufficient for the convergence of martingales bounded in the Orlicz space L_Ψ, is also necessary for this purpose if the Orlicz function Φ satisfies the (Δ_2) condition.

In each instance, the convergence of appropriate classes of amarts *exactly* characterizes the corresponding Vitali condition. This is of particular interest for (V) and (V^O) since there is no corresponding characterization in the classical theory. In general, to nearly every Vitali type of covering condition there corresponds the convergence of an appropriate class of "amarts." The understanding of this fact was helped by new formulations of Vitali conditions in terms of stopping times. Informally, a Vitali condition says that the essential upper limit of a 0-1 valued process (1_{A_t}) can be approximated by the process stopped by appropriate stopping times. This has a clear intuitive meaning even in the case of multivalued stopping times because, as a condition for convergence, the overlap of the values is small, in a precise sense.

The application of martingale theory to derivation theory has been long known, but also amarts come into their own. A derivative of a superadditive set-function is both a supermartingale and an amart (4.2.18). In the classical setting of the derivation theory the Vitali condition (V) holds (4.2.8), but (V^O) does not. So supermartingales need not converge essentially, and the amart theory is needed to prove convergence. Similarly, derivatives of functions of measures are amarts, and the Riesz decomposition sheds some light on their behavior (4.2.19).

Since the condition (V) is not necessary for essential convergence of martingales, is there a covering condition both sufficient and necessary? The answer is yes: There is a condition (C) for this purpose. Condition (C) is sufficient for convergence of L_1-bounded martingales, and also necessary if the index set has a countable cofinal subset. The question of necessity remains open for the general index set. It is also not known whether (V_Φ) is necessary for convergence of L_Ψ-bounded martingales if (Δ_2) fails. Another open question is the existence of a covering condition both sufficient and necessary for convergence of L_1-bounded submartingales.

The conditions (V_Φ) are modeled on similar covering conditions in the theory of derivations (see Chapter 7). Condition (C) has moved in the opposite direction—it was first introduced in the study of convergence of processes indexed by directed sets, then translated into a condition for derivation theory. As with Doob's theory of martingales indexed by $\mathbb{N}$, the main interest and the main difficulties occur in the L_1-bounded case. The greater part of the Chapter is devoted to that case.

4.1. Essential and stochastic convergence

There are two modes of convergence that will be used in this chapter: essential convergence and stochastic convergence.

When the index set is countable, essential convergence coincides with almost everywhere convergence, but when the index set is uncountable, essential convergence is still reasonable, although almost everywhere convergence may not be. Essential convergence is called order convergence in some of the literature.

Stochastic convergence is also known as convergence in probability (or in measure). We will consider the corresponding stochastic upper limit and stochastic lower limit.

Essential convergence

Let $(\Omega, \mathcal{F}, \mathbf{P})$ be a probability space. Let S be the set of all extended-real-valued random variables. Let $P \subseteq S$ be some set of random variables. The random variable $Z \in S$ is called the *essential supremum* of the set P iff

(1) $Z \geq X$ a.s. for all $X \in P$;
(2) if $Y \in S$ and $Y \geq X$ a.s. for all $X \in P$, then $Y \geq Z$ a.s.

(Thus Z is the least upper bound of P in the partial order obtained by identifying functions that agree almost everywhere.)

(4.1.1) Proposition. *(a) Every subset P of S has an essential supremum, $\operatorname{ess\,sup} P$, unique up to null sets. (b) There exists a sequence X_n in P such that $\sup_n X_n = \operatorname{ess\,sup} P$ a.s. (c) If the family P is directed, the sequence X_n may be chosen to be a.s. increasing.*

Proof. Assume first that all $X \in P$ satisfy $0 \leq X \leq 1$. Let P_1 be the set of all countable suprema $\sup_k X_k$ for $X_k \in P$. Write

$$a = \sup \left\{ \mathbf{E}\,[Y] : Y \in P_1 \right\}.$$

(Since $0 \leq Y \leq 1$ for all $Y \in P_1$, we have $0 \leq a \leq 1$.) For each n, choose $Y_n \in P_1$ with $\mathbf{E}[Y_n] \geq a - 2^{-n}$, and define $Z(\omega) = \sup_n Y_n(\omega)$. Then Z is itself a countable supremum of elements of P, so $Z \in P_1$. We have $\mathbf{E}[Z] = a$.

We claim that Z is an essential supremum for the set P. If $X \in P$, then $X \vee Z \in P_1$ and

$$a \geq \mathbf{E}[X \vee Z] \geq \mathbf{E}[Z] = a,$$

so $X \vee Z = Z$ a.s., hence $X \leq Z$ a.s. On the other hand, if Y is a random variable with $Y \geq X$ a.s. for all $X \in P$, then $Y \geq Z$ a.s., since Z is a countable supremum of elements of P.

For the general case, choose a continuous strictly increasing bijection between $[-\infty, \infty]$ and $[0, 1]$, and apply the preceding case.

For the uniqueness, suppose that both Z_1 and Z_2 are essential suprema for the set P. Now $Z_1 \geq X$ a.s. for all $X \in P$, so (since Z_2 is essential supremum), $Z_1 \geq Z_2$ a.s. Similarly, $Z_2 \geq Z_1$ a.s. Therefore $Z_1 = Z_2$ a.s. This completes the proof of (a) and (b).

For (c), note that if the sequence X_n is chosen for (b), we may choose $X'_n \in P$ so that $X'_n \geq X_n$ and $X'_n \geq X'_k$ for $1 \leq k < n$ since P is directed. Then X'_n is increasing and $\sup X'_n = \operatorname{ess\,sup} P$. ∎

We will write $\operatorname{ess\,sup} P$ for the essential supremum of the set P. The *essential infimum* is defined analogously (the greatest lower bound in the partial order obtained by identifying functions that agree almost everywhere), or equivalently in terms of the essential supremum: $\operatorname{ess\,inf} P = -(\operatorname{ess\,sup}\{-X : X \in P\})$.

Note: it is easily verified that if P is countable, then $Z = \operatorname{ess\,sup} P$ is the pointwise supremum,

$$Z(\omega) = \sup\{X(\omega) : X \in P\}.$$

This need not be true for uncountable P. For example, if $(\Omega, \mathcal{F}, \mathbf{P})$ is $[0, 1]$ with Lebesgue measure, and P is the set

$$P = \{\mathbf{1}_{\{x\}} : x \in [0, 1]\},$$

then $\operatorname{ess\,sup} P$ is 0 a.s., but $\sup\{X(\omega) : X \in P\}$ is identically 1. If P is replaced by a nonmeasurable subset, then $\sup\{X(\omega) : X \in P\}$ is not measurable. The essential supremum is better behaved.

(4.1.2) Proposition. *Suppose the set P is a nonempty family of indicator functions of measurable sets. Then $\operatorname{ess\,sup} P$ is also an indicator of a measurable set.*

Proof. Let $Z = \operatorname{ess\,sup} P$. Since P is nonempty, there is $X \in P$, and thus $Z \geq X \geq 0$ a.s. For all $X \in P$, we have $X \leq 1$, so $Z \leq 1$. Let $A = \{Z \geq 1\}$. We claim $Z = \mathbf{1}_A$ a.s. If $X \in P$, then $X \leq Z$ a.s. But X is an indicator function, so it is 0 whenever it is less than 1, and we have

$X \leq 1_A$ a.s. Therefore, by the definition of essential supremum, $Z \leq 1_A$. Clearly $Z \geq 1_A$ a.s., so $Z = 1_A$ a.s. ∎

If $\mathcal{C} \subseteq \mathcal{F}$ is a family of measurable sets, then we will write $B = \operatorname{ess\,sup} \mathcal{C}$ for the set of the proposition; that is,

$$1_B = \operatorname*{ess\,sup}_{A \in \mathcal{C}} 1_A.$$

(By convention, if $\mathcal{C} = \emptyset$, then $\operatorname{ess\,sup} \mathcal{C} = \emptyset$.)

(4.1.3) Definition. Let J be a directed set, and let $(X_t)_{t \in J}$ be a net of random variables. The *essential upper limit* of (X_t) is defined by

$$\operatorname*{e\,lim\,sup}_{t \in J} X_t = \operatorname*{ess\,inf}_{s \in J} \operatorname*{ess\,sup}_{\substack{t \in J \\ t \geq s}} X_t.$$

We will often write X^* for the $\operatorname{e\,lim\,sup}$ of a process (X_t). The *essential lower limit* $\operatorname{e\,lim\,inf} X_t$ is defined analogously:

$$\operatorname*{e\,lim\,inf}_{t \in J} X_t = \operatorname*{ess\,sup}_{s \in J} \operatorname*{ess\,inf}_{\substack{t \in J \\ t \geq s}} X_t,$$

or $\operatorname{e\,lim\,inf} X_t = -\operatorname{e\,lim\,sup}(-X_t)$.

(4.1.4) Proposition. *If (X_t) is a net of random variables, then we have* $\operatorname{e\,lim\,inf} X_t \leq \operatorname{e\,lim\,sup} X_t$ *a.s.*

Proof. For each $s \in J$, $\operatorname{ess\,inf}_{t \geq s} X_t \leq X_s$. Therefore, for each $u \in J$,

$$\operatorname*{ess\,sup}_{s \geq u} \operatorname*{ess\,inf}_{t \geq s} X_t \leq \operatorname*{ess\,sup}_{s \geq u} X_s,$$

hence $\operatorname{e\,lim\,inf}_t X_t \leq \operatorname{ess\,sup}_{s \geq u} X_s$. Therefore,

$$\operatorname*{e\,lim\,inf}_t X_t \leq \operatorname*{ess\,inf}_u \operatorname*{ess\,sup}_{s \geq u} X_s = \operatorname*{e\,lim\,sup}_s X_s. \qquad ■$$

Let (X_t) be a net of random variables. It is said to *converge essentially* to a random variable X_∞ iff

$$X_\infty = \operatorname*{e\,lim\,sup}_{t \in J} X_t = \operatorname*{e\,lim\,inf}_{t \in J} X_t \quad \text{a.s.}$$

Then we write $X_\infty = \operatorname{e\,lim}_{t \in J} X_t$.

We leave to the reader the verification that when J is countable, we have

$$(\operatorname*{e\,lim\,sup}_{t \in J} X_t)(\omega) = \limsup_{t \in J}(X_t(\omega)),$$

$$(\operatorname*{e\,lim\,inf}_{t \in J} X_t)(\omega) = \liminf_{t \in J}(X_t(\omega)),$$

$$(\operatorname*{e\,lim}_{t \in J} X_t)(\omega) = \lim_{t \in J}(X_t(\omega))$$

for almost all ω.

Stochastic convergence

Stochastic convergence (convergence in probability, convergence in measure) will be useful not only in situations when essential convergence fails, but also as a tool in proofs of essential convergence. We begin with the stochastic upper and lower limit. (There does not seem to be a reasonable notion of "stochastic supremum," however.)

Let $(\Omega, \mathcal{F}, \mathbf{P})$ be a probability space.

(4.1.5) Definition. Let $(X_t)_{t \in J}$ be a net of extended-real-valued random variables. A random variable Y such that $\lim_t \mathbf{P}\{X_t > Y\} = 0$ is called *asymptotically greater than X_t in probability*. (The words "in probability" will sometimes be omitted.) The *stochastic upper limit* of (X_t), written $\text{s lim sup}_t X_t$, is the essential infimum of the set of all extended random variables which are asymptotically greater than X_t in probability. The *stochastic lower limit*, written $\text{s lim inf } X_t$, is $\text{s lim inf } X_t = -\text{s lim sup}(-X_t)$. If $\text{s lim sup } X_t = \text{s lim inf } X_t = X_\infty$, then X_∞ is called the *stochastic limit* of (X_t), which is then said to *converge stochastically* (or to *converge in probability*) to X_∞. We write $\text{s lim}_{t \in J} X_t = X_\infty$.

There are relations between stochastic convergence and essential convergence.

(4.1.6) Proposition. *Let $(X_t)_{t \in J}$ be a net of extended-real-valued random variables. Then*

(4.1.6a) $\text{e lim inf } X_t \leq \text{s lim inf } X_t \leq \text{s lim sup } X_t \leq \text{e lim sup } X_t.$

Therefore, if (X_t) converges essentially, then it converges stochastically to the same limit.

Proof. If $s_0 \in J$, then $\text{ess sup}_{t \geq s_0} X_t \geq X_s$ for all $s \geq s_0$. Hence

$$\lim_s \mathbf{P}\left\{\text{ess sup}_{t \geq s_0} X_t < X_s\right\} = 0.$$

That is, $\text{ess sup}_{t \geq s_0} X_t$ is asymptotically greater than the net X_s. Thus

$$\text{ess inf}_{s_0} \text{ess sup}_{t \geq s_0} X_t \geq \text{s lim sup}_s X_s;$$

this holds for all s_0, so

$$\text{e lim sup}_{t \in J} X_t \geq \text{s lim sup}_{t \in J} X_t.$$

Similarly,

$$\text{e lim inf}_{t \in J} X_t \leq \text{s lim inf}_{t \in J} X_t.$$

∎

(4.1.7) There is a natural definition intermediate between essential convergence and stochastic convergence. When three of the four random variables in (4.1.7a) coincide, we say that (X_t) *demiconverges* to their common value. More specifically, we may say that (X_t) *upper demiconverges* to X_∞ iff

$$\operatorname{s\,lim}_t X_t = \operatorname{e\,lim\,sup}_t X_t = X_\infty$$

and that (X_t) *lower demiconverges* to X_∞ iff

$$\operatorname{s\,lim}_t X_t = \operatorname{e\,lim\,inf}_t X_t = X_\infty.$$

Thus demiconvergence implies stochastic convergence. (For some results on demiconvergence, see (7.3.24), (9.1.2), (9.4.4), (9.4.13).)

Let (A_t) be a net of events. Then $\operatorname{s\,lim\,sup} \mathbf{1}_{A_t}$ is an indicator function. Indeed, if Y is asymptotically greater than $(\mathbf{1}_{A_t})$, then

$$0 = \lim_t \mathbf{P}\left\{\mathbf{1}_{A_t} > Y\right\} \geq \mathbf{P}\{0 > Y\},$$

so $Y \geq 0$ a.s. Thus $\left\{\mathbf{1}_{A_t} > Y\right\} \geq \left\{\mathbf{1}_{A_t} > \mathbf{1}_{\{Y \geq 1\}}\right\}$; that is, $\mathbf{1}_{\{Y \geq 1\}}$ is also asymptotically greater than $(\mathbf{1}_{A_t})$, and $\mathbf{1}_{\{Y \geq 1\}} \leq Y$. Thus $\operatorname{s\,lim\,sup} \mathbf{1}_{A_t}$ is the essential infimum of a family of indicator functions, and therefore by (4.1.2) is an indicator function itself:

$$\mathbf{1}_B = \operatorname{s\,lim\,sup} \mathbf{1}_{A_t}.$$

We will call B the *stochastic upper limit* of (A_t), and write $B = \widehat{A} = \operatorname{s\,lim\,sup} A_t$. Translating the definition, we get:

(4.1.7a) $\widehat{A}$ is the smallest set C such that $\lim_t \mathbf{P}(A_t \setminus C) = 0$.

Indeed, $A_t \setminus C = \left\{\mathbf{1}_{A_t} > \mathbf{1}_C\right\}$, hence $\lim_t \mathbf{P}(A_t \setminus C) = 0$ if and only if $\mathbf{1}_C$ is asymptotically greater than $\mathbf{1}_{A_t}$ in probability.

It can be easily verified that if X_∞ is finite a.s., then $\operatorname{s\,lim}_{t \in J} X_t = X_\infty$ is equivalent to the usual definition:

(4.1.7b) $\displaystyle\lim_{t \in J} \mathbf{P}\{|X_\infty - X_t| > \varepsilon\} = 0$

for all $\varepsilon > 0$.

For sets, $B = \operatorname{s\,lim} A_t$ holds if and only if

(4.1.7c) $$\mathbf{P}(A_t \triangle B) \to 0.$$

(4.1.8) Lemma. *Suppose (A_t) is an adapted sequence of sets. Then $\widehat{A} = \operatorname{s\,lim\,sup} A_t$ is the largest set A such that*

(4.1.8a) $$\limsup_t \mathbf{P}(A_t \cap B) > 0$$

for all subsets $B \subseteq A$ of positive probability.

Proof. Let A be a set such that (4.1.8a) holds for every $B \subseteq A$ of positive probability. Then

$$\mathbf{P}\big(A_t \cap (A \setminus \widehat{A})\big) \le \mathbf{P}(A_t \setminus \widehat{A}) \to 0,$$

which implies $\mathbf{P}(A \setminus \widehat{A}) = 0$ and therefore $\widehat{A} \subseteq A$ a.s.

Conversely, let $B \subseteq A$ with $\mathbf{P}(B) > 0$ but suppose that (4.1.8a) fails. We have

$$\mathbf{P}\big(A_t \setminus (\widehat{A} \setminus B)\big) \le \mathbf{P}(A_t \setminus \widehat{A}) + \mathbf{P}(A_t \cap B).$$

Applying (4.1.7a) with $C = \widehat{A}$, we obtain $\mathbf{P}(A_t \setminus \widehat{A}) = 0$. Therefore $\lim \mathbf{P}\big(A_t \setminus (\widehat{A} \setminus B)\big) = 0$, hence $\mathbf{P}(B) = 0$, because $\widehat{A}$ is the smallest set C as in (4.1.7a). ∎

(4.1.9) Lemma. *Let $\widehat{A} = \operatorname{s\,lim\,sup} A_t$, and let (s_n) be a sequence of indices. Then there exists an increasing sequence (t_n) of indices such that $s_n \le t_n$ and $\widehat{A} \subseteq \bigcup_{n=1}^{\infty} A_{t_n}$.*

Proof. On measurable subsets of $\widehat{A}$ define a function γ by

$$\gamma(B) = \limsup_t \mathbf{P}(A_t \cap B).$$

Set $B_1 = \widehat{A}$ and choose $t_1 \ge s_1$ such that $\mathbf{P}\big(A_{t_1} \cap B_1\big) \ge \gamma(B_1)/2$. Then set $B_2 = \widehat{A} \setminus A_{t_1}$ and choose $t_2 \ge t_1$, $t_2 \ge s_2$, such that $\mathbf{P}\big(A_{t_2} \cap B_2\big) \ge \gamma(B_2)/2$. Continue the definition of (t_n) and (B_n) recursively: given $t_1, \cdots, t_n$ and $B_1, \cdots, B_n$, set

$$B_{n+1} = \widehat{A} \setminus \bigcup_{i=1}^{n} A_{t_i}$$

and choose $t_{n+1} \ge t_n$, $t_{n+1} \ge s_{n+1}$, with $\mathbf{P}\big(A_{t_{n+1}} \cap B_{n+1}\big) \ge \gamma(B_{n+1})/2$. Since

$$\bigcap_{n=1}^{\infty} B_n = \widehat{A} \setminus \bigcup_{n=1}^{\infty} A_{t_n},$$

it suffices to show $\mathbf{P}(\bigcap_n B_n) = 0$. The sets $B_n \cap A_{t_n}$ are pairwise disjoint in a finite measure space, and $\gamma(B_n)$ decreases, hence

$$\lim_n \gamma(B_n) = \lim_n \limsup_t \mathbf{P}(A_t \cap B_n) \le 2 \limsup_n \mathbf{P}\big(A_{t_n} \cap B_n\big) = 0.$$

It follows that $\gamma(\bigcap B_n) = 0$, hence by Lemma (4.1.8), $\mathbf{P}(\bigcap B_n) = 0$, so $\widehat{A} \subseteq \bigcup_n A_{t_n}$. ∎

Proofs of convergence in Chapter 1 (for example (1.2.4)) are based on approximation of the a.s. lim sup by the stopped process. We will see below that also on directed sets the stochastic upper limit s lim sup A_t can be approximated by a stopped set $A(\tau)$, with $\tau \in \Sigma$. This was implicitly used in the proof of (1.3.9). (For the *essential* lim sup, e lim sup A_t, such an approximation need not hold; it must be postulated; this is the meaning of the Vitali conditions below.)

In fact, it suffices to stop with ordered stopping times. A *simple ordered stopping time* is a stopping time $\tau \colon \Omega \to J$ such that its set of values is totally ordered (see Section 1.3). We write Σ^O for the set of ordered stopping times.

(4.1.10) Proposition. *Let (A_t) be an adapted family of sets, and write $\widehat{A} = \text{s lim sup } A_t$. For every $\varepsilon > 0$ and every $t_0 \in J$, there is a $\tau \in \Sigma^O$ such that $\tau \geq t_0$ and $\mathbf{P}(\widehat{A} \triangle A(\tau)) < \varepsilon$.*

Proof. Given ε and t_0, choose a sequence $s_n \in J$ such that $s_n \geq t_0$ and $\mathbf{P}(A_t \setminus \widehat{A}) < \varepsilon 2^{-n-1}$ for all $t \geq s_n$. Let (t_n) be the sequence obtained by application of Lemma (4.1.9). Choose k such that $\mathbf{P}(\widehat{A} \setminus \bigcup_{i=1}^{k-1} A_i) \leq \varepsilon/2$. Define $\tau = t_n$ on $A_{t_n} \setminus \bigcup_{i=1}^{n-1} A_{t_i}$ for $n \leq k$ and $\tau = t_{k+1}$ on the rest of Ω. Then $\tau \in \Sigma^O$ and $\mathbf{P}(\widehat{A} \triangle A(\tau)) < \varepsilon$. ∎

The corresponding result for processes is proved next. A process $(X_t)_{t \in J}$ is called *asymptotically uniformly absolutely continuous* iff, for every $\varepsilon > 0$, there exists $s \in J$ and $\delta > 0$ such that if $\mathbf{P}(B) < \delta$ and $t \geq s$, then $\mathbf{E}\left[|X_t \mathbf{1}_B|\right] \leq \varepsilon$.

(4.1.11) Stochastic maximal inequality. *Let $(X_t)_{t \in J}$ be a positive adapted process, let $\lambda > 0$, and define $A = \{\text{s lim sup } X_t \geq \lambda\}$. (a) Then*

$$\mathbf{P}(A) \leq \frac{1}{\lambda} \limsup_{\tau \in \Sigma^O} \mathbf{E}\left[X_\tau\right].$$

(b) Suppose also that $(X_\tau)_{\tau \in \Sigma^O}$ is asymptotically uniformly absolutely continuous. Then

$$\mathbf{P}(A) \leq \frac{1}{\lambda} \limsup_{\tau \in \Sigma^O} \mathbf{E}\left[X_\tau \mathbf{1}_A\right].$$

Proof. (a) Let (X_t) be a positive process and $\lambda > 0$. Fix a number α with $0 < \alpha < \lambda$. Set $B_t = \{X_t > \lambda - \alpha\}$. Then

$$\{\text{s lim sup } X_t \geq \lambda\} \subseteq \text{s lim sup } B_t = \widehat{B}.$$

Indeed, suppose Y is a random variable asymptotically greater than (X_t), and D is an event such that $\mathbf{1}_D$ is asymptotically greater than $(\mathbf{1}_{B_t})$. If

$$Y' = \begin{cases} \lambda - \alpha & \text{on } \Omega \setminus D \\ Y & \text{on } D, \end{cases}$$

then we claim Y' is asymptotically greater than (X_t):

$$\mathbf{P}\{X_t > Y'\} \leq \mathbf{P}\{X_t > Y\} + \mathbf{P}\{B_t \setminus D\} \to 0.$$

Therefore $s \lim \sup X_t \leq Y'$. Thus

$$\{s \lim \sup X_t \geq \lambda\} \subseteq \{Y' \geq \lambda\} \subseteq D.$$

This is true for all events D such that $\mathbf{1}_D$ is asymptotically greater than $(\mathbf{1}_{B_t})$, so

$$\{s \lim \sup X_t \geq \lambda\} \subseteq \widehat{B}.$$

Now given $\varepsilon > 0$, choose $s \in J$ and $\tau \in \Sigma^{\mathrm{O}}$ such that $\tau \geq s$ and $\mathbf{P}(\widehat{B} \triangle B(\tau)) \leq \varepsilon$. This is possible by (4.1.10). Then we have

$$\mathbf{P}(\widehat{B}) \leq \mathbf{P}(B(\tau)) + \varepsilon \leq \frac{1}{\lambda - \alpha} \mathbf{E}\left[X_\tau \mathbf{1}_{B(\tau)}\right] + \varepsilon.$$

The result follows on letting $s \to \infty$, $\alpha \to 0$ and $\varepsilon \to 0$.

(b) Given $\delta > 0$, choose $s \in J$ and $\varepsilon > 0$ with $\varepsilon < \delta$ such that $\mathbf{P}(B) < 2\varepsilon$ and $\tau \geq s$ implies $\mathbf{E}\left[X_\tau \mathbf{1}_B\right] \leq \delta\lambda$. Then let $\alpha < \lambda/2$ be so small that

$$(4.1.11a) \qquad \mathbf{P}\left(s \lim \sup\{X_t > \lambda - \alpha\} \setminus A\right) < \varepsilon.$$

Write $B_t = \{X_t > \lambda - \alpha\}$ and $\widehat{B} = s \lim \sup B_t$. Now choose $\tau \in \Sigma^{\mathrm{O}}$, $\tau \geq s$, such that $\mathbf{P}(\widehat{B} \triangle B(\tau)) \leq \varepsilon$. We have $A \subseteq \widehat{B}$, so $\mathbf{P}(A \setminus B(\tau)) \leq \mathbf{P}(\widehat{B} \setminus B(\tau))$, so by (4.1.11a) we have

$$\mathbf{E}\left[X_\tau \mathbf{1}_{B(\tau)}\right] \leq \mathbf{E}\left[X_\tau \mathbf{1}_A\right] + \delta\lambda.$$

Then:

$$\mathbf{P}(A) \leq \mathbf{P}(\widehat{B}) \leq \mathbf{P}(B(\tau)) + \varepsilon$$
$$\leq \frac{1}{\lambda - \alpha} \mathbf{E}\left[X_\tau \mathbf{1}_A\right] + \frac{\delta\lambda}{\lambda - \alpha} + \varepsilon$$
$$\leq \frac{1}{\lambda - \alpha} \mathbf{E}\left[X_\tau \mathbf{1}_A\right] + 3\delta.$$

The maximal inequality follows on letting $s \to \infty$, $\alpha \to 0$ and $\delta \to 0$. ■

In the previous result, note that if (X_t) is a positive submartingale or positive supermartingale, then the right hand side of the inequality in (a) simplifies to $(1/\lambda) \lim_{t \in J} \mathbf{E}\left[X_t\right]$.

We will now use the stochastic maximal inequality to show that the (essential) maximal inequality is equivalent with convergence of martingales. We state it here in a somewhat abstract form in part (ii); the best known application is $M(\lambda, \varepsilon) = \varepsilon/\lambda$, as in part (i).

(4.1.12) Proposition. *Let $(\mathcal{F}_t)_{t \in J}$ be a stochastic basis. The following are equivalent:*

(i) *For every integrable real random variable X, for every $\lambda > 0$, we have*

$$\mathbf{P}\left\{\operatorname{e\,lim\,sup}_t \left|\mathbf{E}^{\mathcal{F}_t}\left[X\right]\right| \geq \lambda\right\} \leq \frac{1}{\lambda}\mathbf{E}\left[|X|\right].$$

(ii) *There exists $M: \mathbb{R}^+ \times \mathbb{R}^+ \to \mathbb{R}^+$ with (a) $\lim_{\varepsilon \to 0} M(\lambda, \varepsilon) = 0$ for every $\lambda > 0$, and (b) for $\varepsilon, \lambda > 0$ and every positive integrable random variable X with $\mathbf{E}\left[X\right] \leq \varepsilon$, we have*

$$\mathbf{P}\{\operatorname{e\,lim\,sup} \mathbf{E}^{\mathcal{F}_t}\left[X\right] \geq \lambda\} \leq M(\lambda, \varepsilon).$$

(iii) *For every positive integrable random variable X, the martingale $X_t = \mathbf{E}^{\mathcal{F}_t}\left[X\right]$ converges essentially.*

(iv) *For every integrable real random variable X, for every $\lambda > 0$, letting $A = \left\{\operatorname{e\,lim\,sup}\left|\mathbf{E}^{\mathcal{F}_t}\left[X\right]\right| \geq \lambda\right\}$, we have*

$$\mathbf{P}(A) \leq \frac{1}{\lambda}\mathbf{E}\left[|X| \mathbf{1}_A\right].$$

Proof. (i) $\implies$ (ii). Set $M(\lambda, \varepsilon) = \varepsilon/\lambda$.

(ii) $\implies$ (iii). Suppose the function M as in (ii) exists. Let $X \in L_1$. Write $\mathcal{F}_\infty = \sigma(\bigcup \mathcal{F}_t)$. We claim that the martingale $X_t = \mathbf{E}^{\mathcal{F}_t}\left[X\right]$ converges essentially to $\mathbf{E}^{\mathcal{F}_\infty}\left[X\right]$. Since X may be replaced by $\mathbf{E}^{\mathcal{F}_\infty}\left[X\right]$, we may assume that X is $\mathcal{F}_\infty$-measurable. In that case, we must show that $\operatorname{e\,lim} X_t = X$.

Fix $\alpha > 0$ and $\lambda > 0$. Choose $\varepsilon > 0$ so small that $\varepsilon < \alpha\lambda/2$ and $M(\lambda/2, \varepsilon) < \alpha$. Since X is $\mathcal{F}_\infty$-measurable, there must exist $t_0 \in J$ and $Y \in L_1(\mathcal{F}_{t_0})$ such that $\mathbf{E}\left[|X - Y|\right] < \varepsilon$. (See (1.1.12).) Then by the triangle inequality, for $t \geq t_0$,

$$\mathbf{P}\left\{\operatorname{e\,lim\,sup}|X_t - X| \geq \lambda\right\} \leq \mathbf{P}\left\{\operatorname{e\,lim\,sup} \mathbf{E}^{\mathcal{F}_t}\left[|X - Y|\right] \geq \frac{\lambda}{2}\right\}$$
$$+ \mathbf{P}\left\{\operatorname{e\,lim\,sup} \mathbf{E}^{\mathcal{F}_\infty}\left[|X - Y|\right] \geq \frac{\lambda}{2}\right\}.$$

But the first term on the right hand side is at most $M(\lambda/2, \varepsilon) < \alpha$, while by Chebyshev's inequality the second term is at most $2\varepsilon/\lambda < \alpha$. Therefore $\mathbf{P}\left\{\operatorname{e\,lim\,sup}|X_t - X| \geq \lambda\right\} < 2\alpha$ for all $t \geq t_0$. Thus $\operatorname{e\,lim\,sup}|X_t - X| = 0$, and therefore $\operatorname{e\,lim} X_t = X$.

(iii) $\implies$ (iv). Let $X \in L_1$. Let $X_t = \mathbf{E}^{\mathcal{F}_t}\left[X\right]$, so that also $X_\tau = \mathbf{E}^{\mathcal{F}_\tau}\left[X\right]$ for $\tau \in \Sigma$ by the localization theorem (1.4.2). Since X_t converges essentially, we have $\operatorname{s\,lim\,sup}|X_t| = \operatorname{e\,lim}|X_t| = \operatorname{e\,lim\,sup}|X_t|$ and

$$A = \{\operatorname{s\,lim\,sup}|X_t| \geq \lambda\}.$$

Now if $Y_t = |X_t|$, then $(Y_\tau)_{\tau\in\Sigma^o}$ is uniformly absolutely continuous by (2.3.12). Then applying (4.1.11(b)), we obtain

$$\begin{aligned}
\mathbf{P}(A) &\leq \frac{1}{\lambda}\limsup_{\tau\in\Sigma^o}\mathbf{E}\left[Y_\tau \mathbf{1}_A\right]\\
&\leq \frac{1}{\lambda}\limsup_{\tau\in\Sigma^o}\mathbf{E}\left[\mathbf{E}^{\mathcal{F}_\tau}\left[|X|\right]\mathbf{1}_A\right]\\
&= \frac{1}{\lambda}\mathbf{E}\left[|X|\mathbf{1}_A\right].
\end{aligned}$$

The last equality follows because $\mathbf{E}^{\mathcal{F}_\tau}\left[|X|\right]$ converges to $\mathbf{E}^{\mathcal{F}_\infty}\left[|X|\right]$ in L_1. (iv) $\Longrightarrow$ (i) is obvious   ∎

The preceding proof in fact shows more. We need not consider all random variables X, but only a certain subclass. A family $\mathcal{E}$ of processes $(X_t,\mathcal{F}_t)$, where X_t is measurable with respect to $\mathcal{F}_t$, is called *stable* iff for every $(X_t,\mathcal{F}_t)\in\mathcal{E}$ and every $t_0\in J$, the process $(Y_t,\mathcal{G}_t)$ is also in $\mathcal{E}$, where, for $t\geq t_0$, $Y_t = X_t - X_{t_0}$ and $\mathcal{G}_t = \mathcal{F}_t$; and for other t, $Y_t = 0$ and $\mathcal{G}_t = \mathcal{F}_{t_0}$. When we say $(Y_t)\in\mathcal{E}$, without specifying the σ-algebras, we understand that $(Y_t,\mathcal{G}_t)\in\mathcal{E}$, where $\mathcal{G}_t$ is the σ-algebra generated by $\{Y_s : s\leq t\}$.

(4.1.13) Proposition. *Let $(\mathcal{F}_t)_{t\in J}$ be a stochastic basis, and let $\mathcal{E}$ be a stable subfamily of the family of uniformly integrable martingales. Then the following conditions are equivalent:*

(i) *For every $(X_t)\in\mathcal{E}$ and every $\lambda > 0$, we have*

$$\mathbf{P}\left\{\operatorname{e\,lim\,sup}|X_t| \geq \lambda\right\} \leq \frac{1}{\lambda}\mathbf{E}\left[|X_t|\right].$$

(ii) *Each $(X_t)\in\mathcal{E}$ converges essentially.*

Snell envelope

There is one application of the essential supremum that is useful even when the processes are indexed by $\mathbb{N}$. But it works also for processes indexed by any directed set J. We will define a variant of the Snell envelope, restricting stopping times to Σ; this is important in connection with amarts. Let (X_t) be an adapted integrable process. Then the *Snell envelope* of (X_t)

is the process (Z_t) defined by:

$$Z_t = \operatorname*{ess\,sup}_{\substack{\tau \in \Sigma \\ \tau \geq t}} \mathbf{E}^{\mathcal{F}_t}\left[X_\tau\right].$$

(4.1.14) Proposition. *Let (Z_t) be the Snell envelope of the integrable process (X_t). Then:*

(1) *If $\sigma \in \Sigma$, then*

$$Z_\sigma = \operatorname*{ess\,sup}_{\substack{\tau \in \Sigma \\ \tau \geq \sigma}} \mathbf{E}^{\mathcal{F}_\sigma}\left[X_\tau\right].$$

(2) *For $\sigma \in \Sigma$, there exists a sequence $\tau_n \in \Sigma$, $\tau_n \geq \sigma$, such that*

$$\mathbf{E}^{\mathcal{F}_\sigma}\left[X_{\tau_n}\right] \uparrow Z_\sigma.$$

(3) *(Z_t) is an amart and a supermartingale.*

Proof. (1) Let $\sigma \in \Sigma$, and let t be one of the values of σ. If $\tau \in \Sigma$ and $\tau \geq \sigma$, then there is $\tau' \in \Sigma$, $\tau' \geq t$, such that $\tau = \tau'$ on $\{\sigma = t\}$. Indeed, choose t_0 with $t_0 \geq \tau$; and let

$$\tau'(\omega) = \begin{cases} \tau(\omega) & \text{if } \sigma(\omega) = \tau(\omega) \\ t_0 & \text{otherwise.} \end{cases}$$

Similarly, if $\tau' \in \Sigma$ and $\tau' \geq t$, then there is $\tau \in \Sigma$, $\tau \geq \sigma$, such that $\tau = \tau'$ on $\{\sigma = t\}$. So on the set $\{\sigma = t\}$, we have (by the localization theorem (1.4.2))

$$\mathbf{E}^{\mathcal{F}_\sigma}\left[X_\tau\right] = \mathbf{E}^{\mathcal{F}_t}\left[X_\tau\right] = \mathbf{E}^{\mathcal{F}_t}\left[X_{\tau'}\right].$$

Hence on $\{\sigma = t\}$ we have

$$\operatorname*{ess\,sup}_{\tau \geq \sigma} \mathbf{E}^{\mathcal{F}_\sigma}\left[X_\tau\right] = \operatorname*{ess\,sup}_{\tau \geq t} \mathbf{E}^{\mathcal{F}_t}\left[X_\tau\right] = Z_t = Z_\sigma.$$

This is true for all t.

(2) First consider the case $\sigma = t$ constant. For $\tau_1, \tau_2 \geq t$, if we define

$$B = \left\{ \mathbf{E}^{\mathcal{F}_t}\left[X_{\tau_1}\right] \geq \mathbf{E}^{\mathcal{F}_t}\left[X_{\tau_2}\right] \right\},$$

then $B \in \mathcal{F}_t$, and τ defined by

$$\tau = \begin{cases} \tau_1 & \text{on } B \\ \tau_2 & \text{on } \Omega \setminus B \end{cases}$$

belongs to Σ. We have $\tau \geq t$ and $\mathbf{E}^{\mathcal{F}_t}\left[X_\tau\right] \geq \mathbf{E}^{\mathcal{F}_t}\left[X_{\tau_i}\right]$ for $i = 1, 2$. Thus the collection

$$\left\{ \mathbf{E}^{\mathcal{F}_t}\left[X_\tau\right] : \tau \in \Sigma, \tau \geq t \right\}$$

is directed. Apply (4.1.1(c)) to complete the proof of (2) in the case $\sigma = t$.

In the general case, for each $t \in J$ with $\{\sigma = t\} \neq \emptyset$, choose a sequence $\tau_n^{(t)} \in \Sigma$ with $\tau_n^{(t)} \geq t$ such that

$$Z_t = \sup_n \mathbf{E}^{\mathcal{F}_t}\left[X_{\tau_n^{(t)}}\right].$$

Now (4.1.23) there exist stopping times τ_n with $\tau_n = \tau_n^{(t)}$ on $\{\sigma = t\}$. We have $Z_\sigma = Z_t$ on $\{\sigma = t\}$, so we get

$$Z_\sigma = \sup_n \mathbf{E}^{\mathcal{F}_\sigma}\left[X_{\tau_n}\right].$$

(3) Let $\sigma_1 \leq \sigma_2$ be simple stopping times. Choose $\tau_n \geq \sigma_2$ with $\mathbf{E}^{\mathcal{F}_{\sigma_2}}[X_{\tau_n}] \uparrow Z_{\sigma_2}$. Now $\tau_n \geq \sigma_1$, so $\mathbf{E}^{\mathcal{F}_{\sigma_1}}[X_{\tau_n}] \leq Z_{\sigma_1}$. So by the monotone convergence theorem, (Z_t) is a submartingale, and

$$\mathbf{E}\left[Z_{\sigma_1}\right] \geq \lim_n \mathbf{E}\left[\mathbf{E}^{\mathcal{F}_{\sigma_1}}[X_{\tau_n}]\right] = \lim_n \mathbf{E}\left[\mathbf{E}^{\mathcal{F}_{\sigma_2}}[X_{\tau_n}]\right] = \mathbf{E}\left[Z_{\sigma_2}\right].$$

∎

As an application of the Snell envelope, we prove a stronger form of the amart Riesz decomposition (1.4.6). Recall that an *amart potential* is an amart (X_n) such that $\lim_n \mathbf{E}[X_n \mathbf{1}_A] = 0$ for all $A \in \bigcup_{m=1}^{\infty} \mathcal{F}_m$. A *Doob potential* is a positive supermartingale S_n with $\lim_n \mathbf{E}[S_n] = 0$.

(4.1.15) Theorem. *Let $(X_n)_{n \in \mathbb{N}}$ be an adapted process. Then (X_n) is an amart potential if and only if there is a Doob potential (S_n) with $|X_n| \leq S_n$ a.s. for all n.*

Proof. Suppose $|X_n| \leq S_n$ a.s. Then $|X_\sigma| \leq S_\sigma$ for all $\sigma \in \Sigma$. Now $\mathbf{E}[S_\sigma]$ decreases as σ increases. But $\mathbf{E}[S_n] \to 0$, so also $\mathbf{E}[S_\sigma] \to 0$. Then $|\mathbf{E}[X_\sigma]| \leq \mathbf{E}[|X_\sigma|] \leq \mathbf{E}[S_\sigma]$, so (X_n) is an amart and $\mathbf{E}[|X_n|] \to 0$, so it is an amart potential.

Conversely, suppose (X_n) is an amart potential. Then $(|X_n|)$ is also an amart potential. Let (S_n) be the Snell envelope of the process $(|X_n|)$. Then (S_n) is an amart and a supermartingale (1.4.3). Also, $S_n \geq |X_n|$, so $S_n \geq 0$. Finally, $\mathbf{E}[|X_\tau|] \to 0$, so by (4.1.14(2)) we have $\mathbf{E}[S_n] \to 0$. ∎

The Riesz decomposition and (4.1.15) indicate that for $J = \mathbb{N}$, the amart convergence theorem is not likely to have striking applications not possible with martingales and supermartingales. In the vector-valued case (Chapter 5) the situation is similar for uniform amarts (5.2.13), but not for other classes of amarts, since the Riesz decompositions are less restrictive. Also on directed sets (this Chapter) the behavior of amarts cannot be reduced to that of martingales and supermartingales.

Complements

(4.1.16) (Fatou's lemma for stochastic convergence.) Let (X_t) be a net of nonnegative random variables. Then

$$\mathbf{E}\left[\operatorname{s\,lim\,inf} X_t\right] \le \liminf \mathbf{E}\left[X_t\right].$$

(4.1.17) (Uniform integrability and stochastic convergence.) Let (X_t) be a uniformly integrable net in L_1. If $X_t \to X$ stochastically, then $\mathbf{E}\left[X_t\right] \to \mathbf{E}\left[X\right]$. In particular, if $X_t \to X$ essentially, then $\mathbf{E}\left[X_t\right] \to \mathbf{E}\left[X\right]$.

(4.1.18) (Monotone convergence theorem for essential convergence.) Let (X_t) be a net bounded in L_1 that is monotone increasing in the sense that if $s \le t$, then $X_s \le X_t$ a.s. Then $\operatorname{e\,lim} X_t$ exists and $\lim \mathbf{E}\left[X_t\right] = \mathbf{E}\left[\operatorname{e\,lim} X_t\right]$. In particular, $\operatorname{s\,lim} X_t$ exists and $\lim \mathbf{E}\left[X_t\right] = \mathbf{E}\left[\operatorname{s\,lim} X_t\right]$.

(4.1.19) If a net (X_t) converges in L_p norm, for some p $(1 \le p \le \infty)$, then (X_t) converges stochastically.

(4.1.20) Let (X_t) be a net of random variables, and let $\lambda \in \mathbb{R}$. Let $A_t = \{X_t > \lambda\}$ and $B_t = \{X_t \ge \lambda\}$. Then:

$$\{X_s \vee X_t > \lambda\} = A_s \cup A_t \subseteq B_s \cup B_t = \{X_s \vee X_t \ge \lambda\}$$
$$\{X_s \wedge X_t > \lambda\} = A_s \cap A_t \subseteq B_s \cap B_t = \{X_s \wedge X_t \ge \lambda\}$$
$$\{X_{t_1} \vee X_{t_2} \vee \cdots > \lambda\} = A_{t_1} \cup A_{t_2} \cup \cdots \subseteq$$
$$\subseteq B_{t_1} \cup B_{t_2} \cup \cdots \subseteq \{X_{t_1} \vee X_{t_2} \vee \cdots \ge \lambda\}$$
$$\{X_{t_1} \wedge X_{t_2} \wedge \cdots > \lambda\} \subseteq A_{t_1} \cap A_{t_2} \cap \cdots \subseteq$$
$$\subseteq B_{t_1} \cap B_{t_2} \cap \cdots = \{X_{t_1} \wedge X_{t_2} \wedge \cdots \ge \lambda\}.$$

(4.1.21) In the same notation,
(a) $\{\operatorname{ess\,sup} X_t > \lambda\} = \operatorname{ess\,sup} A_t \subseteq \operatorname{ess\,sup} B_t \subseteq \{\operatorname{ess\,sup} X_t \ge \lambda\}$
(b) $\{\operatorname{ess\,inf} X_t > \lambda\} \subseteq \operatorname{ess\,inf} A_t \subseteq \operatorname{ess\,inf} B_t = \{\operatorname{ess\,inf} X_t \ge \lambda\}$
(c) $\{\operatorname{e\,lim\,sup} X_t > \lambda\} \subseteq \operatorname{e\,lim\,sup} A_t \subseteq \operatorname{e\,lim\,sup} B_t \subseteq \{\operatorname{e\,lim\,sup} X_t \ge \lambda\}$
(d) $\{\operatorname{e\,lim\,inf} X_t > \lambda\} \subseteq \operatorname{e\,lim\,inf} A_t \subseteq \operatorname{e\,lim\,inf} B_t \subseteq \{\operatorname{e\,lim\,inf} X_t \ge \lambda\}.$

(4.1.22) Let (X_t) be a net of random variables, and let $\lambda \in \mathbb{R}$. Then

$$\operatorname*{s\,lim\,sup}_{t}\{X_t \ge \lambda\} \subseteq \{\operatorname*{s\,lim\,sup}_{t} X_t \ge \lambda\}.$$

Indeed, suppose Y is a random variable asymptotically above X_t, that is $\mathbf{P}\{X_t > Y\} \to 0$. Write $B_t = \{X_t \ge \lambda\}$. Then

$$\mathbf{P}(B_t \setminus \{Y \ge \lambda\}) = \mathbf{P}\{Y < \lambda \le X_t\} \le \mathbf{P}\{Y < X_t\} \to 0.$$

Therefore $\{Y \ge \lambda\} \supseteq \operatorname{s\,lim\,sup} B_t$, or $Y \ge \lambda$ on $\operatorname{s\,lim\,sup} B_t$. This is true for all such Y, so by the definition of $\operatorname{s\,lim\,sup} X_t$, we have also $\operatorname{s\,lim\,sup} X_t \ge \lambda$ on $\operatorname{s\,lim\,sup} B_t$. That is, $\{\operatorname{s\,lim\,sup} X_t \ge \lambda\} \supseteq \operatorname{s\,lim\,sup} B_t$.

Note that in general the reverse inclusion

$$\text{s}\limsup_{t}\{X_t \geq \lambda\} \supseteq \{\text{s}\limsup_{t} X_t \geq \lambda\}$$

fails. For example, with $J = \mathbb{N}$, if $X_n = 1 - 1/n$ and $\lambda = 1$, then the right side is Ω and the left side is $\emptyset$.

(4.1.23) (Generalized waiting lemma.) Generalize (1.1.5) to directed sets: Let $(\mathcal{F}_t)_{t \in J}$ be a stochastic basis, and let Σ be the corresponding set of simple stopping times. Let $\sigma \in \Sigma$ be given, and for each $t \in J$ with $\{\sigma = t\} \neq \emptyset$, let $\tau^{(t)} \in \Sigma$ be given with $\tau^{(t)} \geq t$ on $\{\sigma = t\}$. Then τ defined by $\tau(\omega) = \tau^{(t)}(\omega)$ on $\{\sigma = t\}$ belongs to Σ and $\tau \geq \sigma$.

Remarks

Stochastic upper and lower limits (also called upper and lower limits in measure) are due to D. E. Menchoff. The definition and treatment were considerably simplified by Goffman & Waterman [1960]. We have used their definition. Propositions (4.1.11) and (4.1.12) are from Millet & Sucheston [1980e]. For processes indexed by $\mathbb{N}$, the connection between maximal inequalities and convergence has been much studied; see for example Burkholder [1964].

Demiconvergence of martingales was first observed by Edgar & Sucheston [1981]. See also Millet & Sucheston [1983] and Frangos & Sucheston [1985].

4.2. The covering condition (V)

Martingales indexed by a directed set converge stochastically. We will now provide an example showing that they need not converge essentially.

(4.2.1) **Example.** Let J be the set of all finite subsets of $\mathbb{N}$, ordered by inclusion. Then J is a countable directed set. Let U_n be independent, identically distributed random variables with

$$\mathbf{P}\{U_n = 1\} = \mathbf{P}\{U_n = -1\} = \frac{1}{2}.$$

For each finite set $t \in J$, let $\mathcal{F}_t$ be the (finite) σ-algebra generated by the random variables U_n, $n \in t$. Thus, if $s \subseteq t$, then $\mathcal{F}_s \subseteq \mathcal{F}_t$. Define

$$X_t = \sum_{n \in t} \frac{1}{n} U_n.$$

First, we verify that the process (X_t) is L_1-bounded. In fact, it is L_2-bounded: since the U_n are orthonormal,

$$\|X_t\|_2^2 = \sum_{n \in t} \left(\frac{1}{n}\right)^2 \leq \sum_{n \in \mathbb{N}} \left(\frac{1}{n}\right)^2 = \frac{\pi^2}{6}.$$

For L_1-boundedness, apply the Schwartz inequality:

$$\|X_t\|_1 = \mathbf{E}\left[|X_t|\right] = \mathbf{E}\left[|X_t|\cdot 1\right] \le \mathbf{E}\left[|X_t|^2\right]^{1/2}\mathbf{E}\left[1^2\right]^{1/2} = \|X_t\|_2.$$

Next, to show that (X_t) is a martingale, we claim: if $s \subseteq t$, then $\mathbf{E}^{\mathcal{F}_s}[X_t] = X_s$. Since t is s plus a finite number of extra elements, it is enough to consider the case where there is one extra element, $t = s \cup \{m\}$, and then apply induction. But U_m is independent of $\mathcal{F}_s$, and therefore

$$\mathbf{E}^{\mathcal{F}_s}[X_t] = \mathbf{E}^{\mathcal{F}_s}\left[X_s + \frac{1}{m}U_m\right] = X_s + \frac{1}{m}\mathbf{E}[U_m] = X_s.$$

This shows that $(X_t)_{t\in J}$ is a martingale.

Now we know by (1.3.1) that (X_t) converges stochastically. (In fact, by elementary Hilbert space theory, it converges in L_2 norm and therefore by (4.1.19) stochastically.) But we claim that it does not converge a.s. Since J is countable, this means also that it does not converge essentially. In fact, we will see that the set of $\omega \in \Omega$ for which the net $(X_t(\omega))_{t\in J}$ converges has probability 0. Indeed, almost all $\omega \in \Omega$ satisfy $|U_n(\omega)| = 1$ for all n. Convergence of $(X_t(\omega))$ to x means: for any $\varepsilon > 0$, there is a finite set $s \subseteq \mathbb{N}$ such that for all finite sets $t \supseteq s$,

$$\left|x - \sum_{n\in t}\frac{1}{n}U_n(\omega)\right| < \varepsilon.$$

But this is equivalent to saying that of the series $\sum_{n\in\mathbb{N}}(1/n)U_n(\omega)$ converges absolutely. Now

$$\sum_{n\in\mathbb{N}}\left|\frac{1}{n}U_n(\omega)\right| = \sum_{n\in\mathbb{N}}\left|\frac{1}{n}\right| = \infty,$$

so the series $\sum_{n\in\mathbb{N}}(1/n)U_n(\omega)$ does not converge absolutely. Therefore $(X_t(\omega))$ does not converge.

Condition (V)

Let $(\mathcal{F}_t)_{t\in J}$ be a stochastic basis. Recall the notation Σ for the set of simple stopping times. For certain stochastic bases (for example a stochastic basis $(\mathcal{F}_n)_{n\in\mathbb{N}}$ indexed by $\mathbb{N}$), all L_1-bounded amarts converge essentially. For other stochastic bases (for example, the one in the preceding example) this is not the case. In this section we study a condition on the stochastic basis that will insure essential convergence. This condition is called the covering condition (V) or the Vitali condition (V).

An *adapted family* of sets is a family $(A_t)_{t\in J}$, where $A_t \in \mathcal{F}_t$ for all $t \in J$. We will often write A^* for e$\limsup A_t$. If τ is a simple stopping time for the stochastic basis $(\mathcal{F}_t)$, we consider a *stopped set* $A(\tau)$ defined by

$$A(\tau) = \bigcup_{t\in J}\left(A_t \cap \{\tau = t\}\right).$$

This union is finite, since $\{\tau = t\}$ is empty except for finitely many values of t. It may be easily verified that $A(\tau) \in \mathcal{F}_\tau$.

The condition (V) states that the essential upper limit of an adapted family of sets may be approximated by a stopped set.

(4.2.2) Definition. The stochastic basis $(\mathcal{F}_t)_{t\in J}$ satisfies the *covering condition* (V) iff, for each adapted family $(A_t)_{t\in J}$ of sets, and each $\varepsilon > 0$, there exists a simple stopping time $\tau \in \Sigma$ with $\mathbf{P}(A^* \setminus A(\tau)) < \varepsilon$.

To understand the meaning of this notion, recall that a.s. convergence of amarts was proved in Chapter 1 by approximation of $\limsup X_t$ by the stopped process X_τ (1.2.4 and 1.2.5). On directed sets this is not possible in general. Condition (V) postulates such an approximation of $\mathrm{e}\limsup$ for zero-one valued processes (A_t). This assumption is crucial: it will be shown below that (V) is necessary and sufficient for convergence of L_1-bounded amarts.

We have stated condition (V) in terms of stopping times since it will be used in that form. But it can also be stated without reference to stopping times: For each adapted family $(A_t)_{t\in J}$ and each $\varepsilon > 0$, there exists an adapted pairwise disjoint family $(B_t)_{t\in J}$, with only finitely many B_t nonempty, such that

$$\mathbf{P}\left(A^* \setminus \bigcup_t (A_t \cap B_t)\right) < \varepsilon.$$

Another variant allows countably many B_t nonempty, and concludes

$$\mathbf{P}\left(A^* \setminus \bigcup_t (A_t \cap B_t)\right) = 0.$$

Another begins with a set $B \subseteq A^*$ and almost covers it by sets $A_t \cap B_t$.

We will now prove the equivalence of a few simple variants of condition (V). To each variant there corresponds an asymptotic version in which, for each t_0, there is an approximating stopping time larger than t_0. The equivalence of an asymptotic version with the corresponding nonasymptotic version follows from the consideration of the process (B_t) defined by $B_t = A_t$ if $t \geq t_0$, $B_t = \emptyset$ otherwise (see the proof of (a) $\Longrightarrow$ (b)). Below, (b) is an asymptotic version of (a); conditions (c) and (d) are in asymptotic forms; their obvious nonasymptotic formulations have been omitted. Any of these equivalent formulations may be referred to as the covering condition (V).

(4.2.3) Proposition. *Let $(\mathcal{F}_t)_{t\in J}$ be a stochastic basis. Then the following formulations of condition* (V) *are equivalent:*

(a) For every adapted family (A_t) and every $\varepsilon > 0$, there exists $\tau \in \Sigma$ such that $\mathbf{P}(A^* \setminus A(\tau)) < \varepsilon$.

(b) For every adapted family (A_t), we have $\liminf_\tau \mathbf{P}(A^* \setminus A(\tau)) = 0$; that is: for every $\varepsilon > 0$, and every $t_0 \in J$, there exists $\tau \in \Sigma$ with $\tau \geq t_0$ and $\mathbf{P}(A^* \setminus A(\tau)) < \varepsilon$.

(c) For every adapted family (A_t), we have $\limsup_\tau \mathbf{P}(A(\tau)) \geq \mathbf{P}(A^*)$; that is: for every $\varepsilon > 0$ and every $t_0 \in J$, there exists $\tau \in \Sigma$ with $\tau \geq t_0$ and $\mathbf{P}(A(\tau)) > \mathbf{P}(A^*) - \varepsilon$.

(d) For every adapted family (A_t), every $t_0 \in J$ and every $\varepsilon > 0$, there is a $\tau \in \Sigma$ with $\tau \geq t_0$ such that $\mathbf{P}(A^* \triangle A(\tau)) < \varepsilon$.

Proof. (a) $\Longrightarrow$ (b): Given $t_0 \in J$, define B_t as follows:

$$B_t = \begin{cases} A_t & \text{if } t \geq t_0 \\ \emptyset & \text{otherwise.} \end{cases}$$

Then $B^* = A^*$. By (a), there is $\tau \in \Sigma$ with $\mathbf{P}(B^* \setminus B(\tau)) < \varepsilon$. Choose $t_1 \geq \tau$, $t_1 \geq t_0$. Define σ by

$$\sigma = \begin{cases} \tau & \text{if } \tau \geq t_0 \\ t_1 & \text{otherwise.} \end{cases}$$

Then $B(\tau) = B(\sigma) \subseteq A(\sigma)$, so $\mathbf{P}(A^* \setminus A(\sigma)) < \varepsilon$.

(b) $\Longrightarrow$ (c): By (b), there is $\tau \geq t_0$ with $\mathbf{P}(A^* \setminus A(\tau)) < \varepsilon$. Thus $\mathbf{P}(A(\tau)) \geq \mathbf{P}(A^*) - \mathbf{P}(A^* \setminus A(\tau)) > \mathbf{P}(A^*) - \varepsilon$.

(c) $\Longrightarrow$ (d). Given $\varepsilon > 0$, choose $s \in J$, $s \geq t_0$, such that

$$\mathbf{P}\left(\operatorname*{ess\,sup}_{t \geq s} A_t \setminus A^* \right) \leq \varepsilon.$$

Let $\tau \geq s$ be given by (c). Then

$$
\begin{aligned}
\mathbf{P}\left(A^* \bigtriangleup A(\tau) \right) &= \mathbf{P}\left(A^* \setminus A(\tau) \right) + \mathbf{P}\left(A(\tau) \setminus A^* \right) \\
&\leq \mathbf{P}\left(\operatorname*{ess\,sup}_{t \geq s} A_t \setminus A(\tau) \right) + \mathbf{P}\left(\operatorname*{ess\,sup}_{t \geq s} A_t \setminus A^* \right) \\
&\leq \mathbf{P}\left(\operatorname*{ess\,sup}_{t \geq s} A_t \right) - \mathbf{P}\left(A(\tau) \right) + \varepsilon \\
&\leq \mathbf{P}(A^*) - \mathbf{P}(A(\tau)) + 2\varepsilon \leq 3\varepsilon.
\end{aligned}
$$

Clearly (d) $\Longrightarrow$ (a). ∎

Condition (V) asserts that $A^* = \mathrm{e}\limsup A_t$ may be "covered" by the stopped set $A(\tau)$. We now show that (V) also holds if only a portion of A^* can be covered, but a *fixed* portion, independent of the choice of (A_t).

(4.2.4) Proposition. *Let $(\mathcal{F}_t)_{t \in J}$ be a stochastic basis. Condition (V) holds if and only if there is a constant α, $0 < \alpha < 1$, such that for each adapted family (A_t) of sets there is a $\tau \in \Sigma$ such that $\mathbf{P}(A^* \cap A(\tau)) \geq \alpha \mathbf{P}(A^*)$.*

Proof. Suppose α exists. Let (A_t) and ε be given. Let $\tau_1 \in \Sigma$ be such that $\mathbf{P}(A^* \cap A(\tau_1)) \geq \alpha \mathbf{P}(A^*)$. Then let $s_2 \in J$, $s_2 > \tau_1$, and set

$$A_t^1 = \begin{cases} A_t \setminus A(\tau_1) & \text{if } t \geq s_1 \\ \emptyset & \text{otherwise.} \end{cases}$$

Since $A^* \setminus A(\tau_1) = \mathrm{e}\limsup_t A_t^1$, there exists $\tau_2 \in \Sigma$ such that $\tau_2 \geq s_2$ and $\mathbf{P}\big((A^* \setminus A(\tau_1)) \cap A(\tau_2)\big) \geq \alpha \mathbf{P}(A^* \setminus A(\tau_1))$. Then $\mathbf{P}\big(A^* \setminus (A(\tau_1) \cup A(\tau_2))\big) \leq$

$(1 - \alpha)^2 \mathbf{P}(A^*)$. Continue inductively to obtain a sequence τ_n of stopping times satisfying for all n the relations $\tau_{n-1} < s_n \leq t_n$ and

$$\mathbf{P}\left(A^* \setminus \bigcup_{j=1}^{n} A(\tau_j)\right) \leq (1 - \alpha)^n \mathbf{P}(A^*).$$

Now we are given $\varepsilon > 0$; choose n so that $(1 - \alpha)^n \mathbf{P}(A^*) < \varepsilon$. Choose $s \geq \tau_n$, and define

$$\tau = \begin{cases} \tau_j & \text{on } A(\tau_j) \setminus \bigcup_{k=1}^{j-1} A(\tau_k), & \text{for } 1 \leq j \leq n, \\ s & \text{on } \Omega \setminus \bigcup_{k=1}^{n} A(\tau_k). \end{cases}$$

Then $\tau \in \Sigma$ and $\mathbf{P}(A^* \setminus A(\tau)) \leq \varepsilon$.

For the converse, suppose (V) holds; let $\alpha = 1/2$ (say). Let (A_t) be given. If $\mathbf{P}(A^*) = 0$, then clearly $\mathbf{P}(A^* \cap A(\tau)) \geq \alpha \mathbf{P}(A^*)$ for any τ. If $\mathbf{P}(A^*) > 0$, use $\varepsilon = P(A^*)/2$ with condition (V) to obtain τ with $\mathbf{P}(A^* \setminus A(\tau)) < \varepsilon$. Then $\mathbf{P}(A^* \cap A(\tau)) < (1/2)\mathbf{P}(A^*)$. ∎

Example: Totally ordered basis

Some simple examples may help explain the covering condition (V). The stochastic basis $(\mathcal{F}_t)_{t \in J}$ is *totally ordered* iff, for any $s, t \in J$, either $\mathcal{F}_s \subseteq \mathcal{F}_t$ or $\mathcal{F}_t \subseteq \mathcal{F}_s$. Note that if J is totally ordered, then the stochastic basis is totally ordered, but the converse is not necessarily true.

(4.2.5) Proposition. *If $(\mathcal{F}_t)_{t \in J}$ is totally ordered, then $(\mathcal{F}_t)$ satisfies condition (V).*

Proof. Let (A_t) be adapted and let $\varepsilon > 0$. Then $A^* = \operatorname{e \, lim \, sup} A_t \subseteq \operatorname{ess \, sup} A_t$, so there is a countable set $\{t_i\}_{i=1}^{\infty} \subseteq J$ such that $A^* \subseteq \bigcup_{i=1}^{\infty} A_{t_i}$. Thus there is $N \in \mathbb{N}$ with $\mathbf{P}\left(A^* \setminus \bigcup_{i=1}^{N} A_{t_i}\right) < \varepsilon$. Renumber the t_i so that $\mathcal{F}_{t_1} \subseteq \mathcal{F}_{t_2} \subseteq \cdots \subseteq \mathcal{F}_{t_N}$. Define τ by:

$$\tau = \begin{cases} t_1 & \text{on } A_{t_1} \\ t_i & \text{on } A_{t_i} \setminus \bigcup_{j=1}^{i-1} A_{t_j} & \text{for } 2 \leq i \leq N \\ t_N & \text{elsewhere.} \end{cases}$$

Then we have $A(\tau) \supseteq \bigcup_{i=1}^{N} A_{t_i}$, and therefore $\mathbf{P}(A^* \setminus A(\tau)) < \varepsilon$. ∎

Example: Finite subsets of $\mathbb{N}$

Consider the directed set J of finite subsets of $\mathbb{N}$, as in (4.2.1). We show now that condition (V) fails for the stochastic basis of the example (4.2.1). See also (4.4.16), where it is shown that this basis also fails the weaker condition (C).

We begin with independent, identically distributed random variables U_n with $\mathbf{P}\{U_n = 1\} = \mathbf{P}\{U_n = -1\} = 1/2$. If J is the set of all finite subsets of $\mathbb{N}$, ordered by inclusion, and $\mathcal{F}_t$ is defined as the least σ-algebra such that U_n ($n \in t$) are measurable, then $(\mathcal{F}_t)_{t \in J}$ is a stochastic basis. If B, C are disjoint finite subsets of $\mathbb{N}$, write $F(B,C)$ for the event

$$\{U_n = 1 \text{ for all } n \in B, U_n = -1 \text{ for all } n \in C\}.$$

Thus $\mathcal{F}_t$ has atoms $F(B,C)$, where $B \cap C = \emptyset$ and $B \cup C = t$. These atoms all have measure 2^{-k}, where k is the number of elements of t.

We claim that condition (V) fails. For $m \in \mathbb{N}$, let $\mathcal{C}_m$ be the set of all m-element subsets of $\{m+1, m+2, \cdots, 4m\}$. For $m \in \mathbb{N}$ and $C \in \mathcal{C}_m$, let $t(m,C) = \{1, 2, \cdots, m\} \cup C \in J$. Define

$$A_t = \begin{cases} F(C, \emptyset) & \text{if } t = t(m,C) \text{ for some } m \in \mathbb{N} \text{ and } C \in \mathcal{C}_m, \\ \emptyset & \text{otherwise.} \end{cases}$$

We will show that (V) fails for the adapted family (A_t).

First, we claim that $e \limsup A_t = \Omega$. Given $s \in J$ and $\varepsilon > 0$, we can choose m so that $s \subseteq \{1, 2, \cdots, m\}$ and

$$(4.2.5a) \qquad 2^{-3m} \sum_{k=m}^{3m} \binom{3m}{k} > 1 - \varepsilon.$$

(This is a simple combinatorial lemma. A probabilistic proof is given in (4.2.19).) Then

$$\bigcup_{t \geq s} A_t \supseteq \bigcup_{C \in \mathcal{C}_m} F(C, \emptyset),$$

which is the event that at least m of the $3m$ random variables U_{m+1}, $U_{m+2}, \cdots, U_{4m}$ are 1. Thus the probability is at least $1 - \varepsilon$ by (4.2.5a). Since s is arbitrary, this shows that $e \limsup A_t = \Omega$.

Next, fix $p \in \mathbb{N}$. Let $\tau \in \Sigma$, $\tau \geq \{1, 2, \cdots, p\}$. Then

$$A(\tau) = \bigcup (A_t \cap \{\tau = t\})$$
$$= \bigcup_{m=p}^{\infty} \bigcup_{C \in \mathcal{C}_m} (F(C, \emptyset) \cap \{\tau = t(m,C)\}).$$

Fix $m \geq p$. The atoms of $\mathcal{F}_{t(m,C)}$ contained in $F(C, \emptyset)$ have the form $F(C \cup B, D)$ where $B \cup D = \{1, \cdots, m\}$, $B \cap C = \emptyset$. Note $\mathbf{P}(F(C \cup B, D)) = 2^{-2m}$. Now two of the sets $F(C \cup B, D)$, $F(C' \cup B, D)$ are not disjoint, since they both contain $F(\{m+1, \cdots, 4m\} \cup B, D)$. Thus, for a fixed pair (B, D), there is at most one C with $F(C \cup B, D) \cap \{\tau = t(m,C)\} \neq \emptyset$. So

$$\mathbf{P}\left(\bigcup_{C \in \mathcal{C}_m} \left(F(C, \emptyset) \cap \{\tau = t(m,C)\} \right) \right)$$
$$\leq \sum_{B \subseteq \{1, \cdots, m\}} 2^{-2m} = 2^m \cdot 2^{-2m} = 2^{-m}.$$

Thus $\mathbf{P}(A(\tau)) \leq \sum_{m=p}^{\infty} 2^{-m} = 2^{-p+1}$. So condition (V) fails by (4.2.3(c)).

Example: Interval partitions

We now discuss a classical situation in which condition (V) holds (Theorem (4.2.8)). It is a result useful in derivation theory. The derivation of set-functions defined on Euclidean space $\mathbb{R}^d$ is closely related to an appropriate directed set consisting of partitions. The topic is treated in a somewhat different setting in Chapter 7.

Let Ω be the d-dimensional cube $[0,1]^d$, and let $\mathbf{P}$ be d-dimensional Lebesgue measure on Ω. A collection $\mathcal{C}$ of open subsets of Ω will be called *substantial* iff there is a constant M such that for every $C \in \mathcal{C}$ there is an open ball B with $C \subseteq B$ and $\mathbf{P}(B) < M\mathbf{P}(C)$. A simple example is the family d-dimensional intervals (rectangular solids with edges parallel to the axes) such that the ratio of the longest edge to the shortest edge is bounded by some constant M'.

First, a Vitali style covering lemma: A stronger version of the Lemma is found in Chapter 7 (7.2.1) . If B is an open ball, we write $r(B)$ for its radius.

(4.2.6) Lemma. *(a) Let $\mathcal{D}$ be a collection of open balls in $[0,1]^d$. Let $W = \bigcup \mathcal{D}$. Then for each $\varepsilon > 0$ there is a finite disjoint subcollection $\mathcal{D}'$ of $\mathcal{D}$ such that*

$$\sum_{B \in \mathcal{D}'} \mathbf{P}(B) > 3^{-d} \left(\mathbf{P}(W) - \varepsilon \right).$$

(b) Let $\mathcal{C}$ be a substantial collection of open sets in $[0,1]^d$ with constant M. Let $W = \bigcup \mathcal{C}$. Then for each $\varepsilon > 0$ there is a finite disjoint subcollection $\mathcal{C}'$ of $\mathcal{C}$ such that

$$\sum_{C \in \mathcal{C}'} \mathbf{P}(C) > M^{-1} 3^{-d} \left(\mathbf{P}(W) - \varepsilon \right).$$

Proof. (a) Let $\varepsilon > 0$ be given. The set W is open, and therefore measurable. So there is a compact set $K \subseteq W$ with $\mathbf{P}(K) > \mathbf{P}(W) - \varepsilon$. Now $\mathcal{D}$ is an open cover of the compact set K, so there is a finite subcover, say $S_1, S_2, \cdots, S_p \in \mathcal{D}$ and $K \subseteq \bigcup_{j=1}^p S_j$. Suppose these sets are ordered in decreasing order of their radii: $r(S_1) \geq r(S_2) \geq \cdots \geq r(S_p)$.

Now we define recursively a sequence $B_1, B_2, \cdots, B_n$ of balls. Let $B_1 = S_1$. Suppose $B_1, \cdots, B_k$ have been defined. Let B_{k+1} be S_j, where j is the least index such that $S_j \cap B_i = \emptyset$ for $1 \leq i \leq k$. If there is no such j, that is every S_j meets some B_i, then the construction stops with B_k. Certainly the construction stops in at most p steps. This completes the definition of the sequence $B_1, B_2, \cdots, B_n$.

Now for each B_i, let B_i' be the ball with the same center as B_i but three times the radius. We claim that

$$\bigcup_{j=1}^p S_j \subseteq \bigcup_{i=1}^n B_i'.$$

Indeed, for each j, the ball S_j meets some B_i with $r(B_i) \geq r(S_j)$, so $S_j \subseteq B_i'$. Thus

$$\mathbf{P}(K) \leq \mathbf{P}\left(\bigcup S_j\right) \leq \mathbf{P}\left(\bigcup B_i'\right) \leq \sum \mathbf{P}(B_i') = 3^d \sum \mathbf{P}(B_i).$$

The required inequality follows.

(b) follows from (a). ∎

We will consider the collection of all countable measurable partitions of Ω. Partitions are ordered by *a.e. refinement*: we write $s \leq t$ iff every atom of s is a union of atoms of t up to sets of measure 0. We will postulate that J is a directed set. This is satisfied in the classical cases. See (7.2.2).

(4.2.7)   Theorem. *Let C be a substantial collection of open subsets of the d-dimensional cube $\Omega = [0,1]^d$. Suppose that the family J of countable partitions of Ω into elements of C is directed by a.e. refinement. For $t \in J$, let $\mathcal{F}_t$ be the σ-algebra generated by the partition t. Then the stochastic basis $(\mathcal{F}_t)_{t \in J}$ satisfies condition (V).*

Proof. Let M be the constant showing that C is substantial. Choose $\varepsilon > 0$ so small that $M^{-1}3^{-d}(1 - 2\varepsilon) - \varepsilon > 0$. We will verify the condition in Proposition 4.2.4 with $\alpha = M^{-1}3^{-d}(1 - 2\varepsilon) - \varepsilon$.

Let (A_t) be a family of sets adapted to the stochastic basis $(\mathcal{F}_t)$ described, and write $A^* = \mathrm{e} \limsup A_t$. We may assume $\mathbf{P}(A^*) > 0$. Write $\varepsilon' = \varepsilon \mathbf{P}(A^*)$. Choose $s \in J$ so that $\mathbf{P}(\mathrm{ess\,sup}_{t \geq s} A_t \setminus A^*) < \varepsilon'$. There exists a sequence $t_k \geq s$ of indices with $\bigcup_k A_{t_k} = \mathrm{ess\,sup}_{t \geq s} A_t$. Decompose each A_{t_k} into atoms C_{kn} of t_k. Then $\mathbf{P}(\bigcup_{k,n} C_{kn}) \geq \mathbf{P}(A^*)$. There is thus a finite set F of pairs (k,n) such that $\mathbf{P}(\bigcup_F C_{kn}) \geq \mathbf{P}(A^*) - \varepsilon'$. By the lemma, there is a subset $F' \subseteq F$ such that the atoms $\{C_{kn} : (k,n) \in F'\}$ are disjoint and $\mathbf{P}(\bigcup_{F'} C_{kn}) > M^{-1}3^{-d}(\mathbf{P}(\bigcup_F C_{kn}) - \varepsilon')$. Now choose an index u larger than all t_k where k occurs as a first coordinate in the finite set F'. By the disjointness of the C_{kn} with $(k,n) \in F'$, we may define a stopping time by

$$\tau(\omega) = \begin{cases} t_k & \text{if } \omega \in C_{kn} \text{ and } (k,n) \in F' \\ u & \text{otherwise.} \end{cases}$$

Thus $A(\tau) \supseteq \bigcup_{F'} C_{kn}$. Finally,

$$\begin{aligned}
\mathbf{P}\big(A^* \cap A(\tau)\big) &\geq \mathbf{P}\big(A(\tau)\big) - \mathbf{P}\big(A(\tau) \setminus A^*\big) \\
&\geq M^{-1}3^{-d}\left(\mathbf{P}\left(\bigcup_F C_{kn}\right) - \varepsilon'\right) - \varepsilon' \\
&\geq M^{-1}3^{-d}\big(\mathbf{P}(A^*) - 2\varepsilon'\big) - \varepsilon' \\
&= \alpha \mathbf{P}(A^*).
\end{aligned}$$

∎

Essential convergence

Here are some consequences of condition (V).

(4.2.8) Proposition. *Let* $(\mathcal{F}_t)_{t\in J}$ *be a stochastic basis. Suppose* (V) *holds. Then:*

(a) If (A_t) is an adapted family of sets, then

$$\operatorname{s\,lim\,sup}_{\tau\in\Sigma} A(\tau) = \operatorname{e\,lim\,sup}_{t\in J} A_t \ \left(= \operatorname{e\,lim\,sup}_{\tau\in\Sigma} A(\tau)\right).$$

(b) If (X_t) is a stochastic process, then

$$\operatorname{s\,lim\,sup}_{\tau\in\Sigma} X_t = \operatorname{e\,lim\,sup}_{t\in J} X_t \ \left(= \operatorname{e\,lim\,sup}_{\tau\in\Sigma} X_\tau\right).$$

That is, for every $\varepsilon > 0$ and every $t_0 \in J$, there is a $\tau \in \Sigma$ with $\tau \geq t_0$ and $\mathbf{P}\{|\operatorname{e\,lim\,sup} X_t - X_\tau| > \varepsilon\} < \varepsilon$.

(c) If (X_t) is a nonnegative process, then for every $\lambda > 0$ and every $t_0 \in J$, there is a $\tau \in \Sigma$ with $\tau \geq t_0$ and

$$\mathbf{P}\{\operatorname{e\,lim\,sup} X_t \geq \lambda\} \leq \frac{1}{\lambda} \operatorname*{lim\,sup}_{\tau\in\Sigma} \mathbf{E}\left[X_\tau\right].$$

(d) If (X_t) is a stochastic process, and σ_n is a sequence of simple stopping times, then there exist $\tau_n \in \Sigma$ with $\tau_n \geq \sigma_n$ and $X_{\tau_n} \to$ e lim sup X_t a.s.

Proof. (a) Write $A^* = \operatorname{e\,lim\,sup} A_t$. To show $\operatorname{s\,lim\,sup}_\tau A(\tau) \supseteq A^*$, we use (4.1.7a). If C is any set with $\lim_\tau \mathbf{P}(A(\tau) \setminus C) = 0$, then we have

$$\mathbf{P}(A^* \setminus C) \leq \mathbf{P}(A^* \setminus A(\tau)) + \mathbf{P}(A(\tau) \setminus C).$$

By (V), we have $\liminf_\tau \mathbf{P}(A^*\setminus A(\tau)) = 0$. Therefore $\mathbf{P}(A^*\setminus C) = 0$, or $C \supseteq A^*$ a.s. This shows $\operatorname{s\,lim\,sup}_\tau A(\tau) \supseteq A^*$, so in fact $\operatorname{s\,lim\,sup}_\tau A(\tau) = A^*$.

(b) We always have $\operatorname{s\,lim\,sup} X_\tau \leq \operatorname{e\,lim\,sup} X_\tau = \operatorname{e\,lim\,sup} X_t$, so we must prove the opposite inequality. Applying part (a), 4.1.21(c) and 4.1.22, we have for any $\lambda > 0$

$$\left\{\operatorname{e\,lim\,sup}_{t\in J} X_t > \lambda\right\} \subseteq \operatorname{e\,lim\,sup}_{t\in J} \{X_t > \lambda\}$$
$$= \operatorname{s\,lim\,sup}_{\tau\in\Sigma} \{X_\tau > \lambda\}$$
$$\subseteq \left\{\operatorname{s\,lim\,sup}_{\tau\in\Sigma} X_\tau \geq \lambda\right\}.$$

Therefore $\operatorname{e\,lim\,sup}_t X_t \leq \operatorname{s\,lim\,sup}_\tau X_\tau$.

(c) Let (X_t) be a nonnegative process, and let $\lambda > 0$. Fix β with $0 < \beta < \lambda$. Define $A_t = \{X_t > \beta\}$. Then $\{\text{e}\limsup X_t > \beta\} \subseteq A^*$. Let $t_0 \in J$ and $\varepsilon > 0$ be given. Then by (V), there is $\tau \in \Sigma$ with $\tau \geq t_0$ and $\mathbf{P}(A^* \setminus A(\tau)) < \varepsilon$. Then

$$\mathbf{E}[X_\tau] = \sum_t \mathbf{E}\left[X_t \, \mathbf{1}_{\{\tau=t\}}\right]$$

$$\geq \beta \sum_t \mathbf{E}\left[\mathbf{1}_{\{\tau=t\}\cap A_t}\right]$$

$$= \beta\mathbf{P}(A(\tau))$$

$$\geq \beta\big(\mathbf{P}(A^*) - \varepsilon\big)$$

$$\geq \beta\big(\mathbf{P}\{\text{e}\limsup X_t > \beta\} - \varepsilon\big).$$

Now t_0 and ε were arbitrary, so

$$\mathbf{P}\{\text{e}\limsup X_t > \beta\} \leq \frac{1}{\beta} \limsup_{\tau\in\Sigma} \mathbf{E}[X_\tau].$$

Finally, let $\beta \to \lambda$ to obtain the result.

(d) By (b), for any $t_0 \in J$ there exists $\tau \geq t_0$ with

$$\mathbf{P}\{|\text{e}\limsup X_t - X_\tau| > \varepsilon\} < \varepsilon.$$

Apply this recursively. ∎

Convergence theorems hold in the presence of condition (V). Our proof will follow the method used in Chapter 1. A *semiamart* is a process $(X_t)_{t\in J}$ with

$$\limsup_{\sigma\in\Sigma} \big|\mathbf{E}[X_\sigma]\big| < \infty.$$

(See, for example, (1.4.26).) Clearly every amart is a semiamart. Note that for $J = \mathbb{N}$ this definition is equivalent to $\sup_{\sigma\in\Sigma} \big|\mathbf{E}[X_\sigma]\big| < \infty$ (Lemma (1.2.1)).

(4.2.9) Lattice property.

(1) If (X_t) is an L_1-bounded semiamart, then (X_t^+) is also a semiamart.
(2) If (X_t) is an L_1-bounded amart, then (X_t^+) is also an amart.

The proof is essentially the same as that of Theorem (1.2.2), and is therefore omitted.

(4.2.10) Theorem (Astbury). Let $(\mathcal{F}_t)_{t\in J}$ be a stochastic basis. The following are equivalent:

(1) Condition (V).
(2) L_∞-bounded amarts converge essentially.
(3) L_1-bounded amarts converge essentially.

Proof. (1) $\implies$ (2). Let (X_t) be an L_∞-bounded amart. For each $n \in \mathbb{N}$, choose $t_n \in J$ so that if $\tau, \sigma \geq t_n$, then $|\mathbf{E}[X_\tau] - \mathbf{E}[X_\sigma]| < 1/n$. Then by Proposition (4.2.8(d)), there exist stopping times τ_n with $\tau_n \geq t_n$, $\tau_{n+1} \geq \tau_n$, and $X_{\tau_n} \to \mathrm{e}\limsup X_t$ a.s. By Proposition (4.2.8(d)) applied to $(-X_t)$, there exist stopping times σ_n with $\sigma_n \geq t_n$, $\sigma_{n+1} \geq \sigma_n$, and $X_{\sigma_n} \to \mathrm{e}\liminf X_t$. Hence

$$0 = \lim_n |\mathbf{E}[X_{\tau_n}] - \mathbf{E}[X_{\sigma_n}]| = \mathbf{E}[\mathrm{e}\limsup X_t - \mathrm{e}\liminf X_t],$$

so $\mathrm{e}\limsup X_t = \mathrm{e}\liminf X_t$.

(1) $\implies$ (3). Suppose (1) holds. Then (as we just proved) also (2) holds. Let (X_t) be an L_1-bounded amart. By the lattice property (4.2.10), if $\lambda > 0$, then the process $((-\lambda) \vee X_t \wedge \lambda)$ is an L_∞-bounded amart. Therefore by (2) it converges essentially. Therefore the original process (X_t) converges essentially on the set $\Omega_\lambda = \{\mathrm{e}\limsup |X_t| \leq \lambda\}$. But the maximal inequality (4.2.8(c)) shows that $\mathbf{P}\{\mathrm{e}\limsup |X_t| < \infty\} = 1$, so Ω is the countable union of sets Ω_λ, hence (X_t) converges essentially.

(3) $\implies$ (2) is easy.

(2) $\implies$ (1). Let (A_t) be an adapted family of sets. For $t \in J$, let X_t be the Snell envelope (4.1.15):

$$X_t = \operatorname*{ess\,sup}_{\tau \geq t} \mathbf{E}^{\mathcal{F}_t}[\mathbf{1}_{A(\tau)}].$$

The net $(X_\sigma)_{\sigma \in \Sigma}$ is decreasing, so (X_t) is an L_∞-bounded amart. So by (2) it converges essentially. Now $X_t \geq \mathbf{1}_{A_t}$ and $\mathbf{1}_{A^*} \leq \mathrm{e}\limsup X_t$, so by the essential convergence, $\mathrm{e}\limsup X_t = \mathrm{s}\limsup X_t$, and thus by (4.1.11(c)),

$$\mathbf{P}(A^*) \leq \mathbf{P}\{\mathrm{e}\limsup X_t \geq 1\} \leq \limsup_{\tau \in \Sigma} \mathbf{E}[X_\tau].$$

Given $\sigma \in \Sigma$, there exists a sequence $\tau_n \geq \sigma$ such that $\mathbf{E}^{\mathcal{F}_\sigma}[\mathbf{1}_{A(\tau_n)}] \uparrow X_\sigma$, so

$$\mathbf{E}[X_\sigma] = \lim_n \mathbf{E}\left[\mathbf{E}^{\mathcal{F}_\sigma}[\mathbf{1}_{A(\tau_n)}]\right] = \lim_n \mathbf{P}(A(\tau_n)).$$

Hence $\limsup_\tau \mathbf{E}[X_\tau] \leq \limsup_\tau \mathbf{P}(A(\tau))$. Thus

$$\mathbf{P}(A^*) \leq \limsup_\tau \mathbf{P}(A(\tau)),$$

so (V) holds. ∎

The next corollary is an immediate consequence of Theorem (4.2.11).

(4.2.11) Corollary. Let $(\mathcal{F}_t)_{t \in J}$ satisfy (V). Then L_1-bounded martingales converge essentially

Note, however, that under condition (V), it it not necessarily true that L_1-bounded submartingales converge essentially (4.2.17). We will see in Section 4.4 that condition (V) is not necessary for convergence of L_1-bounded martingales.

Complements

(4.2.12) (σ-directed set.) Suppose the directed set J has the property that every countable subset has an upper bound. (We say that J is a σ-*directed set*.) Then any stochastic basis $(\mathcal{F}_t)_{t \in J}$ indexed by J satisfies condition (V). To see this, let (A_t) be adapted, and suppose

$$\alpha = \liminf_{\tau \in \Sigma} \mathbf{P}(A^* \setminus A(\tau)) > 0.$$

Next, choose indices $s_n \in J$ with

$$\inf_{\tau \geq s_n} \mathbf{P}(A^* \setminus A(\tau)) > \alpha - \frac{1}{n}$$

for $n = 1, 2, \cdots$. There is $s_\infty \in J$ larger than all s_n, so

$$\liminf_{\tau \geq s_\infty} \mathbf{P}(A^* \setminus A(\tau)) = \inf_{\tau \geq s_\infty} \mathbf{P}(A^* \setminus A(\tau)) = \alpha.$$

Now choose $\tau_1 = s_\infty$, and continue choosing recursively τ_n with $\tau_n \lll \tau_{n+1}$ so that $\mathbf{P}(A^* \setminus A(\tau_n)) \to \alpha$. Choose t_∞ larger than all τ_n. Then define a countably valued stopping time τ_∞ by:

$$\tau_\infty = \begin{cases} \tau_n & \text{on } A(\tau_n) \setminus \bigcup_{j=1}^{n-1} A(\tau_j) \quad \text{for } n = 1, 2, \cdots \\ t_\infty & \text{elsewhere.} \end{cases}$$

So $A(\tau_\infty) \supseteq \bigcup A(\tau_n)$, and thus $\mathbf{P}(A^* \setminus A(\tau_\infty)) = \alpha > 0$. But $A^* \setminus A(\tau_\infty)$ is the e lim sup of the $A_t \setminus A(\tau_\infty)$, so there is $t_1 > t_0$ such that

$$\mathbf{P}\big((A^* \cap A_{t_1}) \setminus A(\tau_\infty)\big) > 0,$$

so $\mathbf{P}\big(A^* \setminus (A(\tau_\infty) \cup A_{t_1}))\big) < \alpha$. Then for large enough n we have also $\mathbf{P}\big(A^* \setminus (A(\tau_n) \cup A_{t_1}))\big) < \alpha$. Then we may construct $\sigma \in \Sigma$ with $\mathbf{P}(A^* \setminus A(\sigma)) < \alpha$, a contradiction.

(4.2.13) (Condition (V).) All the conditions in Theorem (4.2.9) are equivalent with condition (V).

(4.2.14) (Other generalizations.) Let $(X_t)_{t \in J}$ be a stochastic process. For $\sigma, \tau \in \Sigma$, $\sigma \leq \tau$, write

$$H(\sigma, \tau) = X_\sigma - \mathbf{E}^{\mathcal{F}_\sigma}[X_\tau].$$

Then we say that (X_t) is a *pramart* iff

$$\mathrm{s} \lim_{\sigma \leq \tau} H(\sigma, \tau) = 0;$$

a *subpramart* iff

$$s \limsup_{\sigma \le \tau} H(\sigma, \tau) \le 0;$$

a *martingale in the limit* iff

$$e \lim_{s \le t} H(s, t) = 0.$$

Among other conditions equivalent with (V) are:

Every amart is a martingale in the limit.
Every pramart is a martingale in the limit.
Every L_1-bounded pramart (or subpramart) converges essentially.
Every L_1-bounded submartingale amart converges essentially.

Pramarts and subpramarts converge under the following condition (d), which is properly weaker than L_1-boundedness:

(d) $$\liminf \mathbf{E}\left[X_t^+\right] + \liminf \mathbf{E}\left[X_t^-\right] < \infty.$$

(4.2.15) (Abstract difference condition.) Let $(\mathcal{F}_t)_{t \in J}$ be a stochastic basis. Suppose, for $\sigma, \tau \in \Sigma$, we are given a random variable $f(\sigma, \tau)$. Assume that:

(1) For each $s \in J$, $\mathbf{1}_{\{\sigma=s\}} f(\sigma, \tau) = \mathbf{1}_{\{\sigma=s\}} f(s, \tau)$ a.s.
(2) For each $s \in J$, $A \in \mathcal{F}_s$, and $\tau, \tau' \in \Sigma$, if $\tau = \tau'$ on A, then $f(s, \tau) = f(s, \tau')$ on A.

Suppose $\mathcal{F}$ satisfies condition (V). If $f(\sigma, \tau)$ converges stochastically, then it converges essentially (Millet & Sucheston [1980b]; this paper missed the needed second localization condition, as was pointed out by A. Bellow).

(4.2.16) (Submartingale and supermartingale compared to amart.)
    (a) An example of a supermartingale that is not an amart. The stochastic basis satisfies condition (V), yet the L_1-bounded supermartingale does not converge essentially.
    Let $c_i = 2^{i^2}$. Let $J = \{ (i,j) : i \ge 0, 1 \le j \le c_i \}$ be ordered by:

$$(i,j) \le (i', j') \quad \Longleftrightarrow \quad i < i' \text{ or } (i,j) = (i', j').$$

For $(i,j) \in J$, let $\mathcal{F}_{(i,j)}$ be the σ-algebra on $\Omega = [0,1)$ generated by the partition

$$\left\{ \left[\frac{k-1}{c_i}, \frac{k}{c_i} \right) : 1 \le k \le c_i \right\}.$$

Define random variables $X_{(i,j)}$ by:

$$X_{(i,j)} = \begin{cases} i & \text{on } [(j-1)/c_i, j/c_i), \\ 1/i & \text{elsewhere.} \end{cases}$$

Then: (X_t) is an L_1-bounded supermartingale and $(\mathcal{F}_t)_{t \in J}$ satisfies (V) since it is totally ordered. But (X_t) clearly does not converge essentially, so it is not an amart. The controlled Vitali condition (4.2.25) fails.

(b) A submartingale that is an amart, but the net $(\mathbf{E}[X_\tau])_{\tau \in \Sigma}$ is not increasing.

Let $J = \mathbb{N} \times \mathbb{N}$, ordered by $(i_1, j_1) \le (i_2, j_2)$ iff $i_1 \le i_2$ and $j_1 \le j_2$. Let $\Omega = [0,1]$, and let $\mathbf{P}$ be Lebesgue measure. Define $\mathcal{F}_{(0,0)}$ as the trivial σ-algebra $\{\emptyset, \Omega\}$; for all other $(i,j) \in J$, define $\mathcal{F}_{(i,j)}$ as the σ-algebra with atoms $[0,1/2]$ and $(1/2,1]$. Thus $(\mathcal{F}_{(i,j)})$ is a totally ordered stochastic basis. Define a stochastic process $(X_{(i,j)})$ as follows:

$$X_{(0,0)} = 0$$
$$X_{(0,1)} = 2\,\mathbf{1}_{[0,1/2]} - \mathbf{1}_{(1/2,1]}$$
$$X_{(1,0)} = -\mathbf{1}_{[0,1/2]} + 2\,\mathbf{1}_{(1/2,1]}$$
$$X_{(i,j)} = 3 \qquad\qquad \text{for all other } (i,j).$$

For this process, we have $X_s \le \mathbf{E}^{\mathcal{F}_s}[X_t]$ for all $s \le t$ in J, but $\mathbf{E}[X_\sigma] \not\le \mathbf{E}[X_\tau]$, where $\sigma \le \tau$ in Σ are given by

$$\sigma(\omega) = (0,0)$$
$$\tau(\omega) = \begin{cases} (0,1) & \text{if } \omega \in (1/2,1], \\ (1,0) & \text{if } \omega \in [0,1/2]. \end{cases}$$

Also, the net $\mathbf{E}[X_\tau]$ converges to 3, so (X_t) is an amart.

(4.2.17) (Supermartingale amarts in the derivation setting.) Let $\mathcal{A}$ be an algebra of subsets of Ω. A set function $\psi \colon \mathcal{A} \to [0, \infty)$ is called a *charge* iff it is finitely additive; it is called a *supercharge* iff it is finitely superadditive, that is

$$\psi(A \cup B) \ge \psi(A) + \psi(B),$$

for disjoint $A, B \in \mathcal{A}$. We discuss in Chapter 8 the decomposition of a supercharge ψ as $\psi = \psi_{\mathrm{m}} + \psi_{\mathrm{c}} + \psi_{\mathrm{s}}$, where ψ_{m} is a measure, ψ_{c} is a pure charge, and ψ_{s} is a pure supercharge (8.4.1). Here we will discuss the connection of amart theory to the derivation of charges and supercharges. As we will see, we obtain a supermartingale that is also an amart; the supermartingale property is less useful than the amart property, since supermartingales converge essentially only under conditions like (V^O), strictly stronger than condition (V); while by Astbury's theorem (4.2.11), amarts converge under (V). Condition (V) holds in the classical setting of Theorem (4.2.8).

(a) Let $(\Omega, \mathcal{F}, \mathbf{P})$ be a probability space; let J be a set of finite measurable partitions of Ω directed by refinement. For each $t \in J$, let $\mathcal{F}_t$ be the σ-algebra generated by t. Let ψ be a supercharge on the algebra $\mathcal{A} = \bigcup \mathcal{F}_t$. For each $t \in J$, let

$$X_t = \sum_{A \in t} \frac{\psi(A)}{\mathbf{P}(A)}\, \mathbf{1}_A,$$

with the convention that $\psi(A)/\mathbf{P}(A) = 0$ if $\mathbf{P}(A) = 0$. Then X_t is an amart (and a supermartingale).

Proof. Let $\sigma, \tau \in \Sigma$ with $\sigma \leq \tau$. We claim that $\mathbf{E}[X_\tau] \leq \mathbf{E}[X_\sigma]$. Write s_i for the values of σ and t_j for the values of τ. Then

$$\mathbf{E}^{\mathcal{F}_\sigma}[X_\tau] = \sum_i \sum_{\substack{A \in s_i \\ A \subseteq \{\sigma = s_i\}}} \frac{\mathbf{E}[X_\tau 1_A]}{\mathbf{P}(A)}$$

$$= \sum_i \sum_{\substack{A \in s_i \\ A \subseteq \{\sigma = s_i\}}} \frac{1}{\mathbf{P}(A)} \left(\sum_j \sum_{\substack{B \in t_j \\ B \subseteq \{\tau = t_j\} \cap A}} \psi(B) \right) 1_A$$

$$\leq \sum_i \sum_{\substack{A \in s_i \\ A \subseteq \{\sigma = s_i\}}} \frac{1}{\mathbf{P}(A)} \left(\sum_{t_j \geq s_i} \psi(\{\tau = t_j\} \cap A) \right) 1_A$$

$$\leq X_\sigma.$$

Integrate to obtain $\mathbf{E}[X_\tau] \leq \mathbf{E}[X_\sigma]$. ∎

(b) The Riesz decomposition is $X_t = Y_t + Z_t$, where the martingale part is given by

$$Y_t = \sum_{A \in t} \frac{\psi_{\mathrm{m}}(A) + \psi_{\mathrm{c}}(A)}{\mathbf{P}(A)} 1_A.$$

Proof. Let $\varepsilon > 0$. Since ψ_{s} is a pure supercharge, there is a finite partition $A_1, \cdots, A_n$ composed of sets of $\mathcal{A}$ such that $\sum_{i=1}^n \psi_{\mathrm{s}}(A_i) \leq \varepsilon$. Choose $t \in J$ so large that $\mathcal{F}_t$ contains all the sets $A_1, \cdots, A_n$. Now

$$0 \leq \sum_{A \in t} [\psi(A) - \psi_{\mathrm{m}}(A) - \psi_{\mathrm{c}}(A)] \leq \sum_{i=1}^n \psi_{\mathrm{s}}(A_i) \leq \varepsilon.$$

The Riesz decomposition follows. ∎

(c) An analogous theorem (with analogous proof) applies when J is a set of countable measurable partitions directed by refinement. Note that since ψ is superadditive, it is automatically countably superadditive. In this case, the martingale part in the Riesz decomposition is:

$$Y_t = \sum_{A \in t} \frac{\psi_{\mathrm{m}}(A)}{\mathbf{P}(A)} 1_A.$$

(4.2.18) (An example of the Riesz decomposition.) Let $(X_t)_{t \in J}$ be a stochastic basis generated by a family J of partitions of $[0, 1]^d$ satisfying the conditions of Theorem (4.2.8). Let Q be a finite signed measure absolutely continuous with respect to $\mathbf{P}$ on $\mathcal{F}_\infty = \sigma(\bigcup \mathcal{F}_t)$. Let f and g be real functions such that $f(0) = g(0) = 0$, derivatives $f'(0), g'(0)$ exist, and $g'(0) \neq 0$. Then the stochastic process (X_t) defined by

$$X_t = \sum_{A \in t} \frac{f(Q(A))}{f(\mathbf{P}(A))} \mathbf{1}_A, \qquad t \in J$$

is an amart. The process X_t converges essentially to

$$\frac{f'(0)}{g'(0)} \frac{dQ}{d\mathbf{P}},$$

where $dQ/d\mathbf{P}$ is the Radon-Nikodým derivative of Q with respect to $\mathbf{P}$ on $\mathcal{F}_\infty$. The martingale part of the Riesz decomposition of (X_t) is given by

$$Y_t = \sum_{A \in t} \frac{f'(0)Q(A)}{g'(0)\mathbf{P}(A)} \mathbf{1}_A, \qquad t \in J.$$

(4.2.19) (A combinatorial limit.)

$$\lim_{m \to \infty} 2^{-3m} \sum_{k=m}^{3m} \binom{3m}{k} = 1.$$

Let $U_1, \cdots, U_{3m}$ be independent random variables with $\mathbf{P}\{U_i = 1\} = \mathbf{P}\{U_i = -1\} = 1/2$ for all i. Then the expression under the limit sign is the probability that there are at least m ones among $U_1, \cdots, U_{3m}$. Now the sum $S = \sum_{i=1}^{3m} U_i$ has mean $\mathbf{E}[S] = 0$ and variance $\mathbf{E}[S^2] = 3m$, since the U_i are orthonormal. Now we may use Chebyshev's inequality to estimate the probability in question, which is

$$\mathbf{P}\{S \geq -m\} \geq 1 - \mathbf{P}\{|S| > m\}$$
$$= 1 - \mathbf{P}\{S^2 \geq m^2\}$$
$$\geq 1 - \frac{1}{m^2}\mathbf{E}[S^2]$$
$$= 1 - \frac{3m}{m^2}.$$

(4.2.20) (Cofinal optional sampling.) A class $\mathcal{E}$ of processes $(X_t, \mathcal{F}_t, J)$ has the *cofinal optional sampling property* iff, for every $(X_t, \mathcal{F}_t, J) \in \mathcal{E}$ and every cofinal subset J' of Σ, also the process $(X_\tau, \mathcal{F}_\tau, J')$ is in $\mathcal{E}$. The classes of amarts, pramarts, and subpramarts have the cofinal optional sampling property (Millet & Sucheston [1980b]).

(4.2.21) (Monotone optional sampling.) A class $\mathcal{E}$ of processes $(X_n, \mathcal{F}_n)$ indexed by $\mathbb{N}$ has the *monotone optional sampling property* iff for every

$(X_n, \mathcal{F}_n)_{n \in \mathbb{N}} \in \mathcal{E}$, and every increasing sequence τ_k in $\Sigma(\mathcal{F}_n)$, also the process $(X_{\tau_k}, \mathcal{F}_{\tau_k})_{k \in \mathbb{N}}$ belongs to $\mathcal{E}$. The classes of amarts, pramarts, and subpramarts indexed by $\mathbb{N}$ have the monotone optional sampling property (Millet & Sucheston [1980b]).

(4.2.22) (Constant stochastic basis.) Suppose the stochastic basis $\mathcal{F}_t$ satisfies $\mathcal{F}_t = \mathcal{F}_s$ for all $s, t \in J$. Then a process X_t is an amart if and only if X_t converges essentially and there is $s \in J$ so that

$$\mathbf{E}\left[\operatorname*{ess\,sup}_{t \geq s} |X_t|\right] < \infty$$

(Edgar & Sucheston [1976a] for $\mathbb{N}$; Millet & Sucheston [1980b] for directed set; see also (1.2.11).)

(4.2.23) (Ordered Vitali condition (V^O).) A stochastic basis $(\mathcal{F}_t)$ satisfies *condition* (V^O) iff, for each adapted family (A_t), and each $\varepsilon > 0$, there exists $\tau \in \Sigma^O$ with $\mathbf{P}(A^* \setminus A(\tau)) < \varepsilon$. As we know (1.3.1), L_1-bounded ordered amarts converge stochastically. Condition (V^O) is necessary and sufficient for essential convergence of L_1-bounded ordered amarts. Condition (V^O), originally called (V'), is sufficient for convergence of L_1-bounded submartingales (Krickeberg [1959]), but it is not necessary.

Totally ordered stochastic bases need not satisfy condition (V^O). In example (4.2.16(a)), there is a totally ordered stochastic basis, but condition (V^O) fails because L_1-bounded supermartingales may not converge. Millet & Sucheston [1980b] has an example of a totally ordered stochastic basis where L_1-bounded supermartingales converge, but nevertheless (V^O) fails.

(4.2.24) (Controlled Vitali condition.) We say that a stopping time $\tau \in \Sigma$ is *controlled* by $\sigma \in \Sigma^O$ iff $\sigma \leq \tau$ and τ is $\mathcal{F}_\sigma$-measurable. We say that τ is a *controlled stopping time* iff τ is controlled by some $\sigma \in \Sigma^O$.

The stochastic basis $(\mathcal{F}_t)$ satisfies the *controlled Vitali condition* (V^C) iff for every adapted family (A_t) of sets, and $A \in \mathcal{F}_\infty$ with $A \subseteq \operatorname{e\,lim\,sup} A_t$, for every $\varepsilon > 0$ there exists a stopping time $\tau \in \Sigma$ controlled by an ordered stopping time $\sigma \in \Sigma^O$, and a set $B \in \mathcal{F}_\sigma$ such that $B \subseteq A(\tau)$ and $\mathbf{P}(A \setminus B) < \varepsilon$.

Then we have

$$(V^O) \implies (V^C) \implies (V)$$

and the implications are not reversible.

(4.2.25) (Controlled amarts.) Write Σ^C for the set of all controlled stopping times. If $\tau_1, \tau_2 \in \Sigma^C$, write $\tau_1 <_C \tau_2$ iff there exists σ_2 controlling τ_2 such that $\sigma_1 \leq \sigma_2$ for every σ_1 controlling τ_1. Then Σ^C is directed by the relation $<_C$. A stochastic process (X_t) is a *controlled amart* iff the net $(\mathbf{E}[X_\tau])_{\tau \in \Sigma^C}$ converges.

Note that every L_1-bounded supermartingale or submartingale is a controlled amart. If condition (V^C) holds, then every L_1-bounded controlled amart converges essentially.

Remarks

The Vitali condition (V) was obtained by analogy from the related setting of derivations (see Chapter 7). In the derivation setting, it goes back to the lemma of Vitali. In the directed-set setting, condition (V) was introduced by Krickeberg [1956], who proved that martingales converge under (V); the condition was also studied by Y. S. Chow [1960b]. The important fact that condition (V) is equivalent to amart convergence (Theorem (4.2.11)) is due to Astbury [1978].

Millet & Sucheston [1980b] is the source of (4.2.2) to (4.2.4), (4.2.8) to (4.2.9), (4.2.14) to (4.2.19), and (4.2.23) to (4.2.25). They also discuss reversed ordered amarts, and Banach-valued ordered amarts.

4.3. L_Ψ-bounded martingales

Our next concern will be the convergence of L_Ψ-bounded martingales, where Ψ is an Orlicz function. (Orlicz functions, and Orlicz spaces, are discussed in Chapter 2.) In this section, we will limit the setting to a probability space $(\Omega, \mathcal{F}, \mathbf{P})$. This means that $L_\Psi \subseteq L_1$ and any L_Ψ-bounded martingale is L_1-bounded. But, on the other hand, covering conditions that ensure essential convergence of all L_1-bounded martingales may be much stronger than necessary to ensure essential convergence of all L_Ψ-bounded martingales. We will return in the next section to the most important case, namely L_1 itself.

The covering condition most often used in the study of L_Ψ-bounded martingales is stated in terms of the Orlicz function Φ conjugate to Ψ and multivalued stopping times.

Multivalued stopping times

A simple stopping time $\tau \in \Sigma$ defines a finite family of sets $A_t = \{\tau = t\}$ such that $A_t \in \mathcal{F}_t$ and $A_s \cap A_t = \emptyset$ for $s \neq t$. In order to study finite adapted families of sets with overlap, we will use a modification of this sort of stopping time, namely the *multivalued stopping time*. There are two equivalent ways to look at the definition.

Let $(\mathcal{F}_t)_{t \in J}$ be a stochastic basis. Let $\mathcal{J}$ be the family of all finite nonempty subsets of J. A *simple multivalued stopping time* is a function $\tau \colon \Omega \to \mathcal{J}$, with only finitely many values, such that, for each $t \in J$,

$$\{\, \omega \in \Omega : t \in \tau(\omega)\,\} \in \mathcal{F}_t.$$

We will write $\{t \in \tau\}$ for this set. (Some of the literature writes $\{\tau = t\}$ for this set.) We write Σ^M for the set of all simple multivalued stopping times. We may identify Σ with the subset of Σ^M such that $\tau(\omega)$ is a singleton for all $\omega \in \Omega$.

Similarly, a *simple incomplete multivalued stopping time* is a function $\tau \colon \Omega \to \mathcal{J} \cup \{\emptyset\}$, with only finitely many values, such that, for each $t \in J$, $\{t \in \tau\} \in \mathcal{F}_t$. We write Σ^{IM} for the set of all simple incomplete multivalued stopping times.

Consider an adapted family $(A_t)_{t\in J}$ of sets, only finitely many of which are nonempty. There is a unique $\tau \in \Sigma^{IM}$ with $\{t \in \tau\} = A_t$ for all t. So an alternative definition could be given in terms of such families of sets.

Let $\tau \in \Sigma^{IM}$. We write $D(\tau) = \bigcup_t \{t \in \tau\}$, called the *domain* of τ. Thus $D(\tau) = \Omega$ if and only if $\tau \in \Sigma^M$. We write $S(\tau) = \sum_t \mathbf{1}_{\{t\in\tau\}}$ for the *sum* of τ. Then $\tau \in \Sigma$ if and only if $S(\tau) = 1$. The *excess* of τ is $e(\tau) = S(\tau) - \mathbf{1}_{D(\tau)}$. This function can be used to measure the overlap properties of the family $(\{t \in \tau\})_{t\in J}$ of sets.

Let $(X_t)_{t\in J}$ be a stochastic process. If $\tau \in \Sigma^{IM}$, we write

$$X_\tau = \sum_{t\in J} X_t \, \mathbf{1}_{\{t\in\tau\}}.$$

Or, if we think of $\tau(\omega)$ as a finite subset of J,

$$(X_\tau)(\omega) = \sum_{t\in\tau(\omega)} X_t(\omega).$$

If $\tau \in \Sigma$, this coincides with the usual definition of X_τ. Let $(A_t)_{t\in J}$ be an adapted family of sets (finite or infinite). For $\tau \in \Sigma^{IM}$ we write

$$A(\tau) = \bigcup_{t\in J} (\{t \in \tau\} \cap A_t).$$

We will say that the stopping time τ is *subordinate* to the adapted family (A_t) iff $\{t \in \tau\} \subseteq A_t$ for all $t \in J$. Then $D(\tau) = A(\tau)$.

An ordering may be defined on Σ^M in more than one way. For our purposes, they are all equally useful. We have chosen the easiest to define. For $\tau \in \Sigma^{IM}$ and $t_0 \in J$, we write $t_0 \le \tau$ iff $\{t \in \tau\} = \emptyset$ except for $t \ge t_0$; we write $\tau \le t_0$ iff $\{t \in \tau\} = \emptyset$ except for $t \le t_0$. If $\sigma, \tau \in \Sigma^M$, we write $\sigma \lll \tau$ iff there is $t_0 \in J$ with $\sigma \le t_0$ and $t_0 \le \tau$.

Covering condition (V_Φ)

We begin with a covering condition analogous to the Vitali condition (V). We allow multivalued stopping times, but require that the L_Φ norm $\|e(\tau)\|_\Phi$ of the excess (known as the *overlap* of τ) be small.

(4.3.1) Definition. Let Φ be an Orlicz function. The stochastic basis $(\mathcal{F}_t)_{t\in J}$ satisfies the covering condition (V_Φ) iff, for each adapted family $(A_t)_{t\in J}$ of sets, and each $\varepsilon > 0$, there exists $\tau \in \Sigma^{IM}$ with $\mathbf{P}(A^* \setminus A(\tau)) < \varepsilon$ and $\|e(\tau)\|_\Phi < \varepsilon$.

First we will record some useful technical variants of condition (V_Φ). Also, since $\mathbf{P}(A(\tau) \setminus A^*) \to 0$ by the definition of e lim sup, any of the formulations with $\mathbf{P}(A^* \setminus A(\tau)) < \varepsilon$ may equivalently be restated with $\mathbf{P}(A^* \bigtriangleup A(\tau)) < \varepsilon$. Any of these formulations may be called "condition (V_Φ)."

(4.3.2) Lemma. *The following are equivalent.*

(a) (V_Φ): *for every adapted $(A_t)_{t \in J}$ and every $\varepsilon > 0$, there exists $\tau \in \Sigma^{IM}$ with $\mathbf{P}(A^* \setminus A(\tau)) < \varepsilon$ and $\|e(\tau)\|_\Phi < \varepsilon$.*

(b) *For every adapted $(A_t)_{t \in J}$ and every $\varepsilon > 0$, there exists $\tau \in \Sigma^M$ with $\mathbf{P}(A^* \setminus A(\tau)) < \varepsilon$ and $\|e(\tau)\|_\Phi < \varepsilon$.*

(c) *For every adapted $(A_t)_{t \in J}$, every $\varepsilon > 0$, and every $t_0 \in J$, there exists $\tau \in \Sigma^M$ with $\tau \geq t_0$, $\mathbf{P}(A^* \setminus A(\tau)) < \varepsilon$ and $\|e(\tau)\|_\Phi < \varepsilon$.*

(d) *For every adapted $(A_t)_{t \in J}$, every $\varepsilon > 0$, and every $t_0 \in J$, there exists $\tau \in \Sigma^{IM}$ subordinate to $(A_t)_{t \in J}$ with $\tau \geq t_0$, $\mathbf{P}(A(\tau)) > \mathbf{P}(A^*) - \varepsilon$, and $\|e(\tau)\|_\Phi < \varepsilon$.*

Proof. (a) $\Longrightarrow$ (b). Given $\tau \in \Sigma^{IM}$, choose $s \in J$ with $s \geq \tau$, and define τ' by

$$\{t \in \tau'\} = \begin{cases} \{t \in \tau\} & t \neq s \\ \Omega \setminus D(\tau) & t = s. \end{cases}$$

Then $\tau' \in \Sigma^M$, $A^* \setminus A(\tau) \supseteq A^* \setminus A(\tau')$, and $e(\tau) = e(\tau')$.

(b) $\Longrightarrow$ (c). Given $(A_t)_{t \in J}$ and t_0, define (B_t) by $B_t = A_t$ if $t \geq t_0$ and $B_t = \emptyset$ otherwise. Then $B^* = A^*$ and we may apply (b) to the family (B_t) to get (c) for the family (A_t).

(c) $\Longrightarrow$ (d). Given $\tau \in \Sigma^M$, let $\tau' \in \Sigma^{IM}$ be defined by $\{t \in \tau'\} = A_t \cap \{t \in \tau\}$. Then $e(\tau') \leq e(\tau)$ and $A^* \setminus A(\tau) = A^* \setminus A(\tau')$.

(d) $\Longrightarrow$ (a). Let (A_t) be an adapted family, and let $\varepsilon > 0$. Choose t_0 so that $\mathbf{P}\{\operatorname{ess\,sup}_{t \geq t_0} A_t \setminus A^*\} < \varepsilon/2$. By (d) there exists $\tau \in \Sigma^{IM}$ subordinate to (A_t) with $\tau \geq t_0$, $\mathbf{P}(A(\tau)) > \mathbf{P}(A^*) - \varepsilon/2$, and $\|e(\tau)\|_\Phi < \varepsilon/2$. But then

$$\mathbf{P}\big(A^* \setminus A(\tau)\big) = \mathbf{P}(A^*) - \mathbf{P}\big(A(\tau)\big) + \mathbf{P}\big(A(\tau) \setminus A^*\big) < \frac{\varepsilon}{2} + \frac{\varepsilon}{2} = \varepsilon. \qquad \blacksquare$$

It should be noted that if $L_\Phi = L_\infty$, then condition (V_Φ) becomes condition (V). So (V) is sometimes known as (V_∞). Similarly, when $L_\Phi = L_p$, we write (V_p) for (V_Φ).

We will see that under fairly general conditions, (V_Φ) is necessary and sufficient for the convergence of L_Ψ-bounded martingales. The case of L_1-bounded martingales is excluded, but L_q-bounded martingales are included, for $1 < q \leq \infty$.

There is a sort of convergence that is suggested by condition (V_Φ). Suppose $f(\tau) \in \mathbb{R}$ for each $\tau \in \Sigma^M$. We will write

$$V_\Phi \lim f(\tau) = u$$

iff for every $\varepsilon > 0$, there is $t_0 \in J$ and $\delta > 0$ so that, if $\tau \in \Sigma^M$, $\tau \geq t_0$, $\|e(\tau)\|_\Phi < \delta$, then $|f(\tau) - u| < \varepsilon$. Note the use of Σ^M, not Σ^{IM}. This could be considered the Moore-Smith convergence defined by a directed set. The elements of the directed set are pairs (τ, δ), where $\tau \in \Sigma^M$, $\delta > 0$, and $\|e(\tau)\|_\Phi < \delta$. The ordering is given by:

$$(\tau, \delta) \leq (\tau', \delta')$$

iff $\delta \geq \delta'$ and either $\tau \ll \tau'$ or $\tau = \tau'$. Note that if $V_\Phi \lim f(\tau)$ exists, then $\lim_{\tau \in \Sigma} f(\tau)$ exists, and the limits are equal.

There is a corresponding $V_\Phi \lim \sup$. We will write

$$V_\Phi \lim \sup f(\tau) \leq K$$

iff for every $\varepsilon > 0$, there exist $t_0 \in J$, and $\delta > 0$, such that if $\tau \in \Sigma^M$ with $\tau \geq t_0$, $\|e(\tau)\|_\Phi < \delta$, then $f(\tau) < K + \varepsilon$. Clearly there is also a corresponding $V_\Phi \lim \inf$.

There are classes of processes that we may naturally associate with condition (V_Φ).

(4.3.3) Definition. Let $(X_t)_{t \in J}$ be an adapted process. Then

(1) $(X_t)_{t \in J}$ is a V_Φ-*amart* iff $V_\Phi \lim \mathbf{E}[X_\tau]$ exists;

(2) $(X_t)_{t \in J}$ is a V_Φ-*semiamart* iff $V_\Phi \lim \sup |\mathbf{E}[X_\tau]| < \infty$.

(3) $(X_t)_{t \in J}$ is a V_Φ-*potential* iff $V_\Phi \lim \mathbf{E}[|X_\tau|] = 0$.

(4.3.4) Proposition. If $(X_t)_{t \in J}$ is a V_Φ-amart and $\lim_t \mathbf{E}[X_t \mathbf{1}_A] = 0$ for all $A \in \bigcup_t \mathcal{F}_t$, then $(X_t)_{t \in J}$ is a V_Φ-potential.

Proof. A V_Φ-amart is an amart, so (X_t) is an amart potential. In particular, $\mathbf{E}[|X_t|] \to 0$. Now given $\varepsilon > 0$, choose $\delta > 0$ and $t_0 \in J$ so that if $\sigma_1, \sigma_2 \in \Sigma^M$, $\sigma_1, \sigma_2 \geq t_0$, $\|e(\sigma_1)\|_\Phi < \delta$, and $\|e(\sigma_2)\|_\Phi < \delta$, then $|\mathbf{E}[X_{\sigma_1} - X_{\sigma_2}]| < \varepsilon$. Let $\sigma \in \Sigma^M$ with $\|e(\sigma)\|_\Phi < \delta$ and $\sigma \geq t_0$. Choose $t_1 > \sigma$. Let $A = \bigcup_t \{X_t > 0, t \in \sigma\}$, so that $A \in \mathcal{F}_{t_1}$. Define $\tau_1, \tau_2 \in \Sigma^M$ by:

$$\{t \in \tau_1\} = \begin{cases} \{X_t > 0, t \in \sigma\} & \text{if } t \neq t_1 \\ \Omega \setminus A & \text{if } t = t_1 \end{cases}$$

$$\{t \in \tau_2\} = \begin{cases} \{X_t \leq 0, t \in \sigma\} & \text{if } t \neq t_1 \\ A & \text{if } t = t_1. \end{cases}$$

Then

$$|X_\sigma| = \left| \sum_t X_t \mathbf{1}_{\{t \in \sigma\}} \right| \leq \sum |X_t| \mathbf{1}_{\{t \in \sigma\}}$$

$$= \sum_{t \neq t_1} \left(X_t \mathbf{1}_{\{t \in \sigma, X_t > 0\}} - X_t \mathbf{1}_{\{t \in \sigma, X_t \leq 0\}} \right)$$

$$= \sum_{t \neq t_1} X_t \mathbf{1}_{\{t \in \tau_1\}} - \sum_{t \neq t_1} X_t \mathbf{1}_{\{t \in \tau_2\}}$$

$$\leq X_{\tau_1} - X_{\tau_2} + |X_{t_1}|.$$

Thus $\mathbf{E}[|X_\sigma|] \leq \varepsilon + \mathbf{E}[|X_{t_1}|]$. This shows that $V_\Phi \lim_\sigma \mathbf{E}[|X_\sigma|] = 0$. ∎

(4.3.5) Proposition (lattice property).

(a) If $(X_t)_{t \in J}$ is an L_1-bounded V_Φ-semiamart, then (X_t^+) is also a V_Φ-semiamart.

(b) If $(X_t)_{t \in J}$ is an L_1-bounded V_Φ-amart, then (X_t^+) is also a V_Φ-amart.

Proof. Write $U_t = X_t^+$. (We must use caution, since $U_\tau \neq (X_\tau)^+$, in general.)

(a) Let

$$V_\Phi \limsup \mathbf{E}\left[X_\tau\right] = K < \infty.$$

Choose $t_1 \in J$ and $\delta_1 > 0$ so that if $\tau \in \Sigma^M$, $\tau \geq t_1$, and $\|e(\tau)\|_\Phi < \delta_1$, then $\mathbf{E}[X_\tau] < K + 1$. Let $\sup_t \mathbf{E}[\|X_t\|] = M < \infty$. Now suppose $\tau \in \Sigma^M$ with $\tau \geq t_1$ and $\|e(\tau)\|_\Phi < \delta_1$. Choose $t_2 > \tau$, and let $A = \bigcup_t \{t \in \tau, X_t \geq 0\}$. Then $A \in \mathcal{F}_{t_2}$. Now define $\sigma \in \Sigma^M$ by:

$$\{t \in \sigma\} = \begin{cases} \{t \in \tau, X_t \geq 0\} & \text{if } t \neq t_2 \\ \Omega \setminus A & \text{if } t = t_2. \end{cases}$$

Thus $\sigma \geq t_1$ and $e(\sigma) \leq e(\tau)$, so $\|e(\sigma)\|_\Phi < \delta_1$. Then

$$\mathbf{E}[U_\tau] = \mathbf{E}\left[\sum_t X_t^+ \mathbf{1}_{\{t \in \tau\}}\right] = \mathbf{E}\left[\sum X_t \mathbf{1}_{\{t \in \tau, X_t \geq 0\}}\right]$$
$$\leq \mathbf{E}[X_\sigma] + \mathbf{E}[\|X_{t_2}\|] \leq K + M + 1.$$

Thus $V_\Phi \limsup |\mathbf{E}[U_\tau]| \leq K + M + 1 < \infty$.

(b) Let $\varepsilon > 0$ be given. Choose $t_1 \in J$ and $\delta_1 > 0$ so that if $\tau, \sigma \in \Sigma^M$, $\tau \geq t_1$, $\sigma \geq t_1$, $\|e(\tau)\|_\Phi < \delta_1$, and $\|e(\sigma)\|_\Phi < \delta_1$, then

$$|\mathbf{E}[X_\tau] - \mathbf{E}[X_\sigma]| < \varepsilon.$$

Next, using (a), there is $t_2 \geq t_1$, $\delta_2 \leq \delta_1/2$, and $\tau_1 \geq t_2$ with $\|e(\tau_1)\|_\Phi < \delta_2$, so that if $\tau \in \Sigma^M$, $\tau \geq t_2$, and $\|e(\tau)\|_\Phi < \delta_2$, then

$$\mathbf{E}[U_\tau] < \mathbf{E}[U_{\tau_1}] + \varepsilon.$$

Let $t_3 > \tau_1$. Now suppose $\tau \in \Sigma^M$, $\tau \geq t_3$, and $\|e(\tau)\|_\Phi < \delta_2$. Define A as the finite union $\bigcup_t \{t \in \tau_1, X_t < 0\}$. Note that $A \in \mathcal{F}_{t_3}$. Define $\sigma \in \Sigma^M$ by:

$$\{t \in \sigma\} = \begin{cases} \{t \in \tau\} \setminus A & t \geq t_3 \\ \{t \in \tau_1, X_t < 0\} & \text{otherwise.} \end{cases}$$

Now $e(\sigma) \le e(\tau_1) + e(\tau)$, so $\|e(\sigma)\|_\Phi \le \delta_2 + \delta_2 < \delta_1$, and $\sigma \ge t_2$. Therefore $|\mathbf{E}\,[X_{\tau_1}] - \mathbf{E}\,[X_\sigma]| < \varepsilon$. Also, $\mathbf{E}\,[U_\tau] < \mathbf{E}\,[U_{\tau_1}] + \varepsilon$. Now

$$
\begin{aligned}
U_{\tau_1} - U_\tau &= \sum_t X_t \mathbf{1}_{\{t\in\tau_1, X_t \ge 0\}} - \sum_t X_t \mathbf{1}_{\{t\in\tau, X_t \ge 0\}} \\
&= \sum_t X_t \mathbf{1}_{\{t\in\tau_1\}} - \sum_t X_t \mathbf{1}_{\{t\in\tau_1, X_t < 0\}} - \sum_t X_t \mathbf{1}_{\{t\in\tau, X_t \ge 0\}} \\
&= X_{\tau_1} - X_\sigma - \mathbf{1}_A \sum_t X_t \mathbf{1}_{\{t\in\tau_1, X_t \ge 0\}} + \mathbf{1}_{\Omega\setminus A} \sum_t X_t \mathbf{1}_{\{t\in\tau, X_t < 0\}} \\
&\le X_{\tau_1} - X_\sigma.
\end{aligned}
$$

Thus $\mathbf{E}\,[U_{\tau_1}] - \mathbf{E}\,[U_\tau] \le \mathbf{E}\,[X_{\tau_1}] - \mathbf{E}\,[X_\sigma] < \varepsilon$. Also $\mathbf{E}\,[U_{\tau_1}] - \mathbf{E}\,[U_\tau] \ge -\varepsilon$. Thus (U_t) is a V_Φ-amart. ∎

As usual, the covering condition is equivalent to a maximal inequality, to an "approximation by the stopped process," and to amart conditions.

(4.3.6) Theorem. *Let $(\mathcal{F}_t)_{t\in J}$ be a stochastic basis, and let Φ be an Orlicz function. The following are equivalent.*

(a) (V_Φ) *holds: for every adapted family* $(A_t)_{t\in J}$,

$$V_\Phi \liminf \mathbf{P}(A^* \setminus A(\tau)) = 0.$$

(b) *For every nonnegative process* $(X_t)_{t\in J}$, *and for any* $\lambda > 0$,

$$\mathbf{P}(\mathrm{e}\limsup X_t \ge \lambda) \le \frac{1}{\lambda}\, V_\Phi \limsup \mathbf{E}\,[X_\tau].$$

(c) *If* $(X_t)_{t\in J}$ *is adapted,* $\varepsilon > 0$, *and* $t \in J$, *then there is* $\tau \in \Sigma^M$ *with* $\tau \ge t$, $\|e(\tau)\|_\Phi < \varepsilon$, *and* $\mathbf{P}(\mathrm{e}\limsup X_t - X_\tau > \varepsilon) < \varepsilon$.

(d) *Every* L_∞-*bounded* V_Φ-*potential converges essentially (to 0).*

Proof. (a) $\Longrightarrow$ (b). Let $(X_t)_{t\in J}$ be a nonnegative process, and let $\lambda > 0$. Write $X^* = \mathrm{e}\limsup X_t$. Fix β with $0 < \beta < \lambda$. Define $A_t = \{X_t > \beta\}$. Then $\{X^* > \beta\} \subseteq A^*$. Let $t_0 \in J$ and $\varepsilon > 0$ be given. Then by (V_Φ), there is $\tau \in \Sigma^{\mathrm{IM}}$, subordinate to $(A_t)_{t\in J}$, with $\tau \ge t_0$, $\|e(\tau)\|_\Phi < \varepsilon$, and $\mathbf{P}(A^* \setminus A(\tau)) < \varepsilon$. Now

$$
\begin{aligned}
\mathbf{E}\,[X_\tau] = \sum_t \mathbf{E}\,[X_t \mathbf{1}_{\{t\in\tau\}}] &\ge \beta \sum_t \mathbf{E}\,[\mathbf{1}_{\{t\in\tau\}}] \\
&\ge \beta \mathbf{P}(A(\tau)) \ge \beta(\mathbf{P}(A^*) - \varepsilon) \\
&\ge \beta(\mathbf{P}(X^* > \beta) - \varepsilon).
\end{aligned}
$$

Take the limit as $\varepsilon \to 0$:

$$\mathbf{P}(X^* > \beta) \le \frac{1}{\beta}\, V_\Phi \limsup \mathbf{E}\,[X_\tau].$$

Then take the limit as $\beta \to \lambda$:

$$\mathbf{P}(X^* \geq \lambda) \leq \frac{1}{\lambda} \, V_\Phi \limsup \mathbf{E}\,[X_\tau].$$

(b) $\Longrightarrow$ (d). Let $(X_t)_{t \in J}$ be a V_Φ-potential. Then by (b), for any $\lambda > 0$,

$$\mathbf{P}(\mathrm{e}\limsup |X_t| \geq \lambda) \leq \frac{1}{\lambda} \, V_\Phi \limsup \mathbf{E}\,\big[|X_\tau|\big] = 0.$$

Thus $\mathrm{e}\limsup |X_t| = 0$, so $(X_t)_{t \in J}$ converges essentially to 0.

(a) $\Longrightarrow$ (c). Let $(X_t)_{t \in J}$ be a process and $\varepsilon > 0$. Let $\delta > 0$ be such that $\|X\|_\Phi < \delta$ implies $\|X\|_1 < \varepsilon$. Now X^* is $\sigma\left(\bigcup \mathcal{F}_t\right)$-measurable, so there is $t_1 \in J$ and a random variable Y, measurable with respect to $\mathcal{F}_{t_1}$, so that

$$\mathbf{P}\big(|X^* - Y| \geq \varepsilon\big) < \varepsilon.$$

Let $A_t = \{Y - X_t < \varepsilon\}$ for $t \geq t_1$. Then $A^* \supseteq \{|X^* - Y| < \varepsilon\}$. Thus $\mathbf{P}(A^*) > 1 - \varepsilon$. Then there is $\tau \in \Sigma^{\mathrm{IM}}$ with $\tau \geq t_1$, $\mathbf{P}(A(\tau)) > 1 - \varepsilon$, and $\|e(\tau)\|_\Phi < \delta$, so that $\mathbf{E}\,[e(\tau)] < \varepsilon$. Then

$$\mathbf{P}(X^* - X_\tau > 2\varepsilon) \leq \mathbf{P}(|X^* - Y| > \varepsilon) + \mathbf{P}(A^* \setminus A(\tau)) + \mathbf{E}\,[e(\tau)] \leq 3\varepsilon.$$

(c) $\Longrightarrow$ (a). Given an adapted family $(A_t)_{t \in J}$ of sets, let $X_t = \mathbf{1}_{A_t}$. Then $X^* = \mathbf{1}_{A^*}$ and for $\tau \in \Sigma^{\mathrm{IM}}$ and $\varepsilon < 1$, we have

$$\mathbf{P}(X^* - X_\tau > \varepsilon) = \mathbf{P}(A^* \setminus A(\tau)).$$

(d) $\Longrightarrow$ (a). Let (A_t) be an adapted family, and let $A^* = \mathrm{e}\limsup A_t$. (Recall that if τ is subordinate to (A_t), then $A(\tau) = D(\tau)$.) We construct recursively sequences $u_k > 0$ and $\tau_k \in \Sigma^{\mathrm{IM}}$. Let

$$u_0 = \sup\big\{\,\mathbf{P}\big(A(\tau)\big) : \tau \text{ is subordinate to } (A_t) \text{ and } \big\|e(\tau)\big\|_\Phi < 1\,\big\}.$$

Then choose τ_1 subordinate to (A_t) with $\big\|e(\tau)\big\|_\Phi < 1$ and $\mathbf{P}\big(A(\tau_1)\big) > u_0/2$. Next let

$$u_1 = \sup\big\{\,\mathbf{P}\big(A(\tau) \setminus A(\tau_1)\big) : \tau \text{ is subordinate to } (A_t),$$
$$\big\|e(\tau)\big\|_\Phi < 1/2, \text{ and } A(\tau) \supseteq A(\tau_1)\,\big\}.$$

Then choose $\tau_2 \in \Sigma^{\mathrm{IM}}$ subordinate to (A_t) with $\big\|e(\tau_2)\big\|_\Phi < 1/2$, $A(\tau_2) \supseteq A(\tau_1)$, and $\mathbf{P}\big(A(\tau_2) \setminus A(\tau_1)\big) \geq u_1/2$. Continue in this way, so that $A(\tau_k) \supseteq A(\tau_{k-1})$, $\mathbf{P}\big(A(\tau_k) \setminus A(\tau_{k-1})\big) \geq u_{k-1}/2$ and

$$u_k = \sup\big\{\,\mathbf{P}\big(A(\tau) \setminus A(\tau_k)\big) : \tau \text{ is subordinate to } (A_t),$$
$$\big\|e(\tau)\big\|_\Phi < 1/(k+1), \text{ and } A(\tau) \supseteq A(\tau_k)\,\big\}.$$

This completes the recursive construction.

Now if τ is subordinate to (A_t), $\|e(\tau)\|_\Phi < 1/(k+1)$, and $A(\tau) \supseteq A(\tau_k)$, then

$$u_{k-1} \geq \mathbf{P}\big(A(\tau) \setminus A(\tau_k)\big) + \mathbf{P}\big(A(\tau_k) \setminus A(\tau_{k-1})\big)$$
$$\geq \mathbf{P}\big(A(\tau) \setminus A(\tau_k)\big) + u_{k-1}/2,$$

so $u_k \leq u_{k-1}/2$. Therefore $u_k \leq 2^{-k}$.

For each t, let

$$C_t = A_t \setminus \bigcup_{u \leq t} \bigcup_{k=1}^{\infty} \{u \in \tau_k\}.$$

Define $X_t = \mathbf{1}_{C_t}$.

We claim that (X_t) is a V_Φ-potential. Fix k, choose $t_0 \in J$ with $t_0 \geq \tau_1, \tau_2, \cdots, \tau_k$, and let $\delta = (1/k) - \|e(\tau_k)\|_\Phi$. Now for any $\tau \in \Sigma^M$ with $\tau \geq t_0$ and $\|e(\tau)\|_\Phi < \delta$, define $\sigma \in \Sigma^{IM}$ by

$$\{t \in \sigma\} = \{t \in \tau_k\} \cup \big(\{t \in \tau\} \cap C_t\big).$$

Certainly $A(\sigma) \supseteq A(\tau_k)$,

$$\|e(\sigma)\|_\Phi \leq \|e(\tau_k)\|_\Phi + \|e(\tau)\|_\Phi < \frac{1}{k},$$

so $\mathbf{P}\big(A(\sigma) \setminus A(\tau_{k-1})\big) \leq u_{k-1} \leq 2^{-k+1}$. Thus

$$X_\tau = \sum X_t \mathbf{1}_{\{t \in \tau\}} = \sum \mathbf{1}_{C_t \cap \{t \in \tau\}} \leq \mathbf{1}_{A(\sigma) \setminus A(\tau_k)} + e(\tau).$$

There is a constant c so that $\|X\|_1 \leq c\|X\|_\Phi$ (2.1.21). Thus $0 \leq \mathbf{E}\,[X_\tau] \leq 2^{-k+1} + c/k$. This approaches 0 as $k \to \infty$, so we see that (X_t) is a V_Φ-potential.

We conclude by (d) that (X_t) converges essentially. The limit is 0. Thus $\mathbf{P}(\mathrm{e}\limsup C_t) = 0$. Now if $B = \bigcup_{k=1}^{\infty} A(\tau_k)$, then

$$A^* \setminus B \subseteq \mathrm{e}\limsup(A_t \setminus B) = \mathrm{e}\limsup C_t,$$

so $\mathbf{P}(A^* \setminus B) = 0$. Given $\varepsilon > 0$, since the sets $A(\tau_k)$ increase, there is k with $\mathbf{P}\big(A^* \setminus A(\tau_k)\big) < \varepsilon$ and $\|e(\tau_k)\|_\Phi < \varepsilon$. ∎

(4.3.7) Corollary. *Suppose* (V_Φ) *holds. If* $(X_t)_{t \in J}$ *is an adapted process, and* $(t_n) \subseteq J$, *then there exists an increasing sequence* (τ_n) *in* Σ^M *with* $\tau_n \geq t_n$, $\|e(\tau_n)\|_\Phi \to 0$, *and* $X_{\tau_n} \to \mathrm{e}\limsup X_t$ *a.s.*

Proof. Apply (4.3.6(c)) repeatedly. ∎

(4.3.8) Proposition. *An L_Ψ-bounded martingale is a V_Φ-amart.*

Proof. Let $(X_t)_{t \in J}$ be a martingale with $\|X_t\|_\Psi \leq M$ for all t. Since it is a martingale, $L = \mathbf{E}[X_t]$ is independent of t. Let $\varepsilon > 0$ be given, and set $\delta = \varepsilon/(2M+1)$.

Suppose $\tau \in \Sigma^{\mathrm{M}}$ and $\|e(\tau)\|_\Phi < \delta$. Choose $t_0 \in J, t_0 \geq \tau$. Then

$$\mathbf{E}[X_\tau] = \sum_t \mathbf{E}[X_t \mathbf{1}_{\{t \in \tau\}}] = \sum_t \mathbf{E}[X_{t_0} \mathbf{1}_{\{t \in \tau\}}]$$

$$= \mathbf{E}\left[X_{t_0} \sum_t \mathbf{1}_{\{t \in \tau\}}\right] = \mathbf{E}[X_{t_0}(1 + e(\tau))]$$

$$= L + \mathbf{E}[X_{t_0} e(\tau)].$$

Thus, applying Young's inequality in the form (2.2.7), we get

$$\left|\mathbf{E}[X_\tau] - L\right| \leq \mathbf{E}[|X_{t_0}|e(\tau)] \leq 2\|X_{t_0}\|_\Psi \|e(\tau)\|_\Phi \leq 2M\delta < \varepsilon.$$

Therefore $(X_t)_{t \in J}$ is a V_Φ-amart. ∎

Here is a convergence theorem for the covering condition (V_Φ).

(4.3.9) Theorem. *Let $(\mathcal{F}_t)_{t \in J}$ be a stochastic basis, and let Φ be an Orlicz function. Then the following are equivalent:*

(1) *Condition (V_Φ) holds.*
(2) *All L_∞-bounded V_Φ-amarts converge essentially.*
(3) *All L_1-bounded V_Φ-amarts converge essentially.*

Proof. (1) $\Longrightarrow$ (2). Let (X_t) be an L_∞-bounded V_Φ-amart. For each $n \in \mathbb{N}$, choose $t_n \in J$ and $\delta_n > 0$ so that if $\sigma, \tau \in \Sigma^{\mathrm{M}}$, $\sigma, \tau \geq t_n$, $\|e(\sigma)\|_\Phi < \delta_n$, and $\|e(\tau)\|_\Phi < \delta_n$, then $|\mathbf{E}[X_\sigma] - \mathbf{E}[X_\tau]| < 1/n$. Then by Corollary 4.3.7, there exist $\tau_n \in \Sigma$ with $\tau_n \geq t_n$, $\|e(\tau_n)\|_\Phi < \delta_n$, and $X_{\tau_n} \to \mathrm{e}\limsup X_t$ a.s. Similarly, there exist $\sigma_n \in \Sigma$ with $\sigma_n \geq t_n$, $\|e(\sigma_n)\|_\Phi < \delta_n$, and $X_{\sigma_n} \to \mathrm{e}\liminf X_t$ a.s. Hence

$$0 = \lim_n \left|\mathbf{E}[X_{\tau_n}] - \mathbf{E}[X_{\sigma_n}]\right| = \mathbf{E}\left[\mathrm{e}\limsup X_t - \mathrm{e}\liminf X_t\right].$$

Therefore $\mathrm{e}\limsup X_t = \mathrm{e}\liminf X_t$.

(1) $\Longrightarrow$ (3). Suppose (1) holds. Then (as we have just proved) also (2) holds. Let (X_t) be an L_1-bounded V_Φ-amart. By the lattice property (4.3.5), if $\lambda > 0$, then $((-\lambda) \vee X_t \wedge \lambda)$ is an L_∞-bounded V_Φ-amart. By (2), it converges essentially. Thus (X_t) itself converges essentially on the set $\Omega_\lambda = \{\mathrm{e}\limsup |X_t| \leq \lambda\}$. The maximal inequality (4.3.6(b)) shows $\mathbf{P}\{\mathrm{e}\limsup |X_t| < \infty\} = 1$, so Ω is a countable union of sets Ω_λ. Therefore (X_t) converges essentially.

(3) $\Longrightarrow$ (2) is easy.

(2) $\Longrightarrow$ (1). Suppose all L_∞-bounded V_Φ-amarts converge essentially. Then all L_∞-bounded V_Φ-potentials converge essentially. Thus, by (4.3.6), (V_Φ) holds. ∎

(4.3.10) Theorem. *If* (V_Φ) *holds, then all* L_Ψ-*bounded martingales converge essentially.*

Proof. Let (X_t) be an L_Ψ-bounded martingale. By (4.3.8), X_t is a V_Φ-amart. An L_Ψ-bounded process is also L_1-bounded. Hence X_t converges essentially by Theorem (4.3.9).   ∎

Necessity of (V_Φ)

We will prove the converse of Theorem (4.3.10) under the assumption that Φ satisfies condition (Δ_2). It is an open problem whether the theorem is true without (Δ_2).

A *countably valued incomplete multivalued stopping time* is a countable collection of sets, written $(\{t \in \tau\})_{t \in J}$, with $\{t \in \tau\} \in \mathcal{F}_t$ for all t. (Since the collection is countable, $\{t \in \tau\} = \emptyset$ except for countably many t.) We write Σ^{CIM} for the set of all countably valued incomplete multivalued stopping times. We still write $S(\tau) = \sum \mathbf{1}_{\{t \in \tau\}}$ (which may have the value ∞); $e(\tau) = S(\tau) - 1 \wedge S(\tau)$; if $(A_t)_{t \in J}$ is an adapted family of sets,

$$A(\tau) = \bigcup_t \left(\{t \in \tau\} \cap A_t \right).$$

(4.3.11) Theorem. *Let* $(\mathcal{F}_t)_{t \in J}$ *be a stochastic basis, and let* Φ *be an Orlicz function satisfying* (Δ_2) *at* ∞. *If every* L_Ψ-*bounded martingale converges essentially, then* $(\mathcal{F}_t)_{t \in J}$ *satisfies* (V_Φ).

Proof. Since $(\Omega, \mathcal{F}, \mathbf{P})$ is a finite measure space, we may change Φ near 0 without changing finiteness of L_Ψ bounds or smallness of L_Φ norms. Thus we will assume that Φ also satisfies (Δ_2) at 0, and $\varphi(1) > 0$. Thus $\Phi(2u) \le c\Phi(u)$ for all u. Then (2.2.18) $\Psi(\varphi(u)) \le c\Phi(u)$ also.

We first claim: If $\alpha, \beta > 0$, $(A_t)_{t \in J}$ is an adapted family of sets, and Y is $\mathcal{F}_\infty$-measurable and satisfies $Y \ge 0$, $Y \in L_\Psi$, and $\mathbf{P}(A^* \setminus \{Y > 0\}) > 0$, then there exist $t \in J$ and $B \in \mathcal{F}_t$ with $\mathbf{P}(B) > 0$, $B \subseteq A_t$, and

$$\mathbf{E}\left[Y \mathbf{1}_B\right] + \alpha \mathbf{P}(B \setminus A^*) \le \beta \mathbf{P}(B).$$

To see this, consider the random variable

$$X = Y + \alpha \mathbf{1}_{\Omega \setminus A^*} \in L_\Psi,$$

and the corresponding martingale $X_t = \mathbf{E}^{\mathcal{F}_t}[X]$. Now X_t converges to X stochastically, and, by assumption, it converges essentially, so it converges essentially to X. Now $X = 0$ on $A^* \setminus \{Y > 0\}$, which has positive probability. Thus $X \ge \beta \mathbf{1}_{A^*}$ a.s. is false. Hence $\mathbf{P}\left(\{X_t \le \beta \mathbf{1}_{A_t}\} \cap A_t\right) > 0$ for some t. If $B = \{X_t \le \beta \mathbf{1}_{A_t}\} \cap A_t$, then $\mathbf{E}[X \mathbf{1}_B] = \mathbf{E}[X_t \mathbf{1}_B] \le \beta \mathbf{P}(B)$. Thus $\mathbf{E}[Y \mathbf{1}_B] + \alpha \mathbf{P}(B \setminus A^*) \le \beta \mathbf{P}(B)$. This completes the proof of the claim.

Now let $\varepsilon > 0$ and let $(A_t)_{t \in J}$ be an adapted family with $\mathbf{P}(A^*) > \varepsilon$. Choose k so large that $1/\varepsilon < 2^k$, and let $\eta = c^{-k}$. [Thus, if $\mathbf{E}\left[\Phi(X)\right] < \eta$, we will have $\mathbf{E}\left[\Phi(X/\varepsilon)\right] \leq \mathbf{E}\left[\Phi(2^k X)\right] \leq c^k \mathbf{E}\left[\Phi(X)\right] \leq 1$, so that $\|X\|_\Phi \leq \varepsilon$.] Then choose $\beta > 0$ so small that

$$\beta \left(1 - \frac{\beta}{\varphi(1)}\right)^{-1} \leq \eta,$$

and $\beta < \varphi(1)/2$. Let $\alpha = \varphi(1)$.

We next claim: If $\tau \in \Sigma^{\text{CIM}}$ is subordinate to (A_t), $\mathbf{P}(A^* \setminus A(\tau)) > 0$, and $\mathbf{E}\left[\Phi(e(\tau))\right] \leq \eta \mathbf{P}(A^* \cap A(\tau))$, then there is $t \in J$ and $B \in \mathcal{F}_t$, with $\mathbf{P}(B) > 0$, $B \subseteq A_t$, and

(4.3.11a) $\mathbf{E}\left[\varphi(S(\tau)) \mathbf{1}_B\right] + \varphi(1)\mathbf{P}(B \setminus A^*) \leq \beta \mathbf{P}(B).$

This is demonstrated by taking $Y = \varphi(S(\tau))$ in the previous claim. Then $\mathbf{E}\left[\Psi(Y)\right] = \mathbf{E}\left[\Psi\big(\varphi(S(\tau))\big)\right] \leq c\mathbf{E}\left[\Phi(S(\tau))\right] < \infty$, since $S(\tau) \leq e(\tau) + 1$ belongs to $L_\Phi = H_\Phi$.

Now we note that if any t and B satisfy (4.3.11a), and if we define $\bar{\tau}$ by $\{s \in \bar{\tau}\} = \{s \in \tau\}$ for $s \neq t$ and $\{t \in \bar{\tau}\} = \{t \in \tau\} \cup B$, then we still have $\mathbf{E}\left[\Phi(e(\bar{\tau}))\right] \leq \eta \mathbf{P}(A^* \cap A(\bar{\tau}))$. Indeed, $S(\tau) \geq 1$ on $A(\tau)$, so

$$\begin{aligned}
\varphi(1)\mathbf{P}\big(B \setminus (A^* \setminus A(\tau))\big) &\leq \varphi(1)\mathbf{P}(B \cap A(\tau)) + \varphi(1)\mathbf{P}(B \setminus A^*) \\
&\leq \mathbf{E}\left[\varphi(S(\tau)) \mathbf{1}_B\right] + \varphi(1)\mathbf{P}(B \setminus A^*) \\
&\leq \beta \mathbf{P}(B),
\end{aligned}$$

and thus

$$\begin{aligned}
\mathbf{P}(B \cap (A^* \setminus A(\tau))) &= \mathbf{P}(B) - \mathbf{P}\big(B \setminus (A^* \setminus A(\tau))\big) \\
&\geq \left(1 - \frac{\beta}{\varphi(1)}\right)\mathbf{P}(B).
\end{aligned}$$

Therefore

$$\begin{aligned}
\mathbf{E}\left[\varphi(S(\tau)) \mathbf{1}_B\right] &\leq \beta \mathbf{P}(B) \\
&\leq \beta \left(1 - \frac{\beta}{\varphi(1)}\right)^{-1} \mathbf{P}\big(B \cap (A^* \setminus A(\tau))\big) \\
&\leq \eta \mathbf{P}\big(B \cap (A^* \setminus A(\tau))\big).
\end{aligned}$$

Now

$$\begin{aligned}
\mathbf{E}\left[\Phi(e(\bar{\tau}))\right] &\leq \mathbf{E}\left[\Phi(e(\tau)) \mathbf{1}_{\Omega \setminus B}\right] + \mathbf{E}\left[\Phi(S(\tau)) \mathbf{1}_B\right] \\
&= \mathbf{E}\left[\Phi(e(\tau))\right] + \mathbf{E}\left[\big(\Phi(S(\tau)) - \Phi(e(\tau))\big) \mathbf{1}_B\right] \\
&\leq \mathbf{E}\left[\Phi(e(\tau))\right] + \mathbf{E}\left[\varphi(S(\tau)) \mathbf{1}_B\right] \\
&\leq \eta \mathbf{P}(A^* \cap A(\tau)) + \eta \mathbf{P}(B \cap (A^* \setminus A(\tau))) \\
&= \eta \mathbf{P}(A^* \cap A(\bar{\tau})).
\end{aligned}$$

Also note

$$\mathbf{P}(B \cap A(\tau)) \le \frac{1}{\varphi(1)} \mathbf{E}\left[\varphi(S(\tau))\,1_B\right] \le \frac{\beta}{\varphi(1)}\mathbf{P}(B) < \frac{1}{2}\mathbf{P}(B),$$

so

(4.3.11b) $\mathbf{P}(A(\bar{\tau})) \ge \mathbf{P}(A(\tau)) + \mathbf{P}(B)/2.$

Now we will recursively construct a (finite or infinite) sequence

$$\tau_0, \tau_1, \tau_2, \cdots \in \Sigma^{\mathrm{IM}}.$$

First let τ_0 be the empty incomplete multivalued stopping time, $\{s \in \tau_0\} = \emptyset$ for all $s \in J$. Then $e(\tau_0) = 0$, so trivially

$$\mathbf{E}\left[\Phi(e(\tau_0))\right] \le \eta\mathbf{P}(A^* \cap A(\tau_0)).$$

Suppose τ_k has been constructed, and $\mathbf{E}\left[\Phi(e(\tau_k))\right] \le \eta\mathbf{P}(A^* \cap A(\tau_k))$. Then if $\mathbf{P}(A^* \cap A(\tau_k)) \ge \mathbf{P}(A^*) - \varepsilon$, stop the recursive construction. But if $\mathbf{P}(A^* \cap A(\tau_k)) < \mathbf{P}(A^*) - \varepsilon$, construct τ_{k+1} as follows. The class

$$\mathcal{B}_k = \big\{\, B \in \mathcal{F} \text{ :there exists } t \text{ such that } B \in \mathcal{F}_t,\ B \subseteq A_t,\ \mathbf{P}(B) > 0,$$

$$\mathbf{E}\left[\varphi(S(\tau_k))\,1_B\right] + \varphi(1)\mathbf{P}(B \setminus A^*) \le \beta\mathbf{P}(B) \,\big\}$$

is nonempty. Choose $B_k \in \mathcal{B}_k$ with

$$\mathbf{P}(B_k) \ge (1/2)\sup\big\{\, \mathbf{P}(B) : B \in \mathcal{B}_k \,\big\},$$

and use it to construct τ_{k+1} as above, with

$$\mathbf{E}\left[\Phi(e(\tau_{k+1}))\right] \le \eta\mathbf{P}(A^* \cap A(\tau_{k+1}))$$

and $\{s \in \tau_{k+1}\} \supseteq \{s \in \tau_k\}$ for all s.

We claim that the recursive construction ends at some finite stage τ_k. If not, we get an infinite sequence $(\tau_k)_{k=0}^\infty$. Since $\{s \in \tau_{k+1}\} \supseteq \{s \in \tau_k\}$, there is a "limiting" $\tau \in \Sigma^{\mathrm{CIM}}$ defined by $\{s \in \tau\} = \bigcup_k\{s \in \tau_k\}$. We have

$$\mathbf{E}\left[\Phi(e(\tau))\right] \le \lim_k \mathbf{E}\left[\Phi(e(\tau_k))\right] \le \eta\mathbf{P}\big\{A^* \cap A(\tau)\big\} < \infty.$$

Thus there is $t \in J$ and $B \in \mathcal{F}_t$ with $\mathbf{P}(B) > 0$, $B \in \mathcal{B}_k$ for all k. This means $\mathbf{P}(B_k) > (1/2)\mathbf{P}(B)$ for all k; but this is impossible, since by (4.3.11b) we have $\sum \mathbf{P}(B_k) \le 2\mathbf{P}(A(\tau)) \le 2$.

Therefore, we have $\mathbf{P}(A^* \cap A(\tau_k)) > \mathbf{P}(A^*) - \varepsilon$ for some k, so

$$\mathbf{P}(A^* \setminus A(\tau_k)) < \varepsilon.$$

Also, $\mathbf{E}\left[\Phi(e(\tau_k))\right] \le \eta$ implies $\|e(\tau_k)\|_\Phi \le \varepsilon$.   ∎

The equivalence of essential convergence of martingales and covering conditions can be simply stated without stopping times in the case $(\mathbf{V_1})$. The following, due to Krickeberg, is also the earliest result of this type.

(4.3.12) Theorem. *Let* $(\mathcal{F}_t)$ *be a stochastic basis. Then every* L_∞-*bounded martingale converges if and only if: for every adapted family* $(A_t)_{t \in J}$ *and every* $\varepsilon > 0$, *there is an adapted family* (B_t), *empty except for finitely many* t, *with* $B_t \subseteq A_t$, $\mathbf{P}(\bigcup B_t) > \mathbf{P}(A^*) - \varepsilon$, *and* $\sum \mathbf{P}(B_t) \leq \mathbf{P}(\bigcup B_t) + \varepsilon$.

Proof. Apply the theorem using $\Phi = \Phi_\infty$ (see Section 2.2), then let $B_t = \{t \in \tau\}$.   ∎

Functional condition (FV$_\Phi$)

A variant (FV$_\Phi$) of the condition (V$_\Phi$) is obtained by replacing the multivalued stopping time τ by an adapted family of functions $(\xi_t)_{t \in J}$. The advantage of (FV$_\Phi$) is that the equivalence with convergence of L_Ψ-bounded martingales is easily established with no special assumptions on Φ (except finiteness). This disadvantage of (FV$_\Phi$) is that it is not a Vitali-type covering condition.

We will say that an adapted family $(\xi_t)_{t \in J}$ of functions is *subordinate* to the family $(A_t)_{t \in J}$ of sets iff $\{\xi_t \neq 0\} \subseteq A_t$ for every t.

(4.3.13) Definition. The stochastic basis $(\mathcal{F}_t)_{t \in J}$ satisfies *condition* (FV$_\Phi$) iff, for every adapted family $(A_t)_{t \in J}$ of sets and every $\varepsilon > 0$, there exists an adapted family $(\xi_t)_{t \in J}$ of nonnegative bounded functions subordinate to $(A_t)_{t \in J}$, with only finitely many ξ_t nonzero, such that $\mathbf{E}\left[\sum \xi_t\right] \geq \mathbf{P}(A^*) - \varepsilon$ and $\left\| \sum \xi_t - 1 \wedge \sum \xi_t \right\|_\Phi < \varepsilon$.

It is easily verified that condition (V$_\Phi$) implies condition (FV$_\Phi$), by taking ξ_t to be the indicator function of the set $\{t \in \tau\}$.

The variants of (FV$_\Phi$), the corresponding amart-like processes, and other interesting topics, will not be discussed here. We will prove only the martingale convergence theorem. For that purpose, we will need a maximal inequality.

(4.3.14) Proposition. *Suppose* (FV$_\Phi$) *holds. If* $(X_t)_{t \in J}$ *is an* L_Ψ-*bounded positive submartingale, and* $\lambda > 0$, *then*

$$\mathbf{P}(X^* \geq \lambda) \leq \frac{1}{\lambda} \sup_t \mathbf{E}\left[X_t\right].$$

Proof. Suppose $\|X_t\|_\Psi \leq M$ for all t. Let β satisfy $0 < \beta < \lambda$. Let $A_t = \{X_t > \beta\}$. Thus $\{X^* \geq \lambda\} \subseteq A^*$. Fix $\varepsilon > 0$. By (FV$_\Phi$), there is a family $(\xi_t)_{t \in J}$ of nonnegative bounded functions subordinate to $(A_t)_{t \in J}$ with finitely many ξ_t nonzero and $\mathbf{E}[\xi] \geq \mathbf{P}(A^*) - \varepsilon$ and $\|\xi'\|_\Phi < \varepsilon$, where $\xi = \sum \xi_t$, $\xi'' = 1 \wedge \xi$, $\xi' = \xi - \xi''$. Let $t_0 \in J$ be so large that $t_0 > t$

whenever $\mathbf{P}(\xi_t \neq 0) > 0$. Then

$$
\begin{aligned}
\mathbf{P}(A^*) - \varepsilon \leq \mathbf{E}\left[\xi\right] = \mathbf{E}\left[\sum_t \xi_t 1_{A_t}\right] &\leq \frac{1}{\beta}\mathbf{E}\left[\sum_t \xi_t X_t\right] \\
= \frac{1}{\beta}\sum \mathbf{E}\left[\xi_t X_t\right] &\leq \frac{1}{\beta}\sum \mathbf{E}\left[\xi_t X_{t_0}\right] \\
= \frac{1}{\beta}\mathbf{E}\left[\sum \xi_t X_{t_0}\right] &= \frac{1}{\beta}\mathbf{E}\left[\xi' X_{t_0}\right] + \frac{1}{\beta}\mathbf{E}\left[\xi'' X_{t_0}\right] \\
&\leq \frac{2}{\beta}\|\xi'\|_\Phi\|X_{t_0}\|_\Psi + \frac{1}{\beta}\mathbf{E}\left[X_{t_0}\right] \\
&\leq \frac{2\varepsilon M}{\beta} + \frac{1}{\beta}\sup_t \mathbf{E}\left[X_t\right].
\end{aligned}
$$

Note that we have applied (2.2.7). Now $(X_t)_{t\in J}$ is L_1-bounded, since it is L_Ψ-bounded. Let $\varepsilon \to 0$, then $\beta \to \lambda$ to get the result. ∎

(4.3.15) Theorem. *Suppose Φ is a finite Orlicz function with conjugate Ψ, and $(\mathcal{F}_t)_{t\in J}$ is a stochastic basis. Then every L_Ψ-bounded martingale converges essentially if and only if $(\mathcal{F}_t)_{t\in J}$ satisfies* (FV$_\Phi$).

Proof. Suppose (FV$_\Phi$) holds. Then we have a maximal inequality by (4.3.14). Thus we may apply Proposition (4.1.13) using the family of L_Ψ-bounded martingales; we may conclude that these martingales converge.

Conversely, suppose (FV$_\Phi$) fails. There is an adapted family $(A_t)_{t\in J}$ of sets and $\varepsilon > 0$ such that, for any family $(\xi_t)_{t\in J}$ of functions, if

(4.3.15a) $\quad \begin{cases} \xi_t \geq 0,\ \xi_t \text{ is bounded},\ \xi_t = 0 \text{ outside } A_t, \\ \text{only finitely many } \xi_t \text{ are nonzero}, \\ \xi_t \text{ is } \mathcal{F}_t\text{-measurable},\ \mathbf{E}\left[\sum \xi_t\right] \geq \mathbf{P}(A^*) - \varepsilon \end{cases}$

then $\|\sum \xi_t - 1 \wedge \sum \xi_t\|_\Phi \geq \varepsilon$. We consider three subsets of L_Φ:

$$
\begin{aligned}
C_1 &= \left\{ \xi = \sum \xi_t : (\xi_t)_{t\in J} \text{ satisfies (4.3.15a)} \right\} \\
C_2 &= \{ \xi \in L_\Phi : \xi \leq 1 \} \\
C_3 &= \{ \xi \in L_\Phi : \|\xi\|_\Phi < \varepsilon \}.
\end{aligned}
$$

These sets are convex. Since C_3 is open, also $C_2 + C_3$ is open. But $C_1 \cap (C_2 + C_3) = \emptyset$. Hence by the Hahn-Banach theorem, there is $x^* \in L_\Phi^*$ such that $x^*(\xi) \geq 1$ for all $\xi \in C_1$ and $x^*(\xi) < 1$ for all $\xi \in C_2 + C_3$. Now $x^*(\xi) < 1$ for all $\xi \leq 1$, so by (2.2.24) the functional x^* has the form $x^*(\xi) = \mathbf{E}[\xi X]$ for some $X \in L_\Psi$.

Now consider the martingale $X_t = \mathbf{E}^{\mathcal{F}_t}[X]$. Since $X \in L_\Psi$, (X_t) L_Ψ-bounded. We claim that

$$
X_t \geq \frac{1}{\mathbf{P}(A^*) - \varepsilon} \quad \text{a.s. on } A_t.
$$

If not, fix t and $B \in \mathcal{F}_t$ with $B \subseteq A_t$, $\mathbf{P}(B) > 0$, and $X_t < 1/(\mathbf{P}(A^*) - \varepsilon)$ on B. Then

$$\xi = \frac{\mathbf{P}(A^*) - \varepsilon}{\mathbf{P}(B)} \mathbf{1}_B$$

belongs to C_1, so $1 \le x^*(\xi) = \mathbf{E}[\xi X] = \mathbf{E}[\xi X_t] < 1$, a contradiction. It follows that

$$X^* \ge \frac{1}{\mathbf{P}(A^*) - \varepsilon} \quad \text{a.s. on } A^*.$$

Now $\mathbf{E}[X] \le 1$, so we also have

$$\mathbf{P}\left(X_* \ge \frac{1}{\mathbf{P}(A^*) - \varepsilon}\right) \le \mathbf{P}(A^*) - \varepsilon.$$

Thus $\mathbf{P}(X^* \ne X_*) \ge \varepsilon$, and $(X_t)_{t \in J}$ does not converge essentially. ∎

Complements

(4.3.16) (Covering condition (D'_Φ).) Let Φ be an Orlicz function. We say that the stochastic basis $(\mathcal{F}_t)_{t \in J}$ satisfies condition (D'_Φ) iff for each $\varepsilon > 0$ there exists $\eta > 0$ such that for each $\gamma > 0$ and each adapted family $(A_t)_{t \in J}$, there exists $\tau \in \Sigma^{\mathrm{IM}}$ with $\mathbf{P}(A^* \setminus A(\tau)) < \varepsilon$ and $\mathbf{E}[\Phi(\eta e(\tau))] \le \gamma$. It can easily be verified that if Φ satisfies the (Δ_2) condition, then (D'_Φ) is equivalent to (V_Φ).

Suppose J has a countable cofinal set. Then (D'_Φ) holds if and only if every L_Ψ-bounded martingale converges (Talagrand [1986]). This result is more satisfactory than Theorem (4.3.11) in the sense that the hypothesis of condition (Δ_2) is not needed. But the covering condition (D'_Φ) is not as easy to understand or verify as condition (V_Φ).

Talagrand introduced another covering condition (C_Φ) in his study of L_Ψ-bounded martingales (4.4.12).

Remarks

The covering condition (V_Φ) appears first in the derivation setting; see, for example, Hayes [1976]. The martingale material here is based on Krickeberg [1956] and Millet [1978]. Krickeberg proved that (V_p) implies essential convergence of all L_q-bounded martingales, where $1/p + 1/q = 1$. He also proved the necessity in the case $p = \infty$. (This is our (4.2.12).) See Krickeberg & Pauc [1963]. Following the lead of Hayes in the derivation setting, Millet proved in the directed-set setting necessity for (V_p), $1 < p < \infty$, as well as for (V_Φ), where Φ satisfies condition (Δ_2) (4.3.11). Millet & Sucheston [1979a] and [1980d] proved several equivalent formulations of (V_p), including the amart convergence: they used the term "amart for M_p" for our "V_p-amart." These papers also introduced the stopping-time formulation of the Vitali conditions.

Talagrand [1986] contains a theorem stating that condition (V_Φ) is equivalent to the convergence of all L_Ψ-bounded martingales under a much more general hypothesis than (Δ_2); unfortunately, we believe there is a gap in his proof. Professor Talagrand now shares this view. It would be interesting to know exactly which Orlicz functions have this useful property.

The functional condition (FV_Φ) appears first in the important paper, Talagrand [1986]. Theorem (4.3.15) appears there.

4.4. L_1-bounded martingales

In this section we return once again to the question of essential convergence of L_1-bounded martingales.

We know that condition (V) = (V$_\infty$) is sufficient for convergence of L_1-bounded martingales. We know that condition (V$_\Phi$) is necessary and sufficient for convergence of L_Ψ-bounded martingales, if Φ satisfies condition (Δ_2). But the Orlicz function Φ_∞ (see Section 2.2) does not satisfy (Δ_2). We will give an example below (4.4.10) showing that condition (V) is *not* necessary for convergence of L_1-bounded martingales.

We will therefore need to consider another covering condition, condition (C), which is slightly more complicated than (V). Under reasonable hypotheses (the directed set J has a countable cofinal subset) we will prove that (C) is necessary and sufficient for the essential convergence of L_1-bounded martingales.

We will retain the notation used in the preceding section. A probability space $(\Omega, \mathcal{F}, \mathbf{P})$, a directed set J, and a stochastic basis $(\mathcal{F}_t)_{t \in J}$ will be fixed. We write Σ^{IM} for the set of (simple) incomplete multivalued stopping times, and Σ^{M} for the set of multivalued stopping times. If $(X_t)_{t \in J}$ is a stochastic process, then

$$X^* = \operatorname{e\,lim\,sup}_t X_t$$

and if $\tau \in \Sigma^{\mathrm{IM}}$,

$$X_\tau = \sum_{t \in J} X_t \mathbf{1}_{\{t \in \tau\}}.$$

If $(A_t)_{t \in J}$ is an adapted family of sets, then

$$A^* = \operatorname{e\,lim\,sup}_t A_t$$

and if $\tau \in \Sigma^{\mathrm{IM}}$,

$$A(\tau) = \bigcup_{t \in J}(\{t \in \tau\} \cap A_t).$$

For $\tau \in \Sigma^{\mathrm{IM}}$, the *domain* of τ is $D(\tau) = \bigcup_t \{t \in \tau\}$, the *sum* of τ is $S(\tau) = \sum \mathbf{1}_{\{t \in \tau\}}$, and the *excess* of τ is $e(\tau) = S(\tau) - \mathbf{1}_{D(\tau)}$. We say τ is *subordinate* to $(A_t)_{t \in J}$ iff $\{t \in \tau\} \subseteq A_t$ for all $t \in J$; then $A(\tau) = D(\tau)$.

Covering condition (C)

The next covering condition to be introduced is similar to the conditions (V$_\Phi$) of the previous section. But instead of requiring that the overlap should be *small*, it is only required that it be *bounded*.

(4.4.1) Definition. The stochastic basis $(\mathcal{F}_t)_{t \in J}$ satisfies the covering condition (C) iff, for each $\varepsilon > 0$, there is a constant M such that for every adapted family $(A_t)_{t \in J}$ of sets, there exists $\tau \in \Sigma^{\mathrm{IM}}$ with $\mathbf{P}(A^* \setminus A(\tau)) < \varepsilon$ and $S(\tau) \leq M$.

Note that (V) implies (C) even with $M = 1$. The stochastic basis of Example (4.2.1) fails condition (C): see (4.4.16).

Here are some technical variants of condition (C). The proofs are similar to those used for (V$_\Phi$) so they are left to the reader.

(4.4.2) Lemma. *The following are equivalent.*

(a) (C): *For every* $\varepsilon > 0$, *there is* M *such that for any adapted family* $(A_t)_{t \in J}$ *of sets, there exists* $\tau \in \Sigma^{\mathrm{IM}}$ *with* $\mathbf{P}(A^* \setminus A(\tau)) < \varepsilon$ *and* $S(\tau) \le M$.

(b) *For every* $\varepsilon > 0$, *there is* M *such that for any adapted family* $(A_t)_{t \in J}$ *of sets, there exists* $\tau \in \Sigma^{\mathrm{M}}$ *with* $\mathbf{P}(A^* \setminus A(\tau)) < \varepsilon$ *and* $S(\tau) \le M$.

(c) *For every* $\varepsilon > 0$, *there is* M *such that for any adapted family* $(A_t)_{t \in J}$ *of sets, and every* $t_0 \in J$, *there exists* $\tau \in \Sigma^{\mathrm{M}}$ *with* $\tau \ge t_0$, $\mathbf{P}(A^* \setminus A(\tau)) < \varepsilon$ *and* $S(\tau) \le M$.

(d) *For every* $\varepsilon > 0$, *there is* M *such that for any adapted family* $(A_t)_{t \in J}$ *of sets, and every* $t_0 \in J$, *there exists* $\tau \in \Sigma^{\mathrm{IM}}$, *subordinate to* $(A_t)_{t \in J}$, *with* $\tau \ge t_0$, $\mathbf{P}(A(\tau)) > \mathbf{P}(A^*) - \varepsilon$ *and* $S(\tau) \le M$.

(e) *For every* $\varepsilon > 0$, *there exist* M *and* $\alpha > 0$ *such that for any adapted family* $(A_t)_{t \in J}$ *of sets with* $\mathbf{P}(A^*) > \varepsilon$, *and every* $t_0 \in J$, *there exists* $\tau \in \Sigma^{\mathrm{IM}}$, *subordinate to* $(A_t)_{t \in J}$, *with* $\tau \ge t_0$, $\mathbf{P}(A(\tau)) > \alpha$ *and* $S(\tau) \le M$.

See Millet & Sucheston [1980e], where the lemma is proved. The same paper contains a proof that the maximal inequality in Lemma (4.4.3), below, is equivalent to condition (C).

Again, we refrain from studying the processes corresponding to condition (C) and proceed directly to martingale convergence. However, our usual technique of truncating at λ and $-\lambda$ to obtain an L_∞-bounded amart seems difficult to carry out here: we do not know of a suitable amart definition corresponding to condition (C). So we will have to use another method, involving a decomposition theorem for martingales (1.4.17).

We begin with a maximal inequality.

(4.4.3) Lemma. *Suppose condition* (C) *holds. If* $(X_t)_{t \in J}$ *is a nonnegative adapted process, then for every* ε *there exists* M *such that, for every* $\lambda > 0$,

$$\mathbf{P}(X^* \ge \lambda) \le \varepsilon + \frac{1}{\lambda} \sup_{\substack{\sigma \in \Sigma^{\mathrm{IM}} \\ S(\sigma) \le M}} \mathbf{E}\left[X_\sigma\right].$$

Proof. Let $\lambda > 0$ be given. Let β satisfy $0 < \beta < \lambda$. Given $\varepsilon > 0$, choose M as in condition (C). Let $A_t = \{X_t > \beta\}$. Then $\{X^* \ge \lambda\} \subseteq A^*$. Then there is $\tau \in \Sigma^{\mathrm{IM}}$ with $\mathbf{P}(A^* \setminus A(\tau)) < \varepsilon$ and $S(\tau) \le M$. Then we have $A(\tau) = \bigcup_t (\{t \in \tau\} \cap A_t)$, so

$$\mathbf{1}_{A(\tau)} \le \sum_t \mathbf{1}_{A_t} \mathbf{1}_{\{t \in \tau\}} \le \frac{1}{\beta} \sum_t X_t \mathbf{1}_{\{t \in \tau\}} = \frac{1}{\beta} X_\tau.$$

Thus $\mathbf{P}(A(\tau)) \le (1/\beta)\mathbf{E}\left[X_\tau\right]$. Therefore

$$P(X^* \ge \lambda) \le \mathbf{P}(A^*) \le \mathbf{P}(A(\tau)) + \varepsilon$$

$$\le \varepsilon + \frac{1}{\beta}\mathbf{E}\left[X_\tau\right] \le \varepsilon + \frac{1}{\beta} \sup_{\substack{\sigma \in \Sigma^{\mathrm{IM}} \\ S(\sigma) \le M}} \mathbf{E}\left[X_\sigma\right].$$

Now let $\beta \to \lambda$.   ∎

(4.4.4) Proposition. *Let $(\mathcal{F}_t)_{t \in J}$ be a stochastic basis. Suppose condition (C) is satisfied. Then all uniformly integrable martingales converge essentially.*

Proof. Let $X_t = \mathbf{E}^{\mathcal{F}_t}[X]$ be a uniformly integrable martingale. We may assume that X is $\sigma(\bigcup \mathcal{F}_t)$-measurable. Fix $\varepsilon > 0$. Let M be the constant that corresponds to ε by (C). There is $s \in J$ and bounded $\mathcal{F}_s$-measurable function Y such that $\|X - Y\|_1 < \varepsilon^2/M$. Now $Z_t = \mathbf{E}^{\mathcal{F}_t}[|X - Y|]$ is a nonnegative martingale. Thus we have by the maximal inequality (4.4.3): there is $\tau \in \Sigma^{\mathrm{IM}}$ with $\tau \geq s$, $S(\tau) \leq M$, and

$$\mathbf{P}\{Z^* \geq \varepsilon\} \leq 2\varepsilon + \frac{1}{\varepsilon} \mathbf{E}[Z_\tau].$$

Choose $t \geq \tau$. We have

$$\mathbf{P}\{Z^* \geq \varepsilon\} \leq 2\varepsilon + \frac{1}{\varepsilon} \mathbf{E}[Z_t S(\tau)]$$
$$\leq 2\varepsilon + \frac{M}{\varepsilon} \mathbf{E}[Z_t] \leq 2\varepsilon + \frac{M}{\varepsilon} \frac{\varepsilon^2}{M} = 3\varepsilon.$$

Thus

$$\mathbf{P}(\mathrm{e}\limsup X_t - \mathrm{e}\liminf X_t \geq 2\varepsilon)$$
$$\leq \mathbf{P}(\{\mathrm{e}\limsup X_t - Y \geq \varepsilon\} \cup \{\mathrm{e}\liminf X_t - Y \leq -\varepsilon\})$$
$$\leq \mathbf{P}\{Z^* \geq \varepsilon\} \leq 3\varepsilon.$$

Thus X_t converges essentially. Since X_t converges stochastically to X, the essential limit is also X. ∎

Next, we must handle the case of nonuniformly integrable martingales. Recall these definitions: A finitely additive set-function μ defined on a subalgebra $\mathcal{G}$ of $\mathcal{F}$ is called *singular* iff for every $\varepsilon > 0$, there exists $A \in \mathcal{G}$ with $\mathbf{P}(\Omega \setminus A) < \varepsilon$ but variation $|\mu|(A) < \varepsilon$. A martingale $(X_t)_{t \in J}$ is called *singular* iff the finitely additive set-function μ defined on $\mathcal{G} = \bigcup_t \mathcal{F}_t$ by

$$\mu(A) = \lim_t \mathbf{E}[X_t \mathbf{1}_A]$$

is a singular measure. We know (1.4.17) that any L_1-bounded martingale can be written as the sum of a uniformly integrable martingale and a singular martingale.

(4.4.5) Proposition. *Let $(\mathcal{F}_t)_{t \in J}$ be a stochastic basis. Suppose condition (C) is satisfied. Then all singular L_1-bounded martingales converge essentially to 0.*

Proof. Suppose $(\mathcal{F}_t)_{t \in J}$ is a singular L_1-bounded martingale. If (X_t) corresponds to singular measure μ:

$$\mu(A) = \lim_t \mathbf{E}[X_t \mathbf{1}_A]$$

for all $A \in \mathcal{G} = \bigcup_t \mathcal{F}_t$, then the absolute value process $Z_t = |X_t|$ corresponds to the variation of μ:

$$|\mu|(A) = \lim_t \mathbf{E}\left[Z_t \, \mathbf{1}_A\right].$$

Thus μ has bounded variation, since (X_t) is L_1-bounded. (The process (Z_t) is a submartingale.)

Let $\varepsilon > 0$. Let M be the constant corresponding to ε by condition (C). Since μ is singular, there exist $s \in J$ and $B \in \mathcal{F}_s$ with $\mathbf{P}(B) > 1 - \varepsilon$ and $|\mu|(B) < \varepsilon^2/M$. If $\sigma \in \Sigma^{\mathrm{IM}}$ with $\sigma \geq s$ and $S(\sigma) \leq M$, then choose $t \geq \sigma$ and compute

$$\mathbf{E}\left[Z_\sigma \, \mathbf{1}_B\right] \leq M \, \mathbf{E}\left[Z_t \, \mathbf{1}_B\right] \leq \varepsilon^2.$$

Now by the maximal inequality (4.4.3), applied to the process $(Z_t \, \mathbf{1}_B)_{t \geq s}$, we have $\mathbf{P}\{Z^* \, \mathbf{1}_B \geq \varepsilon\} \leq 2\varepsilon$, so $\mathbf{P}(Z^* \geq \varepsilon) \leq 2\varepsilon$. Thus $Z^* = 0$, so that (X_t) converges essentially to 0.   ∎

(4.4.6) Theorem. *Let $(\mathcal{F}_t)_{t \in J}$ be a stochastic basis. Suppose condition (C) is satisfied. Then all L_1-bounded martingales converge essentially.*

Proof. Apply the decomposition (1.4.17), then use Theorems (4.4.4) and (4.4.5).   ∎

Necessity of (C)

We next undertake the proof of the converse: If all L_1-bounded martingales converge essentially, then condition (C) holds. (We assume the existence of a countable cofinal set for this.) In fact, we need only the convergence of the *uniformly integrable* L_1-bounded martingales; that is, martingales of the form $X_t = \mathbf{E}^{\mathcal{F}_t}[X]$ for some $X \in L_1$.

We will say an adapted family $(A_t)_{t \in J}$ of sets is *finite* iff $A_t = \emptyset$ except for finitely many t. (Such an adapted family is thus really the same thing as an element of Σ^{IM}, viewed from a different perspective.) We also say that the family $(A_t)_{t \in J}$ is *supported beyond* t_0 iff $A_t = \emptyset$ for all t except those with $t \geq t_0$. (Thus the corresponding $\tau \in \Sigma^{\mathrm{IM}}$ satisfies $\tau \geq t_0$.)

We begin with an application of the Hahn-Banach theorem.

(4.4.7) Lemma. *Let $(A_t)_{t \in J}$ be a finite adapted family of sets. Let $a > 0$. Suppose, for each adapted family $(\xi_t)_{t \in J}$ of nonnegative bounded functions, subordinate to $(A_t)_{t \in J}$, with $\mathbf{E}\left[\sum \xi_t\right] = 1$, we have*

$$\left\|\sum \xi_t\right\|_\infty \geq a.$$

Then, for each $\gamma > 0$, there is $Y \in L_1$ with $Y \geq 0$, $\mathbf{E}[Y] \leq 1/a$, and

$$\mathbf{P}\left(A_t \setminus \{\mathbf{E}^{\mathcal{F}_t}[Y] > 1/2\}\right) \leq \gamma$$

for all t.

Proof. Fix $\gamma > 0$. Consider the following two subsets of L_∞:

$$C_1 = \left\{ \xi : \xi = \textstyle\sum \xi_t \text{ for some family } (\xi_t)_{t \in J} \text{ of nonnegative} \right.$$

$$\left. \text{functions, subordinate to } (A_t),\ \mathbf{E}\left[\xi\right] = 1,\ \|\xi\|_\infty \le 1/\gamma \right\}$$

$$C_2 = \left\{ \xi \in L_\infty : \xi \le a/2 \right\}.$$

Then C_1 and C_2 are convex sets, and $C_1 \cap C_2 = \emptyset$ by assumption. Also, C_1 is weak-star compact in L_∞ (recall that $(A_t)_{t \in J}$ is finite) and C_2 is weak-star closed. So by the Hahn-Banach theorem (in the weak-star topology of L_∞), there is $Y \in L_1$ with

$$\mathbf{E}\left[\xi Y\right] < \frac{1}{2} \qquad \text{for all } \xi \in C_2$$

$$\mathbf{E}\left[\xi Y\right] > \frac{1}{2} \qquad \text{for all } \xi \in C_1.$$

Now $\mathbf{E}\left[\xi Y\right] < 1/2$ for all $\xi \le 0$ implies $Y \ge 0$; and $\mathbf{E}\left[\xi Y\right] < 1/2$ for all ξ with $\|\xi\|_\infty \le a/2$ implies $\|Y\|_1 \le 1/a$.

Now for given t, let $B = A_t \setminus \{\mathbf{E}^{\mathcal{F}_t}\left[Y\right] > 1/2\}$. If $\mathbf{P}(B) > \gamma$, then

$$\xi = \frac{1}{\mathbf{P}(B)} \mathbf{1}_B$$

belongs to C_1, and thus

$$\frac{1}{2} < \mathbf{E}\left[\xi Y\right] = \frac{1}{\mathbf{P}(B)} \mathbf{E}\left[Y \mathbf{1}_B\right]$$

$$= \frac{1}{\mathbf{P}(B)} \mathbf{E}\left[\mathbf{E}^{\mathcal{F}_t}\left[Y\right] \mathbf{1}_B\right]$$

$$\le \frac{1}{\mathbf{P}(B)} \mathbf{E}\left[\frac{1}{2} \mathbf{1}_B\right] = \frac{1}{2}.$$

This contradiction shows that $\mathbf{P}(B) \le \gamma$. ∎

(4.4.8) Lemma. *Let $(\mathcal{F}_t)_{t \in J}$ be a stochastic basis, where J is a directed set with countable cofinal subset. Suppose every L_1-bounded martingale converges essentially. Then for every $\varepsilon > 0$, there exist $t_0 \in J$ and $N > 0$ so that for all finite adapted families $(A_t)_{t \in J}$ of sets supported beyond t_0, with $\mathbf{P}(\bigcup_t A_t) \ge \varepsilon$, there is an adapted family $(\xi_t)_{t \in J}$ of bounded nonnegative functions, subordinate to $(A_t)_{t \in J}$, with $\mathbf{E}\left[\sum \xi_t\right] = 1$ and $\sum \xi_t \le N$.*

Proof. Suppose (for purposes of contradiction) that this is false. Then there is $\varepsilon > 0$ such that for each $t_0 \in J$ and $N > 0$, there is a finite adapted family $(A_t)_{t \in J}$ supported beyond t_0, with $\mathbf{P}(\bigcup_t A_t) \ge \varepsilon$ such that any $(\xi_t)_{t \in J}$ subordinate to $(A_t)_{t \in J}$ with $\mathbf{E}\left[\sum \xi_t\right] = 1$ has $\|\sum \xi_t\|_\infty > N$.

Let (s_k) be an increasing sequence cofinal in J. For each k, use $N = 2^{k+2}/\varepsilon$ to obtain a finite family $(A_t)_{t \in J}$ supported beyond s_k as above. Say $A_t \neq \emptyset$ for p different values of t. (Note p depends on k.) Then apply Lemma (4.4.7) with $\gamma = 1/(kp)$ and $a = N$ to obtain $Y_k \in L_1$ with $\|Y_k\|_1 \leq \varepsilon/2^{k+2}$ and

$$\mathbf{P}\left(A_t \setminus \left\{\mathbf{E}^{\mathcal{F}_t}[Y_k] > \frac{1}{2}\right\}\right) \leq \frac{1}{kp},$$

so that

$$\mathbf{P}\left(\bigcup_t \left\{\mathbf{E}^{\mathcal{F}_t}[Y_k] > \frac{1}{2}\right\}\right) \geq \varepsilon - \frac{1}{k}.$$

Let $X = \sum_{k=1}^{\infty} Y_k$. Then $\|X\|_1 \leq \varepsilon/4$. Consider the uniformly integrable martingale $X_t = \mathbf{E}^{\mathcal{F}_t}[X]$. Since (s_k) is cofinal, we have

$$\mathbf{P}(\mathrm{e}\limsup X_t \geq 1/2) \geq \varepsilon.$$

But $\mathbf{P}(\mathrm{e}\liminf X_t \geq 1/2) \leq \varepsilon/2$, since $\|X\|_1 \leq \varepsilon/4$. So (X_t) does not converge. This contradiction completes the proof.   ∎

(4.4.9)   Theorem. *Suppose J is a directed set with countable cofinal subset, and $(\mathcal{F}_t)_{t \in J}$ is a stochastic basis. If every uniformly integrable L_1-bounded martingale converges essentially then $(\mathcal{F}_t)_{t \in J}$ satisfies (C).*

Proof. We claim: For every $\varepsilon > 0$, there exist $M > 0$, $\alpha > 0$, $\gamma > 0$, and $t_0 \in J$ such that for any finite adapted family (A_t) supported beyond t_0 with $\mathbf{P}(\bigcup A_t) > \varepsilon$ and $\sum \mathbf{P}(A_t) \leq (1+\gamma)\mathbf{P}(\bigcup A_t)$, there is $\tau \in \Sigma^{\mathrm{IM}}$, subordinate to (A_t) with $\mathbf{P}(A(\tau)) > \alpha$ and $S(\tau) \leq M$.

Indeed, given $\varepsilon > 0$, let N and t_0 be as in Lemma (4.4.8), and let $\alpha = 1/(4N)$, $\gamma = 1/(2N)$, and $M = 4N$. Let $(A_t)_{t \in J}$ be a finite adapted family supported beyond t_0 with $\mathbf{P}(\bigcup A_t) > \varepsilon$ and $\sum \mathbf{P}(A_t) \leq (1+\gamma)\mathbf{P}(\bigcup A_t)$. Overlap occurs on the set

$$C = \bigcup_{s \neq t}(A_s \cap A_t).$$

(It is really a finite union.) Then $\mathbf{P}(C) \leq \sum \mathbf{P}(A_t) - \mathbf{P}(\bigcup A_t) \leq \gamma = 1/(2N)$. There is a family $(\xi_t)_{t \in J}$ subordinate to $(A_t)_{t \in J}$ with $\mathbf{E}\left[\sum \xi_t\right] = 1$ and $\sum \xi_t \leq N$. Now $\mathbf{E}\left[\left(\sum \xi_t\right) 1_C\right] \leq N\mathbf{P}(C) \leq 1/2$, so

$$\mathbf{E}\left[\left(\sum \xi_t\right) 1_{\Omega \setminus C}\right] \geq \frac{1}{2}.$$

Thus if $H = \{\sum \xi_t > 1/4\} \setminus C$, we have $\mathbf{P}(H) \geq 1/(4N)$. Define $\tau \in \Sigma^{\mathrm{IM}}$ by $\{t \in \tau\} = \{\xi_t > 1/4\}$. Then, since the sets A_t do not overlap outside C, we have $\mathbf{P}(A(\tau)) \geq \mathbf{P}(H) \geq 1/(4N) = \alpha$. Also, $S(\tau) \leq 4\sum \xi_t \leq 4N = M$. This proves the claim.

Now since all uniformly integrable L_1-bounded martingales converge, in particular, all L_∞-bounded martingales converge. So (4.3.12) condition (V_1) holds.

Now let $\varepsilon > 0$ and let (B_t) be an adapted family of sets with $\mathbf{P}(B^*) > \varepsilon$. By (V_1), there is a finite adapted family (A_t), subordinate to (B_t), with $\mathbf{P}(\bigcup A_t) > \varepsilon$ and $\sum \mathbf{P}(A_t) < \mathbf{P}(\bigcup A_t) + \gamma\varepsilon \leq (1+\gamma)\mathbf{P}(\bigcup A_t)$. So there is $\tau \in \Sigma^{\mathrm{IM}}$ with $\mathbf{P}(A(\tau)) > \alpha$ and $S(\tau) \leq M$. By Lemma (4.4.2), we see that condition (C) is satisfied. ∎

A counterexample

(4.4.10) We will next consider an example of a stochastic basis $(\mathcal{F}_t)_{t \in J}$ not satisfying condition (V), but for which all L_1-bounded martingales converge essentially.

If we wish, the probability space $(\Omega, \mathcal{F}, \mathbf{P})$ may be $[0,1)$ with Lebesgue measure, and the sets D_t and $I_t(W)$ defined below may be taken to be half-open intervals $[a, b)$. But any continuous probability space will suffice.

Choose integers n_i with $2 < n_1 < n_2 < \cdots$ and $\sum 1/n_i < \infty$. Choose positive numbers α_i with $1/2 > \alpha_1 > \alpha_2 > \cdots$ and $\sum \alpha_i < \infty$. For each i, let $K(i) = \{1, 2, \cdots, n_i\}$, and let $L(i)$ be the collection of all 2-element subsets of $\{1, 2, \cdots, n_i\}$. Thus $L(i)$ has $n_i(n_i - 1)/2$ elements. Let J_i be the Cartesian product $K_1 \times K_2 \times \cdots \times K_i$, and let $J = \bigcup_{i=0}^\infty J_i$. [Write $\emptyset$ for the unique element of J_0.] For $t \in J_i$, we say that t belongs to *level* i, and write $|t| = i$. If $t = (t_1, t_2, \cdots, t_i) \in J_i$ and $p \in K(i+1)$, then we write tp for the element of J_{i+1} defined by $tp = (t_1, t_2, \cdots, t_i, p)$. The set J is directed when the ordering is defined by:

$$s \leq t \quad \text{iff} \quad |s| < |t| \text{ or } s = t.$$

We next define (recursively on the level i):

σ-algebras $\mathcal{G}_i$ $(i \in \mathbb{N})$,
σ-algebras $\mathcal{F}_t$ $(t \in J)$,
sets A_t $(t \in J, |t| \geq 1)$,
sets $I_t(W)$ $(t \in J, W \in L(|t| + 1))$, and
sets D_t $(t \in J)$, with

$$\mathbf{P}(D_t) = \frac{(1 - \alpha_1)(1 - \alpha_2) \cdots (1 - \alpha_i)}{n_1 n_2 \cdots n_i},$$

where $i = |t|$.

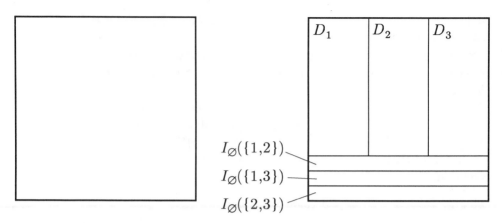

Figure (4.4.10a). σ-algebra $\mathcal{G}_0$.   Figure (4.4.10b). σ-algebra $\mathcal{G}_1$.

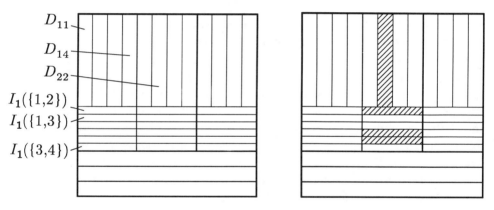

Figure (4.4.10c). σ-algebra $\mathcal{G}_2$.   Figure (4.4.10d). $A_{22} = D_{22} \cup$
$I_2(\{1,2\}) \cup I_2(\{2,3\}) \cup I_2(\{2,4\})$.

You may find the Figures (4.4.10a) to (4.4.10d) helpful when reading the following description.

To begin,

$$\mathcal{G}_0 = \mathcal{F}_0 = \{\Omega, \emptyset\}, \qquad D_\emptyset = \Omega.$$

Suppose for some $i \geq 0$, that $\mathcal{G}_i$, $\mathcal{F}_t$, and D_t (for $|t| = i$) have been defined. Let $t \in J_i$. Subdivide D_t into n_{i+1} sets D_{tp} $(p \in K(i+1))$, each of probability

$$\frac{(1-\alpha_1)(1-\alpha_2)\cdots(1-\alpha_i)(1-\alpha_{i+1})}{n_1 n_2 \cdots n_i n_{i+1}},$$

and $n_{i+1}(n_{i+1} - 1)/2$ sets $I_t(W)$ $(W \in L(i+1))$, each of probability

$$\frac{(1 - \alpha_1)(1 - \alpha_2) \cdots (1 - \alpha_i)\alpha_{i+1} 2}{n_1 n_2 \cdots n_i n_{i+1}(n_{i+1} - 1)}.$$

This subdivision is possible, since the sum of the probabilities of all the subdividing sets listed is exactly the probability

$$\frac{(1 - \alpha_1)(1 - \alpha_2) \cdots (1 - \alpha_i)}{n_1 n_2 \cdots n_i},$$

of D_t. For $p \in K(i+1)$, let

$$A_{tp} = D_{tp} \cup \bigcup \{ I_t(W) : W \in L(i+1), p \in W \}.$$

Then

$$\mathbf{P}(A_{tp}) = \frac{(1 - \alpha_1) \cdots (1 - \alpha_{i+1})}{n_1 \cdots n_{i+1}} + (n_{i+1} - 1)\frac{(1 - \alpha_1) \cdots (1 - \alpha_i)\alpha_{i+1} 2}{n_1 \cdots n_i n_{i+1}(n_{i+1} - 1)}$$

$$= \frac{1 + \alpha_{i+1}}{n_{i+1} - 1}\mathbf{P}(D_t) \leq \frac{2}{n_{i+1}}\mathbf{P}(D_t).$$

Note that $A_{tp} \cap A_{sq} = \emptyset$ if $t \neq s$, but $A_{tp} \cap A_{tq} = I_t(\{p, q\}) \neq \emptyset$ if $p \neq q$. Let $\mathcal{F}_{tp}$ be the σ-algebra generated by $\mathcal{G}_i$ and the single set A_{tp}. Let $\mathcal{G}_{i+1}$ be the σ-algebra with atoms

$$D_{tp} \qquad \text{for } t \in J_i, p \in K(i+1), \text{ and}$$
$$I_t(W) \qquad \text{for } |t| \leq i, W \in L(|t| + 1).$$

Since $D_t \supseteq A_{tp} \supseteq D_{tp}$, we have $\mathcal{G}_i \subseteq \mathcal{F}_{tp} \subseteq \mathcal{G}_{i+1}$. This completes the recursive definition. As required, if $s \leq t$, then $\mathcal{F}_s \subseteq \mathcal{F}_t$.

We claim that $(\mathcal{F}_t)_{t \in J}$ fails condition (V). We use the adapted family $(A_t)_{t \in J}$ for this purpose. For each i,

$$\bigcup_{|t|=i+1} D_t \subseteq \bigcup_{|t| \geq i+1} A_t \subseteq \bigcup_{|t|=i} D_t,$$

so $A^* = \limsup A_t = \bigcap_{i=0}^{\infty} \bigcup_{|t|=i} D_t$. Now

$$\mathbf{P}(A^*) = \lim_{i \to \infty} \mathbf{P}\left(\bigcup_{|t|=i} D_t \right)$$

$$= \lim_{i \to \infty} \frac{(1 - \alpha_1) \cdots (1 - \alpha_i)}{n_1 \cdots n_i} n_1 \cdots n_i$$

$$= \prod_{j=1}^{\infty}(1 - \alpha_j) > 0.$$

Now fix i_0 and let $\tau \in \Sigma$ be any simple stopping time with values in levels i_0 and above. Now A_t is an atom of $\mathcal{F}_t$, so either $\{\tau = t\} \supseteq A_t$ or $\{\tau = t\} \cap A_t = \emptyset$. Because of the overlap properties of the A_t, for each fixed index t we must have $\{\tau = tp\} \cap A_{tp} = \emptyset$ except for at most one value of p, so that

$$\mathbf{P}\left(\bigcup_{|t|=i+1} \{\tau = t\} \cap A_t\right) \leq \sum_{|t|=i} \mathbf{P}(A_{t1})$$

$$\leq n_1 \cdots n_i \frac{2}{n_{i+1}} \mathbf{P}(D_t)$$

$$\leq \frac{2}{n_{i+1}}.$$

Thus

$$\mathbf{P}(A(\tau)) \leq \sum_{i=i_0}^{\infty} \frac{2}{n_i}.$$

Since this approaches 0 as $i_0 \to \infty$, we see that (V) fails.

However, we claim that all L_1-bounded martingales adapted to $(\mathcal{F}_t)_{t \in J}$ converge essentially (or, what is the same thing since J is countable, a.s.). We verify this by proving that condition (C) holds. In fact, a uniform version of (C) is true: If (B_t) is an adapted family of sets and $\varepsilon > 0$, then there is $\tau \in \Sigma^{\text{IM}}$ with $\mathbf{P}(B^* \setminus B(\tau)) < \varepsilon$ and $S(\tau) \leq 2$.

Let (B_t) and ε be given. Fix i. We will construct $\tau_i \in \Sigma^{\text{IM}}$ with values in J_i such that $S(\tau_i) \leq 2$ and $\mathbf{P}\left(\bigcup_{|t|=i} B_t \setminus B(\tau_i)\right) < 2/n_i$. First consider the atoms of $\mathcal{G}_{i-1}$. For $s \in J_{i-1}$, if $D_s \subseteq B_t$ for some $t \in J_i$, then choose one and call it $t = \gamma(s)$. For $|s| \leq i - 2$ and $W \in L(i-1)$, if $I_s(W) \subseteq B_t$ for some $t \in J_i$, then choose one and call it $t = \beta(s, W)$. If some atom D_s, $s \in J_{i-1}$, is *not* contained in any B_t, then consider the sets $H_{sp} = D_s \cap B_{sp}$. This is $\mathcal{F}_{sp}$ measurable, and does not include all of D_s. So it is one of: $\emptyset$, A_{sp}, $D_{sp} \setminus A_{sp}$. For this value of s, if $H_{sp} = D_{sp} \setminus A_{sp}$ for some value of p, then choose one, and write $R_{sp} = H_{sp}$ for that value of p and $R_{sp} = \emptyset$ for the other values of p. If H_{sp} is $\emptyset$ or A_{sp} for all p, then let $R_{sp} = H_{sp}$ for all p. Now define τ_i by: $\{t \in \tau_i\} = \emptyset$ if $|t| \neq i$, and

$$\{t \in \tau_i\} = \bigcup \{D_s : t = \gamma(s)\} \cup \bigcup \{I_s(W) : t = \beta(s, W)\} \cup R_t.$$

Clearly $\{t \in \tau_i\} \subseteq B_t$. Since the sets R_t intersect at most two at a time (and the others are disjoint), $S(\tau_i) \leq 2$.

Now how large is $V_i = \bigcup_{|t|=i} B_t \setminus B(\tau_i)$? Its intersection with most atoms of $\mathcal{G}_{i-1}$ is $\emptyset$. The only exception is an atom D_s, $s \in J_{i-1}$, where $H_{sp} = D_{sp} \setminus A_{sp}$, so $\mathbf{P}(V_i \cap D_s) \leq \mathbf{P}(A_{sp}) \leq (2/n_i)\mathbf{P}(D_s)$. Thus $P(V_i) \leq \sum_s (2/n_i)\mathbf{P}(D_s) = (2/n_i) \sum_s \mathbf{P}(D_s) \leq 2/n_i$.

Now given $\varepsilon > 0$, choose i_0 with $\sum_{i=i_0}^{\infty} 2/n_i < \varepsilon/2$, then choose $i_1 > i_0$ so that

$$\mathbf{P}\left(B^* \setminus \bigcup_{i_0 \leq |t| \leq i_1} B_t\right) < \frac{\varepsilon}{2}.$$

Define $\tau \in \Sigma^{\mathrm{IM}}$ by:

$$\{t \in \tau\} = \{t \in \tau_{|t|}\} \setminus \bigcup_{i_1 \leq |s| < |t|} B_s$$

for $i_0 \leq |t| \leq i_1$, and $\{t \in \tau\} = \emptyset$ otherwise. Then $\mathbf{P}(B^* \setminus B(\tau)) < \varepsilon$ and $S(\tau) \leq 2$. This completes the verification of condition (C).

Complements

(4.4.11) (Conditions between (V) and (C).) Let m be a positive integer. We say that a stochastic basis $(\mathcal{F}_t)$ satisfies *condition* SV(m) iff for every $\varepsilon > 0$ and every adapted family (A_t) of sets there is $\tau \in \Sigma^{\mathrm{IM}}$ such that $\mathbf{P}(A^* \setminus A(\tau)) < \varepsilon$ and $S(\tau) \leq m$. Thus the stochastic basis in the counterexample (4.4.10) satisfies condition SV(2). We say that $(\mathcal{F}_t)$ satisfies *condition* (SV) iff SV(m) holds for some m. We say that $(\mathcal{F}_t)$ satisfies *condition* $(\sigma$SV$)$ iff there exists a sequence of sets Ω_n in the algebra $\bigcup \mathcal{F}_t$ such that $\mathbf{P}(\bigcup \Omega_n) = 1$ and, for each n, the restriction of $(\mathcal{F}_t)$ to Ω_n satisfies (SV).

These conditions are related as follows. None of the implications can be reversed.

$$(\mathrm{V}) \implies \mathrm{SV}(2) \implies \mathrm{SV}(3) \implies \cdots \implies \mathrm{SV}(m) \implies$$
$$\mathrm{SV}(m+1) \implies \cdots \implies (\mathrm{SV}) \implies (\sigma\mathrm{SV}) \implies (\mathrm{C}).$$

Reference: Millet & Sucheston [1980e].

(4.4.12) (Condition (C_Φ).) Let Φ be an Orlicz function. We say that the stochastic basis $(\mathcal{F}_t)_{t \in J}$ satisfies *condition* (C_Φ) iff for each $\varepsilon > 0$ there exists M such that for each adapted family $(A_t)_{t \in J}$ with $\mathbf{P}(A^*) > \varepsilon$, there exists $\tau \in \Sigma^{\mathrm{IM}}$ with $\mathbf{P}(A^* \setminus A(\tau)) < \varepsilon$ and $\|S(\tau)\|_\Phi \leq M$. Then, of course, condition (C) is the special case where $L_\Phi = L_\infty$.

Suppose J has a countable cofinal set and the conjugate Orlicz function Ψ satisfies condition (Δ_2). Then condition (C_Φ) is satisfied if and only if every L_Ψ-bounded martingale converges essentially (Talagrand [1986]).

(4.4.13) (Covering condition (A).) A stochastic basis (A_t) satisfies *covering condition* (A) iff, for every adapted family (A_t) of sets with $\mathbf{P}(A^*) > 0$, there is a constant M such that for all $s \in J$ there exists $\tau \in \Sigma$, $\tau \geq s$, τ subordinate to (A_t), with

$$\|S(\tau)\|_\infty \leq M\|S(\tau)\|_1.$$

If (A) holds, then all L_1-bounded martingales converge (Astbury [1981b]).

(4.4.14) (Uniformly integrable martingales.) Suppose J has a countable cofinal subset. Let $(\mathcal{F}_t)_{t \in J}$ be a stochastic basis. If all uniformly integrable martingales converge essentially, then all L_1-bounded martingales converge essentially (Astbury [1981a]).

(4.4.15) (A counterexample.) Assuming the Continuum Hypothesis, there exists a stochastic basis of finite algebras on $[0,1]$ such that all uniformly integrable martingales converge essentially, but some L_1-bounded martingales fail to converge essentially (Talagrand [1986]). This example shows that the hypothesis of countable cofinal set cannot be removed entirely in Theorem (4.4.9) (or in the preceding result (4.4.14) of Astbury).

(4.4.16) (Failure of condition (C).) Condition (C) fails for the stochastic basis of Example (4.2.1), since the martingale there does not converge essentially. The failure of (C) may be established directly. We use the notation of Proposition (4.2.6). The adapted family A_t constructed there has $\mathbf{P}(A^*) = 1$. For $\varepsilon = 1/2$, we will show that the condition in Lemma (4.4.2(d)) fails. Let M be given, then choose $p \in \mathbb{N}$ so large that $M \cdot 2^{-p} < 1/4$. Continue as in the argument for (V): Let $\tau \in \Sigma^{\mathrm{IM}}$ satisfy $\tau \geq \{1, 2, \cdots, p\}$ and $S(\tau) \leq M$. For $m \geq p$, for a fixed pair (B, D), the sets

$$F(C \cup B, D) \qquad \text{for } C \in \mathcal{C}_m$$

all contain the nonnull set $F(\{m+1, \cdots, 4m\} \cup B, D)$. Since $S(\tau) \leq M$, there are at most M sets C with $F(C \cup B, D) \cap \{t(m, C) \in \tau\} \neq \emptyset$. Thus

$$\mathbf{P}\left(\bigcup_{C \in \mathcal{C}_m} \left(F(C, \emptyset) \cap \{t(m, C) \in \tau\}\right)\right) \leq M \sum_{B \subseteq \{1, \cdots, m\}} 2^{-2m}$$
$$= M \cdot 2^m \cdot 2^{-2m} = M \cdot 2^{-m}.$$

Thus $\mathbf{P}(A(\tau)) \leq \sum_{m=p}^{\infty} M \cdot 2^{-m} = M \cdot 2^{-p+1} < 1/2$. Condition (C) fails.

Remarks

The covering condition (C) was introduced by Millet & Sucheston [1980e]. Astbury [1981b] independently introduced covering condition (A), which he called the "dominated sums property." Millet & Sucheston [1980c] proved (assuming the existence of a countable cofinal set) that (A) is equivalent to condition (C). Millet & Sucheston proved that (C) implies convergence of L_1-bounded martingales, and Astbury independently proved that (A) implies convergence of L_1-bounded martingales. The proof that convergence of L_1-bounded martingales implies (C) is due to Talagrand [1986].

We will see in Chapter 7 that condition (C) can be used profitably in the derivation setting. The interplay between derivation and directed-sets is part of the reason that we feel the two settings should be studied together.

The counterexample (4.4.10) showing that convergence of L_1-bounded martingales does not imply condition (V) is from Millet & Sucheston [1979b], where conditions $SV(m)$ and (SV) were introduced. A similar example is given in Astbury [1981b].

5

Banach-valued random variables

In this chapter, we consider martingales, amarts, and related processes that take values in Banach spaces. They are useful in Banach spaces that occur naturally in mathematics, such as function spaces. They have also played an important role in the understanding of some geometric properties of Banach spaces. The reader who knows nothing of Banach spaces should, of course, skip this chapter; but someone with only a minimal knowledge of Banach space theory should be able to work through most of the chapter, with the exception of Section 5.5.

There is a close connection between martingales with values in a Banach space E and measures with values in E. This is not unexpected. More interesting are the connections of these two topics with the geometric properties of the Banach space. These are explored in Section 5.4.

If a theorem is true in the real-valued case and extends to a more general setting without change of argument, this may be useful, but is hardly exciting. One modern approach to probability in Banach space consists in exactly matching the convergence property to the geometry of the space. Such theorems are not only the best possible in terms of convergence, but they may also shed new light on the structure of the space. The remarkable theorem of A. & C. Ionescu Tulcea and S. D. Chatterji (that L_1-bounded martingales taking values in a Banach space E converge a.s. if and only if E has the Radon-Nikodým property, Theorem (5.3.30); also (5.3.34)) characterizes a geometrical property of Banach spaces in terms of convergence of martingales. It will be seen that there are many such characterizations in terms of amarts. For example, the Radon-Nikodým property of the space, the Radon-Nikodým property of the dual, the reflexivity and even the finite-dimensionality, can be characterized in terms of convergence of appropriate classes of amarts. Some of the more common operator ideals can be characterized as well in terms of convergence of classes of amarts.

These results have various degrees of depth. In some instances, the probabilistic result is an interesting but easy consequence of a deep theorem in functional analysis.

Martingale convergence is applied several times in Section 5.4: the fixed-point theorem of Ryll-Nardzewski, the integral representation theorem of Choquet-Edgar, and two geometric characterizations of the Radon-Nikodým property, (5.4.13) and (5.4.17). Amart investigations led to the following basic result which can be stated without amarts: If for bounded processes scalar convergence implies weak a.s. convergence, then (and only then) the dual of the Banach space is separable (5.5.27). There is also an operator generalization of this result (5.5.26).

5.1. Vector measures and integrals

Theorems of functional analysis

The Banach spaces in this chapter will be understood to be Banach spaces over the field $\mathbb{R}$ of real numbers.

If E is a Banach space, we will write E^* for the *dual space*, that is the set of all bounded linear functionals on E with norm

$$\|x^*\| = \sup \left\{ |x^*(x)| : x \in E, \|x\| \leq 1 \right\}.$$

If $x \in E$ and $x^* \in E^*$, we will sometimes write $\langle x, x^* \rangle$ for $x^*(x)$. The Banach space E will often be identified with a subspace of E^{**} in the natural way.

The norm on a Banach space defines a metric on the space, and therefore a topology. When topological words are used without qualification, they will refer to this norm topology. A Banach space also has a *weak topology*. A net (x_t) in E converges weakly to the vector x iff

$$\langle x_t, x^* \rangle \rightarrow \langle x, x^* \rangle$$

for all $x^* \in E^*$. If $E = F^*$ is a dual space, then it has a *weak-star topology*. A net (x_t) in E converges weak-star to the vector x iff

$$\langle y, x_t \rangle \rightarrow \langle y, x \rangle$$

for all $y \in F$. Details of these and other items concerning Banach spaces can be found for example in Lindenstrauss & Tzafriri [1968] or in Dunford & Schwartz [1958].

The study of the geometry of convex sets in a Banach space frequently uses the Hahn-Banach theorem in two different forms:

(5.1.1) The Hahn-Banach extension theorem. *Let E be a Banach space, and E_1 a closed subspace. If $x_1^* \in E_1^*$, then there exists an extension $x^* \in E^*$ (that is, $x_1^*(x) = x^*(x)$ for all $x \in E_1$) with $\|x^*\| = \|x_1^*\|$.*

(5.1.2) The Hahn-Banach separation theorem. *Let E be a Banach space. Suppose C a closed convex subset, and K a compact convex subset. If $C \cap K = \emptyset$, then there exists a functional $x^* \in E^*$ that strictly separates the two sets (that is, $\sup x^*(C) < \inf x^*(K)$). Suppose C a closed convex subset, and G an open convex subset. If $C \cap G = \emptyset$, then there exists a functional $x^* \in E^*$ that separates the two sets (that is, $\sup x^*(C) \leq \inf x^*(G)$).*

A reference for Hahn-Banach theorems is, for example, Rudin [1973], pp. 55–59.

Another important result from functional analysis that will be used often is the *closed graph theorem*.

(5.1.3) Closed graph theorem. *Let E and F be Banach spaces, and let $T\colon E \to F$ be a linear transformation. If the graph*

$$\{\,(x,y) \in E \times F : y = Tx\,\}$$

is a closed set in the product $E \times F$, then T is a bounded linear transformation.

See Rudin [1973], Theorem 2.15, for a discussion of the closed graph theorem.

Random variables

Let E be a Banach space. By a *random variable* in E, we mean a function measurable in the sense of Bochner. This can be defined as follows. Let $(\Omega, \mathcal{F}, \mathbf{P})$ be a probability space. A *measurable simple function* in E is a function $X\colon \Omega \to E$ of the form

$$X(\omega) = \sum_{j=1}^{n} \mathbf{1}_{A_j}(\omega)\, x_j$$

where $A_j \in \mathcal{F}$ and $x_j \in E$. The *integral* or *expectation* of X is the vector

$$\mathbf{E}\left[X\right] = \sum_{j=1}^{n} \mathbf{P}(A_j)\, x_j.$$

A *random variable* in E is a function $X\colon \Omega \to E$ that is equal almost surely to the limit (in the norm of E) of a sequence X_n of measurable simple functions:

$$\lim_{n\to\infty} \left\| X(\omega) - X_n(\omega) \right\| = 0 \qquad \text{for almost every } \omega.$$

Almost all the values of X lie in a separable subspace of E, namely the closed span of the countable set obtained from the values of all the simple functions X_n.

The random variable X is *Bochner integrable* iff $\mathbf{E}\left[\|X\|\right] < \infty$; the quantity

$$\|X\|_{L_1} = \mathbf{E}\left[\|X\|\right]$$

is called the *Bochner norm* of X. If X is Bochner integrable, then the approximating sequence X_n of simple functions can be chosen so that

$$\lim_{n\to\infty} \mathbf{E}\left[\|X - X_n\|\right] = 0;$$

then the *Bochner integral* of X is defined by

$$\mathbf{E}\left[X\right] = \lim_{n\to\infty} \mathbf{E}\left[X_n\right].$$

It can be shown that this expression does not depend on the choice of the approximating sequence X_n. We write $L_1(\Omega, \mathcal{F}, \mathbf{P}; E)$ for the set of all Bochner integrable random variables. It is a Banach space (when its elements are considered to be equivalence classes). The Bochner integral has many of the properties of the real-valued integral.

(5.1.4) Definition. Let $X: \Omega \to E$ be a random variable. The *Pettis norm* of X is

$$\|X\|_P = \sup_{\substack{x^* \in E^* \\ \|x^*\| \leq 1}} \mathbf{E}\left[|\langle X, x^* \rangle|\right].$$

Note that $\|X\|_P \leq \|X\|_{L_1}$, so every Bochner integrable random variable has finite Pettis norm. The space $L_1(\Omega, \mathcal{F}, \mathbf{P}; E)$ is not complete under the Pettis norm (unless E is finite-dimensional). See Theorem (5.5.34).

The random variable X is *scalarly integrable* iff, for every bounded linear functional $x^* \in E^*$, the composition $\langle X, x^* \rangle$ is a (real-valued) integrable random variable. Of course, if X has finite Pettis norm, then it is scalarly integrable. But in fact the converse is also true. This is an application of the closed graph theorem, which may be carried out as follows: If X is scalarly integrable, define a map $T: E^* \to L_1(\Omega, \mathcal{F}, \mathbf{P})$ by $T(x^*) = \langle X, x^* \rangle$. We claim that the graph of T is closed. If $x_n^* \to x^*$ in E^*, then $\langle X(\omega), x_n^* \rangle \to \langle X(\omega), x^* \rangle$ for each ω, hence $\langle X, x_n^* \rangle \to \langle X, x^* \rangle$ in probability. If, in addition, $\langle X, x_n^* \rangle \to Y$ in L_1 norm, then $\langle X, x_n^* \rangle \to Y$ in probability. Therefore $Y = \langle X, x^* \rangle$ a.s. Thus T is a closed linear transformation defined on a Banach space, so it is bounded. The norm $\|T\|$ is the Pettis norm of X.

When the Banach space is separable, there are many equivalent ways to recognize random variables. Our proofs are based on a simple lemma.

A *closed half-space* in a Banach space E is a set of the form

$$H_\lambda(x^*) = \{ y \in E : \langle y, x^* \rangle \leq \lambda \}$$

for some $x^* \in E^*$ and $\lambda \in \mathbb{R}$.

(5.1.5) Lemma. *Let E be a separable Banach space, and let C be a closed, convex subset of E. Then C is a countable intersection of closed half-spaces.*

Proof. It follows from the Hahn-Banach separation theorem that C is an intersection of closed half-spaces, say

$$C = \bigcap_{\gamma \in \Gamma} H_{\lambda_\gamma}(x_\gamma^*).$$

The set $E \setminus C$ is a Lindelöf space, so C is also the intersection of a countable subcollection of the half-spaces $H_{\lambda_\gamma}(x_\gamma^*)$. This point will now be explained more fully. Let V be a countable set dense in $E \setminus C$, and let $\mathcal{V}$ be the collection of all balls with centers in V and rational radii. $\mathcal{V}$ is a countable collection. The union of all the elements of $\mathcal{V}$ that are disjoint from some $H_{\lambda_\gamma}(x_\gamma^*)$ is exactly $E \setminus C$. Thus C is the intersection of countably many of these half-spaces, one disjoint from each of these balls. ∎

The σ-algebra $\mathcal{B}$ of *Borel sets* in the Banach space E is the σ-algebra generated by the open sets. If E is separable, then that is the same as the σ-algebra generated by the open balls, since each open set is a countable union of open balls. In fact, it is the same as the σ-algebra generated by the closed balls, since each open ball is a countable union of closed balls. Then the previous result will show that $\mathcal{B}$ is the σ-algebra generated by the closed half-spaces. (Note: This may fail in a nonseparable Banach space.)

(5.1.6) **Proposition.** *Let E be a separable Banach space, and let $(\Omega, \mathcal{F}, \mathbf{P})$ be a complete probability space. A function $X \colon \Omega \to E$ is Bochner measurable if and only if it is Borel measurable.*

Proof. A simple function is Borel measurable, and a limit of a sequence of Borel measurable functions (with values in a metric space) is Borel measurable. This is enough to show that any Bochner measurable function is Borel measurable. Now let $X \colon \Omega \to E$ be Borel measurable. Fix a positive integer n. The Banach space E is the union of countably open balls of diameter less than 2^{-n}, so E is the union of a countable pairwise disjoint family $\{D_j\}$ of Borel sets of diameter less than 2^{-n}. By the countable additivity of $\mathbf{P}$, there is m so that

$$\mathbf{P}\left\{X \in \bigcup_{j=1}^{m} D_j\right\} > 1 - 2^{-n}.$$

Let $a_j \in D_j$ and write $A_j = \{X \in D_j\}$. Then

$$X_n = \sum_{j=1}^{m} a_j \, 1_{A_j}$$

is a measurable simple function satisfying

$$\mathbf{P}\{\|X - X_n\| < 2^{-n}\} > 1 - 2^{-n}.$$

Clearly X is Bochner measurable. ∎

(5.1.7) **Pettis measurability theorem.** *Let E be a Banach space, and let $(\Omega, \mathcal{F}, \mathbf{P})$ be a complete probability space. A function $X \colon \Omega \to E$ is Bochner measurable if and only if X is scalarly measurable and there is a separable subspace $E_1 \subseteq E$ with $\mathbf{P}\{X \in E_1\} = 1$.*

Proof. If X is Bochner measurable, then it is scalarly measurable and (almost) separably valued, since simple functions have these properties and they are preserved by pointwise a.s. limits.

For the converse, suppose X is scalarly measurable and has values (a.s.) in the separable space E_1. By ignoring a set of measure zero, we may assume that X has all its values in E_1. Now the collection of all sets $D \subseteq E_1$ such that $\{X \in D\} \in \mathcal{F}$ is a σ-algebra and includes the closed half-spaces. Therefore it includes all Borel sets. This means that X is Borel measurable, so by the preceding result X is Bochner measurable. ∎

A slight variant of this theorem will also be useful. If F is a Banach space and $X \colon \Omega \to F^*$, we will say that X is *weak-star scalarly measurable* iff $\langle y, X \rangle$ is a measurable function on Ω for each $y \in F$.

(5.1.8) Proposition. *Let F be a Banach space with separable dual, and let $X\colon \Omega \to F^*$ be weak-star scalarly measurable. Then X is Bochner measurable.*

This is proved by almost the same method as we have used for the Pettis measurability theorem. We must use *weak-star closed half-spaces* in place of closed half-spaces. The unit ball of F^* is weak-star closed, so it is the intersection of a family of weak-star closed half-spaces. The rest of the proof is the same.

Vector measures

Let E be a Banach space. Suppose $\mathcal{F}$ is a σ-algebra of subsets of a set Ω. A *vector measure* in E is a function $\mu\colon \mathcal{F} \to E$ such that

$$\mu\left(\bigcup_{n=1}^{\infty} A_n\right) = \sum_{n=1}^{\infty} \mu(A_n),$$

for every pairwise disjoint sequence (A_n) in $\mathcal{F}$, where the series converges in the norm of E. The measure μ is *absolutely continuous* with respect to $\mathbf{P}$ iff $\mu(A) = 0$ for all $A \in \mathcal{F}$ with $\mathbf{P}(A) = 0$. (We write $\mu \ll \mathbf{P}$.)

The *variation* of μ on a set $A \in \mathcal{F}$ is

$$|\mu|(A) = \sup \sum_{i=1}^{n} \|\mu(A_i)\|,$$

where the supremum is taken over all finite disjoint sequences $A_1, A_2, \cdots,$ $A_n \subseteq A$ in $\mathcal{F}$. The set-function $|\mu|(\cdot)$ is a (possibly infinite) measure on $\mathcal{F}$. We say that μ has *σ-finite variation* on a set A iff A is a countable union of sets on which μ has finite variation.

Suppose the random variable X is Bochner integrable, and a set function μ is defined by $\mu(A) = \mathbf{E}\left[X\,\mathbf{1}_A\right]$ for $A \in \mathcal{F}$. Then μ has finite variation; in fact the variation of μ on Ω is exactly the Bochner norm $\mathbf{E}\left[\|X\|\right]$.

The *semivariation* of a vector measure μ on a set $A \in \mathcal{F}$ is

$$\|\mu\|(A) = \sup\left\{\,|x^*\mu|(A) : x^* \in E^*, \|x^*\| \le 1\,\right\},$$

where $|x^*\mu|$ is the variation of the real-valued measure $x^*\mu$ (see Section 1.3). The set-function $\|\mu\|(\cdot)$ is not additive, but it is subadditive and has the following property: $\|\mu\|(A) = 0$ if and only if $\mu(B) = 0$ for all $B \subseteq A$. Indeed, we claim that

$$\sup\left\{\,\|\mu(B)\| : B \subseteq A\,\right\} \le \|\mu\|(A) \le 2\sup\left\{\,\|\mu(B)\| : B \subseteq A\,\right\}.$$

To see this, first consider any $B \subseteq A$. Then

$$
\begin{aligned}
\|\mu(B)\| &= \sup\left\{\,|x^*\mu|(B) : x^* \in E^*, \|x^*\| \le 1\,\right\}\\
&\le \sup\left\{\,|x^*\mu|(A) : x^* \in E^*, \|x^*\| \le 1\,\right\} = \|\mu\|(A).
\end{aligned}
$$

Thus $\sup\{\,\|\mu(B)\| : B \subseteq A\,\} \le \|\mu\|(A)$. On the other hand, if $x^* \in E^*$ and $\|x^*\| \le 1$, choose the positive and negative sets B_1 and B_2 for the signed measure $x^*\mu$ (as in (1.3.7)). Then

$$|x^*\mu|(A) = |x^*\mu(A \cap B_1)| + |x^*\mu(A \cap B_2)|$$
$$\le \|\mu(A \cap B_1)\| + \|\mu(A \cap B_2)\|$$
$$\le 2 \sup\{\,\|\mu(B)\| : B \subseteq A\,\}.$$

Therefore

$$\sup\{\,|x^*\mu|(A) : x^* \in E^*, \|x^*\| \le 1\,\} = \|\mu\|(A) \le 2 \sup\{\,\|\mu(B)\| : B \subseteq A\,\}.$$

(5.1.9) If μ has the form $\mu(A) = \mathbf{E}\,[X\,\mathbf{1}_A]$, then the semivariation is the Pettis norm of X (5.1.4):

$$\|X\|_P = \|\mu\|(\Omega).$$

Therefore we have

(5.1.9a) $$\|X\|_P \le 2 \sup\{\,\|\mathbf{E}\,[X\,\mathbf{1}_A]\,\| : A \in \sigma(X)\,\}.$$

Here is an extension theorem for vector measures.

(5.1.10) Theorem. *Let* $(\Omega, \mathcal{F}, \mathbf{P})$ *be a probability space, and let* E *be a Banach space. Let* $\mathcal{A} \subseteq \mathcal{F}$ *be an algebra of sets, and let* $\mathcal{G}$ *be the* σ-algebra *generated by* $\mathcal{A}$. *If* $\mu \colon \mathcal{A} \to E$ *is finitely additive and satisfies*

(5.1.10a) $$\|\mu(A)\| \le \mathbf{P}(A) \quad \text{for all } A \in \mathcal{A},$$

then there is a unique extension of μ *to* $\mathcal{G}$ *that is a countably additive vector measure.*

Proof. First, if $B \in \mathcal{G}$, then for every $\varepsilon > 0$ there is an approximating set $A \in \mathcal{A}$ with $\mathbf{P}(A \triangle B) < \varepsilon$. (This is true because the set of all B with this property is a σ-algebra containing $\mathcal{A}$.) Now if $B \in \mathcal{G}$, we may choose a sequence (A_n) in $\mathcal{A}$ with $\mathbf{P}(A_n \triangle B) \to 0$. But now by (5.1.10a), we have

$$\|\mu(A_m) - \mu(A_n)\| \le |\mathbf{P}(A_m) - \mathbf{P}(A_n)|,$$

so the sequence $(\mu(A_n))$ converges. Call the limit $\mu(B)$. This limit does not depend on the choice of the sequence (A_n), again by (5.1.10a). This extension of μ to $\mathcal{G}$ is countably additive, since it satisfies

$$\|\mu(B)\| \le \mathbf{P}(B) \quad \text{for all } B \in \mathcal{G}.$$

∎

The Radon-Nikodým property

A Banach space E has the *Radon-Nikodým property* iff, for every probability space $(\Omega, \mathcal{F}, \mathbf{P})$ and every measure $\mu \colon \mathcal{F} \to E$ such that μ is absolutely continuous with respect to $\mathbf{P}$ and μ has finite variation on Ω, there is a Bochner integrable random variable $X \colon \Omega \to E$ such that

$$\mu(A) = \mathbf{E}\left[X\, \mathbf{1}_A\right]$$

for all $A \in \mathcal{F}$. The random variable X is called the *Radon-Nikodým derivative* of μ with respect to $\mathbf{P}$, and is denoted

$$X = \frac{d\mu}{d\mathbf{P}}.$$

Here is an elementary, but useful, reformulation of the condition.

(5.1.11) Proposition. *Let E be a Banach space. Then E has the Radon-Nikodým property if and only if for every probability space $(\Omega, \mathcal{F}, \mathbf{P})$ and every measure $\mu \colon \mathcal{F} \to E$ such that $\|\mu(A)\| \leq \mathbf{P}(A)$ for all $A \in \mathcal{F}$, there is a Radon-Nikodým derivative $d\mu/d\mathbf{P}$ and $\|d\mu/d\mathbf{P}\| \leq 1$ a.s.*

Proof. If $\|\mu(A)\| \leq \mathbf{P}(A)$ for all $A \in \mathcal{F}$, then μ has variation at most 1. If E has the Radon-Nikodým property, then it satisfies the condition stated here. Conversely, suppose E satisfies the condition. Let $(\Omega, \mathcal{F}, \mathbf{P})$ be a probability space, and $\mu \colon \mathcal{F} \to E$ a measure, absolutely continuous with respect to $\mathbf{P}$, with finite variation. We may suppose that μ is not identically 0. Define

$$\mathbf{P}'(A) = \frac{|\mu|(A)}{|\mu|(\Omega)}$$

for all $A \in \mathcal{F}$. Then $(\Omega, \mathcal{F}, \mathbf{P}')$ is a probability space. Define

$$\mu'(A) = \frac{\mu(A)}{|\mu|(\Omega)}$$

for $A \in \mathcal{F}$. Then $\mu' \colon \mathcal{F} \to E$ is a vector measure, and $\|\mu'(A)\| \leq \mathbf{P}'(A)$ for all $A \in \mathcal{F}$. Thus there is a Bochner integrable random variable $X' \colon \Omega \to E$ such that

$$\mu'(A) = \mathbf{E}'\left(X'\, \mathbf{1}_A\right)$$

for all $A \in \mathcal{F}$, where $\mathbf{E}'$ is the expectation with respect to $\mathbf{P}'$. Now $\mathbf{P}'$ is absolutely continuous with respect to $\mathbf{P}$, so there exists a scalar-valued Radon-Nikodým derivative $H = d\mathbf{P}'/d\mathbf{P}$. A short calculation shows that the product $X = |\mu|(\Omega) \cdot H \cdot X'$ is the required Radon-Nikodým derivative of μ with respect to $\mathbf{P}$. ∎

There are some special cases under which Radon-Nikodým derivatives exist regardless of the Radon-Nikodým property. One of them occurs when the probability space $(\Omega, \mathcal{F}, \mathbf{P})$ is atomic. The Radon-Nikodým derivative of μ is

$$X = \sum_A \frac{\mu(A)}{\mathbf{P}(A)}\, \mathbf{1}_A$$

where the sum is over a maximal disjoint partition of Ω into atoms. Another such special case is the conditional expectation, which we consider next.

(5.1.12) Definition. Let E be a Banach space, let $(\Omega, \mathcal{F}, \mathbf{P})$ be a probability space, let $X \colon \Omega \to E$ be a Bochner integrable random variable, and let $\mathcal{G} \subseteq \mathcal{F}$ be a σ-algebra. The *conditional expectation* of X given $\mathcal{G}$ is the unique (a.s.) random variable $Y \colon \Omega \to E$, measurable with respect to $\mathcal{G}$, such that

$$\mathbf{E}\left[Y\, \mathbf{1}_A\right] = \mathbf{E}\left[X\, \mathbf{1}_A\right]$$

for all $A \in \mathcal{G}$. We write $Y = \mathbf{E}^{\mathcal{G}}\left[X\right]$ or $\mathbf{E}\left[X \mid \mathcal{G}\right]$.

(5.1.13) Proposition. *Let E be a Banach space, let $(\Omega, \mathcal{F}, \mathbf{P})$ be a probability space, let $X \colon \Omega \to E$ be a Bochner integrable random variable, and let $\mathcal{G} \subseteq \mathcal{F}$ be a σ-algebra. Then the conditional expectation $\mathbf{E}^{\mathcal{G}}\left[X\right]$ exists.*

Proof. The conditional expectation $\mathbf{E}^{\mathcal{G}}\left[X\right]$ exists for simple random variables of the form

$$X = \sum_{j=1}^{n} \mathbf{1}_{A_j}\, x_j,$$

where $A_j \in \mathcal{F}$ and $x_j \in E$, namely

$$\mathbf{E}^{\mathcal{G}}\left[X\right] = \sum_{j=1}^{n} \mathbf{E}^{\mathcal{G}}\left[\mathbf{1}_{A_j}\right] x_j.$$

It is not hard to establish the estimate

$$\mathbf{E}\left[\left\|\mathbf{E}^{\mathcal{G}}\left[X_1\right] - \mathbf{E}^{\mathcal{G}}\left[X_2\right]\right\|\right] \leq \mathbf{E}\left[\left\|X_1 - X_2\right\|\right],$$

where X_1 and X_2 are simple functions. If X is Bochner integrable, and $(X_n)_{n \in \mathbb{N}}$ is a sequence of simple functions that converges to X in the Bochner norm, then the sequence $\mathbf{E}^{\mathcal{G}}\left[X_n\right]$ is Cauchy in the Bochner norm. The limit of this sequence defines $\mathbf{E}^{\mathcal{G}}\left[X\right]$. ∎

The conditional expectation is a kind of "average," so the following result is not unexpected.

(5.1.14) Proposition. *Let X be a Bochner integrable random variable taking values in a closed convex subset C of the Banach space E. Then (a) for any sub-σ-algebra $\mathcal{G}$ of $\mathcal{F}$, the conditional expectation $\mathbf{E}^{\mathcal{G}}\left[X\right]$ has values in C; and (b) for any $A \in \mathcal{F}$ with $\mathbf{P}(A) > 0$, we have $\mathbf{E}\left[X\, \mathbf{1}_A\right]/\mathbf{P}(A) \in C$.*

Proof. Since X is Bochner measurable, its values lie a.s. in a separable subspace of E. So we may assume E itself is separable. By Lemma (5.1.5), C is an intersection of countably many closed half-spaces. That is, there exist linear functionals $x_i^* \in E^*$ and scalars α_i such that

(5.1.14a) $$C = \bigcap_{i=1}^{\infty} \left\{ x \in E : \langle x, x_i^* \rangle \leq \alpha_i \right\}.$$

Now X has its values in C, so for any $A \in \mathcal{F}$ we have

(5.1.14b) $\langle \mathbf{E}\left[X\,\mathbf{1}_A\right], x_i^* \rangle = \mathbf{E}\left[\langle X, x_i^* \rangle\,\mathbf{1}_A\right] \le \mathbf{E}\left[\alpha_i\,\mathbf{1}_A\right] = \alpha_i\,\mathbf{P}(A).$

Therefore, by (5.1.14a), the vector $\mathbf{E}\left[X\,\mathbf{1}_A\right]/|P(A)|$ belongs to C. This proves (b). For the proof of (a), write $Y = \mathbf{E}^{\mathcal{G}}\left[X\right]$. In (5.1.14b) take $A = \{\,\omega \in \Omega : \langle Y(\omega), x_i^* \rangle > \alpha_i\,\}$. Then

$$\langle \mathbf{E}\left[Y\,\mathbf{1}_A\right], x_i^* \rangle = \mathbf{E}\left[\langle Y, x_i^* \rangle\,\mathbf{1}_A\right] \ge \mathbf{E}\left[\alpha_i\,\mathbf{1}_A\right] = \alpha_i\,\mathbf{P}(A).$$

with equality only if $\mathbf{P}(A) = 0$. But $A \in \mathcal{G}$, so $\mathbf{E}\left[Y\,\mathbf{1}_A\right] = \mathbf{E}\left[X\,\mathbf{1}_A\right]$, and equality holds. Thus $\mathbf{P}\{\langle Y, x_i^* \rangle > \alpha_i\} = 0$. The union of countably many events of probability zero still has probability zero, so by (5.1.14a), almost all the values of Y lie in C.   ∎

A corollary of this is the vector-valued version of Jensen's inequality. (For a simplified statement, replace "lower semicontinuous" with "continuous." Recall that the function φ is called *lower semicontinuous* if, for each c, we have $\varphi(c) \le \liminf_{x \to c} \varphi(x)$.)

(5.1.15)   Theorem. *Let E be a Banach space, $(\Omega, \mathcal{F}, \mathbf{P})$ a probability space, $\mathcal{G} \subseteq \mathcal{F}$ a σ-algebra, and $X \colon \Omega \to E$ a Bochner integrable random variable. Let $C \subseteq E$ be a closed convex set, and $\varphi \colon C \to \mathbb{R}$ a convex lower semicontinuous function. If $X \in C$ a.s. and $\mathbf{E}\left[|\varphi(X)|\right] < \infty$, then*

$$\varphi\big(\mathbf{E}^{\mathcal{G}}\left[X\right]\big) \le \mathbf{E}^{\mathcal{G}}\left[\varphi(X)\right]   \text{a.s.}$$

Proof. In the Banach space $\widetilde{E} = E \oplus \mathbb{R}$, let $\widetilde{C} = \{\,(x, t) : x \in C, t \ge \varphi(x)\,\}$. Then $\widetilde{C}$ is a convex set since φ is a convex function and $\widetilde{C}$ is closed since φ is lower semicontinuous. Define $\widetilde{X} \colon \Omega \to \widetilde{E}$ by $\widetilde{X}(\omega) = \big(X(\omega), \varphi(X(\omega))\big)$. Then $\widetilde{X}$ is Bochner integrable, with values in $\widetilde{C}$, so the conditional expectation $\mathbf{E}[\widetilde{X}\,|\mathcal{G}]$ also has values in $\widetilde{C}$. But

$$\mathbf{E}^{\mathcal{G}}\big[\widetilde{X}\big] = \big(\mathbf{E}^{\mathcal{G}}\left[X\right], \mathbf{E}^{\mathcal{G}}\left[\varphi(X)\right]\big),$$

so we have $\mathbf{E}^{\mathcal{G}}\left[\varphi(X)\right] \ge \varphi(\mathbf{E}^{\mathcal{G}}\left[X\right])$ a.s.   ∎

Jensen's inequality shows clearly that the conditional expectation is a contraction on the vector-valued L_p-spaces. If $p \ge 1$ then the function $\varphi(x) = \|x\|^p$ is convex, so

$$\big\|\mathbf{E}^{\mathcal{G}}\left[X\right]\big\|^p \le \mathbf{E}^{\mathcal{G}}\left[\|X\|^p\right]   \text{a.s.}$$

Integrate both sides and raise to the power $1/p$ to obtain

$$\mathbf{E}\left[\big\|\mathbf{E}^{\mathcal{G}}\left[X\right]\big\|^p\right]^{1/p} \le \mathbf{E}\left[\|X\|^p\right]^{1/p}.$$

(For the corresponding result in Orlicz spaces, see (5.1.22)).

We include here a few examples that illustrate the Radon-Nikodým property. The Banach space l_1 is the set of all sequences

$$(x_1, x_2, x_3, \cdots)$$

of real numbers with

$$\sum_{i=1}^{\infty} |x_i| < \infty.$$

It is a Banach space when given the norm $\sum |x_i|$. The Banach space c_0 is the set of all sequences

$$(x_1, x_2, x_3, \cdots)$$

of real numbers with

$$\lim_{i \to \infty} x_i = 0.$$

It is a Banach space when given the norm $\max_i |x_i|$.

(5.1.16) Proposition. (a) *The space l_1 has the Radon-Nikodým property.* (b) *The space c_0 fails the Radon-Nikodým property.*

Proof. (a) Let $(\Omega, \mathcal{F}, \mathbf{P})$ be a probability space, and let $\mu \colon \mathcal{F} \to l_1$ satisfy $\|\mu(A)\| \leq \mathbf{P}(A)$ for all $A \in \mathcal{F}$. When $\mu(A)$ is written in terms of its components,

$$\mu(A) = (\mu_1(A), \mu_2(A), \mu_3(A), \cdots),$$

each $\mu_i(\cdot)$ is a measure (since the map that selects the ith component is linear and continuous). Each $\mu_i(\cdot)$ is a scalar-valued measure, absolutely continuous with respect to $\mathbf{P}$, so there exists a Radon-Nikodým derivative $X_i \colon \Omega \to \mathbb{R}$ such that

$$\mu_i(A) = \mathbf{E}\,[X_i\, \mathbf{1}_A]$$

for all $A \in \mathcal{F}$. Now $\mathbf{E}\,[\|X_i\|] = |\mu_i|(\Omega)$, and $\sum_{i=1}^{\infty} |\mu_i|(\Omega) < \infty$, so for almost all $\omega \in \Omega$, the combined random variable

$$X(\omega) = (X_1(\omega), X_2(\omega), X_3(\omega), \cdots)$$

has its values in l_1. This combined random variable satisfies

$$\mu(A) = \mathbf{E}\,[X\, \mathbf{1}_A]$$

for all $A \in \mathcal{F}$.

(b) Let $(\Omega, \mathcal{F}, \mathbf{P})$ be $[0, 1]$ with Lebesgue measure. Define $\mu \colon \mathcal{F} \to c_0$ by

$$\mu(A) = \left(\int_A \sin \omega \, d\omega, \int_A \sin 2\omega \, d\omega, \int_A \sin 3\omega \, d\omega, \cdots \right).$$

By the Riemann-Lebesgue lemma, $\mu(A) \in c_0$ for all $A \in \mathcal{F}$. Also,

$$\|\mu(A)\| \leq \mathbf{P}(A)$$

for all $A \in \mathcal{F}$. But we claim that μ has no Radon-Nikodým derivative with respect to $\mathbf{P}$. As in part (a), the continuity of the coordinate functionals on c_0 shows that if a Radon-Nikodým derivative X did exist, we would have

$$X(\omega) = (\sin \omega, \sin 2\omega, \sin 3\omega, \cdots)$$

for almost all $\omega \in [0,1]$. But this $X(\omega)$ is in c_0 for (almost) no ω. This shows that c_0 fails the Radon-Nikodým property. ∎

Complements

(5.1.17) (Banach space L_1.) The space $L_1 = L_1([0,1])$ fails the Radon-Nikodým property. This can be seen using the probability space $(\Omega, \mathcal{F}, \mathbf{P}) = [0,1]$ with Lebesgue measure and the vector measure $\mu \colon \mathcal{F} \to L_1$ defined by:

$$\mu(A) = \mathbf{1}_A \qquad \text{for all } A \in \mathcal{F}.$$

(5.1.18) (Pettis integral.) There is another kind of integral sometimes used for random variables with values in a Banach space. Suppose the random variable X is scalarly integrable. We say that X is *Pettis integrable* on the set $A \in \mathcal{F}$ iff there is a vector $x_A \in E$ satisfying

$$x^*(x_A) = \mathbf{E}\left[x^*(X)\,\mathbf{1}_A\right]$$

for all $x^* \in E^*$. We say that X is *Pettis integrable* iff X is Pettis integrable on each set in $\mathcal{F}$. The vector x_A is called the *Pettis integral* of X on A, and we will usually write it using the same notation as the Bochner integral:

$$x_A = \mathbf{E}\left[X\,\mathbf{1}_A\right].$$

(5.1.19) (Pettis vs. Bochner.) Let X be a Bochner integrable random variable. Then X is also Pettis integrable, and the two integrals agree.

(5.1.20) (Pettis vs. Bochner.) Define a random variable $X \colon [0,1] \to l_2$ as follows. Let e_n be an orthonormal sequence in l_2, and let A_n be disjoint sets in $[0,1]$ with Lebesgue measures $\mathbf{P}(A_n) = 2^{-n}$ for $n = 1, 2, 3, \cdots$. Define

$$X(\omega) = \sum_{n=1}^{\infty} \frac{2^n}{n}\,\mathbf{1}_{A_n}(\omega)\,e_n.$$

Then X is Pettis integrable but not Bochner integrable.

(5.1.21) (Variant definition.) The Banach space E has the Radon-Nikodým property if and only if for every probability space $(\Omega, \mathcal{F}, \mathbf{P})$ and every vector measure $\mu \colon \mathcal{F} \to E$ with σ-finite variation, absolutely continuous with respect to $\mathbf{P}$, there is a Pettis integrable random variable X such that

$$\mu(A) = \mathbf{E}\left[X\,\mathbf{1}_A\right]$$

for all $A \in \mathcal{F}$.

(5.1.22) (Vector-valued Orlicz spaces.) Let Φ be an Orlicz function. For E-valued random variables, let

$$\|X\|_\Phi = \inf\left\{ a > 0 : \mathbf{E}\left[\Phi\left(\frac{\|X\|}{a}\right)\right] \leq 1 \right\}.$$

The set $L_\Phi(\Omega, \mathcal{F}, \mathbf{P}; E)$ of all (equivalence classes of) random variables X with $\|X\|_\Phi < \infty$ is a Banach space with norm $\|X\|_\Phi$.

(5.1.23) (Orlicz norm and variation.) If $\mu \colon \mathcal{F} \to E$ is a vector measure, then its Φ-*variation* of μ with respect to $\mathbf{P}$ is

$$\sup \sum_i \Phi\left(\frac{\|\mu(A_i)\|}{\mathbf{P}(A_i)}\right) \mathbf{P}(A_i),$$

where the supremum is over all finite disjoint sequences $(A_1, A_2, \cdots, A_n)$ in $\mathcal{F}$. If X is a Bochner integrable random variable, and $\mu(A) = \mathbf{E}[X\,\mathbf{1}_A]$ for all A, then the Φ-variation of μ is equal to the Orlicz modular $\mathbf{E}\left[\Phi(\|X\|)\right]$ of X.

Remarks

Additional material covering measurable subsets of a Banach space is in Edgar [1979b] and Talagrand [1984].

A more thorough discussion of vector-valued measures and the Bochner integral can be found in Diestel & Uhl [1977], or Bourgin [1983]. These two books also contain much more material on the geometry of Banach spaces with the Radon-Nikodým property.

The proof [Proposition (5.1.16(a))] that l_1 has the Radon-Nikodým property actually shows that any Banach space with a boundedly complete basis has the Radon-Nikodým property. (For the definition, see for example Lindenstrauss & Tzafriri [1968], page 13.) This fact is a special case of the result proved below (5.3.32) that a separable dual Banach space has the Radon-Nikodým property.

5.2. Martingales and amarts

In this section we begin the discussion of martingales and related processes with values in a Banach space. Then we discuss some difference properties and Riesz decompositions for vector-valued processes. Convergence theorems will be treated in the next section, because of the close connection with the Radon-Nikodým property.

Let $(\Omega, \mathcal{F}, \mathbf{P})$ be a probability space, let $(\mathcal{F}_n)_{n\in\mathbb{N}}$ be a stochastic basis on Ω, and let Σ be the set of all simple stopping times for $(\mathcal{F}_n)_{n\in\mathbb{N}}$. Let E be a Banach space, and let $(X_n)_{n\in\mathbb{N}}$ be a sequence of Bochner integrable random variables with values in E, adapted to $(\mathcal{F}_n)_{n\in\mathbb{N}}$. These data will be fixed throughout Section 5.2.

Elementary properties

(5.2.1) Definition. The E-valued adapted process $(X_n)_{n\in\mathbb{N}}$ is a *martingale* iff the net $(\mathbf{E}\,[X_\sigma])_{\sigma\in\Sigma}$ is constant.

One shows (applying a linear functional, then using the scalar case (1.4.3)) that $(X_n)_{n\in\mathbb{N}}$ is a martingale if and only if

$$\mathbf{E}^{\mathcal{F}_m}\,[X_n] = X_m$$

for $m \leq n$. Also, if $(X_n)_{n\in\mathbb{N}}$ is a martingale, and $\sigma \leq \tau$ in Σ, then $\mathbf{E}^{\mathcal{F}_\sigma}\,[X_\tau] = X_\sigma$. The optional sampling theorem for martingales follows from this.

(5.2.2) Optional sampling theorem. *Suppose $(X_n)_{n\in\mathbb{N}}$ is a martingale with respect to the stochastic basis $(\mathcal{F}_n)_{n\in\mathbb{N}}$. Let $\tau_1 \leq \tau_2 \leq \cdots$ be an increasing sequence in Σ. Then the process*

$$Y_k = X_{\tau_k}$$

is a martingale with respect to the stochastic basis $(\mathcal{G}_k)_{k\in\mathbb{N}}$ defined by

$$\mathcal{G}_k = \mathcal{F}_{\tau_k}.$$

There are several different vector-valued analogs of the scalar-valued amarts. One that inherits most of their properties is the "uniform amart."

(5.2.3) Definition. The process $(X_n)_{n\in\mathbb{N}}$ is a *uniform amart* iff the following difference property is satisfied: For every $\varepsilon > 0$, there is $m_0 \in \mathbb{N}$ such that for all $\sigma, \tau \in \Sigma$ with $m_0 \leq \sigma \leq \tau$,

$$\mathbf{E}\left[\left\|\mathbf{E}^{\mathcal{F}_\sigma}\,[X_\tau] - X_\sigma\right\|\right] < \varepsilon.$$

It should be noted that (unlike the scalar case (1.4.21)) this difference property is not equivalent to the one in which σ is replaced by an integer m (5.2.35).

(5.2.4) Definition. The process $(X_n)_{n\in\mathbb{N}}$ is a *quasimartingale* iff

$$\sum_{n=1}^{\infty}\mathbf{E}\left[\left\|\mathbf{E}^{\mathcal{F}_n}\,[X_{n+1}] - X_n\right\|\right] < \infty.$$

Clearly, every martingale is a quasimartingale. It can be proved (as in (1.4.4)) that every quasimartingale is a uniform amart.

(5.2.5) Definition. The process $(X_n)_{n\in\mathbb{N}}$ is an *amart* iff the net $(\mathbf{E}\,[X_\sigma])_{\sigma\in\Sigma}$ converges according to the norm of E.

Since the norm on E defines a metric topology, we know by the sequential sufficiency theorem (1.1.3) that $(X_n)_{n\in\mathbb{N}}$ is an amart iff, for every sequence (σ_n) in Σ converging to ∞, the sequence $(\mathbf{E}\,[X_{\sigma_n}])$ converges in E. Amarts can also be characterized by:

(5.2.6) Pettis norm difference property. *Let $(X_n)_{n\in\mathbb{N}}$ be a process with values in the Banach space E. Then the following are equivalent:*

(1) (X_n) *is an amart.*
(2) $\mathbf{E}^{\mathcal{F}_\sigma}[X_\tau] - X_\sigma \to 0$ *in Pettis norm, that is: For every $\varepsilon > 0$, there is $m_0 \in \mathbb{N}$ such that*

$$\left\| \mathbf{E}^{\mathcal{F}_\sigma}[X_\tau] - X_\sigma \right\|_P < \varepsilon$$

for all $\sigma, \tau \in \Sigma$ with $m_0 \le \sigma \le \tau$.

Proof. If $m_0 \le \sigma \le \tau$, then

$$
\begin{aligned}
\left\| \mathbf{E}[X_\tau] - \mathbf{E}[X_\sigma] \right\| &= \left\| \mathbf{E}\left[\mathbf{E}^{\mathcal{F}_\sigma}[X_\tau] - X_\sigma \right] \right\| \\
&= \sup\left\{ x^*\left(\mathbf{E}\left[\mathbf{E}^{\mathcal{F}_\sigma}[X_\tau] - X_\sigma \right] \right) : x^* \in E^*, \|x^*\| \le 1 \right\} \\
&\le \left\| \mathbf{E}^{\mathcal{F}_\sigma}[X_\tau] - X_\sigma \right\|_P.
\end{aligned}
$$

Thus, if the difference property (2) holds, the net $\mathbf{E}[X_\sigma]$ is Cauchy, and therefore convergent.

For the converse, again let $m_0 \le \sigma \le \tau$. Note that if $A \in \mathcal{F}_\sigma$, then (as in (1.4.5)) there are stopping times $\sigma_1, \tau_1 \ge m_0$ with

$$
\begin{aligned}
\mathbf{E}[X_{\tau_1} - X_{\sigma_1}] &= \mathbf{E}\left[(X_\tau - X_\sigma)\, \mathbf{1}_A \right] \\
&= \mathbf{E}\left[\left(\mathbf{E}^{\mathcal{F}_\sigma}[X_\tau] - X_\sigma \right) \mathbf{1}_A \right].
\end{aligned}
$$

By (5.1.9a) the Pettis norm satisfies

$$\left\| \mathbf{E}^{\mathcal{F}_\sigma}[X_\tau] - X_\sigma \right\|_P \le 2 \sup_{A \in \mathcal{F}_\sigma} \left\| \mathbf{E}\left[\left(\mathbf{E}^{\mathcal{F}_\sigma}[X_\tau] - X_\sigma \right) \mathbf{1}_A \right] \right\|,$$

so if (X_n) is an amart, this Pettis norm converges to 0. ∎

(5.2.7) Corollary. *Every uniform amart is an amart.*

Proof. The Pettis norm is dominated by the Bochner norm. ∎

The converse of this proposition is not true in general. In fact (5.5.11), only finite-dimensional spaces E have the property that amarts and uniform amarts coincide.

The optional sampling theorem for amarts is proved as in the real-valued case.

(5.2.8) Optional sampling theorem. *Let $(X_n)_{n\in\mathbb{N}}$ be an amart in E adapted to $(\mathcal{F}_n)_{n\in\mathbb{N}}$. If $\sigma_1 \le \sigma_2 \le \cdots$ is an increasing sequence in Σ, then the process $Y_k = X_{\sigma_k}$ is an amart adapted to the stochastic basis $(\mathcal{G}_k)$ defined by $\mathcal{G}_k = \mathcal{F}_{\sigma_k}$.*

(5.2.9) Restriction theorem. *Let* $(X_n)_{n\in\mathbb{N}}$ *be an amart in* E, *and let* $A \in \mathcal{F}_m$. *Then the process* $(X_n \mathbf{1}_A)_{n=m}^{\infty}$ *is also an amart in* E. *In particular, the limit*

$$\lim_{n\to\infty} \mathbf{E}[X_n \mathbf{1}_A]$$

exists.

Proof. Let (X_n) be an amart, and $A \in \mathcal{F}_m$. Given $\sigma, \tau \geq m$, choose $n \geq \sigma, \tau$ and define

$$\sigma'(\omega) = \begin{cases} \sigma(\omega) & \text{if } \omega \in A \\ n & \text{otherwise} \end{cases}$$

$$\tau'(\omega) = \begin{cases} \tau(\omega) & \text{if } \omega \in A \\ n & \text{otherwise.} \end{cases}$$

Then $\sigma', \tau' \in \Sigma$ and $\mathbf{E}[\mathbf{1}_A X_\sigma] - \mathbf{E}[\mathbf{1}_A X_\tau] = \mathbf{E}[X_{\sigma'}] - \mathbf{E}[X_{\tau'}]$. Therefore $(\mathbf{E}[\mathbf{1}_A X_\sigma])_{\sigma\in\Sigma}$ is a Cauchy net. So $(\mathbf{1}_A X_n)$ is an amart. ∎

(5.2.10) Definition. The process $(X_n)_{n\in\mathbb{N}}$ is a *weak amart* iff the net $(\mathbf{E}[X_\sigma])_{\sigma\in\Sigma}$ converges in the weak topology of E; and $(X_n)_{n\in\mathbb{N}}$ is a *weak sequential amart* iff, for every increasing sequence (σ_n) in Σ, the sequence $(\mathbf{E}[X_{\sigma_n}])$ converges weakly in E.

The weak topology in a Banach space is not metrizable in general, so it is not true in general that a weak amart is a weak sequential amart. The optional sampling theorem and the restriction theorem may fail for weak amarts, but they both remain correct for weak sequential amarts. The optional sampling theorem is immediate from the definition:

(5.2.11) Optional sampling theorem. *Let* $(X_n)_{n\in\mathbb{N}}$ *be a weak sequential amart in* E *adapted to* $(\mathcal{F}_n)_{n\in\mathbb{N}}$. *If* $\sigma_1 \leq \sigma_2 \leq \cdots$ *is an increasing sequence in* Σ, *then the process* $Y_k = X_{\sigma_k}$ *is a weak sequential amart adapted to the stochastic basis* $(\mathcal{G}_k)$ *defined by* $\mathcal{G}_k = \mathcal{F}_{\sigma_k}$.

The restriction theorem requires a proof:

(5.2.12) Restriction theorem. *Let* $(X_n)_{n\in\mathbb{N}}$ *be a weak sequential amart in* E, *and let* $A \in \mathcal{F}_m$. *Then the process* $(X_n \mathbf{1}_A)_{n=m}^{\infty}$ *is also a weak sequential amart in* E. *In particular, the limit*

$$\lim_{n\to\infty} \mathbf{E}[X_n \mathbf{1}_A]$$

exists in the weak topology of E.

Proof. Let σ_n be an increasing sequence in Σ. Write $\sigma = \lim_n \sigma_n$. Then σ is a (possibly infinite) stopping time. Now for each n, let

$$\tau_n(\omega) = \begin{cases} m & \text{if } \omega \in A \text{ and } m \leq \sigma_n(\omega) \\ \sigma_n(\omega) & \text{otherwise.} \end{cases}$$

Then τ_n is an increasing sequence of bounded stopping times, and $\lim_n \tau_n = \sigma \wedge m$ on A. Then

$$\mathbf{E}\left[X_{\sigma_n}\, \mathbf{1}_A\right] = \mathbf{E}\left[X_{\sigma_n}\right] - \mathbf{E}\left[X_{\tau_n}\right] + \mathbf{E}\left[X_{\sigma \wedge m}\, \mathbf{1}_A\right] + \mathbf{E}\left[\left(X_{\tau_n} - X_{\sigma \wedge m}\right)\mathbf{1}_A\right].$$

The first two terms on the right converge weakly, and the third term is constant, so all that remains is the proof that the last term converges weakly (to zero). Now on A, both τ_n and $\sigma \wedge m$ take values between 1 and m, so

$$\left\| \left(X_{\tau_n} - X_{\sigma \wedge m}\right)\mathbf{1}_A \right\| \le 2 \sup_{1 \le k \le m} \|X_k\|,$$

which is an integrable function. But $\left(X_{\tau_n} - X_{\sigma \wedge m}\right)\mathbf{1}_A$ converges pointwise a.s. to 0, so the expectation converges in norm to 0.   ∎

Uniform amart Riesz decomposition

We next take up the Riesz decomposition for uniform amarts.

(5.2.13) Riesz decomposition. *Let $(X_n)_{n \in \mathbb{N}}$ be a uniform amart. Then there is a unique decomposition $X_n = Y_n + Z_n$, where (Y_n) is a martingale and (Z_σ) converges to 0 a.s. and in Bochner norm.*

Proof. Fix $n \in \mathbb{N}$. Then we have for $n \le \sigma \le \tau$

$$\mathbf{E}\left[\left\|\mathbf{E}^{\mathcal{F}_n}\left[X_\tau\right] - \mathbf{E}^{\mathcal{F}_n}\left[X_\sigma\right]\right\|\right] \le \mathbf{E}\left[\left\|\mathbf{E}^{\mathcal{F}_\sigma}\left[X_\tau\right] - X_\sigma\right\|\right],$$

so the net $\left(\mathbf{E}^{\mathcal{F}_n}\left[X_\sigma\right]\right)_{\sigma \in \Sigma}$ is Cauchy in $L_1(\Omega, \mathcal{F}, \mathbf{P}; E)$. We will write Y_n for its limit. Now for $m \le n$, we have

$$\mathbf{E}\left[\left\|\mathbf{E}^{\mathcal{F}_m}\left[Y_n\right] - Y_m\right\|\right]$$
$$= \lim_{p \to \infty} \mathbf{E}\left[\left\|\mathbf{E}^{\mathcal{F}_m}\left[\mathbf{E}^{\mathcal{F}_n}\left[X_p\right]\right] - \mathbf{E}^{\mathcal{F}_m}\left[X_p\right]\right\|\right]$$
$$= 0.$$

So $(Y_n)_{n \in \mathbb{N}}$ is a martingale. Let $Z_n = X_n - Y_n$. For $\sigma \le \tau$ in Σ we have

$$\mathbf{E}^{\mathcal{F}_\sigma}\left[Z_\tau\right] = \mathbf{E}^{\mathcal{F}_\sigma}\left[X_\tau - Y_\tau\right] = \mathbf{E}^{\mathcal{F}_\sigma}\left[X_\tau\right] - Y_\sigma.$$

Thus

$$\lim_{\tau \ge \sigma} \mathbf{E}\left[\left\|\mathbf{E}^{\mathcal{F}_\sigma}\left[Z_\tau\right]\right\|\right] = 0.$$

But

$$\|Z_\sigma\| \le \left\|Z_\sigma - \mathbf{E}^{\mathcal{F}_\sigma}\left[Z_\tau\right]\right\| + \left\|\mathbf{E}^{\mathcal{F}_\sigma}\left[Z_\tau\right]\right\|,$$

so

$$\lim_{\sigma \in \Sigma} \mathbf{E}\left[\|Z_\sigma\|\right] = 0.$$

Then by (1.4.22), we have $\|Z_n\| \to 0$ a.s.   ∎

This proposition enables us to reduce much of the theory of uniform amarts to the martingale case.

(5.2.14) Lemma. *Let C be a closed convex subset of a Banach space E.
Let $(X_n)_{n \in \mathbb{N}}$ be a uniform amart with values in C. Let $X_n = Y_n + Z_n$ be
its Riesz decomposition. Then Y_n has its values a.s. in C.*

Proof. Now X_n has its values in C, and C is closed and convex, so by
(5.1.14) $\mathbf{E}^{\mathcal{F}_m}[X_n]$ has its values in C. But Y_m is the pointwise limit of a
subsequence of such conditional expectations, so Y_m has its values in C.
∎

The convergence theorems that will be proved in this chapter utilize
certain boundedness conditions. Many of them are essentially the same as
the definitions used in the scalar valued case.

(5.2.15) Definition. The set R of E-valued random variables is said to
be L_1-*bounded* iff it is bounded in the Bochner norm:

$$\sup_{X \in R} \mathbf{E}\left[\|X\|\right] < \infty.$$

A process $(X_n)_{n \in \mathbb{N}}$ is accordingly said to be L_1-bounded iff

$$\sup_n \mathbf{E}\left[\|X_n\|\right] < \infty.$$

The set R is said to be *Pettis bounded* iff

$$\sup_{X \in R} \|X\|_P < \infty.$$

The set R is said to be *uniformly integrable* iff for every $\varepsilon > 0$ there exists
$\lambda > 0$ so that for all $X \in R$,

$$\mathbf{E}\left[\|X\| \mathbf{1}_{\{\|X\| > \lambda\}}\right] < \varepsilon;$$

or, if the probability space is atomless, equivalently [see Proposition
(2.3.2(3))]: for every $\varepsilon > 0$ there exists $\delta > 0$ so that for all $X \in R$
and all $A \in \mathcal{F}$ with $\mathbf{P}(A) < \delta$,

$$\mathbf{E}\left[\|X\| \mathbf{1}_A\right] < \varepsilon.$$

The set R is said to be *uniformly absolutely continuous* iff, for every $\varepsilon > 0$
there exists $\delta > 0$ so that for all $X \in R$ and all $A \in \mathcal{F}$ with $\mathbf{P}(A) < \delta$,

$$\left\|\mathbf{E}\left[X \mathbf{1}_A\right]\right\| < \varepsilon.$$

The process $(X_n)_{n \in \mathbb{N}}$ satisfies *condition* (B), or is of *class* (B) iff

$$\sup_{\sigma \in \Sigma} \mathbf{E}\left[\|X_\sigma\|\right] < \infty.$$

(5.2.16) Proposition. *Let R be a set of Pettis integrable E-valued random variables. Then R is uniformly absolutely continuous if and only if the set*

$$R_1 = \left\{ \langle X, x^* \rangle : X \in R, x^* \in E^*, \|x^*\| \leq 1 \right\}$$

of scalar-valued random variables is uniformly absolutely continuous.

Proof. Suppose R is uniformly absolutely continuous. Let $\varepsilon > 0$. There is δ such that $\|\mathbf{E}[X \, \mathbf{1}_A]\| < \varepsilon$ if $\mathbf{P}(A) < \delta$. Let $x^* \in E^*$ with $\|x^*\| \leq 1$. Then for A with $\mathbf{P}(A) < \delta$, the two sets $A_1 = A \cap \{\langle X, x^* \rangle \geq 0\}$ and $A_2 = A \cap \{\langle X, x^* \rangle < 0\}$ also have probability less than δ. So

$$\mathbf{E}\left[|\langle X, x^* \rangle| \, \mathbf{1}_A\right] = \left|\mathbf{E}[\langle X, x^* \rangle \, \mathbf{1}_{A_1}]\right| + \left|\mathbf{E}[\langle X, x^* \rangle \, \mathbf{1}_{A_2}]\right| \leq 2\varepsilon.$$

This shows that the set R_1 is uniformly absolutely continuous.

Conversely, suppose that R_1 is uniformly absolutely continuous. Let $\varepsilon > 0$ be given. There is $\delta > 0$ so that $\mathbf{E}\left[|\langle X, x^* \rangle| \, \mathbf{1}_A\right] < \varepsilon$ if $\mathbf{P}(A) < \delta$ and $\|x^*\| \leq 1$. Then, for all $x^* \in E^*$ with $\|x^*\| \leq 1$,

$$\left|\langle \mathbf{E}[X \, \mathbf{1}_A], x^* \rangle\right| \leq \mathbf{E}\left[|\langle X, x^* \rangle| \, \mathbf{1}_A\right] \leq \varepsilon,$$

so that $\|\mathbf{E}[X \, \mathbf{1}_A]\| \leq \varepsilon$. ∎

This proposition means that many of the properties of uniform absolute continuity in the scalar case carry over to vector case. For example: If (X_n) is uniformly integrable, then it is bounded in Bochner norm. If (X_n) is uniformly absolutely continuous, then it is bounded in Pettis norm. If (X_n) is uniformly absolutely continuous, and $\mathcal{G} \subseteq \mathcal{F}$ is a σ-algebra, then $(\mathbf{E}^{\mathcal{G}}[X_n])$ is also uniformly absolutely continuous.

(5.2.17) Proposition. *Let (X_n) be a uniformly absolutely continuous amart. Then*

$$\mu(A) = \lim_{n \to \infty} \mathbf{E}[X_n \, \mathbf{1}_A]$$

exists for all $A \in \mathcal{F}$, and defines a vector measure $\mu \colon \mathcal{F} \to E$.

Proof. By (5.2.9), $\mu(A)$ exists for $A \in \bigcup \mathcal{F}_n$. Let $A \in \mathcal{F}_\infty = \sigma(\bigcup \mathcal{F}_n)$. Let $\varepsilon > 0$. There is $\delta > 0$ so that $\mathbf{P}(D) < \delta$ implies $\|\mathbf{E}[\mathbf{1}_D X_n]\| < \varepsilon$ for all n. Then there is $k_1 \in \mathbb{N}$ and $B \in \mathcal{F}_{k_1}$ with $\mathbf{P}(A \triangle B) < \delta$. There is $k_2 \geq k_1$ so that $\|\mathbf{E}[\mathbf{1}_B X_n] - \mathbf{E}[\mathbf{1}_B X_m]\| < \varepsilon$ for all $n, m \geq k_2$. Thus

$$\begin{aligned}
\|\mathbf{E}[\mathbf{1}_A X_n] - \mathbf{E}[\mathbf{1}_A X_m]\| \leq {} & \|\mathbf{E}[\mathbf{1}_B X_n] - \mathbf{E}[\mathbf{1}_B X_m]\| \\
& + \|\mathbf{E}[\mathbf{1}_{A \setminus B} X_n]\| + \|\mathbf{E}[\mathbf{1}_{A \setminus B} X_m]\| \\
& + \|\mathbf{E}[\mathbf{1}_{B \setminus A} X_n]\| + \|\mathbf{E}[\mathbf{1}_{B \setminus A} X_m]\| \\
& \leq 5\varepsilon.
\end{aligned}$$

This shows that $\lim_{n \to \infty} \mathbf{E}[\mathbf{1}_A X_n]$ exists.

Next we claim that if $Y \in L_\infty(\Omega, \mathcal{F}_\infty, \mathbf{P})$, the limit $\lim_{n\to\infty} \mathbf{E}[YX_n]$ exists. To see this, note that (X_n) is bounded in Pettis norm, say $\|X_n\|_P \leq a$ for all n. For $\varepsilon > 0$, there is a simple function $Y' \in L_\infty(\mathcal{F}_\infty)$ with $|Y - Y'| < \varepsilon/a$ everywhere. But by the preceding, $\lim_n \mathbf{E}[Y'X_n]$ exists. So there is $k \in \mathbb{N}$ so that $\|\mathbf{E}[Y'X_n] - \mathbf{E}[Y'X_m]\| < \varepsilon$ for $n, m \geq k$. But

$$\|\mathbf{E}[YX_n] - \mathbf{E}[Y'X_n]\| \leq \frac{\varepsilon}{a} a = \varepsilon.$$

Then we have for $n, m \geq k$

$$\|\mathbf{E}[YX_n] - \mathbf{E}[YX_m]\| < 3\varepsilon,$$

so $\lim_n \mathbf{E}[YX_n]$ exists.

Now consider $A \in \mathcal{F}$. Then

$$\mathbf{E}[1_A X_n] = \mathbf{E}[\mathbf{E}^{\mathcal{F}_\infty}[1_A] X_n],$$

so it converges as $n \to \infty$. Let $\mu(A)$ be the limit.

We must show that μ is countably additive. Suppose $A_k \downarrow \emptyset$. Given $\varepsilon > 0$, there is $\delta > 0$ so that $\mathbf{P}(D) < \delta$ implies $\|\mathbf{E}[1_D X_n]\| < \varepsilon$. Now $\mathbf{P}(A_k) \to 0$, so $\mathbf{P}(A_k) < \delta$ for k large enough, and thus $\|\mathbf{E}[1_{A_k} X_n]\| \leq \varepsilon$ for all n. Therefore $\|\mu(A_k)\| \leq \varepsilon$. This shows $\lim_k \|\mu(A_k)\| = 0$. Thus μ is countably additive. ∎

The preceding result is also true for weak sequential amarts.

Next we consider condition (B), namely $\sup_{\sigma \in \Sigma} \mathbf{E}[\|X_\sigma\|] < \infty$. It was not necessary to emphasize condition (B) in the scalar case, because an L_1-bounded scalar amart automatically satisfies condition (B). This is no longer true for vector-valued amarts (5.5.29). However it does remain true for uniform amarts.

(5.2.18) Proposition. *A uniform amart is L_1-bounded if and only if it satisfies condition (B).*

Proof. By the Riesz decomposition (5.2.13), it is enough to prove the result when (X_n) is a martingale. But then $(\|X_n\|)_{n\in\mathbb{N}}$ is a submartingale. For every σ in Σ there is an $n \in \mathbb{N}$ with $\sigma \leq n$, and thus $\mathbf{E}[\|X_\sigma\|] \leq \mathbf{E}[\|X_n\|]$. Therefore

$$\sup_{\sigma \in \Sigma} \mathbf{E}[\|X_\sigma\|] = \sup_{n \in \mathbb{N}} \mathbf{E}[\|X_n\|].$$

∎

The following is obtained by applying (1.1.7) to the real process $(\|X_n\|)$.

(5.2.19) Maximal inequality. *Let $(X_n)_{n\in\mathbb{N}}$ be an adapted sequence in the Banach space E, and let λ be a positive real number. Then*

$$\mathbf{P}\{\sup \|X_n\| \geq \lambda\} \leq \frac{1}{\lambda} \sup_{\sigma \in \Sigma} \mathbf{E}[\|X_\sigma\|].$$

This maximal inequality is a good illustration of the usefulness of condition (B). Another illustration of its usefulness is the next stopping result.

(5.2.20) Definition. Let $(X_n)_{n\in\mathbb{N}}$ be a stochastic process with values in a Banach space E, and let C be a subset of E. We say that the process (X_n) *stops outside* C iff $X_n(\omega) \notin C$ implies $X_n(\omega) = X_{n+1}(\omega)$.

(5.2.21) Definition. Let $(X_n)_{n\in\mathbb{N}}$ be a process adapted to the stochastic basis $(\mathcal{F}_n)_{n\in\mathbb{N}}$. Suppose $(X_n)_{n\in\mathbb{N}}$ has values in a Banach space E, and let C be a Borel subset of E. The *first entrance time* of (X_n) in $E \setminus C$ is the (possibly infinite) stopping time σ defined as follows. Let $D = \{X_n \in C \text{ for all } n\}$. Then

$$\sigma(\omega) = \begin{cases} \min\{n : X_n(\omega) \notin C\} & \text{for } \omega \notin D \\ \infty & \text{for } \omega \in D. \end{cases}$$

The process (X_n) *stopped outside* C is the process (Y_n) defined by $Y_n = X_{n\wedge\sigma}$.

(5.2.22) Proposition. *Let C be a closed bounded subset of the Banach space E. Suppose the process $(X_n)_{n\in\mathbb{N}}$ in E satisfies condition (B). Let (Y_n) be the process (X_n) stopped outside C. Then*

$$\mathbf{E}\left[\sup_n \|Y_n\|\right] < \infty.$$

Proof. Let D and σ be as in the definition. Outside the set D, we have convergence $Y_n \to X_\sigma$, so by Fatou's lemma,

$$\mathbf{E}\left[\|X_\sigma\|\,\mathbf{1}_{\Omega\setminus D}\right] \leq \liminf_n \mathbf{E}\left[\|X_{n\wedge\sigma}\|\,\mathbf{1}_{\Omega\setminus D}\right]$$
$$\leq \liminf_n \mathbf{E}\left[\|X_{n\wedge\sigma}\|\right]$$
$$\leq \sup_{\tau\in\Sigma} \mathbf{E}\left[\|X_\tau\|\right]$$
$$< \infty.$$

Thus, if $\lambda = \sup_{x\in C}\|x\|$, we have

$$\mathbf{E}\left[\sup_n \|Y_n\|\right] \leq \mathbf{E}\left[\|X_\sigma\|\,\mathbf{1}_{\Omega\setminus D}\right] + \mathbf{E}\left[\lambda\,\mathbf{1}_\Omega\right] < \infty. \qquad\blacksquare$$

(5.2.23) Corollary. *Let C be a closed bounded subset of the Banach space E. Suppose $(X_n)_{n\in\mathbb{N}}$ is an L_1-bounded uniform amart. Let (Y_n) be the process (X_n) stopped outside C. Then*

$$\mathbf{E}\left[\sup_n \|Y_n\|\right] < \infty.$$

Proof. By (5.2.18), an L_1-bounded uniform amart satisfies condition (B).

$\blacksquare$

(5.2.24) Corollary. An L_1-bounded uniform amart (X_n) that stops outside a bounded set satisfies

$$\mathbf{E}\left[\sup_n \|X_n\|\right] < \infty.$$

Proof. In the previous Corollary, set $X_n = Y_n$. ∎

Amart Riesz decomposition

We next consider the Riesz decomposition for amarts. We begin with some preliminaries.

(5.2.25) Definition. An *amart potential* is an amart $(X_n)_{n\in\mathbb{N}}$ such that

$$\lim_{n\to\infty} \left\|\mathbf{E}\left[X_n \, \mathbf{1}_A\right]\right\| = 0$$

for all $A \in \bigcup_{m\in\mathbb{N}} \mathcal{F}_m$.

(5.2.26) Lemma. Let $(X_n)_{n\in\mathbb{N}}$ be an amart potential in a Banach space. Then

$$\lim_{\sigma\in\Sigma} \|X_\sigma\|_P = 0.$$

Proof. Let $\varepsilon > 0$ be given. There exists $n_0 \in \mathbb{N}$ so that

$$\left\|\mathbf{E}\left[X_n\right]\right\| < \varepsilon \qquad \text{for } n \geq n_0;$$
$$\left\|\mathbf{E}\left[X_\sigma - X_{\sigma'}\right]\right\| < \varepsilon \qquad \text{for } \sigma, \sigma' \geq n_0.$$

Fix $m \geq n_0$. Choose $D \in \mathcal{F}_m$ so that

$$\left\|\mathbf{E}\left[X_m \, \mathbf{1}_D\right]\right\| \geq \sup_{A\in\mathcal{F}_m} \left\|\mathbf{E}\left[X_m \, \mathbf{1}_A\right]\right\| - \varepsilon.$$

Then choose $n \geq m$ so that

$$\left\|\mathbf{E}\left[X_n \, \mathbf{1}_{\Omega\setminus D}\right]\right\| < \varepsilon.$$

Define $\tau \in \Sigma$ by

$$\tau = \begin{cases} m & \text{on } D \\ n & \text{on } \Omega \setminus D. \end{cases}$$

Now $\|\mathbf{E}\left[X_\tau\right]\| \leq \|\mathbf{E}\left[X_\tau - X_m\right]\| + \|\mathbf{E}\left[X_m\right]\| \leq 2\varepsilon$, so

$$\begin{aligned}
\left\|\mathbf{E}\left[X_m \, \mathbf{1}_D\right]\right\| &= \left\|\mathbf{E}\left[X_\tau\right] - \mathbf{E}\left[X_n \, \mathbf{1}_{\Omega\setminus D}\right]\right\| \\
&\leq \left\|\mathbf{E}\left[X_\tau - X_m\right]\right\| + \left\|\mathbf{E}\left[X_m\right]\right\| + \left\|\mathbf{E}\left[X_n \, \mathbf{1}_{\Omega\setminus D}\right]\right\| \\
&\leq 3\varepsilon.
\end{aligned}$$

Therefore
$$\sup_{A \in \mathcal{F}_m} \| \mathbf{E} [X_m \, \mathbf{1}_A] \| \le 4\varepsilon$$

for any $m \ge n_0$.

Now if $\sigma \ge n_0$ and $A \in \mathcal{F}_\sigma$, choose $m \ge \sigma$ and let
$$\sigma' = \begin{cases} m & \text{on } A \\ \sigma & \text{on } \Omega \setminus A, \end{cases}$$

so that $\sigma' \in \Sigma$ and $\sigma' \ge n_0$. Then
$$\| \mathbf{E} [X_\sigma \, \mathbf{1}_A] \| \le \| \mathbf{E} [X_m \, \mathbf{1}_A] \| + \| \mathbf{E} [X_\sigma - X_{\sigma'}] \| \le 4\varepsilon + \varepsilon = 5\varepsilon.$$

Thus we get
$$\| X_\sigma \|_P \le 2 \sup_{A \in \mathcal{F}_\sigma} \| \mathbf{E} [X_\sigma \, \mathbf{1}_A] \| \le 10\varepsilon. \qquad \blacksquare$$

The goal of the Riesz decomposition is to decompose an amart $(X_n)_{n \in \mathbb{N}}$ into a martingale and an amart potential. As in the case of the uniform amart, the martingale (Y_m) should be obtained as a limit
$$Y_m = \lim_{n \to \infty} \mathbf{E}^{\mathcal{F}_m} [X_n].$$

The existence of the limit presents more of a problem this time. The sequence $(\mathbf{E}^{\mathcal{F}_m} [X_n])_{n \in \mathbb{N}}$ is clearly Cauchy in the Pettis norm (by the difference property (5.2.6)), but that alone does not guarantee convergence, since the Pettis norm is usually not complete. So some care must be taken in the following proof.

(5.2.27) Vector Riesz decomposition. *Let E be a Banach space with the Radon-Nikodým property, and let $(X_n)_{n \in \mathbb{N}}$ be an L_1-bounded amart in E. Then (X_n) can be written uniquely as $X_n = Y_n + Z_n$, where (Y_n) is a martingale, and (Z_n) is an amart potential. Furthermore,*
$$\mathbf{E} [\|Y_n\|] \le \sup_{n \in \mathbb{N}} \mathbf{E} [\|X_n\|]$$

and
$$\lim_{\sigma \in \Sigma} \| Z_\sigma \|_P = 0.$$

Proof. The uniqueness is easy: If $X_n = Y_n + Z_n = Y_n' + Z_n'$, where (Y_n) and (Y_n') are martingales and (Z_n) and (Z_n') are amart potentials, then the difference $Y_n - Y_n' = Z_n' - Z_n$ is both a martingale and an amart potential. Clearly this difference is identically 0.

Fix an integer n_0. By (5.2.9), the limit
$$\mu(A) = \lim_{\sigma \in \Sigma} \mathbf{E} [X_\sigma \, \mathbf{1}_A]$$

exists for all $A \in \mathcal{F}_{n_0}$. Clearly μ is absolutely continuous with respect to $\mathbf{P}$. We claim that μ has bounded variation on $\mathcal{F}_{n_0}$. Let $A_1, A_2, \cdots, A_k \in \mathcal{F}_{n_0}$ be disjoint, and let $M = \liminf_{n \in \mathbb{N}} \mathbf{E}\left[\|X_n\|\right]$. Given $\varepsilon > 0$, choose $m \in \mathbb{N}$ such that

$$\left\|\mathbf{E}\left[X_m \, \mathbf{1}_{A_i}\right] - \mu(A_i)\right\| \leq \frac{\varepsilon}{k} \quad \text{for } i = 1, 2, \cdots, k$$

and

$$\mathbf{E}\left[\|X_m\|\right] \leq M + \varepsilon.$$

Then

$$\sum_{i=1}^{k}\left\|\mu(A_i)\right\| \leq \sum_{i=1}^{k}\left\|\mu(A_i) - \mathbf{E}\left[X_m \, \mathbf{1}_{A_i}\right]\right\| + \sum_{i=1}^{k}\left\|\mathbf{E}\left[X_m \, \mathbf{1}_{A_i}\right]\right\|$$

$$\leq \varepsilon + \mathbf{E}\left[\sum_{i=1}^{k}\|X_m\| \, \mathbf{1}_{A_i}\right]$$

$$\leq \varepsilon + \mathbf{E}\left[\|X_m\|\right]$$

$$\leq M + 2\varepsilon.$$

Therefore the variation of μ on Ω is at most $M < \infty$.

Now, by the Radon-Nikodým property, there is a random variable $Y_{n_0} \in L_1(\Omega, \mathcal{F}_{n_0}, \mathbf{P}; E)$ with

$$\mu(A) = \mathbf{E}\left[Y_{n_0} \, \mathbf{1}_A\right] \qquad \text{for all } A \in \mathcal{F}_{n_0}.$$

This can be done for each $n_0 \in \mathbb{N}$, to produce a process (Y_n). But this process clearly satisfies $\mathbf{E}^{\mathcal{F}_m}\left[Y_n\right] = Y_m$ for $m \leq n$, so it is a martingale. Also, $\mathbf{E}\left[\|Y_n\|\right] \leq \|\mu\|(\Omega) \leq M$.

Let $Z_n = X_n - Y_n$. Then

$$\lim_{n \in \mathbb{N}} \left\|\mathbf{E}\left[Z_n \, \mathbf{1}_A\right]\right\| = 0$$

for all $A \in \bigcup_{n \in \mathbb{N}} \mathcal{F}_n$, so that (Z_n) is an amart potential. ∎

It should be noted that L_1-boundedness may be replaced by the weaker condition $\liminf_{n \to \infty} \mathbf{E}\left[\|X_n\|\right] < \infty$. Then if $M = \liminf \mathbf{E}\left[\|X_n\|\right]$, we have

(5.2.27a) $\mathbf{E}\left[\|Y_n\|\right] \leq M.$

It should also be noted that the Riesz decomposition preserves many important properties.

(5.2.28) Proposition. *Let $X_n = Y_n + Z_n$ be the Riesz decomposition of the L_1-bounded amart (X_n).*

(1) *If Φ is an Orlicz function, then*

$$\sup_n \|Y_n\|_\Phi \leq \sup_n \|X_n\|_\Phi;$$

so if (X_n) is L_Φ-bounded, then so are (Y_n) and (Z_n).

(2) *If (X_n) is uniformly integrable, then so are (Y_n) and (Z_n).*

(3) *If (X_n) is uniformly absolutely continuous, then so are (Y_n) and (Z_n).*

(4) *If (X_n) satisfies condition (B), then so do (Y_n) and (Z_n).*

Proof. Note first: $\mathbf{E}[Y_m \mathbf{1}_A] = \lim_n \mathbf{E}[X_n \mathbf{1}_A]$ for all $A \in \mathcal{F}_m$.

(1) Let $M = \sup_n \|X_n\|_\Phi$. If $M = \infty$, then the inequality is trivial, so suppose $M < \infty$. Fix m. We will compute the L_Φ norm of Y_m using the Φ-variation of the indefinite integral (5.1.23). Let (A_i) be a finite disjoint sequence of sets in $\mathcal{F}_m$. Then we have (since Φ is convex)

$$\sum_i \Phi\left(\left\|\frac{\mathbf{E}[Y_m \mathbf{1}_{A_i}]}{M\mathbf{P}(A_i)}\right\|\right) \mathbf{P}(A_i) = \lim_n \sum_i \Phi\left(\frac{\|\mathbf{E}[X_n \mathbf{1}_{A_i}]\|}{M\mathbf{P}(A_i)}\right) \mathbf{P}(A_i)$$

$$\leq \liminf_n \mathbf{E}\left[\Phi\left(\frac{\|X_n\|}{M}\right)\right] \leq 1.$$

Therefore $\mathbf{E}[\Phi(\|Y_m\|/M)] \leq 1$, so $\|Y_m\|_\Phi \leq M$.

(2) If (X_n) is uniformly integrable, then the sequence $(\|X_n\|)$ of scalar random variables is also uniformly integrable. By (2.3.5), there is a finite Orlicz function Φ with $\Phi(u)/u \to \infty$ such that $(\|X_n\|)$ is L_Φ-bounded. By (1), we see that $(\|Y_n\|)$ is also L_Φ-bounded, so (Y_n) is uniformly integrable.

(3) If (X_n) is uniformly absolutely continuous, then the collection

$$\left\{ \langle X_n, x^* \rangle : n \in \mathbb{N}, x^* \in E^*, \|x^*\| \leq 1 \right\}$$

of scalar random variables is uniformly absolutely continuous. By the uniqueness of the scalar Riesz decomposition (1.4.22), we see that for each x^*, the Riesz decomposition of $\langle X_n, x^* \rangle$ is

$$\langle X_n, x^* \rangle = \langle Y_n, x^* \rangle + \langle Z_n, x^* \rangle.$$

Thus $\{ \langle Y_n, x^* \rangle : n \in \mathbb{N}, x^* \in E^*, \|x^*\| \leq 1 \}$ is uniformly absolutely continuous, and (Y_n) is uniformly absolutely continuous.

(4) If (X_n) satisfies condition (B), then it is L_1-bounded. So Y_n is L_1-bounded, and therefore satisfies condition (B). Finally, $Z_n = X_n - Y_n$ satisfies condition (B). ∎

Complements

(5.2.29) (Counterexample: The Radon-Nikodým property cannot be omitted in the amart Riesz decomposition theorem (5.2.27)). Let $E = c_0$. Let e_n^i, $n \in \mathbb{N}$, $1 \leq i \leq 2^n$ be the standard basis in some order. For each n, let measurable sets A_n^i, $1 \leq i \leq 2^n$ be disjoint and $\mathbf{P}(A_n^i) = 2^{-n}$. Let

$$X_n = \sum_{k=1}^{n} \sum_{i=1}^{2^k} e_k^i \, 1_{A_k^i}.$$

Then the Riesz decomposition fails to exist. Indeed, since each coordinate projection is continuous, the Riesz decomposition must agree with the scalar Riesz decomposition in each coordinate. The values of the martingale part Y_m lie in $E^{**} = l_\infty$, not in $E = c_0$.

(5.2.30) Definition. Let s be a positive integer. Write Σ_s for the set of all stopping times with at most s values. A process $(X_n)_{n \in \mathbb{N}}$ in a Banach space E is an *amart of order s* iff the net $(\mathbf{E}[X_\sigma])_{\sigma \in \Sigma_s}$ converges in the norm of E.

(5.2.31) (Amart potential of order $s + 1$.) Suppose that (X_n) is an amart of order $s + 1$ and $\lim_{n \to \infty} \mathbf{E}[X_n \, 1_A] = 0$ for all $A \in \bigcup_{m \in \mathbb{N}} \mathcal{F}_m$. Then

$$\lim_{\sigma \in \Sigma_s} \|X_\sigma\|_P = 0.$$

This is proved as in (5.2.26), above.

(5.2.32) (Riesz decomposition.) The proof given above for the amart Riesz decomposition can be used to establish the following interesting result of Lu'u [1981]. The process $(X_n)_{n \in \mathbb{N}}$ is called an *amart of finite order* iff (X_n) is an amart of order s for all $s \in \mathbb{N}$. Let (X_n) be an L_1-bounded amart of finite order with values in the Banach space E with the Radon-Nikodým property. Then (X_n) has a decomposition as $X_n = Y_n + Z_n$, where (Y_n) is a martingale, and

$$\lim_{\sigma \in \Sigma} \|Z_\sigma\|_P = 0.$$

(5.2.33) (Converse Riesz decomposition.) Let the process $(X_n)_{n \in \mathbb{N}}$ be adapted and Bochner integrable. Suppose $X_n = Y_n + Z_n$, where (Y_n) is a martingale and

$$\lim_{\sigma \in \Sigma} \|Z_\sigma\|_P = 0.$$

Then (X_n) is an amart of finite order, but it need not be an amart (Lu'u [1981]).

(5.2.34) (Vector version of (1.4.25).) Let F be a Banach space, and suppose $E = F^*$ is separable. Then any adapted L_1-bounded sequence X_n of E-valued random variables satisfies

$$\mathbf{E}\left[\limsup_{n,m \in \mathbb{N}} \|X_n - X_m\| \right] \leq 2 \limsup_{\substack{\sigma, \tau \in \Sigma \\ \sigma \geq \tau}} \mathbf{E}\left[\left\| \mathbf{E}^{\mathcal{F}_\tau}[X_\sigma] - X_\tau \right\| \right].$$

(Edgar [1979a]; see also Bellow & Egghe [1982].) Convergence of uniform amarts in $E = F^*$ clearly follows from this. The inequality may fail in a space E with Radon-Nikodým property.

(5.2.35) (Difference property.) Let $\{e_1, e_2, \cdots\}$ be an orthonormal sequence in the Banach space l^2. Let $r_k = 1! + 2! + \cdots + k!$. In the measure space $[0, 1]$ with Lebesgue measure, define sets A_n by

$$A_n = \left[\frac{n - r_{k-1}}{r_k}, \frac{n - r_{k-1} + 1}{r_k} \right)$$

if $r_{k-1} \leq n \leq r_k - 1$. Then the process $X_n = e_n \mathbf{1}_{A_n}$, with natural stochastic basis $\mathcal{F}_n = \sigma(A_1, A_2, \cdots, A_n)$, satisfies

$$\lim_{m \in \mathbb{N}} \sup_{\substack{\tau \geq m \\ \tau \in \Sigma}} \mathbf{E}\left[\left\| \mathbf{E}^{\mathcal{F}_m}[X_\tau] - X_m \right\| \right] = 0$$

but

$$\limsup_{\sigma \in \Sigma} \sup_{\tau \geq \sigma} \mathbf{E}\left[\left\| \mathbf{E}^{\mathcal{F}_\sigma}[X_\tau] - X_\sigma \right\| \right] > 0$$

(Edgar [1979a], Example 8, corrected).

(5.2.36) (Variants of the maximal inequality.) If $(X_n)_{n \in \mathbb{N}}$ is a martingale, then

$$\mathbf{P}\left\{ \sup_n \|X_n\| \geq \lambda \right\} \leq \frac{1}{\lambda} \lim_n \mathbf{E}\left[\|X_n\| \right] = \frac{1}{\lambda} \sup_n \mathbf{E}\left[\|X_n\| \right].$$

If $X_n = \mathbf{E}^{\mathcal{F}_n}[X]$ for all n, then

$$\mathbf{P}\left\{ \sup_n \|X_n\| \geq \lambda \right\} \leq \frac{1}{\lambda} \mathbf{E}\left[\|X\| \right].$$

This follows from (1.4.18), since $\|X_n\|$ is a positive submartingale.

(5.2.37) (Maximal inequality for reversed processes.) Let $(\mathcal{F}_n)_{n \in -\mathbb{N}}$ be a reversed stochastic basis. Then for any process $(X_n)_{n \in -\mathbb{N}}$,

$$\mathbf{P}\left\{ \sup_n \|X_n\| \geq \lambda \right\} \leq \sup_{\sigma \in \Sigma} \frac{1}{\lambda} \mathbf{E}\left[\|X_\sigma\| \right].$$

(For the maximum of a finite number of terms, apply the direct maximal inequality. For the complete supremum, take a limit.) If (X_n) is a reversed martingale, then

$$\mathbf{P}\left\{ \sup_n \|X_n\| \geq \lambda \right\} \leq \frac{1}{\lambda} \mathbf{E}\left[\|X_{-1}\| \right].$$

(5.2.38) (Riesz decomposition for weak sequential amarts.) If (X_n) is a weak sequential amart with $\liminf \mathbf{E}\left[\|X_n\| \right] < \infty$ and E has the Radon-Nikodým property, then $X_n = Y_n + Z_n$, where (Y_n) is a martingale and (Z_n) is a weak sequential potential, that is, the weak limit $\mathbf{E}[\mathbf{1}_A X_n]$ is 0 for each $A \in \bigcup_n \mathcal{F}_n$ (Brunel & Sucheston [1976b]).

Remarks

Banach-valued amarts were defined by Chacon & Sucheston [1975]. The notion of uniform amart is due to Bellow [1978a]. The Riesz decomposition for uniform amarts (5.2.13) is due to Ghoussoub & Sucheston [1978]. The Riesz decomposition for amarts (5.2.27) is due to Edgar & Sucheston [1976c]. Weak and weak sequential amarts were defined by Brunel & Sucheston [1976a], and their properties are investigated in [1976b].

Astbury [1978] characterized amarts by the Pettis norm difference property. He also used the difference property to prove the Riesz decomposition (5.2.27) for (scalar-valued and vector-valued) amarts.

S. N. Bagchi [1983], [1985] discusses convergence of set-valued processes. In a dual Banach space, the right notion is the weak* amart. It converges only weak* a.s., but even set-valued martingales cannot do any better (Neveu [1972]).

5.3. The Radon-Nikodým property

The content of this section can be viewed in two different ways. One could consider it as an exposition of the convergence theorems for vector-valued martingales (and related processes). The Radon-Nikodým property is an important hypothesis in many of the convergence theorems. One could also consider the material in this section as a characterization of a certain class of Banach spaces (the spaces with the Radon-Nikodým property) using martingale or amart convergence.

Let $(\Omega, \mathcal{F}, \mathbf{P})$ be a probability space; let $(\mathcal{F}_n)_{n \in \mathbb{N}}$ be a stochastic basis on $(\Omega, \mathcal{F}, \mathbf{P})$; and let Σ be the set of all bounded stopping times for $(\mathcal{F}_n)_{n \in \mathbb{N}}$. Let E be a Banach space. We will keep this notation throughout Section 5.3.

Scalar and Pettis norm convergence

Singling out a few ideas in advance will help in understanding the convergence proofs.

(5.3.1) Definition. Let $\mu \colon \mathcal{F} \to E$ be a vector measure. The *average range* of μ (with respect to $\mathbf{P}$) is the set of vectors

$$\left\{ \frac{\mu(A)}{\mathbf{P}(A)} : A \in \mathcal{F}, \ \mathbf{P}(A) > 0 \right\}.$$

(5.3.2) Definition. Let C be a nonempty closed convex subset of the Banach space E. We say that C is a *Radon-Nikodým set* iff, for every probability space $(\Omega, \mathcal{F}, \mathbf{P})$ and every absolutely continuous vector measure $\mu \colon \mathcal{F} \to E$ with finite variation and average range contained in C, the Radon-Nikodým derivative $d\mu/d\mathbf{P}$ exists.

On first consideration, it might seem more natural to consider measures μ with range in C rather than average range in C. For linear spaces C, this is, of course, equivalent. But consider, for example, in $E = L_1[0, 1]$ the set

$$C = \{ f \in L_1 : f \ge 0, \mathbf{E}[f] = 1 \}.$$

This set vacuously satisfies the property that measures with range in C have derivatives, since no nontrivial measure has range in C. But it has none of the useful properties of Radon-Nikodým sets to be proved below: for example, a martingale with values in C need not converge.

(5.3.3) Proposition. *A closed nonempty convex set C is a Radon-Nikodým set if and only if every closed bounded convex nonempty subset of C is a Radon-Nikodým set. A Banach space E has the Radon-Nikodým property if and only if E is a Radon-Nikodým set, which holds if and only if the unit ball $\{\, x \in E : \|x\| \le 1 \,\}$ is a Radon-Nikodým set.*

Proof. Suppose every closed nonempty convex subset of C is a Radon-Nikodým set. Let μ be an absolutely continuous vector measure with finite variation and average range contained in C. The Radon-Nikodým derivative $h = d|\mu|/d\mathbf{P}$ is a.e. finite, and μ has bounded average range on each of the sets $A_\lambda = \{h \le \lambda\}$, so by assumption $d\mu/d\mathbf{P}$ exists on each of the sets A_λ. So $d\mu/d\mathbf{P}$ exists. ∎

(5.3.4) Definition. The sequence $(X_n)_{n\in\mathbb{N}}$ of random variables in the Banach space E *converges scalarly* to a random variable X iff, for each $x^* \in E^*$, the sequence $(\langle X_n, x^*\rangle)$ of random variables in $\mathbb{R}$ converges a.s. to $\langle X, x^*\rangle$.

If we say that a sequence (X_n) converges scalarly, we are asserting more than the a.s. convergence of each sequence $(\langle X_n, x^*\rangle)$. We are asserting the existence of a limit random variable X, Bochner measurable, with values in E.

It should be emphasized that for scalar convergence, the exceptional null set where $(\langle X_n, x^*\rangle)$ does not converge to $\langle X, x^*\rangle$ may depend on x^*. A related, but stronger sort of convergence is defined next.

(5.3.5) Definition. The sequence $(X_n)_{n\in\mathbb{N}}$ of random variables *converges weakly a.s.* to the random variable X iff, for almost every $\omega \in \Omega$, the sequence $X_n(\omega)$ converges weakly to $X(\omega)$.

In this case, there is a single exceptional null set.

For scalar-valued processes, in the presence of uniform integrability, pointwise convergence implies L_1 convergence. The corresponding condition for scalar convergence is uniform absolute continuity (5.2.15).

(5.3.6) Proposition. *Let $(X_n)_{n\in\mathbb{N}}$ be a process with values in a Banach space E. If X_n converges scalarly to X and $\{X_n\}$ is uniformly absolutely continuous, then $X_n - X$ converges to 0 in Pettis norm.*

Proof. Let $\varepsilon > 0$. Then there exists $\delta > 0$ so that $\|\mathbf{E}\,[X_n\, 1_A]\,\| < \varepsilon$ if $\mathbf{P}(A) < \delta$. Let $x^* \in E^*, \|x^*\| \le 1$. If $\mathbf{P}(A) < \delta$ then $\mathbf{E}\,[\,|\langle X_n, x^*\rangle|\, 1_A] < 2\varepsilon$. By Fatou's lemma, if $\mathbf{P}(A) < \delta$, then

$$\mathbf{E}\,[\,|\langle X, x^*\rangle|\, 1_A] \le \liminf_n \mathbf{E}\,[\,|\langle X_n, x^*\rangle|\, 1_A] < 2\varepsilon.$$

Now $\langle X_n, x^*\rangle \to \langle X, x^*\rangle$ a.s., so there is N such that for all $n \ge N$,

$$\mathbf{P}\Big\{|\langle X_n - X, x^*\rangle| > \varepsilon\Big\} < \delta.$$

Write $A_n = \{|\langle X_n - X, x^* \rangle| > \varepsilon\}$. Then

$$\mathbf{E}\left[|\langle X_n - X, x^* \rangle|\right] \leq \mathbf{E}\left[|\langle X_n - X, x^* \rangle| \, \mathbf{1}_{A_n}\right] + \mathbf{E}\left[|\langle X_n - X, x^* \rangle| \, \mathbf{1}_{\Omega \setminus A_n}\right]$$
$$\leq 4\varepsilon + \varepsilon = 5\varepsilon.$$

Therefore $\|X_n - X\|_P \leq 5\varepsilon$. Thus $X_n - X \to 0$ in Pettis norm. ∎

Our first convergence theorem will be proved for bounded sets, and then generalized to unbounded sets.

(5.3.7) Proposition. *Let E be a Banach space, and $C \subseteq E$ a closed bounded convex Radon-Nikodým set. If $(X_n)_{n \in \mathbb{N}}$ is an amart in C, then (X_n) converges scalarly and in Pettis norm.*

Proof. For each $A \in \bigcup_{n=1}^{\infty} \mathcal{F}_n$, the restriction theorem (5.2.9) implies that the limit

$$\mu(A) = \lim_{n \to \infty} \mathbf{E}\left[X_n \, \mathbf{1}_A\right]$$

exists in E, and $\mu(A)/\mathbf{P}(A) \in C$ by (5.1.14b). By (5.1.9a), μ can be extended to a countably additive vector measure μ on the σ-algebra $\mathcal{F}_\infty$ generated by $\bigcup_{n=1}^{\infty} \mathcal{F}_n$, and

$$\frac{\mu(A)}{\mathbf{P}(A)} \in C \qquad \text{for } A \in \mathcal{F}_\infty, \mathbf{P}(A) > 0.$$

But C is a Radon-Nikodým set, so there is a random variable X with $\mathbf{E}[X \, \mathbf{1}_A] = \mu(A)$ for all $A \in \mathcal{F}_\infty$.

We claim that X_n converges scalarly to X. If $x^* \in E^*$, then $\langle X_n, x^* \rangle$ is a scalar-valued amart, and hence converges a.s. To see that X has values in C, write $C = \bigcap \{ x \in E : \langle x, x_i^* \rangle \leq \alpha_i \}$ as in (5.1.5); then $\mathbf{P}\{\langle X, x_i \rangle > \alpha_i\} \leq \liminf_n \mathbf{P}\{\langle X_n, x_i \rangle > \alpha_i\} = 0$. Thus $X \in C$ a.s. Now X_n and X all have values in C, a bounded set, so for all $A \in \mathcal{F}_\infty$, we have, by the dominated convergence theorem,

$$\mathbf{E}\left[\left(\langle X, x^* \rangle - \lim \langle X_n, x^* \rangle\right) \mathbf{1}_A\right] = \langle \mu(A), x^* \rangle - \langle \mu(A), x^* \rangle = 0,$$

so $\langle X_n, x^* \rangle \to \langle X, x^* \rangle$ a.s. This proves scalar convergence. The process is uniformly bounded, so in particular uniformly absolutely continuous, which proves Pettis norm convergence by (5.3.6). ∎

The converse of this result is also true: If C is a closed bounded convex set, and every amart in C converges scalarly, then C is a Radon-Nikodým set (5.3.28).

The conclusion of (5.3.7) can be proved with weak sequential amarts instead of amarts. See (5.3.14), below.

(5.3.8) Proposition. *Let C be a closed bounded convex Radon-Nikodým set. Suppose (X_n) is an amart that stops outside C. If either*

(a) *(X_n) satisfies condition (B); or*
(b) *(X_n) is uniformly absolutely continuous;*

then (X_n) converges scalarly and hence in Pettis norm.

Proof. If (a) holds, then by (5.2.24), we have $\mathbf{E}\left[\sup\|X_n\|\right] < \infty$, which implies that (X_n) is uniformly absolutely continuous. Thus it is enough to consider case (b). Now by translation, we may assume that $0 \in C$. Let $G_n = \{X_n \in C\}$. Since (X_n) stops outside C, we have $G_n \supseteq G_{n+1}$. Now $Y_n = X_n \mathbf{1}_{G_n}$ is an adapted process with its values in C.

We claim that (Y_n) is an amart. Let $\varepsilon > 0$ be given. There is $\delta > 0$ so that if $\mathbf{P}(A) < \delta$, then $\|\mathbf{E}[X_n \mathbf{1}_A]\| < \varepsilon$. Write $G = \bigcap_{n=1}^{\infty} G_n$. There is an $N \in \mathbb{N}$ with $\mathbf{P}(G_N \setminus G) < \delta$. Since (X_n) is an amart, there is $M \in \mathbb{N}$ with

$$\|\mathbf{E}[X_\sigma - X_\tau]\| < \varepsilon \quad \text{for } \sigma, \tau \in \Sigma; \sigma, \tau \geq M.$$

Note that

$$X_n - Y_n = \begin{cases} 0 & \text{on } G_n \\ X_n & \text{on } \Omega \setminus G_n, \end{cases}$$

so that if $m < n$, then

$$(X_n - X_m) - (Y_n - Y_m) = \begin{cases} 0 & \text{on } G_n \\ X_n & \text{on } G_m \setminus G_n \\ 0 & \text{on } \Omega \setminus G_m. \end{cases}$$

Also, if $p > n$ then $X_p = X_n$ outside G_n. Now consider simple stopping times σ, τ with $\max\{M, N\} \leq \sigma \leq \tau$. There is an integer $p \geq \tau$. Let $A = G_\sigma \setminus G_\tau$. This is a subset of $G_N \setminus G$, so $\mathbf{P}(A) < \delta$. Therefore $(X_\tau - X_\sigma) - (Y_\tau - Y_\sigma) = X_\tau \mathbf{1}_A$, hence

$$\|\mathbf{E}[(X_\tau - X_\sigma) - (Y_\tau - Y_\sigma)]\| = \|\mathbf{E}[X_\tau \mathbf{1}_A]\| = \|\mathbf{E}[X_p \mathbf{1}_A]\| \leq \varepsilon,$$

so that $\|\mathbf{E}[Y_\tau] - \mathbf{E}[Y_\sigma]\| \leq 2\varepsilon$. This shows that (Y_n) is an amart.

By the preceding proposition, (Y_n) converges scalarly and in Pettis norm, say to Y. Let σ be the first entrance time of (X_n) in $E \setminus C$. If

$$X = \begin{cases} Y & \text{on } G \\ X_\sigma & \text{on } \Omega \setminus G \end{cases}$$

then (X_n) converges to X scalarly and in Pettis norm. ∎

Next we can go to unbounded sets.

(5.3.9) Theorem. *Let C be a Radon-Nikodým set. If (X_n) is an amart in C satisfying condition* (B), *then (X_n) converges scalarly.*

Proof. Fix a real number $\lambda > 0$. Let (Y_n) be the process obtained by stopping (X_n) outside

$$C_\lambda = \{\, x \in C : \|x\| \leq \lambda \,\}.$$

The set C_λ is a bounded Radon-Nikodým set, so by the preceding result, (Y_n) converges scalarly. On the set

$$A_\lambda = \{\|X_n\| \leq \lambda \text{ for all } n\}$$

the two processes (X_n) and (Y_n) agree, so $X_n \mathbf{1}_{A_\lambda}$ converges scalarly. But by the maximal inequality (5.2.19),

$$\lim_{\lambda \to \infty} \mathbf{P}(A_\lambda) = 1,$$

so (X_n) itself converges scalarly. ∎

(5.3.10) Theorem. *Let E be a Banach space with the Radon-Nikodým property. Then every amart in E satisfying condition* (B) *converges scalarly.*

Proof. Take $E = C$ in the previous result. ∎

The converse of this result is also true (5.3.29), so this is a characterization of the Radon-Nikodým property. In general, the amart in the preceding proposition does not converge in Pettis norm. The Pettis norm convergence theorem of Uhl is proved below (5.3.24).

To state the next corollary, we define: X_n converges scalarly to X on the set A iff, for every $x^* \in E^*$, there is a null set N (depending on x^*) such that $x^*(X_n(\omega)) \to x^*(X(\omega))$ for all $\omega \in A \setminus N$.

(5.3.11) Corollary. *Let E be a Banach space with the Radon-Nikodým property. If (X_n) is a uniformly absolutely continuous amart, then X_n converges scalarly on the set where $\sup \|X_n\| < \infty$.*

Proof. For $\lambda > 0$, define $C_\lambda = \{\, x \in E : \|x\| \leq \lambda \,\}$. By (5.3.8(b)), the process X_n stopped outside C_λ converges scalarly. So X_n itself converges scalarly on $\{\sup \|X_n\| \leq \lambda\}$. Therefore it converges scalarly on a countable union $\bigcup_{n \in \mathbb{N}} C_n = \{\sup_n \|X_n\| < \infty\}$. ∎

Weak a.s. convergence

Scalar convergence is a weak conclusion for a theorem. Stronger sorts of convergence may be obtained by strengthening the hypotheses of the theorem. We first consider strengthening the restrictions on the Banach space E. Later in this section we will consider specializing the process $(X_n)_{n \in \mathbb{N}}$.

(5.3.12) Theorem. Let (X_n) be an amart in the Radon-Nikodým set C in the Banach space E. Suppose (X_n) satisfies condition (B) and E^* is separable. Then (X_n) converges weakly a.s.

Proof. By (5.3.9), we know that (X_n) converges scalarly, say to X. Let $(x_k^*)_{k \in \mathbb{N}}$ be a countable set dense in E^*. There exist events $A_k \in \mathcal{F}$ with $\mathbf{P}(A_k) = 0$ such that

$$\lim_n \langle X_n(\omega), x_k^* \rangle = \langle X, x^* \rangle \qquad \text{for all } \omega \in \Omega \setminus A_k.$$

By the maximal inequality, there is a set $N \in \mathcal{F}$ with $\mathbf{P}(N) = 0$ such that $\sup_n \|X_n\| < \infty$ on $\Omega \setminus N$. Let $A = N \cup \bigcup_{k \in \mathbb{N}} A_k$. Then $\mathbf{P}(A) = 0$. Fix $\omega \in \Omega \setminus A$. We claim that $X_n(\omega)$ converges weakly to $X(\omega)$. Since $\omega \notin N$, we have

$$\lambda = \sup \|X_n(\omega)\| < \infty.$$

Now let $x^* \in E^*$, and let $\varepsilon > 0$. Choose $k \in \mathbb{N}$ so that

$$\|x^* - x_k^*\| \le \frac{\varepsilon}{4\lambda + 1}.$$

For this k, we have $\omega \notin A_k$, so $\lim_n \langle X_n(\omega), x_k^* \rangle = \langle X(\omega), x^* \rangle$. Choose $m \in \mathbb{N}$ so that for $n \ge m$

$$\left| \langle X_n(\omega) - X(\omega), x_k^* \rangle \right| < \frac{\varepsilon}{2}.$$

Then for $n \ge m$ it follows that

$$
\begin{aligned}
\left| \langle X_n(\omega) - X(\omega), x^* \rangle \right| \\
\le \left| \langle X_n(\omega), x^* \rangle - \langle X_n(\omega), x_k^* \rangle \right| + \left| \langle X_n(\omega) - X(\omega), x_k^* \rangle \right| \\
+ \left| \langle X(\omega), x^* \rangle - \langle X(\omega), x_k^* \rangle \right| \\
\le \|x^* - x_k^*\| \left(\|X_n(\omega)\| + \|X(\omega)\| \right) + \varepsilon \\
\le \frac{\varepsilon}{4\lambda + 1} 2\lambda + \frac{\varepsilon}{2} \le \varepsilon.
\end{aligned}
$$

This shows that $\langle X_n(\omega), x^* \rangle \to \langle X(\omega), x^* \rangle$. This is true for any $x^* \in E^*$, so $X_n(\omega)$ converges weakly to $X(\omega)$. ∎

(5.3.13) Theorem. Let $(\Omega, \mathcal{F}, \mathbf{P})$ be a probability space, let E be a Banach space, and let $(X_n)_{n \in \mathbb{N}}$ be an amart in E. Assume:

(1) $(X_n)_{n \in \mathbb{N}}$ satisfies condition (B);
(2) E has the Radon-Nikodým property;
(3) E^* is separable.

Then $(X_n)_{n \in \mathbb{N}}$ converges weakly a.s.

Proof. Take $C = E$ in the preceding result. ∎

In this theorem, the conditions on the space E are necessary as well as sufficient. A Banach space E has the Radon-Nikodým property if and only if every L_1-bounded martingale in E converges weakly a.s. (5.3.30). Suppose E is a separable Banach space. Then the dual E^* is separable if and only if every amart potential in E converges weakly a.s. (5.5.27).

Weak sequential amarts

The preceding convergence theorems prove only weak convergence. This makes the definition of amarts (involving norm convergence) seem too strong. These convergence theorems fail in general for weak amarts. But they remain true for weak sequential amarts. Recall that a *weak sequential amart* is a process (X_n) such that, for every increasing sequence $\sigma_n \in \Sigma$, the sequence of vectors $\mathbf{E}[X_{\sigma_n}]$ converges weakly.

(5.3.14) Proposition. *Let E be a Banach space, and $C \subseteq E$ a closed bounded convex Radon-Nikodým set. If $(X_n)_{n\in\mathbb{N}}$ is a weak sequential amart in C, then (X_n) converges scalarly.*

Proof. By (5.2.12), for each $A \in \bigcup_{n=1}^{\infty} \mathcal{F}_n$, the sequence $\mathbf{E}[X_n \mathbf{1}_A]$ converges in the weak topology of E. Let $\mu(A)$ be its limit. Each $\mathbf{E}[X_n \mathbf{1}_A]/\mathbf{P}(A)$ is in C, and C is weakly closed, so $\mu(A)/\mathbf{P}(A) \in C$. The remainder of the proof is the same as in (5.3.7). ∎

(5.3.15) Proposition. *Let C be a closed bounded convex Radon-Nikodým set. Every weak sequential amart satisfying condition (B) that stops outside C converges scalarly.*

Proof. Let $(X_n)_{n\in\mathbb{N}}$ be a weak sequential amart that satisfies condition (B) and stops outside C. Let G_n and Y_n be as in the proof of (5.3.8). We claim that (Y_n) is a weak sequential amart. Now (X_{σ_n}) is a weak sequential amart for any increasing sequence (σ_n) in Σ, so it is enough to prove that $\mathbf{E}[Y_n]$ converges weakly.

Let σ be the first entrance time of (X_n) in the set $E \setminus C$. Then we have $\mathbf{E}[\|X_\sigma\| \mathbf{1}_{\Omega\setminus G}] < \infty$ by (5.2.22). The difference $X_n - Y_n$ converges a.s. (in norm) to $X_\sigma \mathbf{1}_{\Omega\setminus G}$, so $\mathbf{E}[X_n - Y_n]$ converges in norm to $\mathbf{E}[X_\sigma \mathbf{1}_{\Omega\setminus G}]$. So the weak limit of $\mathbf{E}[Y_n]$ is the difference of the weak limit of $\mathbf{E}[X_n]$ minus $\mathbf{E}[X_\sigma \mathbf{1}_{\Omega\setminus G}]$.

The remainder of the proof is the same as (5.3.8). ∎

(5.3.16) Proposition. *Let E be a Banach space with the Radon-Nikodým property. Then every weak sequential amart in E satisfying condition (B) converges scalarly.*

The proof is the same as that given for (5.3.9).

(5.3.17) Theorem. *Let $(\Omega, \mathcal{F}, \mathbf{P})$ be a probability space, let E be a Banach space, and let $(X_n)_{n\in\mathbb{N}}$ be a weak sequential amart in E. Assume:*

(1) *$(X_n)_{n\in\mathbb{N}}$ satisfies condition (B);*
(2) *E has the Radon-Nikodým property;*
(3) *E^* is separable.*

Then $(X_n)_{n\in\mathbb{N}}$ converges weakly a.s.

This proof, too, is the same as given above for the amart case (5.3.13).

One reason that the convergence theorems for weak sequential amarts are interesting is that there is no larger class with the same convergence properties (in bounded sets).

(5.3.18) Theorem. *Let E be a Banach space, let C be a bounded subset, and let $(X_n)_{n \in \mathbb{N}}$ have values in C. If (X_n) converges scalarly, then (X_n) is a weak sequential amart.*

Proof. Suppose $X_n \to X$ scalarly. Let σ_n be an inc `asing sequence in Σ. Let $\sigma = \lim_n \sigma_n$. Then σ is a (possibly infinite) stopping time. Write $A = \{\sigma = \infty\}$. We claim that $\mathbf{E}[X_{\sigma_n}]$ converges weakly to $\mathbf{E}\left[X_\sigma \mathbf{1}_{\Omega \setminus A} + X \mathbf{1}_A\right]$. Indeed, if $x^* \in E^*$, then $\langle X_{\sigma_n}, x^* \rangle$ converges a.s. to $\langle X_\sigma \mathbf{1}_{\Omega \setminus A} + X \mathbf{1}_A, x^* \rangle$; so the convergence of the expectations follows from the dominated convergence theorem. ∎

One could observe that the preceding result is not really probabilistic in nature, since the stochastic basis $(\mathcal{F}_n)_{n \in \mathbb{N}}$ does not matter. If X_n converges scalarly, then (X_n) is a weak sequential amart with respect to the stochastic basis $\mathcal{F}_n = \mathcal{F}$.

Strong convergence

(5.3.19) Definition. *Let $(X_n)_{n \in \mathbb{N}}$ be a process with values in a Banach space. We will say that (X_n) converges a.s. iff the sequence $(X_n(\omega))$ converges in the norm of E for almost every $\omega \in \Omega$.*

Next we take up an easy norm convergence theorem. It does not require the Radon-Nikodým property.

(5.3.20) Proposition. *Let $(\Omega, \mathcal{F}, \mathbf{P})$ be a probability space, let $(\mathcal{F}_n)_{n \in \mathbb{N}}$ be a stochastic basis, and let E be a Banach space. Write $\mathcal{F}_\infty$ for the σ-algebra generated by $\bigcup_{n=1}^\infty \mathcal{F}_n$. For every $X \in L_1(\Omega, \mathcal{F}, \mathbf{P}; E)$, the martingale*

$$X_n = \mathbf{E}^{\mathcal{F}_n}[X]$$

converges a.s. and in Bochner norm to $\mathbf{E}^{\mathcal{F}_\infty}[X]$.

Proof. First suppose X is $\mathcal{F}_m$-measurable for some $m \in \mathbb{N}$. Then $X_n = X$ for all $n \geq m$, so certainly X_n converges a.s. to $X = \mathbf{E}^{\mathcal{F}_\infty}[X]$.

Now consider a general $X \in L_1(\Omega, \mathcal{F}, \mathbf{P}; E)$. Let ε be given, $0 < \varepsilon < 1$. There is $m_0 \in \mathbb{N}$ and $Y \in L_1(\Omega, \mathcal{F}_{m_0}, \mathbf{P}; E)$ such that

$$\mathbf{E}\left[\|\mathbf{E}^{\mathcal{F}_\infty}[X] - Y\|\right] < \varepsilon^2.$$

Set $Y_n = \mathbf{E}^{\mathcal{F}_n}[Y]$; by the maximal inequality (5.2.19), applied to

$$X_n - Y_n = \mathbf{E}^{\mathcal{F}_n}\left[\mathbf{E}^{\mathcal{F}_\infty}[X] - Y\right],$$

we have

$$\mathbf{P}\left\{\sup_n \|X_n - Y_n\| > \varepsilon\right\} < \frac{1}{\varepsilon}\varepsilon^2 = \varepsilon.$$

Now $Y_n \to Y$ a.s., so there is $m_1 \geq m_0$ with

$$\mathbf{P}\{\|Y_n - Y\| > \varepsilon \quad \text{for some } n \geq m_1\} < \varepsilon.$$

Then, with the triangle inequality, we have

$$\mathbf{P}\{\|X_n - \mathbf{E}^{\mathcal{F}_\infty}[X]\| > 3\varepsilon \text{ for some } n \geq m_1\} < 3\varepsilon.$$

Since ε was arbitrary, $X_n \to \mathbf{E}^{\mathcal{F}_\infty}[X]$ a.s.

For proof of convergence in Bochner norm, observe that $\mathbf{E}^{\mathcal{F}_n}[X]$ is uniformly integrable. ■

Here is the observation that will be used to strengthen the scalar convergence results proved above to strong convergence results. If a martingale (or even a uniform amart) converges in a very weak sense, then in fact it converges in a much stronger sense.

(5.3.21) Proposition. *Let $(X_n)_{n\in\mathbb{N}}$ be a uniform amart in E.*

(1) *Suppose (X_n) is L_1-bounded. Then X_n converges scalarly if and only if X_n converges a.s.*

(2) *Suppose (X_n) is L_∞-bounded. Then X_n converges scalarly if and only if X_n converges in the Bochner norm.*

(3) *Suppose (X_n) satisfies $\mathbf{E}\left[\sup_n \|X_n\|\right] < \infty$. Then X_n converges scalarly if and only if X_n converges in the Bochner norm.*

Proof. First observe that it is enough in all parts to prove the result in the case that (X_n) is a martingale by the Riesz decomposition (5.2.13). Observe that a.s. convergence implies scalar convergence. Also, for a martingale (X_n), Bochner norm convergence (to X_∞) implies mean convergence $\mathbf{E}\left[\langle X_n, x^*\rangle - \langle X_\infty, x^*\rangle\right] \to 0$; these are scalar-valued martingales, so they converge a.s.; thus X_n converges scalarly to X_∞. (2) follows trivially from (3).

We begin with (3). Let (X_n) be a martingale with

$$\mathbf{E}\left[\sup_n \|X_n\|\right] < \infty.$$

Suppose X_n converges scalarly to X. Now if $A \in \mathcal{F}_m$ for some m, then by the dominated convergence theorem, we have

$$\lim_{n\to\infty} \mathbf{E}\left[\left(\langle X_n, x^*\rangle - \langle X, x^*\rangle\right)\mathbf{1}_A\right] = 0$$

for all $x^* \in E^*$, so that $\mathbf{E}[X_n\mathbf{1}_A] \to \mathbf{E}[X\mathbf{1}_A]$ weakly. But for $n \geq m$ the sequence $\mathbf{E}[X_n\mathbf{1}_A]$ is constant. Thus $\mathbf{E}[X\mathbf{1}_A] = \mathbf{E}[X_m\mathbf{1}_A]$. This is true for all $A \in \mathcal{F}_m$, so that

$$\mathbf{E}^{\mathcal{F}_m}[X] = X_m.$$

Then by the preceding result, X_n converges a.s. and in Bochner norm to X.

For part (1), we will use the same stopping technique as before. Fix $\lambda > 0$ and let Y_n be obtained by stopping X_n outside

$$C_\lambda = \{x \in E : \|x\| \leq \lambda\}.$$

Then the process (Y_n) is a martingale and satisfies $\mathbf{E}[\sup \|Y_n\|] < \infty$. Hence by part (2), (Y_n) converges a.s. Thus X_n converges a.s. on $\{\sup \|X_n\| \leq \lambda\}$. The maximal inequality (5.2.19) then shows that X_n converges a.s. ■

These results, together with the scalar convergence results proved above, produce some convergence theorems in Banach space.

(5.3.22) Theorem. *Suppose E is a Banach space and $C \subseteq E$ is a bounded Radon-Nikodým set.*

(1) *A uniform amart in C converges a.s. and in Bochner norm.*

(2) *An L_1-bounded uniform amart that stops outside C converges a.s. and in Bochner norm.*

(5.3.23) Theorem. *Let E be a Banach space with the Radon-Nikodým property. Let $(X_n)_{n \in \mathbb{N}}$ be an L_1-bounded uniform amart in E. Then $(X_n)_{n \in \mathbb{N}}$ converges a.s.*

Here is the Pettis norm convergence theorem of Uhl:

(5.3.24) Theorem. *Let E be a Banach space with the Radon-Nikodým property. Let $(X_n)_{n \in \mathbb{N}}$ be a uniformly absolutely continuous amart such that*

$$\liminf \mathbf{E} \left[\|X_n\| \right] < \infty.$$

Then X_n converges in Pettis norm.

Proof. Consider the Riesz decomposition $X_n = Y_n + Z_n$. The potential Z_n converges to 0 in Pettis norm. The martingale Y_n is L_1-bounded by (5.2.27a), and therefore a.s. norm convergent by (5.3.23). But Y_n is uniformly absolutely continuous, so it converges in Pettis norm by (5.3.6). Thus the sum $Y_n + Z_n$ converges in Pettis norm.   ∎

T-convergence

Let E be a Banach space and E^* its dual. A subset T of E^* is called *total* iff $x^*(x) = 0$ for all $x^* \in T$ implies $x = 0$. Let (X_n) be an amart of class (B) and X a (Bochner measurable) random variable. If the Radon-Nikodým property is not assumed, it is natural to ask how small can be the class T of functionals x^* such that a.s. convergence of $x^*(X_n)$ to $x^*(X)$ for each $x^* \in T$ implies scalar convergence of X_n (necessarily to X). In fact, T can be any total subset of E^*.

For martingales, the conclusion is even more dramatic, because scalar convergence implies strong almost sure convergence. Since an L_1-bounded martingale is an amart of class (B), it follows that an L_1-bounded martingale X_n such that $x^*(X_n)$ converges to $x^*(X)$ for each x^* in a total set T, converges to X strongly a.s.

(5.3.25) Lemma. *Let (X_n) be an amart with $\sup_n \|X_n\| \in L_1$. Let X be a random variable, and let T be a total subset of E^*. Assume that $x^*(X_n)$ converges to $x^*(X)$ almost surely for each $x^* \in T$.*

(i) *If $A \in \mathcal{F}$ and $\mathbf{E} \left[1_A \|X\| \right] < \infty$, then $\lim_n \mathbf{E} \left[1_A X_n \right] = \mathbf{E} \left[1_A X \right]$.*

(ii) *If X is Bochner integrable, then X_n converges scalarly to X.*

Proof. The process (X_n) is uniformly absolutely continuous, since we have $\sup \|X_n\| \in L_1$. Thus by (5.2.17), the limit

$$\mu(A) = \lim \mathbf{E}\left[\mathbf{1}_A X_n\right]$$

exists for all $A \in \mathcal{F}$ and defines a measure on $\mathcal{F}$.

(i) Now suppose $\mathbf{1}_A X$ is Bochner integrable. For each $x^* \in T$, we have

$$x^*\big(\mu(A)\big) = \lim_n \mathbf{E}\left[\mathbf{1}_A x^*(X_n)\right] = \mathbf{E}\left[\mathbf{1}_A x^*(X)\right] = x^*\big(\mathbf{E}\left[\mathbf{1}_A X\right]\big).$$

Since T is total, it follows that $\mu(A) = \mathbf{E}\left[\mathbf{1}_A X\right]$, as required.

(ii) If $y^* \in E^*$, then $y^*(X_n)$ is a real-valued amart with integrable supremum; therefore $y^*(X_n)$ converges a.s. to a real random variable, say Z_{y^*}. For each $A \in \mathcal{F}$, we have by (i) $\mu(A) = \mathbf{E}\left[\mathbf{1}_A X\right]$, so that

$$\mathbf{E}\left[\mathbf{1}_A y^*(X)\right] = y^*\big(\mu(A)\big) = \lim_n \mathbf{E}\left[\mathbf{1}_A y^*(X_n)\right] = \mathbf{E}\left[\mathbf{1}_A Z_{y^*}\right].$$

Therefore $y^*(X) = Z_{y^*}$ a.s. Thus X is the scalar limit of X_n.   ∎

The lemma is not the best result, because it is not necessary to assume that X is Bochner integrable.

(5.3.26) Theorem. *Let (X_n) be an E-valued amart of class (B), and let X be an E-valued random variable. Let T be a total subset of E^*. Assume, for each $x^* \in T$, that $x^*(X_n)$ converges almost surely to $x^*(X)$. Then X_n converges scalarly to X.*

Proof. Assume first that $\sup_n \|X_n\| \in L_1$. For $k \in \mathbb{N}$, let $A_k = \{\|X\| \le k\}$. Then for any $A \in \mathcal{F}$, by (5.3.25(i)),

$$\lim_n \mathbf{E}\left[\mathbf{1}_{A \cap A_k} X_n\right] = \mathbf{E}\left[\mathbf{1}_{A \cap A_n} X\right].$$

Therefore $\mathbf{E}\left[\mathbf{1}_{A_k} \|X\|\right] = |\mu|(A_k) \le \mathbf{E}\left[\sup_n \|X_n\|\right]$. Let $k \to \infty$; so

$$\mathbf{E}\left[\|X\|\right] \le \mathbf{E}\left[\sup_n \|X_n\|\right] < \infty,$$

and X is Bochner integrable. By 5.3.25(ii), X_n converges scalarly to X.

For the general case, we may apply the usual stopping argument: If $x^*(X_n) \to x^*(X)$ a.s. and $\tau = \inf\{n : \|X_n\| \ge k\}$, then $x^*(X_{n \wedge \tau}) \to x^*(Y)$ a.s., where

$$Y = \begin{cases} X_\tau & \text{on } \{\tau < \infty\} \\ X & \text{on } \{\tau = \infty\}. \end{cases}$$

   ∎

The preceding result also holds for weak sequential amarts. The conclusion is stronger for martingales.

(5.3.27) Theorem. *Let (X_n) be an E-valued L_1-bounded martingale, and X an E-valued random variable. Let T be a total subset of E^*. Assume, for each $x^* \in T$, that $x^*(X_n)$ converges almost surely to $x^*(X)$. Then X_n converges almost surely to X.*

Proof. An L_1-bounded martingale is necessarily an amart of class (B). Therefore Theorem (5.3.26) applies. Now it is easy to see that scalar convergence implies strong a.s. convergence for martingales. Indeed, after reducing to the case $\sup_n \|X_n\| \in L_1$ as usual, the relation

$$\mu(A) = \mathbf{E}\left[\mathbf{1}_A X\right]$$

proved above, applied to the terms X_n of a martingale and sets A in $\mathcal{F}_m$, shows that each X_m is $\mathbf{E}^{\mathcal{F}_m}[X]$. This reduces the problem to closed martingales, which converge without the Radon-Nikodým property (5.3.20). ∎

Converses

In most cases, the Radon-Nikodým property cannot be omitted from convergence theorems. In fact, it is necessary and sufficient for convergence of martingales in E. This converse result has a version for bounded sets and a version for unbounded sets. Recall that a set $C \subseteq E$ is a Radon-Nikodým set iff, for every probability space $(\Omega, \mathcal{F}, \mathbf{P})$ and every absolutely continuous vector measure $\mu : \mathcal{F} \to E$ with finite variation and average range contained in C, the Radon-Nikodým derivative $d\mu/d\mathbf{P}$ exists.

(5.3.28) Theorem. *Let C be a nonempty closed bounded convex set in a Banach space E. The following are equivalent:*

(1) *C is a Radon-Nikodým set;*
(2) *every martingale in C converges a.s.;*
(3) *every martingale in C converges in Bochner norm;*
(4) *every martingale in C converges in Pettis norm;*
(5) *every martingale in C converges scalarly;*
(6) *every amart in C converges scalarly.*

Proof. The proof outline is: $(1) \Longrightarrow (6) \Longrightarrow (5) \Longrightarrow (3) \Longrightarrow (1) \Longrightarrow (2)$ $\Longrightarrow (3) \Longrightarrow (4) \Longrightarrow (5)$. Seven of the eight implications are easily deduced from what has already been done.

$(1) \Longrightarrow (6)$ by (5.3.7). $(6) \Longrightarrow (5)$ since every martingale is an amart. $(5) \Longrightarrow (3)$ by (5.3.21(2)). $(3) \Longrightarrow (4)$ since the Bochner norm dominates the Pettis norm. $(1) \Longrightarrow (2)$ by (5.3.22). $(2) \Longrightarrow (3)$ by the dominated convergence theorem.

$(4) \Longrightarrow (5)$. Pettis norm convergence (to X_∞) implies mean convergence of scalar-valued martingales $\langle X_n, x^* \rangle \to \langle X_\infty, x^* \rangle$, so they converge a.s.; thus X_n converges scalarly to X_∞.

$(3) \Longrightarrow (1)$. Suppose, then, that every martingale in C converges in Bochner norm. Then any martingale in C with directed index set also converges in Bochner norm. This follows from the sequential sufficiency theorem (1.1.3) and the directed-set version of the optional sampling theorem (5.2.2). Let $(\Omega, \mathcal{F}, \mathbf{P})$ be a probability space, and let $\mu \colon \mathcal{F} \to E$ be a vector measure with average range contained in C. Consider (as in (1.3.2)) the directed set D of partitions of Ω into finitely many sets of $\mathcal{F}$ of positive measure, directed by essential refinement. For each $t = \{A_1, A_2, \cdots, A_k\}$ in D, let

$$X_t = \sum_{i=1}^{k} \frac{\mu(A_i)}{\mathbf{P}(A_i)} \, \mathbf{1}_{A_i}.$$

Then $(X_t)_{t \in D}$ is a martingale indexed by D. Let $X \in L_1(\Omega, \mathcal{F}, \mathbf{P}; E)$ be the limit in the Bochner norm. We claim that X is the Radon-Nikodým derivative of μ with respect to $\mathbf{P}$.

Let $A \in \mathcal{F}$. (We may assume $0 < \mathbf{P}(A) < 1$.) For any $t \in D$ that essentially refines the partition $s = \{A, \Omega \setminus A\}$, we have

$$\mathbf{E}\,[X_t \, \mathbf{1}_A] = \mathbf{E}\,[X_s \, \mathbf{1}_A] = \mu(A).$$

But $\mathbf{E}\,[X_t \, \mathbf{1}_A]$ converges to $\mathbf{E}\,[X \, \mathbf{1}_A]$, so $\mathbf{E}\,[X \, \mathbf{1}_A] = \mu(A)$. Thus X is the Radon-Nikodým derivative of μ.   ∎

The version for unbounded sets is an easy consequence.

(5.3.29) Theorem. *Let C be a nonempty closed convex set in a Banach space E. The following are equivalent:*

(1) *C is a Radon-Nikodým set;*
(2) *every L_1-bounded martingale in C converges a.s.;*
(3) *every L_∞-bounded martingale in C converges in Bochner norm;*
(4) *every L_∞-bounded martingale in C converges in Pettis norm;*
(5) *every L_1-bounded martingale in C converges scalarly;*
(6) *every amart in C satisfying condition (B) converges scalarly;*
(7) *every L_∞-bounded amart in C converges scalarly.*

Proof. Each of the parts holds for an unbounded set C if and only if it holds for all nonempty closed bounded convex subsets of C. For (1) this is (5.3.3), and for the others the usual stopping argument (as in (5.3.9)) will suffice.   ∎

The most commonly used version of this result is the theorem of the Ionescu Tulceas and Chatterji, which is (1) $\Longleftrightarrow$ (2) of the following result. For a simple proof of (1) $\Longrightarrow$ (2), see (5.3.34).

(5.3.30) Theorem. *Let E be a Banach space. The following are equivalent:*

(1) *E has the Radon-Nikodým property;*
(2) *every L_1-bounded martingale in E converges a.s.;*
(3) *every L_∞-bounded martingale in E converges in Bochner norm;*
(4) *every L_∞-bounded martingale in E converges in Pettis norm;*
(5) *every L_1-bounded martingale in E converges scalarly;*
(6) *every amart in E satisfying condition (B) converges scalarly;*
(7) *every L_∞-bounded amart in E converges scalarly.*

These theorems can be used to establish some of the usual results on the Radon-Nikodým property in Banach spaces.

(5.3.31) Proposition. *Let E be a Banach space and let (X_n) be a sequence of E-valued random variables.*

(i) *Suppose that there exists an event $\Omega_0 \subseteq \Omega$ with $\mathbf{P}(\Omega_0) = 1$ such that for each $\omega \in \Omega_0$, the sequence $X_n(\omega)$ is relatively weakly sequentially compact in E (that is, each subsequence contains a subsequence that converges weakly to a point in E). Suppose also that for each $x^* \in E^*$, the limit $\lim\langle X_n, x^*\rangle$ exists a.s. Then X_n converges scalarly.*

(ii) *Assume $E = F^*$ is the dual of a Banach space F. Suppose there is a separable set $E_1 \subseteq E$ and an event $\Omega_0 \subseteq \Omega$ with $\mathbf{P}(\Omega_0) = 1$ such that for each $\omega \in \Omega_0$, the sequence $X_n(\omega)$ is relatively weak-star sequentially compact in E_1 (that is, every subsequence contains a subsequence that converges weak-star to an element of E_1). Suppose also that for each $y \in F$, the limit $\lim\langle y, X_n\rangle$ exists a.s. Then there is a Bochner measurable random variable X such that $\langle y, X_n\rangle \to \langle y, X\rangle$ for each $y \in F$.*

Proof. (i) For each $x^* \in E^*$, denote by R_{x^*} the a.s. limit of $\langle X_n, x^*\rangle$. For each $\omega \in \Omega$, choose a subsequence $X_{n_k}(\omega)$ that converges weakly. (This choice depends on ω, possibly in a nonmeasurable way.) Let $X(\omega)$ be the weak limit of $X_{n_k}(\omega)$. Now for each $x^* \in E^*$,

$$R_{x^*}(\omega) = \lim_{k\to\infty} \langle X_{n_k}(\omega), x^*\rangle = \langle X(\omega), x^*\rangle \quad \text{a.s.}$$

But R_{x^*} is measurable, so X is scalarly measurable. Note that the values $X(\omega)$ all belong to the closed span of the set of values of all the random variables X_n (since the weak closure of a convex set is equal to its norm closure, see Dunford & Schwartz [1958], p. 422). So the set of values of X is separable. Therefore by the Pettis measurability theorem (5.1.7), X is Bochner measurable. For each $x^* \in E^*$,

$$\lim_{n\to\infty} \langle X_n(\omega), x^*\rangle = \langle X(\omega), x^*\rangle = R_{x^*}(\omega) \quad \text{a.s.}$$

Hence X is the scalar limit of X_n.

(ii) The proof is similar. The weak-star limit $X(\omega)$ of a subsequence $X_{n_k}(\omega)$ must be chosen in the separable set E_1. Then for each $y \in F$,

$$\langle y, X_n(\omega)\rangle \to \langle y, X(\omega)\rangle \quad \text{a.s.}$$

and X is weak-star measurable, so it is Bochner measurable by (5.1.8). ∎

(5.3.32) Proposition. *Let F be a Banach space, and let $C \subseteq F^*$ be a separable, weak* closed, bounded, convex set. Then C is a Radon-Nikodým set.*

Proof. First note that countably many linear functionals $y \in F$ separate points of C: if $\{x_n : n \in \mathbb{N}\}$ is a dense set of distinct points in C, choose $y_{mn} \in F$ with $\|y_{mn}\| \leq 1$ and $|\langle y_{mn}, x_n \rangle - \langle y_{mn}, x_m \rangle| > (1/2)\|x_n - x_m\|$. Now if $u, v \in C$ and $u \neq v$, then there exist n and m such that $\|x_n - u\| < (1/6)\|u - v\|$ and $\|x_m - v\| < (1/6)\|u - v\|$. Thus

$$
\begin{aligned}
|\langle y_{nm}, u - v \rangle| &\geq |\langle y_{nm}, x_n - x_m \rangle| - |\langle y_{nm}, x_n - u \rangle| - |\langle y_{nm}, x_m - v \rangle| \\
&> \frac{1}{2}\|x_n - x_m\| - \|x_n - u\| - \|x_m - v\| \\
&\geq \frac{1}{2}\|u - v\| - \frac{3}{2}\|x_n - u\| - \frac{3}{2}\|x_m - v\| \\
&> \frac{1}{2}\|u - v\| - \frac{1}{4}\|u - v\| - \frac{1}{4}\|u - v\| \geq 0.
\end{aligned}
$$

So $\langle y_{nm}, u - v \rangle \neq 0$.

Let $(X_n)_{n \in \mathbb{N}}$ be a martingale in C. For $y \in F$, the scalar process $(\langle y, X_n \rangle)$ is an L_∞-bounded martingale, so it converges a.s. Write R_y for the a.s. limit. Then $\mathbf{E}^{\mathcal{F}_n}[R_y] = \langle y, X_n \rangle$ a.s. For each $\omega \in \Omega_0$, choose a subsequence $X_{n_k}(\omega)$ that converges weak-star. (This is possible by the theorem of Alaoglu, for example Rudin [1973], p. 66.) Proposition (5.3.31(ii)) can now be applied to conclude that there exists a random variable X such that $\langle y, X_n \rangle \to \langle y, X \rangle$ a.s. For each $y \in F$ we have $\langle y, \mathbf{E}^{\mathcal{F}_n}[X] \rangle = \mathbf{E}^{\mathcal{F}_n}[R_y] = \langle y, X_n \rangle$ a.s. Now since countably many linear functionals separate points of C, we have $\mathbf{E}^{\mathcal{F}_n}[X] = X_n$ a.s. Therefore the martingale (X_n) converges a.s. This shows that C has the Radon-Nikodým property. ∎

(5.3.33) Proposition. *(i) A separable dual Banach space has the Radon-Nikodým property. (ii) A weakly compact convex subset of a Banach space is a Radon-Nikodým set.*

Proof. (i) The unit ball of the space satisfies the conditions for C in the preceding proposition. Now apply Proposition (5.3.3).

(ii) follows from (5.3.31(i)), since a weakly compact set is weakly sequentially compact (Eberlein-Smulian theorem; see, for example, Dunford & Schwartz [1958], p. 466). ∎

The preceding result will show that any reflexive Banach space has the Radon-Nikodým property. Indeed, E is reflexive if and only if the unit ball of E is weakly compact (Dunford & Schwartz [1958], p. 430).

Complements

(5.3.34) (Convergence of L_1-bounded martingales under Radon-Nikodým property.) We sketch a direct short proof of the important theorem of the Ionescu Tulceas. Most arguments have appeared before.

(i) Martingales of the form $X_n = \mathbf{E}^{\mathcal{F}_n}[X]$ converge a.s. (5.3.20).

(ii) If (X_n) is uniformly integrable, then $X_n = \mathbf{E}^{\mathcal{F}_n}[X]$ for some $X \in L_1(E)$. Indeed, $\mu(A) = \lim_n \mathbf{E}[1_A X_n]$ defines a countably additive measure of bounded variation on $\sigma(\bigcup \mathcal{F}_n)$. Let X be the Radon-Nikodým derivative $d\mu/d\mathbf{P}$.

(iii) For a constant $\lambda \geq 0$, define a stopping time σ as the first n such that $\|X_n\| > \lambda$ if such an n exists; otherwise $\sigma = \infty$. Then $(X_{n \wedge \sigma})$ is a martingale (5.2.2). Let $Z = \sup_n \|X_{n \wedge \sigma}\|$. We assert that $Z \in L_1$. Indeed, $\|Z\| \leq \lambda$ on the set $\{\sigma = \infty\}$; and on $A = \{\sigma < \infty\}$, we have $X_{n \wedge \sigma} \to X_\sigma$; so by Fatou and since $(\|X_n\|)$ is a submartingale,

$$\mathbf{E}\left[1_A \|X_\sigma\|\right] \leq \liminf \mathbf{E}\left[\|1_A X_{n \wedge \sigma}\|\right] \leq \sup_n \mathbf{E}\left[\|X_n\|\right].$$

But since $\|X_{n \wedge \sigma}\| \leq \|X_\sigma\|$ on A, we have $\mathbf{E}\left[\|Z\|\right] \leq \lambda + \sup \mathbf{E}\left[\|X_n\|\right] < \infty$.

By (i) and (ii), $X_{n \wedge \sigma}$ converges a.s.

(iv) By the maximal inequality (5.2.19), Ω is a union of sets

$$\Omega_\lambda = \{\sup_n \|X_n\| \leq \lambda\},$$

say $\lambda = 1, 2, 3, \cdots$. On Ω_λ, the process X_n coincides with $X_{n \wedge \sigma}$, hence it converges a.e. by (iii). Therefore X_n converges a.e. on Ω.

(5.3.35) (Reversed amarts.) Let $(\mathcal{F}_n)_{n \in -\mathbb{N}}$ be a reversed stochastic basis (1.1.18). Let E be a Banach space. Suppose E has the Radon-Nikodým property and E^* is separable. If $(X_n)_{n \in -\mathbb{N}}$ is an amart, then $X_n(\omega)$ converges weakly for almost all $\omega \in B = \{\sup \|X_n\| < \infty\}$ (and diverges weakly for all $\omega \notin B$) (Edgar & Sucheston [1976a]). It is possible that $B = \Omega$ (so this result applies on Ω) even if (X_n) is not L_1-bounded. If X_n is of class (B), then $\sup \|X_n\| < \infty$ a.s., so X_n converges a.s.

(5.3.36) (Reversed martingales.) We now discuss convergence of reversed martingales. Unlike direct martingales, or direct and reversed amarts, they converge a.s. also if E does not have the Radon-Nikodým property. A reversed martingale $(X_n)_{n \in -\mathbb{N}}$ has the form $X_n = \mathbf{E}^{\mathcal{F}_n}[X]$. In fact, X may be taken to be X_{-1}. So (X_n) is uniformly integrable. We claim that such a martingale converges a.s. to $\mathbf{E}^{\mathcal{F}_{-\infty}}[X]$, where

$$\mathcal{F}_{-\infty} = \bigcap_{n \in -\mathbb{N}} \mathcal{F}_n.$$

To prove this convergence, consider the set G of all $X \in L_1(\Omega, \mathcal{F}, \mathbf{P}; E)$ such that $\mathbf{E}^{\mathcal{F}_n}[X] = \mathbf{E}^{\mathcal{F}_{-\infty}}[X]$ for some finite n. Then we claim that the closed linear span of G is all of $L_1(E)$: to see this, it suffices to show that if $Y \in L_\infty(E^*)$ and $Y \perp G$, then $Y = 0$. But first G contains all random variables of the form $X - \mathbf{E}^{\mathcal{F}_n}[X]$, so $Y \perp G$ implies Y is $\mathcal{F}_n$-measurable. This is true for all n, so Y is $\mathcal{F}_{-\infty}$-measurable. On the other hand, G contains all $\mathcal{F}_{-\infty}$-measurable random variables X, so $Y \perp G$ implies $Y = 0$.

Next, we must show that the set of all $X \in L_1(E)$ for which convergence $\mathbf{E}^{\mathcal{F}_n}[X] \to \mathbf{E}^{\mathcal{F}_{-\infty}}[X]$ holds is a closed linear space. This is proved from the maximal inequality (5.2.37), as in (5.3.20). Since convergence is clear for $X \in G$, it follows that convergence holds for all $X \in L_1(E)$.

(5.3.37) (Directed index set.) Let J be a directed set, and let $(\mathcal{F}_t)_{t \in J}$ be a stochastic basis. Bochner norm convergence and Pettis norm convergence are metric, so the sequential sufficiency theorem applies to generalize results for convergence of these types.

(a) Let E be a Banach space, and $C \subseteq E$ a closed bounded convex Radon-Nikodým set. If $(X_t)_{t \in J}$ is an amart in C, then (X_t) converges in Pettis norm (5.3.7).

(b) Let C be a closed bounded convex Radon-Nikodým set. Suppose (X_t) is an amart that stops outside C [that is: if $s \le t$ and $X_s \notin C$, then $X_s = X_t$]. If either (X_t) satisfies condition (B); or (X_t) is uniformly absolutely continuous; then (X_t) converges in Pettis norm (5.3.8).

(c) Let E be a Banach space with the Radon-Nikodým property. Let (X_t) be a uniformly absolutely continuous amart such that

$$\sup_t \mathbf{E}\left[\|X_t\|\right] < \infty.$$

Then X_t converges in Pettis norm (5.3.24).

(d) Write $\mathcal{F}_\infty$ for the σ-algebra generated by $\bigcup_{t \in J} \mathcal{F}_t$. For every random variable X in $L_1(\Omega, \mathcal{F}, \mathbf{P}; E)$, the martingale

$$X_t = \mathbf{E}^{\mathcal{F}_t}[X]$$

converges in Bochner norm to $\mathbf{E}^{\mathcal{F}_\infty}[X]$ (5.3.20).

(e) Let (X_t) be a reversed martingale. Write $\mathcal{F}_{-\infty}$ for the σ-algebra generated by $\bigcap_{t \in J} \mathcal{F}_t$. Then X_n converges in Bochner norm to $\mathbf{E}^{\mathcal{F}_{-\infty}}[X_{-1}]$ (5.3.36).

(5.3.38) (P_0 convergence.) Let $(\mathcal{F}_t)_{t \in J}$ be a stochastic basis and let E be a Banach space. The process (X_t) of E-valued random variables *converges in probability* to X_∞ iff

$$\mathbf{P}\{\|X_t - X_\infty\| > \varepsilon\} \to 0$$

for every $\varepsilon > 0$. A weaker notion is P_0 convergence, related to convergence in probability as Pettis norm convergence is related to Bochner norm convergence. The process (X_n) *converges in P_0* to the random variable X_∞ iff:

for every $\varepsilon > 0$ there exists $t_0 \in J$ such that for all $t \geq t_0$ and all $x^* \in E^*$ with $\|x^*\| \leq 1$,

$$\mathbf{P}\Big\{|\langle x^*, X_t - X_\infty\rangle| > \varepsilon\Big\} < \varepsilon.$$

Suppose the Banach space E has the Radon-Nikodým property. If X_t is an L_1-bounded E-valued amart, then X_t converges in P_0. To prove this, apply the directed-set version of the Riesz decomposition (5.2.27): $X_t = Y_t + Z_t$, where Y_t is a martingale and $Z_t \to 0$ in Pettis norm. Then Y_t converges in probability and Z_t converges in Pettis norm. Both of these modes of convergence imply convergence in P_0 (Edgar [1982]).

(5.3.39) (Related processes.) As in the scalar case, there are in the vector case many additional generalizations and variants of the class of amarts. Such classes of processes may be more or less general, and correspondingly have less or more desirable properties. We will mention here only a few of the many results along this line.

Let E be a Banach space with the Radon-Nikodým property. If $(X_n)_{n \in \mathbb{N}}$ is a process with values in E, for positive integers $m \leq n$, write

$$H_{mn} = \mathbf{E}^{\mathcal{F}_m}[X_n] - X_m.$$

Similarly, for bounded stopping times $\sigma \leq \tau$, write

$$H_{\sigma\tau} = \mathbf{E}^{\mathcal{F}_\sigma}[X_\tau] - X_\sigma.$$

The process (X_n) is a *pramart* iff $H_{\sigma\tau} \to 0$ in probability, i.e.

$$\lim_{\substack{\sigma \in \Sigma \\ \tau \geq \sigma \\ \tau \in \Sigma}} \sup \mathbf{P}\Big\{\|H_{\sigma\tau}\| > \varepsilon\Big\} = 0$$

a.s. for every $\varepsilon > 0$. Millet & Sucheston [1980b] introduced pramarts and proved that real-valued pramarts converge a.s. if they satisfy condition (d):

$$\liminf \|X_n^+\|_1 + \liminf \|X_n^-\|_1 < \infty.$$

For pramarts, this condition is strictly weaker than L_1-boundedness (but not for martingales or amarts). Pramarts have the optional sampling property, and therefore there is a continuous parameter theory (see Frangos [1985]). Millet & Sucheston [1980b], Egghe [1981], Slaby[1982] and Frangos [1985] considered almost sure convergence of Banach valued pramarts, and Banach lattice valued "subpramarts." The Banach valued case is included in the result of Talagrand stated below.

The process (X_n) is a *martingale in the limit* iff $H_{mn} \to 0$ a.s., that is,

$$\lim_{m \to \infty} \sup_{n \geq m} \|H_{mn}\| = 0 \text{ a.s.}$$

Every pramart is a martingale in the limit. This definition is due to Mucci [1973] in the scalar-valued case. Bellow & Dvoretzky (unpublished) and Edgar [1979a] proved partial results on the a.s. convergence of martingales in the limit. These results were also improved by the result of Talagrand, which is stated below.

The process (X_n) is a *Talagrand mil* iff $H_{\sigma n} \to 0$ in probability, that is, for every $\varepsilon > 0$ there exists N such that for all bounded stopping times $\sigma \geq N$ and all integers $n \geq \sigma$, we have $\mathbf{P}\{\|H_{\sigma n}\| > \varepsilon\} < \varepsilon$. Every martingale in the limit (and therefore every pramart) is a Talagrand mil. Talagrand [1985] showed that (if E has the Radon-Nikodým property) all L_1-bounded Talagrand mils converge a.s.

A large class of processes is the *game which becomes fairer with time* (GFT). The process (X_n) is a *GFT* iff $H_{mn} \to 0$ in probability, that is, for every $\varepsilon > 0$ there is N so that if $n \geq m \geq N$, then $\mathbf{P}\{\|H_{mn}\| > \varepsilon\} < \varepsilon$. The definition is due to Blake [1970] in the scalar case. This very general class contains all those mentioned above. Mucci [1973] and Subramanian [1973] proved independently (in the scalar case) that a uniformly integrable GFT converges in mean. The vector case can be proved from a Riesz decomposition as follows.

(5.3.40)   Theorem. *Let E be a Banach space with the Radon-Nikodým property. Let $(X_n)_{n\in\mathbb{N}}$ be a uniformly integrable GFT. Then X_n converges in Bochner norm.*

Proof. Write $H_{mn} = \mathbf{E}^{\mathcal{F}_m}[X_n] - X_m$ for $m \leq n$. Since $H_{mn} \to 0$ in probability, and this double sequence is uniformly integrable, we get $H_{mn} \to 0$ in Bochner norm. Thus, for a fixed m, the sequence $(\mathbf{E}^{\mathcal{F}_m}[X_n])_{n=m}^{\infty}$ is Cauchy in Bochner norm. Therefore it converges in Bochner norm. Write Y_m for the limit, and $Z_m = X_m - Y_m$. Thus we have $X_n = Y_n + Z_n$, where (Y_n) is a uniformly integrable martingale, and (Z_n) converges to 0 in Bochner norm. By the Radon-Nikodým property, (Y_n) converges in Bochner norm. ∎

(5.3.41)   (Failure of strong convergence of amarts.) Let $E = l_2$ and let

$$\{\, e_{ni} : n \in \mathbb{N}, 1 \leq i \leq 2^n \,\}$$

be an orthonormal set in E. Let $(\Omega, \mathcal{F}, \mathbf{P})$ be $[0, 1)$ with Lebesgue measure. Write $A_{ni} = [(i-1)/2^n, i/2^n)$, and let $\mathcal{F}_n$ be the σ-algebra with atoms A_{ni}, for $1 \leq i \leq 2^n$. Then the process

$$X_n = \sum_{i=1}^{2^n} e_{ni} \mathbf{1}_{A_{ni}}$$

is an amart potential, but does not converge a.s. (Chacon & Sucheston [1975]). In fact, Bellow [1976a] showed that norm convergence of all bounded amarts in a Banach space E implies that E is finite-dimensional ((5.5.11(2))), below).

(5.3.42) (Condition (B) cannot be replaced by L_1-boundedness.) Let $(\Omega, \mathcal{F}, \mathbf{P})$, e_{ni}, A_{ni} be as in the previous exercise. For $1 \leq k \leq 2^n$, define

$$Y_{nk} = n e_{nk} \mathbf{1}_{A_{nk}} + \sum_{\substack{1 \leq i \leq 2^n \\ i \neq k}} e_{ni} \mathbf{1}_{A_{ni}}.$$

Given $m \in \mathbb{N}$, define $X_m = Y_{nk}$, where $m = 2^{n-1} + k - 1$. Then (X_m) is an L_1-bounded amart potential, but does not converge weakly a.s. (This example, due to W. J. Davis, appeared in Chacon & Sucheston [1975].) Again in this case, it is true that if E is a Banach space, and every L_1-bounded amart converges weakly a.s., then E is finite-dimensional (5.5.11(3)).

(5.3.43) (Scalar convergence and weak a.s. convergence.) Let $\{e_n\}$ be the usual unit vectors in the Banach space l_1, and let Z_n be an independent sequence of real random variables with $\mathbf{P}\{Z_n = 1\} = \mathbf{P}\{Z_n = -1\} = 1/2$. Define

$$X_n = 2^{-n} \sum_{j=0}^{2^n - 1} Z_{2^n + j} e_{2^n + j}.$$

Then (X_n) is an amart potential with respect to its natural stochastic basis $\mathcal{F}_n = \sigma(X_1, X_2, \cdots, X_n)$, it converges scalarly to 0, but it converges weakly for (almost) no ω. The last point can be verified from the fact that (in l_1) weak and norm convergence are equivalent for sequences (Davis & Johnson [1977]). In fact, there is such an example in a separable Banach space E if and only if E^* is not separable (5.5.26).

(5.3.44) (Submartingales in Banach lattices.) A Banach lattice has the Radon-Nikodým property if and only if all L_1-bounded positive submartingales converge a.s. The proof of necessity uses the Riesz decomposition (5.2.38) (Heinich [1978b]).

Remarks

The Radon-Nikodým property has become an important idea in the study of the geometry of Banach spaces. This began with Rieffel [1968] and culminated in the books by Diestel & Uhl [1977] and Bourgin [1983].

S. D. Chatterji [1960] proved almost sure convergence of martingales of the form $\mathbf{E}^{\mathcal{F}_n}[X]$, where X is Bochner integrable. Convergence of L_1-bounded martingales taking values in a reflexive Banach space E was proved by F. Scalora [1961]. The same result, assuming only that E has the Radon-Nikodým property appears as a remark in A. & C. Ionescu Tulcea [1963] (page 121; see also Bellow [1984]). The converse, in a stronger form—namely that convergence in Bochner norm of uniformly integrable martingales implies the Radon-Nikodým property—was proved by Rønnow [1967] and Chatterji [1968]. Chatterji in 1968 was apparently unaware of the Ionescu Tulcea convergence result, which he reproved. His article was influential in that it emphasized the importance of the Radon-Nikodým property.

The convergence theorem for uniform amarts (5.3.23) is due to Bellow [1978a]. The Pettis norm convergence of L_1-bounded uniformly absolutely continuous amarts (5.3.24) is due to Uhl [1977]. The result that a separable dual has the Radon-Nikodým property is due to Dunford & Pettis [1940]. Proposition (5.3.31) is due to Brunel & Sucheston [1976b].

The amart convergence theorem in Banach space (5.3.10) is due to Chacon & Sucheston [1975]. Theorem (5.3.27) is from Davis, Ghoussoub, Johnson, Kwapien, Maurey [1990]. Theorem (5.3.26) is due to Marraffa [1988].

5.4. Geometric properties

This section, even more than the last one, is an application of the martingale (and related) results to the geometry of Banach spaces. Martingales have found many varied uses in the study of the geometry of Banach spaces, and we will only describe a few of them here.

We will consider extreme points for closed bounded convex sets, and the integral representation that is available for sets with the Radon-Nikodým property. Another use of Radon-Nikodým sets is in connection with the Ryll-Nardzewski fixed-point theorem. We will discuss a geometric characterization of the Radon-Nikodým property in terms of "dentability." Next we will provide a martingale proof for the "strongly exposed point" characterization of the Radon-Nikodým property.

The following result is a simple consequence of the Hahn-Banach theorem.

(5.4.1) Proposition. *Let E be a Banach space, and suppose $x^*, y^* \in E^*$ with $\|x^*\| = \|y^*\| = 1$. Let*

$$\alpha = \sup \left\{ \, |x^*(x)| : \|x\| \leq 1, y^*(x) = 0 \, \right\}.$$

Then either $\|x^* - y^*\| \leq 2\alpha$ or $\|x^* + y^*\| \leq 2\alpha$.

Proof. Let $E_1 = \{\, x \in E : y^*(x) = 0 \,\}$. Then E_1 is a subspace of E. The definition of α shows that the norm of the restriction of x^* to E_1 is α. Let $z^* \in E^*$ be a Hahn-Banach extension of it, so that $\|z^*\| = \alpha$. Now $x^* - z^*$ vanishes on the nullspace E_1 of y^*, so $x^* - z^* = \beta y^*$ for some $\beta \in \mathbb{R}$. Then

$$\big|1 - |\beta|\big| = \big|\|x^*\| - \|x^* - z^*\|\big| \leq \|z^*\| \leq \alpha.$$

If $\beta \geq 0$, then

$$\|y^* - x^*\| = \big\|(1 - \beta)y^* - z^*\big\| \leq |1 - \beta| + \|z^*\| \leq 2\alpha,$$

and if $\beta < 0$, then

$$\|y^* + x^*\| = \big\|(1 + \beta)y^* + z^*\big\| \leq |1 + \beta| + \|z^*\| \leq 2\alpha. \qquad \blacksquare$$

We will need a selection theorem such as the following variant of the Yankov-von Neumann selection theorem. An expository account of it can be found in Section 8.5 of Cohn [1980].

(5.4.2) Theorem. *Let $(\Omega, \mathcal{F}, \mathbf{P})$ be a complete probability space, let R and S be complete separable metric spaces, and let $X \colon \Omega \to S$ be a Borel measurable random variable. Suppose $D \subseteq R$ is a nonempty Borel set, and $f \colon D \to S$ is a Borel function. Then $\{X \in f(D)\} \in \mathcal{F}$ and there is a Borel measurable random variable $Y \colon \Omega \to R$ such that $f(Y) = X$ on the set $\{X \in f(D)\}$.*

Some processes appearing below are very close to martingales. By the following lemma, they converge almost surely if martingales do.

(5.4.3) Lemma. *Let (X_n) be an E-valued Bochner integrable process such that $\left\|\mathbf{E}^{\mathcal{F}_n}[X_{n+1}] - X_n\right\| \leq 2^{-n}$ for all n, and let $Y_n = \lim_m \mathbf{E}^{\mathcal{F}_n}[X_{n+m}]$. Then (Y_n) is a martingale, and $\|X_n - Y_n\| \leq 2^{-n+1}$ for all n.*

Proof.

$$\mathbf{E}^{\mathcal{F}_n}[X_{n+m}] - X_n = \mathbf{E}^{\mathcal{F}_n}\left[\mathbf{E}^{\mathcal{F}_{n+m-1}}[X_{n+m}] - X_{n+m-1}\right]$$
$$+ \mathbf{E}^{\mathcal{F}_n}\left[\mathbf{E}^{\mathcal{F}_{n+m-2}}[X_{n+m-1}] - X_{n+m-2}\right]$$
$$+ \cdots + \mathbf{E}^{\mathcal{F}_n}\left[\mathbf{E}^{\mathcal{F}_n}[X_{n+1}] - X_n\right].$$

Since the conditional expectation is a contraction on L_∞,

$$\left\|\mathbf{E}^{\mathcal{F}_n}[X_{n+m}] - X_n\right\| \leq 2^{-n-m+1} + 2^{-n-m+2} + \cdots + 2^{-n} < 2^{-n+1}.$$

A similar computation shows that $Y_n = \lim_{m\to\infty} \mathbf{E}^{\mathcal{F}_n}[X_{n+m}]$ exists. Now the conditional expectation is continuous on $L_1(E)$, so

$$\mathbf{E}^{\mathcal{F}_n}[Y_{n+1}] = \mathbf{E}^{\mathcal{F}_n}\left[\lim_m \mathbf{E}^{\mathcal{F}_{n+1}}[X_m]\right] = \lim \mathbf{E}^{\mathcal{F}_n}\left[\mathbf{E}^{\mathcal{F}_{n+1}}[X_m]\right] = Y_n,$$

hence Y_n is a martingale. ∎

The study of convex sets is often coupled with the study of extreme points.

Let C be a convex set in a vector space E. The point $x \in C$ is called an *extreme point* of C iff

$$x = \frac{1}{2}(y + z) \text{ and } y, z \in C \quad \text{imply} \quad x = y = z.$$

We will write ex C for the set of all extreme points of the set C.

The Choquet-Edgar theorem

A theorem of Minkowski states that if C is a closed bounded set in a finite-dimensional Euclidean space, and $x_0 \in C$, then x_0 can be written as a convex combination of extreme points of C. While this is no longer strictly true in an infinite-dimensional space, there is a theorem of Choquet that shows that one can still sometimes write a point in a convex set as a generalized average of extreme points of the convex set. Here is the variant of Choquet's theorem due to Edgar. Notice the method used to construct the martingale. A similar method of construction will be used again below.

(5.4.4) Theorem. *Let E be a Banach space. Let C be a separable closed bounded convex set with the Radon-Nikodým property. Let $x_0 \in C$. Then there is a random variable $X \colon [0, 1] \to C$ such that*

(1) *$X(t) \in \mathrm{ex}\, C$ for almost all t;*

(2) *$\int_0^1 X(t)\, dt = x_0$.*

Proof. The proof is in three parts: (a) construction of a martingale; (b) the martingale converges; (c) the limit has the properties required.

(a) First, there is a *strictly convex* bounded continuous function $\psi \colon C \to \mathbb{R}$. (That is, $\psi(\frac{1}{2}(x + y)) \le \frac{1}{2}(\psi(x) + \psi(y))$, with equality only if $x = y$.) For example, since C is separable, there is a countable set $\{x_i^*\} \in E^*$ of norm-one functionals that separates the points of C. Then

$$\psi(x) = \sum_{i=1}^{\infty} 2^{-i} \left(x_i^*(x) \right)^2$$

is strictly convex.

We will construct a martingale (X_n) in C recursively, using the function ψ to measure how "close to the extreme points" a random variable is. The probability space $(\Omega, \mathcal{F}, \mathbf{P})$ will be the interval $[0, 1)$ with Lebesgue measure. The σ-algebra $\mathcal{F}$ will be the class of Lebesgue measurable sets, and for each $n \in \mathbb{N}$, $\mathcal{F}_n$ will be a finite σ-algebra with atoms which are subintervals of $[0, 1)$.

We begin with $X_1(t) = x_0$ for all $t \in [0, 1)$. Let $\mathcal{F}_1 = \{\emptyset, \Omega\}$. Now we want to write x_0 as $\frac{1}{2}(y_0 + z_0)$ for some $y_0, z_0 \in C$ (if x_0 is an extreme point of C, the only way to do this is $y_0 = z_0 = x_0$). Among all possible ways of doing this, we want to choose one representation $x_0 = y_0 + z_0$ that makes ψ large in the following way:

$$\frac{1}{2} \left(\psi(y_0) + \psi(z_0) \right) - \psi(x_0)$$
$$\ge \sup \left\{ \frac{1}{2} \left(\psi(y) + \psi(z) \right) - \psi(x_0) : y, z \in C, \ x_0 = \frac{1}{2}(y + z) \right\} - 2^{-1}.$$

Let

$$X_2 = y_0 \, \mathbf{1}_{[0,1/2)} + z_0 \, \mathbf{1}_{[1/2,1)},$$

and let $\mathcal{F}_2$ be the σ-algebra with the two atoms $[0, 1/2)$ and $[1/2, 1)$. Then clearly

$$\mathbf{E}^{\mathcal{F}_1} [X_2] = X_1.$$

Now suppose that X_n has been defined, for some $n \ge 2$, as

$$X_n = \sum_{i=1}^{2^n} x_i \, \mathbf{1}_{[(i-1)/2^n, i/2^n)},$$

where $x_i \in C$, and that $\mathcal{F}_n$ is the σ-algebra with atoms $[(i-1)/2^n, i/2^n)$ for $1 \le i \le 2^n$. For each x_i, choose y_i and z_i in C so that $x_i = \frac{1}{2}(y_i + z_i)$ and

$$\frac{1}{2}\left(\psi(y_i) + \psi(z_i)\right) - \psi(x_i)$$

$$\ge \sup\left\{ \frac{1}{2}\left(\psi(y) + \psi(z)\right) - \psi(x_i) : \right.$$

(5.4.4a)

$$\left. y, z \in C, \; x_i = \frac{1}{2}(y+z) \right\} - 2^{-n}.$$

Then write

$$X_{n+1} = \sum_{i=1}^{2^n}\left(y_i\, \mathbf{1}_{[(i-1)/2^n,(2i-1)/2^{n+1})} + z_i\, \mathbf{1}_{[(2i-1)/2^{n+1},i/2^n)}\right),$$

and take $\mathcal{F}_{n+1}$ as the σ-algebra with atoms given by the sets in the subscripts. A little thought will show that $\mathbf{E}^{\mathcal{F}_n}[X_{n+1}] = X_n$. This completes the recursive construction of the martingale (X_n).

(b) The martingale just constructed has its values in the set C. Since C is a Radon-Nikodým set, the martingale converges a.s. and in Bochner norm (5.3.22). Write $X = \lim_{n\to\infty} X_n$.

(c) We claim that the random variable X has the properties stated in the theorem. Since (X_n) is a uniformly bounded martingale, we have

$$\mathbf{E}[X] = \mathbf{E}\left[\mathbf{E}^{\mathcal{F}_1}[X]\right] = \mathbf{E}[X_1] = x_0.$$

The proof of property (1) is more involved.

Suppose (for purposes of contradiction) that $X(t)$ is not a.s. extreme. Write $\Delta = \{(x,x) \in C \times C : x \in C\}$. The set $D = C \times C \setminus \Delta$ is a Borel set in the complete separable metric space $C \times C$, and the map $f : D \to C \setminus \mathrm{ex}\, C$ defined by $f(x,y) = \frac{1}{2}(x+y)$ is continuous. Thus by the selection theorem (5.4.2), there is a random variable $X' : \Omega \to C \times C$ such that $f(X'(t)) = X(t)$ a.s. on the set $\{X \notin \mathrm{ex}\, C\}$, such that if we write $X'(t) = (Y(t), Z(t))$, then $X(t) = \frac{1}{2}(Y(t) + Z(t))$ and $Y(t) = Z(t)$ only if $X(t) \in \mathrm{ex}\, C$. Therefore $\mathbf{P}\{Y \ne Z\} > 0$. But then $\mathbf{P}\{\frac{1}{2}(\psi(Y) + \psi(Z)) > \psi(X)\} > 0$, so

(5.4.4b) $$\mathbf{E}\left[\frac{1}{2}\left(\psi(Y) + \psi(Z)\right)\right] > \mathbf{E}\left[\psi(X)\right] + 2^{-m}$$

for some $m \in \mathbb{N}$. We may define $Y_n = \mathbf{E}^{\mathcal{F}_n}[Y]$ and $Z_n = \mathbf{E}^{\mathcal{F}_n}[Z]$ to obtain martingales (Y_n) and (Z_n) in C with $X_n = \frac{1}{2}(Y_n + Z_n)$. But Y_n converges to Y and Z_n converges to Z, so by the continuity (and boundedness) of ψ, (5.4.4b) implies that there is $n_0 > m$ such that for all $n \ge n_0$,

$$\mathbf{E}\left[\frac{1}{2}\left(\psi(Y_n) + \psi(Z_n)\right)\right] > \mathbf{E}\left[\psi(X_n)\right] + 2^{-m}.$$

By (5.4.4a), we have

$$\mathbf{E}^{\mathcal{F}_n}\left[\psi(X_{n+1})\right] - \psi(X_n) \geq \frac{1}{2}\left(\psi(Y_n) + \psi(Z_n)\right) - \psi(X_n) - 2^{-n},$$

so $\mathbf{E}\left[\psi(X_{n+1}) - \psi(X_n)\right] > 2^{-m} - 2^{-n}$. But X_n converges, and ψ is bounded and continuous, so by the dominated convergence theorem

$$\mathbf{E}\left[\psi(X_{n+1}) - \psi(X_n)\right] \to 0.$$

This is a contradiction. ∎

Common fixed points for noncommuting maps

The Schauder fixed-point theorem asserts that a weakly continuous map of a weakly compact convex set C into itself has a fixed point. The Markov-Kakutani fixed-point theorem asserts that a *commuting* family of continuous affine maps of a weakly compact convex set C into itself has a common fixed point. The Ryll-Nardzewski fixed-point theorem is a generalization to certain noncommuting families of maps. A martingale proof of a key element of the theorem is given here.

(5.4.5) Definition. Let C be a convex set. A map $S \colon C \to C$ is *affine* iff

$$S\left(\sum_{i=1}^{n} t_i\, x_i\right) = \sum_{i=1}^{n} t_i\, S x_i$$

for $x_i \in C, t_i \geq 0, \sum t_i = 1$.

(5.4.6) Definition. A family $\mathcal{S}$ of maps from C to itself is *distal* iff for any pair $x, y \in C$ with $x \neq y$, we have $\inf_{S \in \mathcal{S}} \|Sx - Sy\| > 0$.

(5.4.7) Theorem. *Let C be a closed bounded convex Radon-Nikodým set in a Banach space E. Let $S_1, S_2, \cdots$ be (not necessarily continuous) affine maps from C into itself. Suppose that $\{S_1, S_2, \cdots\}$ generates a distal semigroup. Let $x_0 \in C$ be a fixed point for the map $S = \sum_{i=1}^{\infty} 2^{-i} S_i$. Then x_0 is also a fixed point for each of the maps S_i.*

Proof. Let (U_n) be an independent sequence of random variables with values in the countable set $\{S_1, S_2, \cdots\}$ and $\mathbf{P}\{U_n = S_i\} = 2^{-i}$ for all i. Note that $\mathbf{E}\left[U_n x\right] = Sx$ for $x \in C$. Let $\mathcal{F}_n$ be the σ-algebra generated by $U_1, U_2, \cdots, U_n$. Define $X_n \colon \Omega \to C$ by

$$X_n(\omega) = U_1(\omega)U_2(\omega)\cdots U_n(\omega)x_0.$$

We claim that (X_n) is a martingale. (In the following calculations, measurability is clear, since the σ-algebras $\mathcal{F}_n$ are atomic.)

$$\begin{aligned}
\mathbf{E}^{\mathcal{F}_n}\left[X_{n+1}\right] &= \mathbf{E}^{\mathcal{F}_n}\left[U_1 U_2 \cdots U_n U_{n+1} x_0\right] \\
&= U_1 U_2 \cdots U_n \mathbf{E}^{\mathcal{F}_n}\left[U_{n+1} x_0\right] \\
&= U_1 U_2 \cdots U_n S x_0 \\
&= U_1 U_2 \cdots U_n x_0 = X_n.
\end{aligned}$$

Suppose $S_i x_0 \neq x_0$ for some i. There is $\delta > 0$ so that $\|SS_i x_0 - Sx_0\| \geq \delta$ for all S in the semigroup generated by $\{S_1, S_2, \cdots\}$. Now $\mathbf{P}\{U_{n+1} = S_i\} = 2^{-i}$ and on the set $\{U_{n+1} = S_i\}$, we have

$$\|X_{n+1} - X_n\| = \|U_1 U_2 \cdots U_n (S_i x_0 - x_0)\| \geq \delta,$$

so $\mathbf{P}\{\|X_{n+1} - X_n\| \geq \delta\} > 2^{-i}$. By Egorov's theorem, the martingale (X_n) does not converge a.s. This contradicts the fact that C is a Radon-Nikodým set. ∎

The conclusion can be extended to include uncountable sets of maps by the usual abstract considerations. We will outline them briefly.

(5.4.8) Corollary. *Let E be a Banach space, and let $C \subseteq E$ be closed, bounded, and convex. Assume:*

(1) *C has the affine fixed-point property; that is, any weakly continuous affine map of C into itself has a fixed point.*

(2) *C has Corson's property (C); that is, a family of closed convex subsets of C with the finite intersection property has nonempty intersection.*

(3) *C is a Radon-Nikodým set.*

Then any distal semigroup $\mathcal{S}$ of weakly continuous affine maps from C to itself has a common fixed point.

Proof. For each $S \in \mathcal{S}$, let $F(S) = \{x \in C : Sx = x\}$. By the affine fixed-point property, each $F(S)$ is nonempty. Clearly, each $F(S)$ is closed and convex. Now C is a Radon-Nikodým set, so by Theorem (5.4.7), the family $\{F(S) : S \in \mathcal{S}\}$ has the countable intersection property. Therefore, by property (C), the family has a nonempty intersection, that is, $\mathcal{S}$ has a common fixed point. ∎

Finally, we can use a single-map fixed-point theorem to deduce the Ryll-Nardzewski fixed-point theorem itself.

(5.4.9) Theorem. *Let C be a weakly compact convex set, and let $\mathcal{S}$ be a distal semigroup of weakly continuous affine maps from C into itself. Then $\mathcal{S}$ has a common fixed point.*

Proof. Weakly compact sets have property (C) [since closed convex sets are weakly closed] and are Radon-Nikodým sets (5.3.33). The fixed point property is a version of the Schauder fixed-point theorem (Dunford & Schwartz [1958], V.10.5). Thus, Corollary (5.4.8) applies to this situation. ∎

Dentability

In the geometric arguments below we will use the following notations. Let E be a Banach space. The *closed ball* about a point x with radius r is denoted

$$B(x, r) = \{y \in E : \|y - x\| \leq r\}.$$

Let $C \subseteq E$ be a nonempty bounded set. A *slice* of C is a set of the form

$$S(C, x^*, \alpha) = \{\, y \in C : x^*(y) \geq \sup x^*(C) - \alpha \,\}$$

for some $x^* \in E^*$ and some $\alpha > 0$. Here $\sup x^*(C) = \sup \{\, x^*(y) : y \in C \,\}$ by definition.

Dentability is a purely geometric concept that turns out to be related to the Radon-Nikodým property and to martingale convergence.

(5.4.10) Definition. Let C be a nonempty closed bounded subset of a Banach space E. We say that C is *dentable* iff for every $\varepsilon > 0$, there is a point $x_0 \in C$ such that

$$x_0 \notin \operatorname{cl} \operatorname{conv} (C \setminus B(x_0, \varepsilon)).$$

The Hahn-Banach separation theorem shows how slices are related to convexity. This proof is given in great detail. Future uses of the Hahn-Banach theorem will be less verbose.

(5.4.11) Proposition. *Let E be a Banach space, and let $C \subseteq E$ be a closed, bounded, and nonempty set. Then C is dentable if and only if C admits slices of arbitrarily small diameter.*

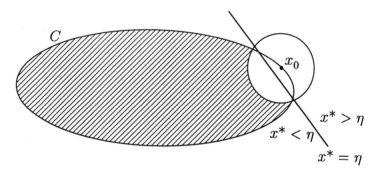

Figure (5.4.11) A slice of a dentable set.

Proof. Suppose C is dentable. Let $\varepsilon > 0$ be given. Then there is $x_0 \in C$ so that

$$x_0 \notin \operatorname{cl} \operatorname{conv} (C \setminus B(x_0, \varepsilon)).$$

By the Hahn-Banach separation theorem, there is $x^* \in E^*$ and $\gamma \in \mathbb{R}$ such that $\sup x^*(C \setminus B(x_0, \varepsilon)) < \gamma < x^*(x_0)$ (see Figure (5.4.11)). But then

$$\{\, y \in C : x^*(y) \geq \gamma \,\}$$

is a slice of C with diameter less than 2ε.

Conversely, suppose C has slices of arbitrarily small diameter. Let $\varepsilon > 0$ be given. Then there is a slice $S(C, x^*, 2\alpha)$ of C with diameter less than ε. Let x_0 be any point in this slice. Then

$$\left(C \setminus B(x_0, \varepsilon)\right) \cap S(C, x^*, \alpha) = \emptyset,$$

so that

$$\left(C \setminus B(x_0, \varepsilon)\right) \subseteq \{ y \in C : x^*(y) \leq \sup x^*(C) - \alpha \}.$$

So also

$$\text{cl conv} \left(C \setminus B(x_0, \varepsilon)\right) \subseteq \{ y \in C : x^*(y) \leq \sup x^*(C) - \alpha \},$$

and x_0 is outside the closed convex hull.   ∎

It will be useful to observe that a nonempty bounded set C is dentable if its closed convex hull is dentable, since every slice of cl conv C contains a slice of C.

The next theorem is probably the most often used geometric characterization of the Radon-Nikodým property.

(5.4.12) Theorem. *Let E be a Banach space, and $C \subseteq E$ a nonempty closed bounded convex set. Then C is a Radon-Nikodým set if and only if every nonempty closed convex subset of C is dentable.*

Proof. Suppose that there is a nonempty closed bounded set $D \subseteq C$ that is not dentable. Then cl conv $D \subseteq C$ is not dentable. So we may assume D is convex. We will construct a quasimartingale on $[0,1)$ with values in $D \subseteq C$ that diverges everywhere, yet is very close to a martingale and should converge a.s. by Lemma (5.4.3) together with (5.3.34). (Alternatively, the uniform amart convergence theorem (5.3.22) could be applied, since every quasimartingale is a uniform amart.)

Since D is not dentable, there is $\varepsilon > 0$ so that every slice of D has diameter greater than ε. Thus if $x \in D$, then $D \cap B(x, \varepsilon/2)$ contains no slice of D. This means that

$$\text{cl conv} \left(D \setminus B(x, \varepsilon/2)\right) = D.$$

Let $\Omega = [0,1)$, $\mathbf{P}$ = Lebesgue measure, and $\mathcal{F}$ = Borel sets. A process $(X_n)_{n \in \mathbb{N}}$ on $(\Omega, \mathcal{F}, \mathbf{P})$ will be constructed recursively. (The construction is similar to that of the martingale constructed in (5.4.4).) Choose $x_0 \in D$ (this is possible since D is nonempty), let $\mathcal{F}_1 = \{ \emptyset, \Omega \}$, and define $X_1(\omega) = x_0$. Now suppose $X_1, X_2, \cdots, X_n$ have been defined, and X_n is a simple function measurable with respect to a σ-algebra $\mathcal{F}_n$ with atoms

$$[0, t_1), [t_1, t_2), \cdots, [t_{N-1}, 1).$$

Let x be one of the values of X_n; say $X_n(\omega) = x$ for $\omega \in [t_{j-1}, t_j)$. Now

$$D = \text{cl conv}\left(D \setminus B(x, \varepsilon/2)\right),$$

so there exist $x_1, x_2, \cdots, x_k \in D$ with $\|x - x_i\| > \varepsilon/2$ for all i, and scalars $\alpha_1, \alpha_2, \cdots, \alpha_k > 0$ with $\sum_{i=1}^{k} \alpha_i = 1$ such that

$$\left\| x - \sum_{i=1}^{k} \alpha_i\, x_i \right\| \leq 2^{-n}.$$

Subdivide the interval $[t_{j-1}, t_j)$ into disjoint subintervals with lengths proportional to the α_i, and define X_{n+1} on these intervals to have the values x_i. Thus:

$$\|X_n(\omega) - X_{n+1}(\omega)\| > \frac{\varepsilon}{2} \qquad \text{for all } \omega \in \Omega$$

and

$$\left\|X_n - \mathbf{E}^{\mathcal{F}_n}\left[X_{n+1}\right]\right\| \leq 2^{-n} \qquad \text{a.s.}$$

This completes the recursive construction of the process $(X_n)_{n \in \mathbb{N}}$. Now (5.4.3) shows that C fails the Radon-Nikodým property.

Conversely, suppose that every nonempty closed convex subset of C is dentable. Then in fact every nonempty subset of C is dentable. We will show that C has the Radon-Nikodým property. Let $(\Omega, \mathcal{F}, \mathbf{P})$ be a probability space, and let $\mu \colon \mathcal{F} \to E$ be an absolutely continuous vector measure with average range in C. For each set $M \in \mathcal{F}$ with $\mathbf{P}(M) > 0$ we will write the *average range* of μ on M this way:

$$\alpha(M) = \left\{ \frac{\mu(W)}{\mathbf{P}(W)} : W \in \mathcal{F}, W \subseteq M, \mathbf{P}(W) > 0 \right\}.$$

We will prove first:

(5.4.12a)
$$\begin{cases} \text{for every } \varepsilon > 0, \text{ and every } M \in \mathcal{F} \text{ with} \\ \mathbf{P}(M) > 0, \text{ there is } F \in \mathcal{F} \text{ with } F \subseteq M, \\ \mathbf{P}(F) > 0, \text{ and } \text{diam}(\alpha(F)) \leq \varepsilon. \end{cases}$$

Now $\alpha(M) \subseteq C$, so it is dentable. Thus there is a slice

$$S = \left\{ y \in \alpha(M) : x^*(y) \geq \gamma \right\}$$

with $S \neq \emptyset$ and $\text{diam}\, S \leq \varepsilon$. Now $x^* \circ \mu$ is a scalar measure and $x^* \circ \mu \ll \mathbf{P}$, so by the scalar Radon-Nikodým theorem, there is $Z \in L_1(\mathbf{P})$ so that $x^*(\mu(W)) = \mathbf{E}\left[Z\, \mathbf{1}_W\right]$ for all $W \in \mathcal{F}$. Let $F = \{Z > \gamma\} \cap M$. Then $\mathbf{P}(F) > 0$ since $S \neq \emptyset$. And $\alpha(F) \subseteq S$: Indeed, for $W \subseteq F$ with $\mathbf{P}(W) > 0$ we have $\mu(W)/\mathbf{P}(W) \in \alpha(M)$ and

$$x^*\left(\frac{\mu(W)}{\mathbf{P}(W)}\right) = \frac{\mathbf{E}\left[Z\, \mathbf{1}_W\right]}{\mathbf{P}(W)} \geq \frac{\mathbf{E}\left[\gamma\, \mathbf{1}_W\right]}{\mathbf{P}(W)} = \gamma,$$

so $\mu(W)/\mathbf{P}(W) \in S$. This completes the proof of (5.4.12a).

Now, take a maximal pairwise disjoint collection $\{M_i\}$ in $\mathcal{F}$ such that $\operatorname{diam} \alpha(M_i) \leq \varepsilon$. We have now proved

(5.4.12b) $\quad \begin{cases} \text{for every } \varepsilon > 0, \text{there is a countable partition} \\ \{M_i\} \text{ of } \Omega \text{ with } \operatorname{diam} \alpha(M_i) \leq \varepsilon \text{ for all } i. \end{cases}$

Thus, we can choose a sequence $(\{M_i^n\}_{i=1}^\infty)_{n=1}^\infty$ of partitions, increasing with respect to refinement, such that $\operatorname{diam} \alpha(M_i^n) < 2^{-n-1}$. Define

$$X_n = \sum_{i=1}^\infty \frac{\mu(M_i^n)}{\mathbf{P}(M_i^n)} \mathbf{1}_{M_i^n}.$$

Now if $M \in \mathcal{F}$, we have

$$\left| \mathbf{E}\left[X_n \mathbf{1}_M - \mu(M)\right] \right|$$
$$= \left| \mathbf{E}\left[\sum_i \frac{\mu(M_i^n)}{\mathbf{P}(M_i^n)} \mathbf{1}_{M_i^n \cap M} - \sum_i \frac{\mu(M_i^n \cap M)}{\mathbf{P}(M_i^n \cap M)} \mathbf{1}_{M_i^n \cap M} \right] \right|$$
$$\leq \mathbf{E}\left[\sum_i 2^{-n} \mathbf{1}_{M_i^n \cap M} \right] \leq 2^{-n-1}.$$

Thus $\mathbf{E}[X_n \mathbf{1}_M] \to \mu(M)$. Now the estimates we have will be used to show that the process $(X_n)_{n \in \mathbb{N}}$ converges uniformly. If $\omega \in \Omega$ and $n \in \mathbb{N}$, then $\omega \in M_i^n$ for some i and $\omega \in M_j^{n+1}$ for some j. But

$$X_n(\omega) = \frac{\mu(M_i^n)}{\mathbf{P}(M_i^n)} \in \alpha(M_i^n),$$

and, since $M_j^{n+1} \subseteq M_i^n$,

$$X_{n+1}(\omega) = \frac{\mu(M_j^{n+1})}{\mathbf{P}(M_j^{n+1})} \in \alpha(M_i^n).$$

Therefore $\|X_n(\omega) - X_{n+1}(\omega)\| \leq 2^{-n-1}$. Let Y be the uniform limit of the sequence X_n. Then, for any $M \in \mathcal{F}$ we have both $\mathbf{E}[X_n \mathbf{1}_M] \to \mu(M)$ and $\mathbf{E}[X_n \mathbf{1}_M] \to \mathbf{E}[Y \mathbf{1}_M]$, so $\mu(M) = \mathbf{E}[Y \mathbf{1}_M]$. $\blacksquare$

(5.4.13) **Theorem.** *Let E be a Banach space. Then E has the Radon-Nikodým property if and only if every nonempty closed bounded convex subset of E is dentable.*

Proof. Combine Theorem (5.4.12) with Proposition (5.3.3). $\blacksquare$

Strongly exposed points

There is a more precise characterization of Radon-Nikodým sets in terms of strongly exposed points.

(5.4.14) Definition. Let C be a nonempty closed bounded convex set. A point $x_0 \in C$ is a *strongly exposed point* of C iff there is $x^* \in E^*$ (the *strongly exposing functional*) such that $x^*(x_0) = \sup x^*(C)$ and if $x_n \in C$ is a sequence with $\lim_{n\to\infty} x^*(x_n) = x^*(x_0)$, then $\lim_{n\to\infty} \|x_n - x_0\| = 0$.

Clearly, a strongly exposed point of C is an extreme point of C. If C has a strongly exposed point, then C is dentable. The geometric criterion we will prove is due to Phelps and Bourgain: a set C is a Radon-Nikodým set if and only if every nonempty closed bounded convex subset of C is the closed convex hull of its strongly exposed points. To begin the proof, we have a lemma, which we prove using Lemma (5.4.3). Recall the notation for slices:

$$S(C, x^*, \alpha) = \{\, y \in C : x^*(y) \geq \sup x^*(C) - \alpha \,\}.$$

(5.4.15) Lemma. *Let E be a Banach space, and let $C \subseteq E$ be a nonempty closed bounded convex Radon-Nikodým set. Let $x^* \in E^*$ with $\|x^*\| = 1$, let $\varepsilon > 0$, and $0 < \eta < \alpha$. Write $M = \sup x^*(C)$. Then there is $x_0 \in S(C, x^*, \eta)$ such that*

$$x_0 \notin \mathrm{cl\ conv}\ \big[(C \setminus B(x_0, \varepsilon)) \cup \{\, x : \|x - x_0\| \leq 1,\ x^*(x) < M - \alpha \,\}\big].$$

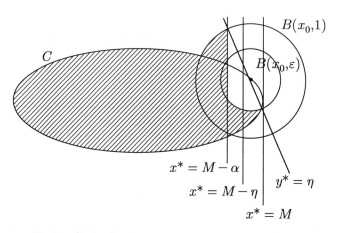

Figure (5.4.15) Illustration for Lemma (5.4.15).

Proof. Let $W = (C \setminus B(x_0, \varepsilon)) \cup \{\, x : \|x - x_0\| \leq 1,\ x^*(x) < M - \alpha \,\}$. (The shaded portion in Figure (5.4.15).) Assume the conclusion is false. Let $x_0 \in C$ be such that $x^*(x_0) > M - \eta$. Let $\delta = x^*(x_0) - M + \eta > 0$. Then $x_0 \in \mathrm{cl\ conv}\, W$, so x_0 is as close as we like to a convex combination of elements $x_1, x_2, \cdots, x_k \in W$, so for each x_i, either $\|x_0 - x_i\| \geq \varepsilon$ or both $\|x_0 - x_i\| \leq 1$ and $x^*(x_i) < M - \alpha$. If the second case holds, then $\alpha - \eta < x^*(x_0) - x^*(x_i) \leq \|x^*\| \|x_0 - x_i\|$. Hence for all x_i we have $\|x_i - x_0\| \geq \min\{\varepsilon, \alpha - \eta\}$.

The above observation may be used to construct a bounded process (X_n) on $\Omega = [0,1]$ that stops when it leaves C and satisfies:

(1) $X_1(\omega) = x_0$ a.s.;

(2) $\|\mathbf{E}^{\mathcal{F}_n}[X_{n+1}] - X_n\| < \delta\, 2^{-n}$;

(3) if $x^*(X_n(\omega)) \geq M - \eta$, then we have $X_n(\omega) \in C$ and either $\|X_{n+1}(\omega) - X_n(\omega)\| \geq \varepsilon$ or both $x^*(X_{n+1}(\omega)) < M - \alpha$ and $\|X_n(\omega) - X_{n+1}(\omega)\| \leq 1$;

(4) if $x^*(X_n(\omega)) < M - \eta$, then $X_{n+1}(\omega) = X_n(\omega)$.

Now if $x^*(X_n(\omega)) \geq M - \eta$, then $\|X_{n+1}(\omega) - X_n(\omega)\| \geq \min\{\varepsilon, \alpha - \eta\}$. But C has the Radon-Nikodým property, so $X_n(\omega)$ converges a.s. by (5.4.3). Write $X_\infty(\omega)$ for the limit, so that $x^*(X_\infty(\omega)) < M - \eta$ a.s. So $\mathbf{E}[x^*(X_\infty)] < M - \eta$. But

$$\left\|\mathbf{E}[X_\infty] - x_0\right\| \leq \sum_{n=1}^{\infty} \left\|\mathbf{E}[X_{n+1}] - \mathbf{E}[X_n]\right\| \leq \delta,$$

so that $M - \eta > x^*(\mathbf{E}[X_\infty]) \geq x^*(x_0) - \delta > M - \eta$, a contradiction. ∎

The preceding lemma can be used to prove a geometric consequence.

(5.4.16) Lemma. *Let E be a Banach space, let $C \subseteq E$ be a nonempty closed bounded convex Radon-Nikodým set, and let $\beta > 0$. Then the set*

$$A_\beta = \left\{\, x^* \in E^* : \|x^*\| = 1 \text{ and } \operatorname{diam} S(C, x^*, \alpha) < \beta \text{ for some } \alpha > 0 \,\right\}$$

is dense in the unit sphere of E^*.

Proof. Let $x^* \in E^*$ with $\|x^*\| = 1$, and let $\delta > 0$. We will show that there is $y^* \in A_\beta$ with $\|x^* - y^*\| < \delta$. We may assume that C is contained in the unit ball and is not just a single point. We may assume that x^* is not constant on C, since functionals not constant on C are dense in the unit sphere of E^*.

Let $\varepsilon > 0$ be so small that

$$2\varepsilon < \beta, \quad 4\varepsilon < \delta, \quad 11\varepsilon < \sup x^*(C) - \inf x^*(C).$$

Write $\eta = \varepsilon$, $\alpha = 2\varepsilon$, and $M = \sup x^*(C)$. Then apply the preceding lemma. Thus: there is $x_0 \in S(C, x^*, \varepsilon)$ such that $x_0 \notin \operatorname{cl} \operatorname{conv} W$, where

$$W = \left(C \setminus B(x_0, \varepsilon)\right) \cup \{\, x : \|x - x_0\| \leq 1, x^*(x) < M - \alpha \,\}.$$

Now by the Hahn-Banach separation theorem there is $y^* \in E^*$, $\|y^*\| = 1$, with $\sup y^*(W) < \gamma < y^*(x_0)$. (The hyperplane $\{y^* = \gamma\}$ is in the illustration.) Then for all $x \in C$ with $y^*(x) > \gamma$ we have $\|x - x_0\| < \varepsilon$. That is, $y^* \in A_\beta$.

It remains only to show that $\|x^* - y^*\| < \delta$. We will do this using (5.4.1). Now if $x \in E$ with $y^*(x) = 0$ and $\|x\| \leq 1$, then $y^*(x_0 + x) = y^*(x_0) > \gamma$ and

$\|(x_0+x)-x_0\| = \|x\| \leq 1$, so $x_0+x \notin W$ and therefore $x^*(x_0+x) > M-\alpha$. But that means

$$x^*(x) > M - \alpha - x^*(x_0) \geq M - \alpha - M = -\alpha.$$

Similarly, using $x_0 - x$ in place of $x_0 + x$, we see that $x^*(x) < \alpha$. Thus $|x^*(x)| < \alpha$ for all $x \in E$ with $\|x\| \leq 1$ and $y^*(x) = 0$. So by Proposition (5.4.1), either $\|x^* + y^*\| \leq 2\alpha$ or $\|x^* - y^*\| \leq 2\alpha$. Since $2\alpha = 4\varepsilon < \delta$, it remains only to show $\|x^* + y^*\| > 4\varepsilon$.

Suppose (for purposes of contradiction) that $\|x^* + y^*\| \leq 4\varepsilon$. Now we have $11\varepsilon < \sup x^*(C) - \inf x^*(C)$ and $M = \sup x^*(C)$, so there is $x_1 \in C$ with $x^*(x_1) < M - 11\varepsilon$. Then

$$y^*(x_1) = -x^*(x_1) + (x^* + y^*)(x_1) \geq -x^*(x_1) - \|x^* + y^*\|$$
$$\geq -M + 11\varepsilon - 4\varepsilon = -M + 7\varepsilon.$$

But $\sup y^*(C) \leq y^*(x_0) + \varepsilon$, hence

$$y^*(x_1) \leq y^*(x_0) + \varepsilon = -x^*(x_0) + (x^* + y^*)(x_0) + \varepsilon$$
$$\leq -M + \varepsilon + 4\varepsilon + \varepsilon = -M + 6\varepsilon.$$

So we have obtained the contradiction $-M + 7\varepsilon \leq -M + 6\varepsilon$.   ∎

(5.4.17)   Theorem. *Let E be a Banach space, and let $C \subseteq E$ be a nonempty closed bounded convex set. If C has the Radon-Nikodým property, then C is the closed convex hull of its strongly exposed points.*

Proof. Suppose C has the Radon-Nikodým property. We may assume that C is contained in the unit ball $B(0,1)$ of E. Let $A =$

$$\{\, x^* \in E^* : \|x^*\| = 1 \text{ and } x^* \text{ is strongly exposing linear functional for } C \,\}.$$

Now if A_β is defined as in Lemma (5.4.16), then we have $A = \bigcap_{n=1}^\infty A_{1/n}$. By Lemma (5.4.16), each $A_{1/n}$ is dense (and open) in the unit sphere of E^*, so by the Baire category theorem, we know that A is itself dense in the unit sphere of E^*.

Let W be the closed convex hull of the strongly exposed points of C. Suppose (for purposes of contradiction) that $C \neq W$. Then there is a linear functional $y^* \in E^*$ with $\|y^*\| = 1$ and $\sup y^*(W) < \sup y^*(C) = M$. Now let $\varepsilon > 0$ satisfy $\sup y^*(W) + 2\varepsilon < M$, and choose $x^* \in A$ with $\|x^* - y^*\| < \varepsilon$. Say x^* strongly exposes $x_0 \in C$. Then $y^*(x_0) > x^*(x_0) - \varepsilon$ and $x^*(x_0) > M - \varepsilon$, so $y^*(x_0) > M - 2\varepsilon > \sup y^*(W)$. Thus $x_0 \notin W$, a contradiction.   ∎

There exists a converse of the preceding result. Suppose, for each nonempty closed convex subset D of C, the set D is the closed convex hull of its strongly exposed points. Then C is a Radon-Nikodým set. In fact, by (5.4.12), it is enough if each nonempty convex $D \subseteq C$ has at least one strongly exposed point, since a set with a strongly exposed point is necessarily dentable.

Complements

(5.4.18) (Locally convex space.) Suppose that F is a locally convex Hausdorff topological vector space. Let $K \subseteq F$ be a nonempty compact convex metrizable subset. Then there is a linear subspace $F_0 \subseteq F$ containing K, a Banach space E, and a continuous linear transformation $T: F_0 \to E$ such that T is a homeomorphism on K.

(5.4.19) (Choquet's theorem.) The preceding can be used to prove Choquet's theorem in one of its original forms: Suppose that F is a locally convex topological vector space. Let $K \subseteq F$ be a nonempty compact convex metrizable subset. If $x_0 \in K$, then there is a random variable $X: \Omega \to K$ such that $\mathbf{P}\{X \in \text{ex } K\} = 1$ and the Pettis integral $\mathbf{E}[X]$ exists and is equal to x_0.

(5.4.20) (An application of Choquet's theorem.) A function $f: \mathbb{Z} \to \mathbb{C}$ is called *positive definite* iff

$$\sum_{j,k=-\infty}^{\infty} t_j \overline{t_k} f(j - k) \geq 0$$

for any choice of complex numbers $\{t_j\}_{j=-\infty}^{\infty}$ with all but finitely many equal to 0. We will outline a use of Choquet's theorem to prove the following theorem of Herglotz: If $f: \mathbb{Z} \to \mathbb{C}$ is positive definite, then there exists a finite measure μ on $[0, 2\pi)$ such that

$$f(n) = \int_0^{2\pi} e^{in\theta} \, d\mu(\theta)$$

for all $n \in \mathbb{Z}$. The following steps are used for the proof.

(a) Using sequences $\{t_j\}_{j=-\infty}^{\infty}$ with only one nonzero term, it can be seen that $f(0) \geq 0$, and using sequences with only two nonzero terms, it can be seen that $|f(n)| \leq f(0)$ for all n.

(b) The set $K = \{ f \in l_\infty(\mathbb{Z}) : f \text{ positive definite}, f(0) = 1 \}$ is a compact convex metrizable subset of the locally convex space $l_\infty(\mathbb{Z})$ with its weak* topology.

(c) For $\theta \in [0, 2\pi)$ let

$$g_\theta(n) = e^{in\theta}.$$

Then $g_\theta \in \text{ex } K$.

(d) Any extreme point f of K is of the form g_θ. This can be seen as follows: Let $\alpha = \text{Re } f(1)$, and show that $f \pm g \in K$, where

$$g(n) = \frac{1}{4}f(n+1) - \frac{1}{2}\alpha f(n) + \frac{1}{4}f(n-1),$$

so that $g = 0$. Let $\beta = \text{Im } f(1)$. Then $f \pm h \in K$, where

$$h(n) = \frac{1}{4i}f(n+1) - \frac{1}{2}\beta f(n) - \frac{1}{4i}f(n-1),$$

so that $h = 0$. Finally, $f(n+1) = f(1)f(n)$ for all n, so that f has the form g_θ.

(This proof is similar to a proof in Edgar [1983].)

(5.4.21) (Krein-Milman property.) We say that a Banach space E has the *Krein-Milman property* iff every closed bounded convex subset is the closed convex hull of its extreme points. (Or, equivalently, every nonempty closed bounded convex subset has at least one extreme point; see Bourgin [1983], Proposition 3.1.1.) Comparing this to Theorem (5.4.17), one might naturally conjecture that the Krein-Milman property is equivalent to the Radon-Nikodým property. This is still an open problem.

Remarks

The selection theorem (5.4.2) is due independently to Yankov [1941] and von Neumann [1949].

Choquet's theorem (for compact sets) is due to Choquet [1956]. The proof by two-point dilations follows Loomis [1975]. The generalization to Radon-Nikodým sets is due to Edgar [1975]. For further reading on Choquet's theorem, we recommend Phelps [1966].

Ryll-Nardzewski [1967] gave a proof for his fixed-point theorem using a "differentiation" argument. Namioka & Asplund [1967] realized that a condition like dentability could be used in the proof. The use of property (C) in our proof may be new. Property (C) holds, for example, in separable Banach spaces, in weakly compactly generated Banach spaces, and in many others. See Corson [1961], Pol [1980]. The Radon-Nikodým type of argument used here to prove the fixed-point theorem will also prove this variant (Namioka & Phelps [1975], Theorem 15): If C is a separable weak-star closed bounded convex subset of the dual F^* of a Banach space F, then any distal semigroup of weak-star continuous affine maps of C into itself admits a common fixed point.

The geometric characterization (5.4.13) was one of the first results showing that the Radon-Nikodým property is relevant in the study of the geometry of Banach spaces. First Rieffel [1968] showed that the Radon-Nikodým property follows from dentability of every subset. (This step was simplified by Girardi & Uhl [1990].) Then the converse was proved in small steps by Maynard [1973], Davis & Phelps [1974], Huff [1974]. The geometric characterization (5.4.17) is due to Phelps [1974] and Bourgain [1977]; the use of martingales for the proof was suggested by Kunen & Rosenthal [1982].

5.5. Operator ideals

This section shows a few more of the connections between our subject matter and the geometric theory of Banach spaces. More knowledge of Banach space theory is required for an understanding of this section than was required in the previous sections.

One concept that has been much used in recent years in the study of Banach spaces is the "operator ideal." We will consider here primarily the following operator ideals: the absolutely summing operators, the Radon-Nikodým operators, the Asplund operators.

Absolutely summing operators

The first ideal that we will consider here, because of its connections with amarts, is the ideal of absolutely summing operators.

(5.5.1) Definition. Let E be a Banach space, and let $\{x_i\}_{i=1}^{\infty}$ be a sequence of vectors in E. We say that the series $\sum x_i$ is *convergent* iff there is a vector y such that

$$\lim_{n\to\infty} \left\| y - \sum_{i=1}^{n} x_i \right\| = 0.$$

The series $\sum x_i$ is *absolutely convergent* iff

$$\sum_{i=1}^{\infty} \|x_i\| < \infty.$$

The series $\sum x_i$ is *unconditionally convergent* iff the series $\sum \theta_i x_i$ converges for all choices of scalars θ_i with $|\theta_i| \leq 1$. (Or, equivalently, the series $\sum \theta_i x_i$ converges for all choices of scalars $\theta_i = \pm 1$.) The series $\sum x_i$ is *weakly unconditionally Cauchy* iff, for all $x^* \in E^*$,

$$\sum_{i=1}^{\infty} |x^*(x_i)| < \infty.$$

An application of the closed graph theorem (5.1.3) shows that if $\sum x_i$ is weakly unconditionally Cauchy, then there is a constant A with

$$\sum_{i=1}^{\infty} |x^*(x_i)| \leq A \|x^*\|$$

for all $x^* \in E^*$. Indeed, the operator $T\colon E^* \to l_1$ defined by

$$T(x^*) = \bigl(x^*(x_i)\bigr)_{i=1}^{\infty}$$

has closed graph, and is therefore bounded.

In a Banach space, every absolutely convergent series is unconditionally convergent, and every unconditionally convergent series is weakly unconditionally Cauchy. (The inverse implications hold if and only if E is finite-dimensional; see (5.5.8).)

Consider the measure space $\mathbb{N}$ with counting measure, and a function $f\colon \mathbb{N} \to E$ defined by $f(n) = x_n$. Then the series $\sum x_n$ is absolutely convergent if and only if f is Bochner integrable (clear); and the series $\sum x_n$ is unconditionally convergent if and only if f is Pettis integrable (see Lindenstrauss & Tzafriri [1977], Proposition 1.c.1(ii)).

(5.5.2) Definition. Let E and F be Banach spaces. The operator $T\colon E \to F$ is said to be *absolutely summing* iff there is a constant C such that for all finite sequences $\{x_1, x_2, \cdots, x_n\}$ in E,

$$\sum_{i=1}^{n} \|Tx_i\| \le C \sup\left\{ \sum_{i=1}^{n} |x^*(x_i)| : x^* \in E^*,\ \|x^*\| \le 1 \right\}.$$

(Stated another way: for $f \in l_1(E)$, the Bochner norm of $T \circ f$ is $\le$ a constant C times the Pettis norm of f.) The smallest such constant C is called the *absolutely summing norm* of T, and will be written $\pi_1(T)$.

Note that

$$\sup_{\|x^*\| \le 1} \sum_{i=1}^{n} |x^*(x_i)| = \sup\left\{ \left\| \sum_{i=1}^{n} \theta_i x_i \right\| : \theta_i = \pm 1 \right\},$$

so that (by the closed graph theorem) T is absolutely summing if and only if the unconditional convergence of a series $\sum x_i$ implies the absolute convergence of $\sum Tx_i$.

A generalization of absolutely summing operators is stated next.

(5.5.3) Definition. Let p be a positive real number. The operator $T\colon E \to F$ is called *p-absolutely summing* iff there is a constant C such that for all finite sequences $\{x_1, x_2, \cdots, x_n\}$ in E,

$$\left(\sum \|Tx_i\|^p \right)^{1/p} \le C \sup\left\{ \left(\sum |x^*(x_i)|^p \right)^{1/p} : x^* \in E^*,\ \|x^*\| \le 1 \right\}.$$

The smallest such constant C is called the *p-summing norm* of T, and will be written $\pi_p(T)$.

The collection of p-absolutely summing operators is a *Banach operator ideal* in the sense given in the next proposition. An operator $T\colon E \to F$ is said to have *rank one* iff it has the form

$$T(x) = x^*(x)y, \quad \text{for all } x \in E,$$

for some $x^* \in E^*$ and $y \in F$.

(5.5.4) Proposition. *Let $p \ge 1$.*

(1) *For any pair E, F of Banach spaces, the set $\Pi_p(E, F)$ of all p-absolutely summing operators from E to F is a Banach space under the norm π_p.*

(2) *For any linear transformation T, we have $\|T\| \le \pi_p(T)$, with equality for rank one operators. In particular, if T is p-absolutely summing, then T is bounded.*

(3) *If $T\colon E \to F$ is p-absolutely summing, $Q\colon E_1 \to E$ and $R\colon F \to F_1$ are bounded operators, then the composition $RTQ\colon E_1 \to F_1$ is p-absolutely summing, and $\pi_p(RTQ) \le \|R\|\, \pi_p(T)\, \|Q\|$.*

Proof. (2) Take $n = 1$ in the definition. If $x \in E$, then

$$\|Tx\| \leq \pi_p(T) \sup \left\{ |\langle x, x^* \rangle| : x^* \in E^*, \|x^*\| \leq 1 \right\}.$$

Therefore $\|Tx\| \leq \pi_p(T) \|x\|$. So we have $\|T\| \leq \pi_p(T)$.

If $T \colon E \to F$ has rank 1, then

$$T(x) = x_0^*(x) \, y_0$$

for some $x_0^* \in E^*$ and $y_0 \in F$, $\|x_0^*\| = 1$. Thus $\|T\| = \|y_0\|$, and for $x_1, x_2, \cdots, x_n \in E$, we have

$$\sum_{i=1}^{n} \|Tx_i\|^p = \sum |x_0^*(x_i)|^p \|y_0\|^p$$

$$\leq \|T\|^p \sup \left\{ \left(\sum |x^*(x_i)|^p \right) : x^* \in E^*, \|x^*\| \leq 1 \right\}.$$

Thus $\pi_p(T) \leq \|T\|$ in this case, so $\pi_p(T) = \|T\|$.

(1) The only nontrivial assertion is the completeness. Suppose (T_m) is a sequence of p-absolutely summing operators, Cauchy in the norm π_p. Then $\pi_p(T_m)$ converges, say to a. By part (2), the sequence (T_m) is also Cauchy in the operator norm. So there is an operator $T \colon E \to F$ with $\|T_m - T\| \to 0$. We claim that $T_m \to T$ also in the norm π_p. Now $T_m x \to Tx$ for all $x \in E$. If the finite set $\{x_1, x_2, \cdots, x_n\}$ is given, then we have

$$\left(\sum_{i=1}^{n} \|Tx_i\|^p \right)^{1/p} = \lim_m \left(\sum_{i=1}^{n} \|T_m x_i\|^p \right)^{1/p}$$

$$\leq \lim_m \pi_p(T_m) \sup_{\|x^*\| \leq 1} \left(\sum_{i=1}^{n} |\langle x_i, x^* \rangle|^p \right)^{1/p}$$

$$= a \sup \left(\sum_{i=1}^{n} |\langle x_i, x^* \rangle|^p \right)^{1/p}.$$

So T is p-absolutely summing. Given $\varepsilon > 0$, we may choose m_0 so that $\pi_p(T_{m_0} - T_m) \leq \varepsilon$ for all $m \geq m_0$. The argument just used shows that $\pi_p(T_{m_0} - T) \leq \varepsilon$. So we have proved that $\pi_p(T_m - T) \to 0$.

(3) follows from the definition. ∎

One of the most useful results on absolutely summing operators is:

(5.5.5) Pietsch factorization. *Let $p \geq 1$, and let E, F be Banach spaces. Write K for the unit ball of E^* with its weak* topology. Then the operator $T \colon E \to F$ is p-absolutely summing if and only if there is a probability measure μ on K and a constant C such that*

$$\|Tx\| \leq C \left(\int_K |x^*(x)|^p \, d\mu(x^*) \right)^{1/p} \qquad \text{for all } x \in E.$$

Moreover, the smallest such constant C is $\pi_p(T)$.

Proof. Suppose that $T\colon E \to F$ is p-absolutely summing, and $\pi_p(T) = 1$. Then $\|T\| \leq 1$. For $x \in E$, define $g_x \in C(K)$ by $g_x(x^*) = |x^*(x)|^p$. Consider two subsets of $C(K)$ defined by

$$F_1 = \left\{ f \in C(K) : \sup_{x^* \in K} f(x^*) < 1 \right\}$$
$$F_2 = \operatorname{conv} \left\{ g_x : x \in E, \ \|Tx\| = 1 \right\}.$$

Then F_1 and F_2 are convex sets and F_1 is open. Since $\pi_p(T) = 1$, we have $F_1 \cap F_2 = \emptyset$. So the two sets can be separated by a linear functional on $C(K)$. By the Riesz representation theorem, there is a positive constant λ and a signed measure μ on K with variation 1 so that

$$f \in F_1 \quad \text{implies} \quad \int_K f(x^*)\, d\mu(x^*) \leq \lambda$$
$$f \in F_2 \quad \text{implies} \quad \int_K f(x^*)\, d\mu(x^*) \geq \lambda.$$

Since F_1 contains all nonpositive functions, the measure μ is a positive measure. Since F_1 contains the open unit ball of $C(K)$, we have $\lambda \geq 1$. Thus for any $x \in E$ with $\|Tx\| = 1$ we have $\int_K |x^*(x)|^p\, d\mu(x^*) \geq 1$, and for every $x \in E$ we have $\int_K |x^*(x)|^p\, d\mu(x^*) \geq \|Tx\|^p$.

For the converse, suppose that μ and C exist. Then if $x_1, x_2, \cdots, x_n$ are in E, we have

$$\sum_{i=1}^{n} \|T x_i\|^p \leq C^p \int_K \sum |x^*(x_i)|^p\, d\mu(x^*)$$
$$\leq C^p \sup \left\{ \sum |x^*(x_i)|^p : x^* \in K \right\}. \quad \blacksquare$$

An easy consequence of the Pietsch factorization is the following comparison theorem.

(5.5.6) Proposition. *Let* $p \leq q$, *and suppose that* $T\colon E \to F$ *is a* p-*absolutely summing operator. Then* T *is also* q-*absolutely summing and* $\pi_q(T) \leq \pi_p(T)$.

Proof. By the Pietsch factorization theorem (5.5.5),

$$\|Tx\| \leq \pi_p(T) \left(\int_K |x^*(x)|^p\, d\mu(x^*) \right)^{1/p}.$$

The function $t^{q/p}$ is convex, since $p \leq q$, so by Jensen's inequality (2.3.10), we have

$$\left(\int_K |x^*(x)|^p\, d\mu(x^*) \right)^{1/p} \leq \left(\int_K |x^*(x)|^q\, d\mu(x^*) \right)^{1/q}.$$

Combining these two inequalities, we obtain

$$\|Tx\| \le \pi_p(T) \left(\int_K |x^*(x)|^q \, d\mu(x^*) \right)^{1/q}.$$

Therefore T is q-absolutely summing and $\pi_q(T) \le \pi_p(T)$.   ∎

The Pietsch factorization theorem shows the existence of an actual "factorization" when certain projections exist. Since every closed linear subspace of a Hilbert space is the range of a projection, the case of 2-absolutely summing operators is particularly simple.

(5.5.7) Corollary. *Let $T: E \to F$ be a 2-absolutely summing operator. Then there is a compact space K and a probability measure μ on K such that the operator T has a factorization*

$$
\begin{array}{ccc}
E & \xrightarrow{\ T\ } & F \\
{\scriptstyle I}\big\downarrow & & \big\uparrow{\scriptstyle S} \\
C(K) & \xrightarrow[\ J_\mu\ ]{} & L_2(K,\mu)
\end{array}
$$

where $I: E \to C(K)$ is an isometry, and $J_\mu: C(K) \to L_2(K,\mu)$ is the formal identity map.

Proof. Let K and μ be as in the Pietsch factorization theorem. The canonical isometry $I: E \to C(K)$ is defined by $I(x)(x^*) = x^*(x)$ for $x^* \in K \subseteq E^*$. The inequality from the theorem shows that the map sending $J_\mu I(x)$ to Tx is a bounded map on the set $J_\mu I(E) \subseteq L_2(K,\mu)$. So it can be extended by continuity to the closure. Composing with the orthogonal projection onto that closure extends it to a map S defined on all of $L_2(K,\mu)$.   ∎

We obtain from this a version of the Dvoretzky-Rogers lemma.

(5.5.8) Theorem. *Let E be a Banach space. The identity operator on E is absolutely summing if and only if E is finite-dimensional. Every unconditionally convergent series in E is absolutely convergent if and only if E is finite-dimensional.*

Proof. If E is finite-dimensional, then the series property follows from the well known one-dimensional case, so the identity operator is absolutely summing.

Suppose every unconditionally convergent series in E is absolutely convergent. That means the identity map $I: E \to E$ is absolutely summing. Then I is also 2-absolutely summing (5.5.6), so by the previous corollary it factors through a Hilbert space. That means that E is linearly homeomorphic to a closed subspace of the Hilbert space. It therefore remains only to show that an infinite-dimensional Hilbert space contains a series that converges unconditionally but not absolutely. If $\{e_i\}$ is an infinite orthonormal sequence, then the series

$$\sum_{i=1}^{\infty} \frac{1}{i} e_i$$

converges unconditionally (Bessel's inequality), but does not converge absolutely.

∎

The absolutely summing operators illustrate the difference between the Bochner and Pettis norms. Since the two norms are related by the inequality $\|X\|_P \le \mathbf{E}\left[\|X\|\right]$, we will have for any operator $T\colon E \to F$ the inequality $\|TX\|_P \le \|T\|\, \mathbf{E}\left[\|X\|\right]$. The reverse inequality (with a multiplicative constant) holds only for the absolutely summing operators.

(5.5.9) Proposition. *Let $T\colon E \to F$ be an operator. Then T is absolutely summing if and only if there is a constant C such that, for every random variable X in E, we have*

$$\mathbf{E}\left[\|TX\|\right] \le C\|X\|_P.$$

The smallest such constant C is $\pi_1(T)$.

Proof. Suppose T is absolutely summing. Let K and μ be as in the Pietsch factorization theorem (5.5.5). Then if $X\colon \Omega \to E$ is a random variable, we have by Fubini's theorem

$$\mathbf{E}\left[\|TX\|\right] \le \pi_1(T)\,\mathbf{E}\left[\int_K |x^*(X)|\, d\mu(x^*)\right]$$

$$= \pi_1(T) \int_K \mathbf{E}\left[|x^*(X)|\right]\, d\mu(x^*)$$

$$\le \pi_1(T) \sup_{x^* \in K} \mathbf{E}\left[|x^*(X)|\right]$$

$$= \pi_1(T)\|X\|_P.$$

Conversely, suppose there is a constant C so that

$$\|TX\|_1 \le C\,\|X\|_P.$$

Let $x_1, x_2, \cdots, x_n \in E$. Let A_i, $i = 1, 2, \cdots, n$ be disjoint sets in $[0,1]$ with measure $1/n$. Define a random variable $X\colon [0,1] \to E$ by

$$X = \sum_{i=1}^{n} n x_i\, \mathbf{1}_{A_i}.$$

Thus

$$\sum_{i=1}^{n} \|Tx_i\| = \sum_{i=1}^{n} \|T(nx_i)\| \, \mathbf{P}(A_i)$$

$$= \mathbf{E}\left[\|TX\|\right]$$

$$\leq C\,\|X\|_P$$

$$= C \sup_{\|x^*\|\leq 1} \mathbf{E}\left[|x^*(X)|\right]$$

$$= C \sup_{\|x^*\|\leq 1} \sum_{i=1}^{n} n\,|x^*(x_i)|\,\mathbf{P}(A_i)$$

$$= C \sup_{\|x^*\|\leq 1} \sum_{i=1}^{n} |x^*(x_i)|. \qquad \blacksquare$$

Absolutely summing operators also characterize certain amart properties.

(5.5.10) Theorem. Let $T\colon E \to F$ be a bounded linear operator. Then the following are equivalent:

(1) T is absolutely summing;
(2) if (X_n) is an amart in E, then (TX_n) is a uniform amart in F;
(3) if (X_n) is an L_1-bounded amart potential in E, then (TX_n) converges a.s. in F (to 0);
(4) if (X_n) is an L_1-bounded amart potential in E, then (TX_n) converges weakly a.s. in F;
(5) if (X_n) is an L_1-bounded amart potential in E, then $\|TX_n\|$ is bounded on a set of positive measure.

Proof. (1) $\Longrightarrow$ (2). Apply the difference properties characterizing amarts (5.2.6) and uniform amarts (5.2.3). Assume that T is absolutely summing. Then (2) follows from Proposition (5.5.9).

(2) $\Longrightarrow$ (3). This is a consequence of the Riesz decomposition theorem (5.2.13).

(3) $\Longrightarrow$ (4) and (4) $\Longrightarrow$ (5) are trivial.

(5) $\Longrightarrow$ (1). Suppose T is not absolutely summing. We will show that there is an L_1-bounded amart (X_n) in E such that (TX_n) diverges weakly a.s. There is an unconditionally convergent series $\sum x_n$ in E such that $\sum Tx_n$ is not absolutely convergent in F. By the convergence of $\sum x_n$, we have $\|x_n\| \to 0$, so $\|Tx_n\| \to 0$. Now $\sum \|Tx_n\| = \infty$. Let β_n be such that $\beta_n \to 0$ but $\sum \|Tx_n\| \beta_n = \infty$. We may assume that $\|Tx_n\| \beta_n < 1$ for all n.

Let $(\Omega, \mathcal{F}, \mathbf{P})$ be $[0,1)$ with Lebesgue measure. Let C_n be independent events with $\mathbf{P}(C_n) = \|Tx_n\| \beta_n$. Define

$$X_n = \frac{x_n}{\mathbf{P}(C_n)}\,\mathbf{1}_{C_n}.$$

The process (X_n) is L_1-bounded, since

$$\mathbf{E}\left[\|X_n\|\right] = \frac{\|x_n\|}{\mathbf{P}(C_n)}\mathbf{P}(C_n) = \|x_n\| \to 0.$$

Let $\mathcal{F}_n$ be the σ-algebra generated by $\{X_1, X_2, \cdots, X_n\}$. The process (X_n) is adapted to the stochastic basis $(\mathcal{F}_n)$. In fact, (X_n) is an amart potential. Indeed, given $\varepsilon > 0$, there is N such that

$$\left\| \sum_{n=N}^{\infty} \alpha_n x_n \right\| < \varepsilon$$

for any choice of scalars α_n with $|\alpha_n| \leq 1$. If $\sigma \in \Sigma$ and $\sigma \geq N$, write $D_n = C_n \cap \{\sigma = n\}$, so that

$$\mathbf{E}\left[X_\sigma\right] = \sum_{n=N}^{\infty} \frac{x_n}{\mathbf{P}(C_n)}\mathbf{P}(D_n).$$

But $\mathbf{P}(D_n)/\mathbf{P}(C_n) \leq 1$, so $\|\mathbf{E}\left[X_\sigma\right]\| < \varepsilon$. Thus (X_n) is an amart potential.

Next, we must show that (TX_n) is not bounded on any set of positive measure. Since $\sum \mathbf{P}(C_n) = \infty$, the Borel-Cantelli lemma tells us that for almost all $\omega \in \Omega$, we have $\omega \in C_n$ for infinitely many n. For those n,

$$\|TX_n(\omega)\| = \frac{\|Tx_n\|}{\mathbf{P}(C_n)} = \frac{1}{\beta_n}.$$

But $\beta_n \to 0$, so the sequence $TX_n(\omega)$ is unbounded.   ∎

As a corollary we obtain some interesting characterizations of finite-dimensional spaces.

(5.5.11) Theorem.

(1) A Banach space is finite-dimensional if and only if every amart in E is a uniform amart.

(2) A Banach space E is finite-dimensional if and only if every L_1-bounded amart in E converges strongly a.s.

(3) A Banach space E is finite-dimensional if and only if every L_1-bounded amart in E converges weakly a.s.

(4) A Banach space E is finite-dimensional if and only if every L_1-bounded amart potential in E is bounded on a set of positive measure.

Proof. In finite dimensional spaces, the convergence properties follow from the one-dimensional cases: the projection on a one-dimensional space is continuous, and a sequence in $\mathbb{R}^n$ converges if and only if each of its coordinates converges.

The converse directions follow from the previous theorem, together with Theorem (5.5.8).   ∎

Radon-Nikodým operators

The next operator ideal to be considered is the ideal of Radon-Nikodým operators.

(5.5.12) Definition. Let E and F be Banach spaces, and let $T: E \to F$ be a bounded linear operator. We say that T is a *Radon-Nikodým operator* (or T has the *Radon-Nikodým property*) iff, for every probability space $(\Omega, \mathcal{F}, \mathbf{P})$ and every vector measure $\mu: \mathcal{F} \to E$ that is absolutely continuous with respect to $\mathbf{P}$ and has finite variation, the measure $T\mu$ has a Radon-Nikodým derivative in $L_1(\Omega, \mathcal{F}, \mathbf{P}; F)$.

This is motivated by the definition of the Radon-Nikodým property for a Banach space E. In fact, the Banach space E has the Radon-Nikodým property if and only if the identity operator on E is a Radon-Nikodým operator. Most of the elementary part of the theory of Banach spaces with the Radon-Nikodým property can be reproduced in terms of Radon-Nikodým operators, simply by inserting the operator into the proofs in appropriate places. For example, as in (5.3.3), it is sufficient to use vector measures μ with bounded average range in E. Or: an operator is a Radon-Nikodým operator if and only if its restriction to each separable subspace is a Radon-Nikodým operator.

The next result states that the collection of Radon-Nikodým operators is a Banach operator ideal. The "Radon-Nikodým norm" of T is simply the operator norm $\|T\|$.

(5.5.13) Proposition.

(1) *For any pair E, F of Banach spaces, the set of all Radon-Nikodým operators $T: E \to F$ is a Banach space under the operator norm $\|T\|$.*

(2) *If $T: E \to F$ is a Radon-Nikodým operator, $Q: E_1 \to E$ and $R: F \to F_1$ are bounded operators, then the composition $RTQ: E_1 \to F_1$ is also a Radon-Nikodým operator.*

Proof. (1) To see that the set of all Radon-Nikodým operators from E to F is a vector space, note that

$$\frac{d(\alpha T\mu)}{d\mathbf{P}} = \alpha \, \frac{dT\mu}{d\mathbf{P}}$$
$$\frac{d(T_1 + T_2)\mu}{d\mathbf{P}} = \frac{dT_1\mu}{d\mathbf{P}} + \frac{dT_2\mu}{d\mathbf{P}}.$$

Next consider completeness. Suppose operators $T_n: E \to F$ are Radon-Nikodým operators, and T_n converges to T in the operator norm. By taking a subsequence, we may assume

$$\sum_{n=1}^{\infty} \|T_{n+1} - T_n\| < \infty.$$

Now each difference $T_{n+1} - T_n$ is a Radon-Nikodým operator, so there exist random variables $X_n: \Omega \to F$ so that $X_n = d(T_{n+1} - T_n)\mu/d\mathbf{P}$. (Let $T_0 = 0$.) Now the variation satisfies

$$|(T_{n+1} - T_n)\mu|(\Omega) \leq \|T_{n+1} - T_n\| \, |\mu|(\Omega),$$

and $\mathbf{E}\left[\|X_n\|\right] \leq \|T_{n+1} - T_n\| \, |\mu|(\Omega)$, so the series $\sum X_n$ converges in $L_1(\Omega; F)$, say to X. For $A \in \mathcal{F}$,

$$\mathbf{E}\left[X\,\mathbf{1}_A\right] = \lim_{n\to\infty} \mathbf{E}\left[\left(\sum_{i=1}^{n} X_i\right)\mathbf{1}_A\right] = \lim_{n\to\infty} T_n\big(\mu(A)\big) = T\big(\mu(A)\big).$$

Thus $X = dT\mu/d\mathbf{P}$, so T is a Radon-Nikodým operator.

(2) Let $(\Omega, \mathcal{F}, \mathbf{P})$ be a probability space, and let $\mu\colon \mathcal{F} \to E_1$ be a vector measure, absolutely continuous with respect to $\mathbf{P}$, with finite variation. Then $Q\mu\colon \mathcal{F} \to E$ is a vector measure, absolutely continuous with respect to $\mathbf{P}$, with variation at most $\|Q\| \, |\mu|(\Omega) < \infty$. Thus the Radon-Nikodým derivative $X = dTQ\mu/d\mathbf{P}$ exists. But then the Radon-Nikodým derivative $RX = dRTQ\mu/d\mathbf{P}$ exists. Thus, the composition RTQ is a Radon-Nikodým operator. ∎

Arguments almost identical to those used above (5.3.29 and 5.4.12) prove the next result.

(5.5.14) Proposition. *Let E and F be Banach spaces, and let $T\colon E \to F$ be a bounded linear operator. Then the following are equivalent:*

(1) *T has the Radon-Nikodým property;*

(2) *every closed bounded nonempty set $C \subseteq E$ has slices S with image $T(S)$ of arbitrarily small diameter in F;*

(3) *for every L_1-bounded martingale $(X_n)_{n\in\mathbb{N}}$ in E, the image (TX_n) converges a.s. in F;*

(4) *for every amart $(X_n)_{n\in\mathbb{N}}$ in E satisfying condition (B), the image (TX_n) converges scalarly in F.*

The Radon-Nikodým operators are related to the Riesz representable operators. Before we make this more precise, let us discuss the representable operators.

(5.5.15) Definition. Let $(\Omega, \mathcal{F}, \mathbf{P})$ be a probability space, and let E be a Banach space. Then an operator $T\colon L_1(\Omega, \mathcal{F}, \mathbf{P}) \to E$ is said to be *representable* iff there is a random variable $X\colon \Omega \to E$ such that

$$T(Z) = \mathbf{E}\left[Z\,X\right] \quad \text{for all } Z \in L_1(\Omega, \mathcal{F}, \mathbf{P}).$$

(Necessarily $X \in L_\infty(\Omega, \mathcal{F}, \mathbf{P}; E)$.)

(5.5.16) Proposition. *Let E be a Banach space. Then E has the Radon-Nikodým property if and only if every operator $T\colon L_1 \to E$ is representable.*

Proof. There is a one-to-one correspondence between the set of all operators $T\colon L_1(\Omega, \mathcal{F}, \mathbf{P}) \to E$ and the set of all absolutely continuous vector measures $\mu\colon \mathcal{F} \to E$ with bounded average range. The measure μ corresponding to the operator T is defined by $\mu(A) = T(\mathbf{1}_A)$. The operator T is representable by a random variable X if and only if the measure μ has Radon-Nikodým derivative X. ∎

Analogous (roughly speaking) to the Pietsch factorization is the Lewis-Stegall factorization, which is proved next.

(5.5.17) Theorem. *Let $(\Omega, \mathcal{F}, \mathbf{P})$ be a probability space, and let E be a Banach space. Then the operator $T\colon L_1(\Omega, \mathcal{F}, \mathbf{P}) \to E$ is representable if and only if it factors through the space l_1, that is, there exist operators $S\colon L_1 \to l_1$ and $R\colon l_1 \to E$ such that $T = RS$.*

$$L_1 \xrightarrow{\ \ S\ \ } l_1$$
$$\downarrow R$$
$$E$$

The range of a representable operator $T\colon L_1(\Omega, \mathcal{F}, \mathbf{P}) \to E$ is separable.

Proof. Suppose first that T has a factorization $T = RS$. Then S is representable, since l_1 has the Radon-Nikodým property (5.1.16). Say $S(Z) = \mathbf{E}\,[Z\,X]$ for all $Z \in L_1$. But R is continuous and linear, so we have $RS(Z) = \mathbf{E}\,[Z \cdot R(X)]$, so that RS is also representable.

Conversely, suppose that T is representable. Then there exists $X \in L_\infty(\Omega, \mathcal{F}, \mathbf{P}; E)$ so that $T(Z) = \mathbf{E}\,[Z\,X]$ for all $Z \in L_1$. Now X has separable range, so T also has separable range. We may assume that E is a separable space. Let $\varepsilon > 0$. For each positive integer n, the space E can be covered by countably many balls of radius $\varepsilon 2^{-n-1}$, so there is a random variable $Y_n\colon \Omega \to E$ with countably many values such that $\|X - Y_n\|_\infty < \varepsilon 2^{-n-1}$. Let $X_1 = Y_1$ and $X_n = Y_n - Y_{n-1}$ for $n \geq 2$, so that X_n has countably many values and $\|X - \sum_{m=1}^{n} X_m\| < \varepsilon 2^{-n-1}$. Then X_n has the form

$$X_n = \sum_{k=1}^{\infty} x_{nk}\, \mathbf{1}_{E_{nk}},$$

where $x_{nk} \in E$ and $E_{nk} \in \mathcal{F}$ with $E_{nk} \cap E_{nk'} = \emptyset$ (for $k \neq k'$) and $\|x_{nk}\| < 3\varepsilon 2^{-n+1}$ if $n \geq 2$. Define $S\colon L_1(\Omega, \mathcal{F}, \mathbf{P}) \to l_1(\mathbb{N} \times \mathbb{N})$ by

$$S(Z)(n, k) = \|x_{nk}\|\, \mathbf{E}\,[Z\,\mathbf{1}_{E_{nk}}]$$

for $Z \in L_1$. Then we have

$$\|S(Z)\|_{l_1} = \sum_{n=1}^{\infty} \sum_{k=1}^{\infty} \|x_{nk}\| \left| \mathbf{E} \left[Z \, \mathbf{1}_{E_{nk}} \right] \right|$$

$$\leq \sum_{k=1}^{\infty} \|x_{1k}\| \, \mathbf{E} \left[|Z| \, \mathbf{1}_{E_{1k}} \right] + \sum_{n=2}^{\infty} \sum_{k=1}^{\infty} 3\varepsilon 2^{-n+1} \mathbf{E} \left[|Z| \, \mathbf{1}_{E_{nk}} \right].$$

But $\|X - X_1\| < \varepsilon/4$ and $\|X\|_{\infty} = \|T\|$, so $\|x_{1k}\| < \|T\| + \varepsilon/4$. Then

$$\|S(Z)\|_{l_1} \leq \left(\|T\| + \frac{\varepsilon}{4} \right) \mathbf{E} \left[|Z| \right] + \frac{3\varepsilon}{4} \mathbf{E} \left[|Z| \right]$$

$$= \left(\|T\| + \varepsilon \right) \mathbf{E} \left[|Z| \right].$$

Thus $\|S\| \leq \|T\| + \varepsilon$. Then define $R\colon l_1(\mathbb{N} \times \mathbb{N}) \to E$ by

$$R(h) = \sum_{n=1}^{\infty} \sum_{k=1}^{\infty} h(n,k) \frac{x_{nk}}{\|x_{nk}\|},$$

with the convention $0/0 = 0$. Then $\|R(h)\| \leq \sum \sum |h(n,k)| = \|h\|$, so $\|R\| \leq 1$. Also

$$RS(Z) = \sum_{n=1}^{\infty} \sum_{k=1}^{\infty} \|x_{nk}\| \, \mathbf{E} \left[Z \, \mathbf{1}_{E_{nk}} \right] \frac{x_{nk}}{\|x_{nk}\|}$$

$$= \sum_{n=1}^{\infty} \mathbf{E} \left[\left(\sum_{k=1}^{\infty} x_{nk} \, \mathbf{1}_{E_{nk}} \right) Z \right]$$

$$= \sum_{n=1}^{\infty} \mathbf{E} \left[X_n \, Z \right]$$

$$= \mathbf{E} \left[X \, Z \right]$$

$$= T(Z).$$

Thus T has been factored, and $\|R\| \, \|S\| \leq \|T\| + \varepsilon$.

If T is a representable operator, then it factors through l_1. Since l_1 is separable, the range of T must be separable as well. ∎

The connections between the representable operators and the Radon-Nikodým operators are illustrated by the next two results.

(5.5.18) Proposition. *Let* $(\Omega, \mathcal{F}, \mathbf{P})$ *be a probability space, and let* E *be a Banach space. Then the operator* $T\colon L_1(\Omega, \mathcal{F}, \mathbf{P}) \to E$ *is a Radon-Nikodým operator if and only if it is representable.*

Proof. Suppose that T is a Radon-Nikodým operator. Define $\mu\colon \mathcal{F} \to L_1$ by $\mu(A) = \mathbf{1}_A$. Then μ is absolutely continuous with respect to $\mathbf{P}$, and

$|\mu|(A) = \mathbf{P}(A)$ for all $A \in \mathcal{F}$, so that μ has average range bounded by 1. Since T is a Radon-Nikodým operator, there exists a Radon-Nikodým derivative $X = dT\mu/d\mathbf{P}$. Now if $Z = \mathbf{1}_A$, then we have

$$\mathbf{E}\,[Z\,X] = \mathbf{E}\,[X\,\mathbf{1}_A] = T\mu(A) = T(\mathbf{1}_A) = T(Z).$$

The equation $\mathbf{E}\,[Z\,X] = T(Z)$ holds for simple functions Z, and each side is a continuous linear function of $Z \in L_1$, so the equation is true for all $Z \in L_1$. Therefore T is representable.

Conversely, suppose T is representable. Then T factors through l_1; say $T = RS$, where $S\colon L_1 \to l_1$ and $R\colon l_1 \to E$. If I is the identity operator on l_1, then I is a Radon-Nikodým operator, since the space l_1 has the Radon-Nikodým property (5.1.16).

$$
\begin{array}{ccc}
L_1 & \xrightarrow{\;\;T\;\;} & E \\[2pt]
{\scriptstyle S}\downarrow & & \uparrow{\scriptstyle R} \\[2pt]
l_1 & \xrightarrow[\;\;I\;\;]{} & l_1
\end{array}
$$

Therefore, by the ideal property (5.3.13(2)), $T = RIS$ is a Radon-Nikodým operator. ∎

(5.5.19) Theorem. *Suppose E and F are Banach spaces, and suppose $T\colon E \to F$ is a bounded linear operator. Then T is a Radon-Nikodým operator if and only if, for every probability space $(\Omega, \mathcal{F}, \mathbf{P})$ and every operator $S\colon L_1(\Omega, \mathcal{F}, \mathbf{P}) \to E$, the composition TS is representable.*

Proof. If T is a Radon-Nikodým operator, then so is TS, and therefore TS is representable.

Conversely, suppose TS is representable for any S. Then if $\mu\colon \mathcal{F} \to E$ is a vector measure with bounded average range, there is a unique operator

$$S\colon L_1(\Omega, \mathcal{F}, \mathbf{P}) \to E$$

such that $S(\mathbf{1}_A) = \mu(A)$ for all $A \in \mathcal{F}$. But TS is representable, say by $X\colon \Omega \to F$, and then $X = dT\mu/d\mathbf{P}$. ∎

In light of the Lewis-Stegall factorization, this result can be restated like this: $T\colon E \to F$ is a Radon-Nikodým operator if and only if, for every operator $S\colon L_1(\Omega, \mathcal{F}, \mathbf{P}) \to E$, there is a factorization of TS through l_1:

$$
\begin{array}{ccc}
L_1 & \xrightarrow{\quad\quad} & l_1 \\[2pt]
{\scriptstyle S}\downarrow & & \downarrow \\[2pt]
E & \xrightarrow[\;\;T\;\;]{} & F
\end{array}
$$

It is not hard to see that there is a relation between the ideal of Radon-Nikodým operators and the ideals of compact and weakly compact operators: An operator $T\colon E \to F$ is *compact* iff the closure of the image of the unit ball of E

$$\overline{T(B_E)}$$

is a compact set. Equivalently, for every bounded sequence (x_n) in E, there is a subsequence (x_{n_k}) such that $T(x_{n_k})$ converges in F. The operator $T\colon E \to F$ is *weakly compact* iff the closure of the image of the unit ball of E is a weakly compact set. Equivalently, for every bounded sequence (x_n) in E, there is a subsequence (x_{n_k}) such that $T(x_{n_k})$ converges weakly in F.

(5.5.20) Proposition. *Every weakly compact operator is a Radon-Nikodým operator.*

Proof. Let $T\colon E \to F$ be weakly compact. Let $(X_n)_{n \in \mathbb{N}}$ be a martingale in the unit ball of E. Then the martingale (TX_n) has values in a weakly compact convex set, namely the closure of the image of that unit ball under T. Now weakly compact convex sets are Radon-Nikodým sets (5.3.33). So (TX_n) converges. This shows that T is a Radon-Nikodým operator. ∎

(5.5.21) Corollary. *Every compact operator is a Radon-Nikodým operator. A weakly compact operator $T\colon L_1 \to X$ has separable range and is representable.*

Proof. Any compact operator is weakly compact.

Suppose $T\colon L_1(\Omega, \mathcal{F}, \mathbf{P}) \to E$ is weakly compact. Then T is a Radon-Nikodým operator, so it is representable. Thus there is a Lewis-Stegall factorization $T = RS$, where $S\colon L_1 \to l_1$ and $R\colon l_1 \to E$. But l_1 is separable, and the range of T is a subset of the range of R, so it is separable. ∎

Asplund operators

The next operator ideal to be considered is the ideal of *Asplund operators*. The main result of concern here is the connection between scalar convergence and weak almost sure convergence.

(5.5.22) Definition. Let E and F be Banach spaces, and let $T\colon E \to F$ be an operator. Then T is an *Asplund operator* iff T^* is a Radon-Nikodým operator.

When defined in this way the ideal properties follow easily from those of the Radon-Nikodým operators.

(5.5.23) Proposition.

(1) For any pair E, F of Banach spaces, the set of all Asplund operators $T\colon E \to F$ is a Banach space under the operator norm $\|T\|$.

(2) If $T\colon E \to F$ is an Asplund operator, $Q\colon E_1 \to E$ and $R\colon F \to F_1$ are bounded operators, then the composition $RTQ\colon E_1 \to F_1$ is also an Asplund operator.

As with the ideals considered above, this one has close connections with factorization conditions. In order to properly state these conditions, we first consider the *Haar operator*.

The *Cantor set* is the countable product topological space

$$\Delta = \{0,1\}^{\mathbb{N}}.$$

It is compact and metrizable. It is made up of made up of parts

$$\{\, \Delta_{ni} : n \in \mathbb{N}, 0 \le i < 2^n \,\}$$

so that $\Delta_{n+1,2i} \cup \Delta_{n+1,2i+1} = \Delta_{ni}$ and $\Delta_{n+1,2i} \cap \Delta_{n+1,2i+1} = \emptyset$. Each set Δ_{ni} is open and closed. To do this, let Δ_{ni} be the set of all sequences in Δ such that the first n terms are the digits of the base two expansion of i:

$$\Delta_{10} = \{\, (x_i) \in \Delta : x_1 = 0 \,\}$$
$$\Delta_{11} = \{\, (x_i) \in \Delta : x_1 = 1 \,\}$$
$$\Delta_{20} = \{\, (x_i) \in \Delta : x_1 = 0, x_2 = 0 \,\}$$
$$\Delta_{21} = \{\, (x_i) \in \Delta : x_1 = 0, x_2 = 1 \,\}$$
$$\Delta_{22} = \{\, (x_i) \in \Delta : x_1 = 1, x_2 = 0 \,\}$$
$$\Delta_{23} = \{\, (x_i) \in \Delta : x_1 = 1, x_2 = 1 \,\}$$

and so on.

The Cantor set supports a probability measure μ with $\mu(\Delta_{ni}) = 2^{-n}$; this is the product measure where the factors assign measure $1/2$ to each point of $\{0,1\}$.

(5.5.24) Definition. The *Haar functions* $h_{ni} \colon \Delta \to \mathbb{R}$ are defined by

$$h_{ni} = 1_{\Delta_{n+1,2i}} - 1_{\Delta_{n+1,2i+1}}.$$

If $\{\, e_{ni} : n \in \mathbb{N}, 0 \le i < 2^n \,\}$ is an enumeration of the unit vector basis of l_1, then the *Haar operator* $H \colon l_1 \to L_\infty(\Delta,\mu)$ is defined by

$$H(e_{ni}) = h_{ni}.$$

Here is Stegall's theorem characterizing Asplund operators. For the proof, see Stegall [1981], p. 515, or Bourgin [1983], Chapter 5. A function $\varphi \colon E \to \mathbb{R}$ is *Fréchet differentiable* at the point $x \in E$ iff there is a linear functional $x^* \in E^*$ such that

$$\lim_{\|y\| \to 0} \frac{|\varphi(x+y) - \varphi(x) - x^*(y)|}{\|y\|} = 0.$$

(5.5.25) Theorem. *Let $U: E \to F$ be an operator. Then the following are equivalent.*

(1) U *is an Asplund operator.*

(2) U^* *factors through a space with the Radon-Nikodým property.*

(3) U^* *maps closed bounded convex subsets of F^* to Radon-Nikodým subsets of E^*.*

(4) *If $\varphi: F \to \mathbb{R}$ is continuous and convex, then φU is Fréchet differentiable on a dense subset of E.*

(5) U *is not a factor of the Haar operator H:*

(6) U *factors through a space, every separable subspace of which has separable dual.*

A Banach space where the identity operator is an Asplund operator is called an *Asplund space*. Then E is an Asplund space if and only if E^* has the Radon-Nikodým property.

The interest here of Asplund operators is the following result.

(5.5.26) Theorem. *Let E and F be Banach spaces, and let $U: E \to F$ be an operator. The following are equivalent.*

(a) U *is an Asplund operator.*

(b) *If X_n is a sequence of E-valued Bochner measurable functions, converging scalarly to 0, with $\sup_n \|X_n\| < \infty$ a.s., then the sequence (UX_n) almost surely converges weakly to 0.*

(c) *If $(X_n)_{n \in \mathbb{N}}$ is a weak sequential amart potential in E satisfying condition (B), then (UX_n) almost surely converges weakly to 0.*

(d) *If $(X_n)_{n \in \mathbb{N}}$ is an amart potential in E satisfying condition (B), then (UX_n) almost surely converges weakly to 0.*

Proof. Three of the four parts of the proof are easy.

(a) $\Longrightarrow$ (b). Suppose U is an Asplund operator. Then by Theorem (5.5.25(6)), U can be factored, $U = WV$, where $V: E \to E_0$, $W: E_0 \to F$, and E_0 is a Banach space, every separable subspace of which has separable dual. Let $X_n: \Omega \to E$ be Bochner measurable, $\sup_n \|X_n\| < \infty$ a.s., and suppose X_n converges scalarly to 0. Then by (5.1.7) there is a separable subspace E_1 of E_0 such that $\mathbf{P}\{VX_n \in E_1\} = 1$ for all n. Now E_1^* is separable; choose a countable dense set $\{x_k^*\}_{k=1}^{\infty}$. For each k there is a null set $\Omega_k \subseteq \Omega$ such that $\langle VX_n, x_k^* \rangle \to 0$ on $\Omega \setminus \Omega_k$. Now since $\{x_k^*\}$ is dense and $\{\|VX_n(\omega)\|\}$ is bounded, we have $\langle VX_n, x^* \rangle \to 0$ for all $x^* \in E_1^*$ and all $\omega \in \Omega \setminus \bigcup \Omega_k$. But then $UX_n = WVX_n$ converges weakly to 0 a.s.

(b) $\implies$ (c). Let $(X_n)_{n\in\mathbb{N}}$ be a weak sequential potential in E satisfying condition (B). If $x^* \in E^*$, then $\langle X_n, x^* \rangle$ is a scalar amart potential, and thus $\langle X_n, x^* \rangle \to 0$ a.e. That is, (X_n) converges scalarly to 0. By the maximal inequality (5.2.19), $\sup_n \|X_n\| < \infty$ a.s. Therefore, $UX_n \to 0$ weakly a.s.

(c) $\implies$ (d). Any amart potential is a weak sequential potential.

(d) $\implies$ (a). Suppose U is not an Asplund operator. By Theorem (5.5.25(5)), U is a factor of the Haar operator H:

$$
\begin{array}{ccc}
E & \xrightarrow{\ U\ } & F \\[4pt]
{\scriptstyle S}\big\uparrow & & \big\downarrow{\scriptstyle T} \\[4pt]
l_1 & \xrightarrow[\ H\ ]{} & L_\infty(\Delta,\mu)
\end{array}
$$

Now we claim that it will suffice to exhibit an amart potential $(X_n)_{n\in\mathbb{N}}$ in l_1 satisfying condition (B) such that HX_n does not converge weakly a.s. to 0 in L_∞. Indeed, then (SX_n) will be a strong amart potential in E satisfying condition (B), but (USX_n) will not converge weakly to 0 a.s. in F.

For this example, we let $\Omega = \Delta$ be the Cantor set, and $\mathbf{P} = \mu$ the natural measure on Δ. We will also use the other notation in Definition (5.5.24). Define $X_n \colon \Delta \to l_1$ by

$$
X_n(\omega) = \frac{1}{n} \sum_{m=1}^{n} \sum_{i=0}^{2^m-1} h_{mi}(\omega) e_{mi}.
$$

We will prove that (X_n) has the required properties.

First, observe that (X_n) is L_∞-bounded, and hence satisfies condition (B), since for any $\omega \in \Omega$,

$$
\left\| X_n(\omega) \right\|_{l_1} = \sum_{m=1}^{n} \sum_{i=0}^{2^m-1} \left| \frac{1}{n} h_{mi}(\omega) \right| = \sum_{m=1}^{n} \frac{1}{n} = 1.
$$

We claim next that HX_n does not converge weakly a.s. to 0 in $L_\infty(\Delta)$. Since the range of H is contained in the subspace $C(\Delta)$, it suffices to show that HX_n does not converge weakly to 0 in $C(\Delta)$. For each $\omega \in \Delta$, we have

$$
HX_n(\omega)(\omega) = \frac{1}{n} \sum_{m=1}^{n} \sum_{i=0}^{2^m-1} h_{mi}(\omega)\, h_{mi}(\omega) = \frac{1}{n} \sum_{m=1}^{n} 1 = 1,
$$

so that $HX_n(\omega)$ does not converge weakly to 0.

Now we claim that if $2 < p < \infty$, there is a constant C_p such that for any scalars α_{ni} with $|\alpha_{ni}| \leq 1$,

$$(5.5.26a) \qquad \mathbf{E}\left[\left|\sum_{m=1}^{n}\sum_{i=0}^{2^m-1}\alpha_{mi}\,h_{mi}\right|^p\right] \leq C_p\,n^{p/2}.$$

This is a classical inequality. It is implicit from Paley [1932]. A modern proof might observe that the Haar functions, taken in the proper order, constitute a martingale difference sequence. The square function S associated with (5.5.26a) satisfies

$$S^2 = \sum_{m=1}^{n}\sum_{i=0}^{2^m-1}\alpha_{mi}^2\,h_{mi}^2 \leq \sum_{m=1}^{n} 1 = n,$$

so $S \leq n^{1/2}$. Since the L_p norm of a martingale and its square function differ at most by a constant factor (6.3.6), we obtain (5.5.26a).

The next claim is that

$$\sup\left\{\,\mathbf{E}\left[|x^*X_\sigma|\right] : x^* \in l_1^*, \|x^*\| \leq 1\,\right\} \to 0 \quad \text{as } \sigma \text{ increases in } \Sigma.$$

Choose p with $2 < p < \infty$, and let $0 < \alpha < (p-2)/(2p-2)$, so that $p/2 - \alpha(p-1) > 1$; for example, take $p = 7, \alpha = 1/4$. Let $\varepsilon_n = n^{-\alpha}$. Given $x^* \in l_1^*$ with $\|x^*\| \leq 1$, write $\alpha_{ni} = x^*(e_{ni})$, so that $|\alpha_{ni}| \leq 1$. By (5.5.26a),

$$\left\|\frac{1}{n}\sum_{m=1}^{n}\sum_{i=0}^{2^m-1}\alpha_{mi}\,h_{mi}\right\|_p^p \leq \frac{C_p}{n^{p/2}},$$

and hence for $\lambda > 0$,

$$\mathbf{P}\left\{\left|\frac{1}{n}\sum_{m=1}^{n}\sum_{i=0}^{2^m-1}\alpha_{mi}\,h_{mi}\right| > \lambda\right\} \leq \frac{C_p}{n^{p/2}\lambda^p}.$$

Let $\sigma \in \Sigma, \sigma \geq N$. Then

$$\mathbf{E}\left[|x^*X_\sigma|\right] = \sum_{n=N}^{\infty}\mathbf{E}\left[\left|\frac{1}{n}\sum_{m=1}^{n}\sum_{i=0}^{2^m-1}\alpha_{mi}\,h_{mi}\right|\mathbf{1}_{\{\sigma=n\}}\right]$$

$$= \sum_{n=N}^{\infty}\int_0^\infty \mathbf{P}\left\{\left|\frac{1}{n}\sum_{m=1}^{n}\sum_{i=0}^{2^m-1}\alpha_{mi}\,h_{mi}\right|\mathbf{1}_{\{\sigma=n\}} > \lambda\right\}d\lambda$$

$$= \sum_{n=N}^{\infty}\left(\int_0^{\varepsilon_n}\mathbf{P}\{\cdots\}d\lambda + \int_{\varepsilon_n}^{\infty}\mathbf{P}\{\cdots\}d\lambda\right)$$

$$\leq \sum_{n=N}^{\infty} \left(\varepsilon_n \mathbf{P}\{\sigma = n\} + \frac{C_p}{n^{p/2}} \int_{\varepsilon_n}^{\infty} \frac{1}{\lambda^p} \, d\lambda \right)$$

$$= \sum_{n=N}^{\infty} n^{-\alpha} \mathbf{P}\{\sigma = n\} + \sum_{n=N}^{\infty} \frac{C_p}{n^{p/2} \varepsilon_n^{p-1}(p-1)}$$

$$\leq \sum_{n=N}^{\infty} N^{-\alpha} \mathbf{P}\{\sigma = n\} + \frac{C_p}{p-1} \sum_{n=N}^{\infty} \frac{1}{n^{(p/2)-\alpha(p-1)}}$$

$$= N^{-\alpha} + \frac{C_p}{p-1} \sum_{n=N}^{\infty} \frac{1}{n^{(p/2)-\alpha(p-1)}}.$$

This tends to 0 as $N \to \infty$, since $\alpha > 0$ and $p/2 - \alpha(p-1) > 1$.

Now it is easy to verify that (X_n) is an amart:

$$\left\| \mathbf{E}\left[X_\sigma\right] \right\| = \sup_{\|x^*\| \leq 1} \left| \langle \mathbf{E}\left[X_\sigma\right], x^* \rangle \right|$$

$$\leq \sup_{\|x^*\| \leq 1} \mathbf{E}\left[|\langle X_\sigma, x^* \rangle| \right],$$

which tends to 0.

Finally, (X_n) is a potential, since the Pettis norm

$$\sup_{\|x^*\| \leq 1} \mathbf{E}\left[|\langle X_n, x^* \rangle| \right]$$

tends to 0 as $n \to \infty$. ∎

As a consequence, we obtain a convergence result not involving amarts.

(5.5.27) Theorem. *Let E be a separable Banach space. Then E^* is separable if and only if for any sequence of random variables (X_n) in E with $\sup_n \|X_n\| < \infty$ a.s., scalar convergence of X_n to 0 implies weak a.s. convergence of X_n to 0.*

Proof. By Theorem (5.5.26), the convergence condition is satisfied if and only if the identity operator on E is an Asplund operator. By (5.5.25), that happens if and only if every separable subspace of E has separable dual. If E itself is separable, this means E has separable dual. ∎

The preceding material on Asplund operators will enable us to prove the following characterization of weak amart convergence.

(5.5.28) Theorem. *Let E be a Banach space. The following are equivalent.*

(a) *E is reflexive.*

(b) *If (X_n) is an L_∞-bounded weak amart in E, then (X_n) converges weakly a.s.*

(c) *If (X_n) is a weak amart of class (B) in E, then (X_n) converges weakly a.s.*

Proof. (a) $\Longrightarrow$ (b). Suppose E is reflexive. Let (X_n) be a weak amart with values in the unit ball of E. In the proof that X_n converges weakly a.s., we may assume that E is separable. For each $x^* \in E^*$, the sequence $\langle X_n, x^* \rangle$ converges a.s. For each $\omega \in \Omega$, the sequence $X_n(\omega)$ has a subsequence that converges weakly. The method used in the proof of (5.3.31) shows that X_n converges weakly a.s. to a Bochner measurable limit.

(b) $\Longrightarrow$ (c). A process of class (B) satisfies the usual maximal inequality. Thus we may apply the stopping time argument as in (5.3.9) to reduce to the bounded case.

(c) $\Longrightarrow$ (a). Suppose E is not reflexive. We will consider two cases. First, suppose E is not an Asplund space; that is, the identity operator on E is not an Asplund operator. Then (5.5.26) there is a (strong) amart (X_n) in E such that (X_n) does not converge weakly a.s.

Next suppose E is an Asplund space. Then by Stegall's theorem (5.5.25), every separable subspace of E has separable dual. Since E is not reflexive, there is a sequence x_n in E with $\|x_n\| \le 1$ such that no subsequence of x_n converges weakly. The vectors x_n are contained in a separable subspace E_1 of E. Since E_1^* is separable, a diagonalization procedure shows that there is a subsequence of x_n that is weakly Cauchy. (We will continue to write x_n for that subsequence.) We will now define a rudimentary weak amart (X_n) on $[0,1)$ as follows: $X_n = x_n \left(\mathbf{1}_{[0,1/2)} - \mathbf{1}_{[1/2,1)} \right)$. To verify that (X_n) is a weak amart, let $x^* \in E^*$, and $\varepsilon > 0$. The sequence x_n is weakly Cauchy, so there is $c \in \mathbb{R}$ and N such that $|\langle x_n, x^* \rangle - c| < \varepsilon$ for all $n \ge N$. So, if $\sigma \in \Sigma$ and $\sigma \ge N$, then

$$\left| \mathbf{E} \left[\langle X_\sigma, x^* \rangle \right] \right| < (1/2)(c + \varepsilon) - (1/2)(c - \varepsilon) < \varepsilon.$$

So in fact, (X_n) is a weak amart potential. But (X_n) does not converge weakly a.s., since (x_n) does not converge weakly. ∎

Complements

(5.5.29) A Banach space E is finite-dimensional if and only if every L_1-bounded amart in E satisfies condition (B).

(5.5.30) The Pietsch factorization theorem (5.5.5) also implies other properties of the absolutely summing operators. If $T \colon E \to F$ is p-absolutely summing for some p, then T is *completely continuous*; that is, if $x_n \to x$ weakly in E, then $Tx_n \to Tx$ in norm in F. (Apply the dominated convergence theorem in the inequality in (5.5.5).) If T is p-absolutely summing, then T is a weakly compact operator. (In fact, T factors through a subspace of an L_p-space with $1 < p < \infty$.)

(5.5.31) Theorem (5.5.11) was generalized in a different way by L. Egghe. He replaced the Banach space E by a Fréchet space. Some of his conditions are as follows:

Theorem. *Let E be a Fréchet space. Then the following are equivalent.*

(1) *E is nuclear;*

(2) *every mean-bounded amart in E converges strongly a.e.;*

(3) *every mean-bounded amart in E satisfies condition* (B);

(4) *every mean-bounded uniformly integrable amart converges in mean.*

See Egghe [1980b] and Egghe [1982a] for the details.

(5.5.32) (Asplund operators.) It would be interesting to prove Theorem (5.5.26) without using Stegall's factorization (5.5.25(5)). Can the Radon-Nikodým property in E^* somehow be connected with an amart potential in E? Is there a direct way to connect them with Fréchet differentiability of φU for convex functions φ?

(5.5.33) (Another ideal.) Clearly the collection of operators that send L_∞-bounded weak amarts to weakly a.s. convergent processes is an operator ideal. Which ideal is it? Theorem (5.5.28) leads to the natural conjecture that it is the ideal of weakly compact operators. But that conjecture is wrong. If J is the quasireflexive Banach space of James (see Lindenstrauss & Tzafriri [1977], Example 1.d.2), there is a sequence e_n in J that is weakly Cauchy but not weakly convergent. Now J is a separable dual, so it has the Radon-Nikodým property, and J^* is separable, so J is an Asplund space. Let $T\colon l_1 \to J$ send the canonical unit vectors in l_1 to the sequence e_n in J. Then T is not weakly compact, since e_n has no weakly convergent subsequence. If (X_n) is a weak amart in l_1, then it is in fact a strong amart (by the Schur property; see Dunford & Schwartz [1958], IV.8.14). But then X_n converges scalarly, so TX_n converges scalarly, and (since J is an Asplund space) TX_n converges weakly a.s.

(5.5.34) (Incompleteness of the Pettis norm.) Let E be a Banach space. Then $L^1(\Omega, \mathcal{F}, \mathbf{P}; E)$ is complete under the Pettis norm $\|\cdot\|_P$ if and only if E is finite-dimensional. To see this, apply Theorem (5.5.9) to the identity operator $T\colon E \to E$. If $\|\cdot\|_P$ is complete, then by the closed graph theorem, $X \mapsto TX$ is a bounded operator in both directions between $\|\cdot\|_P$ and $\|\cdot\|_{L_1}$.

Remarks

The brief treatment of p-absolutely summing operators, the Pietsch factorization, and the Dvoretzky-Rogers lemma follows Lindenstrauss & Tzafriri [1979]. For further reading on operator ideals and their applications to Banach space theory, see Pisier [1986]. Bellow [1976a] proved that L_1-bounded vector amarts converge strongly only if the Banach space is finite-dimensional (Theorem (5.5.11(2))). Edgar & Sucheston [1977a] proved that L_1-bounded vector amarts converge weakly only if the Banach space is finite-dimensional (Theorem (5.5.11(3))). Ghoussoub [1979b] showed how absolutely summing operators come into the problem.

The operator ideal of Radon-Nikodým operators was introduced by Reinov [1975] and Linde [1976], who independently developed its important properties. The Lewis-Stegall factorization is from Lewis & Stegall [1973].

Stegall [1981] discusses the ideal of Asplund operators, but the name is from Edgar [1980], which contains Theorem (5.5.26).

The characterization of weak amart convergence (5.5.28) is due to Brunel & Sucheston [1976b]. Theorem (5.5.27) (the dual is separable if and only if scalar convergence implies weak convergence) is due to Brunel & Sucheston [1977].

6

Martingales

In this chapter we will look at martingales more carefully. In many cases, submartingales and supermartingales will be included. We will consider some of the many interesting results known for these processes, and some of the applications of them. We discuss many well known results concerning maximal inequalities, L_p laws of large numbers for martingale differences (extended to a class of mixing processes), decompositions, convergence of transforms of martingales, Burkholder's square-function inequalities. The Maharam lifting theorem is derived from the martingale theorem.

6.1. Maximal inequalities for supermartingales

In this section we will discuss various maximal inequalities for martingales and supermartingales. We will also prove a law of large numbers for martingale differences. As an application of this, we present a proof of the law of large numbers under the condition of star-mixing.

Let $(\Omega, \mathcal{F}, \mu)$ be a σ-finite measure space, let $(\mathcal{F}_n)_{n \in \mathbb{N}}$ be a stochastic basis, and write $\mathcal{F}_\infty = \sigma(\bigcup \mathcal{F}_i)$. We will write E_i for the conditional expectation: if X is a measurable function,

$$E_i X = \mathbf{E}^{\mathcal{F}_i}[X].$$

This may or may not exist: see (2.3.8). We will also write $\mathbf{P}_i$ for the corresponding "conditional probability": If A is a measurable set,

$$\mathbf{P}_i(A) = E_i \mathbf{1}_A.$$

In particular, E_1 will play the role usually reserved for the expectation $\mathbf{E}$. A reader not interested in this generality may assume that μ is a probability measure, and $E_1 = \mathbf{E}$. However, the $\mathcal{F}_1$ variant has applications: see (6.1.3) below.

A maximal inequality

Let $\lambda \geq 0$ be a random variable, measurable with respect to $\mathcal{F}_1$. Let (X_i) be an adapted sequence of random variables. Fix a value of $n \in \mathbb{N}$. For $1 \leq i \leq n$ write $B_i = \{X_1 < \lambda, \cdots, X_i < \lambda\}$, and

$$A = \Omega \setminus B_n = \left\{ \sup_{1 \leq i \leq n} X_i \geq \lambda \right\}.$$

Now $B_i \in \mathcal{F}_i$, so we may define a stopping time $\sigma \in \Sigma$ by

$$\sigma(\omega) = \begin{cases} \inf\{i : 1 \leq i \leq n, X_i(\omega) \geq \lambda(\omega)\} & \text{if } \omega \in A, \\ n & \text{if } \omega \in B_n. \end{cases}$$

(6.1.1) Lemma. *On the set $\{X_n \geq 0\}$, we have*

$$(6.1.1a) \qquad \lambda \mathbf{1}_A \leq X_1 \mathbf{1}_{B_1} + \sum_{i=1}^{n-1} (X_{i+1} - X_i) \mathbf{1}_{B_i} = X_\sigma.$$

Proof. If $j < n$, then on $\{\sigma = j\}$ we have $\omega \in B_i$ if and only if $i < j$, so

$$X_1 \mathbf{1}_{B_1} + \sum_{i=1}^{n-1} (X_{i+1} - X_i) \mathbf{1}_{B_i} = X_j = X_\sigma.$$

Therefore the formula is: $\lambda \mathbf{1}_A = \lambda \leq X_j = X_\sigma$. On $\{\sigma = n\} \cap A$, we have $\omega \in B_i$ for all $i < n$, so

$$X_1 \mathbf{1}_{B_1} + \sum_{i=1}^{n-1} (X_{i+1} - X_i) \mathbf{1}_{B_i} = X_n = X_\sigma.$$

On $\{\sigma = n\} \cap A$, we have $\lambda \mathbf{1}_A = \lambda \leq X_n = X_\sigma$. On the other hand, on the set $\{\sigma = n\} \cap \{X_n \geq 0\} \cap B_n$ we have $\lambda \mathbf{1}_A = 0 \leq X_n = X_\sigma$. ∎

Now assume that $X_n \geq 0$ a.e. and that $E_1 X_i = \mathbf{E}^{\mathcal{F}_1}[X_i]$ exists for all i. It follows that $E_j X_i$ exists for all j, and that $E_j X_\sigma$ exists for $\sigma \in \Sigma$. Now if we apply E_1 to both sides of (6.1.1a), we obtain:

$$(6.1.1b) \quad \lambda \mathbf{P}_1 \left\{ \sup_{1 \leq i \leq n} X_i \geq \lambda \right\} \leq X_1 + \sum_{i=1}^{n-1} E_1[(X_{i+1} - X_i) \mathbf{1}_{B_i}] = E_1 X_\sigma.$$

A process (X_n) is *eventually positive* iff there is $n_0 \in \mathbb{N}$ so that $X_n \geq 0$ for all $n \geq n_0$. Assume (X_n) is eventually positive. When we let $n \to \infty$ in (6.1.1b), we obtain

$$\lambda \mathbf{P}_1 \left\{ \sup_{1 \leq i < \infty} X_i > \lambda \right\} \leq \sup_{\tau \in \Sigma} E_1 X_\tau.$$

Replacing λ by $\lambda - 1/k$ and letting $k \to \infty$, we obtain

$$(6.1.1c) \qquad \lambda \mathbf{P}_1 \left\{ \sup_{1 \leq i < \infty} X_i \geq \lambda \right\} \leq \sup_{\tau \in \Sigma} E_1 X_\tau.$$

This is the $\mathcal{F}_1$ variant of the maximal inequality (1.1.7).

Another application of (6.1.1b) is a maximal inequality for eventually positive supermartingales. We have $B_i \in \mathcal{F}_i$, so by the supermartingale property,

$$E_i[(X_{i+1} - X_i) \mathbf{1}_{B_i}] = \mathbf{1}_{B_i}[E_i(X_{i+1}) - X_i] \leq 0.$$

Applying E_1 and using (6.1.1b), we obtain a maximal inequality for a fixed n; then we let $n \to \infty$, and we obtain

$$\lambda \mathbf{P}_1 \left\{ \sup_{i \geq 1} X_i \geq \lambda \right\} \leq X_1.$$

Now a conditional probability is bounded by 1, so we obtain:

(6.1.2) Proposition. *If* (X_n) *is an eventually positive supermartingale, then for each positive* $\mathcal{F}_1$*-measurable random variable* λ*, we have*

$$(6.1.2a) \qquad \mathbf{P}_1 \left\{ \sup_{i \geq 1} X_i \geq \lambda \right\} \leq \min \left\{ \frac{X_1}{\lambda}, 1 \right\}.$$

(6.1.3) Corollary. *Let* (X_n) *be an eventually positive supermartingale, let* $r \in \mathbb{N}$ *be fixed, and let* λ *be an* $\mathcal{F}_r$*-measurable random variable. If* $\lambda \leq \sup_{i \geq 1} X_i$ *a.s., then* $\lambda \leq \sup_{1 \leq i \leq r} X_i$ *a.s.*

Proof. For $r = 1$, the assertion follows from Proposition (6.1.2). The general case is obtained by applying the case $r = 1$ to the supermartingale (X'_n) defined by $X'_1 = \sup_{1 \leq i \leq r} X_i$ and $X'_n = X_{r+n-1}$ for $n > 1$. ∎

The inequality (6.1.1b) also renders the Hájek-Rényi-Chow inequality a variant of Kolmogorov's inequality more directly applicable to certain laws of large numbers; examples will be given below. A sequence (a_i) of random variables is called *predictable* iff each a_i is measurable with respect to $\mathcal{F}_{i-1}$.

(6.1.4) Proposition (Hájek-Rényi-Chow). *Let* (X_i) *be a submartingale, let* (a_i) *be a nonnegative increasing predictable sequence, and let* $\lambda > 0$ *be an* $\mathcal{F}_1$*-measurable random variable. Then for each* n*,*

$$(6.1.4a) \qquad \lambda \mathbf{P}_1 \left\{ \sup_{1 \leq i \leq n} \frac{X_i}{a_i} \geq \lambda \right\} \leq \frac{X_1^+}{a_1} + \sum_{i=1}^{n-1} E_1 \left[\frac{X_{i+1}^+ - X_i^+}{a_{i+1}} \right].$$

Proof. First, note that (X_i^+) is also a submartingale. Indeed, we have $E_i X_{i+1}^+ \geq E_i X_{i+1} \geq X_i$ and also $E_i X_{i+1}^+ \geq 0$, so $E_i X_{i+1}^+ \geq X_i^+$. Therefore we have $(E_i X_{i+1}^+ - X_i^+)/a_{i+1} \geq 0$. Now in order to apply (6.1.1b) to $Y_i = X_i^+/a_i$, write $B_i = \{Y_1 < \lambda, Y_2 < \lambda, \cdots Y_i < \lambda\}$. Then

$$\lambda \mathbf{P}_1 \left\{ \sup_{1 \leq i \leq n} \frac{X_i}{a_i} \geq \lambda \right\} = \lambda \mathbf{P}_1 \left\{ \sup_{1 \leq i \leq n} \frac{X_i^+}{a_i} \geq \lambda \right\} = \lambda \mathbf{P}_1 \left\{ \sup_{1 \leq i \leq n} Y_i \geq \lambda \right\}$$

$$\leq Y_1 + \sum_{i=1}^{n-1} E_1 \left[(Y_{i+1} - Y_i) \mathbf{1}_{B_i} \right]$$

$$= \frac{X_1^+}{a_1} + \sum_{i=1}^{n-1} E_1 \left[E_i \left(\left(\frac{X_{i+1}^+}{a_{i+1}} - \frac{X_i^+}{a_i} \right) \mathbf{1}_{B_i} \right) \right]$$

$$= \frac{X_1^+}{a_1} + \sum_{i=1}^{n-1} E_1 \left[\left(\frac{E_i X_{i+1}^+}{a_{i+1}} - \frac{X_i^+}{a_i} \right) \mathbf{1}_{B_i} \right]$$

$$\leq \frac{X_1^+}{a_1} + \sum_{i=1}^{n-1} E_1 \left[\frac{E_i X_{i+1}^+ - X_i^+}{a_{i+1}} \mathbf{1}_{B_i} \right]$$

$$\leq \frac{X_1^+}{a_1} + \sum_{i=1}^{n-1} E_1 \left[\frac{E_i X_{i+1}^+ - X_i^+}{a_{i+1}} \right].$$

∎

Note that the special case $a_i = 1$ is the original Doob inequality (1.4.18): If $\lambda > 0$ is $\mathcal{F}_1$-measurable, then

(6.1.4b) $$\lambda \mathbf{P}_1 \left\{ \sup_{1 \leq i \leq n} X_i \geq \lambda \right\} \leq E_1 X_n^+.$$

A smaller expression on the right will be useful for strong inequalities in L_p and in Orlicz spaces.

(6.1.5) Proposition. *Let* (X_i) *be a positive submartingale, and let* λ *be positive and* $\mathcal{F}_1$*-measurable. Then for each* $n \in N$,

$$\lambda \mathbf{P}_1 \left\{ \sup_{1 \leq i \leq n} X_i \geq \lambda \right\} \leq E_1 \left[\mathbf{1}_{\{\sup_{1 \leq i \leq n} X_n \geq \lambda\}} X_n^+ \right].$$

Proof. Let the (possibly infinite) stopping time σ be defined as usual: $\sigma = \inf \{ i : X_i \geq \lambda \}$. By the localization theorem (1.4.2), if $k \leq n$ we have on the set $\{\sigma = k\}$,

$$X_\sigma = X_k \leq \mathbf{E}^{\mathcal{F}_k} [X_n] = \mathbf{E}^{\mathcal{F}_\sigma} [X_n].$$

Therefore

$$X_\sigma \leq \mathbf{E}^{\mathcal{F}_\sigma} [X_n]$$

on the set $\{\sigma \leq n\}$. Write $A = \{\sup_{1 \leq i \leq n} X_i \geq \lambda\} = \{\sigma \leq n\}$. Then:

$$\lambda \mathbf{P}_1 (A) = E_1 [\mathbf{1}_A \lambda] \leq E_1 [\mathbf{1}_A X_\sigma]$$
$$\leq E_1 \left[\mathbf{1}_A \mathbf{E}^{\mathcal{F}_\sigma} [X_n] \right] = E_1 [\mathbf{1}_A X_n],$$

which proves the Proposition.   ∎

A law of large numbers

We will prove a generalization of the classical law of large numbers for independent random variables. The partial sums of independent mean zero random variables form a martingale, so it is natural to generalize to other martingales, or even submartingales (X_i). The analog of the classical hypothesis would involve powers of differences $|X_{i+1} - X_i|^p$. A result with this sort of hypothesis is obtained later for $p \leq 2$. For $p > 2$ the exact analog fails, and the theorem will be shown to hold with the exponent $1 + p/2$ replacing p.

We will use this result:

(6.1.6) Kronecker's lemma. *Let (x_i) be a sequence of real numbers, and let (c_i) be a positive unbounded increasing sequence. If $\sum_{i=1}^{\infty}(x_i/c_i)$ converges, then*

$$\lim_{m \to \infty} \frac{1}{c_m} \sum_{i=1}^{m} x_i = 0.$$

Proof. Write $c_0 = 0$. Let $\varepsilon > 0$. There is $p \in \mathbb{N}$ such that $\left|\sum_{i=j}^{m}(x_i/c_i)\right| < \varepsilon$ for $p \le j \le m$. Note that

$$\sum_{i=1}^{m} x_i = \sum_{j=1}^{m}(c_j - c_{j-1}) \sum_{i=j}^{m} \frac{x_i}{c_i}.$$

Therefore if $p < m$ we have

$$\left| \frac{1}{c_m} \sum_{i=1}^{m} x_i \right| \le \frac{1}{c_m} \left| \sum_{j=1}^{p}(c_j - c_{j-1}) \sum_{i=j}^{m} \frac{x_i}{c_i} \right| + \frac{1}{c_m} \sum_{j=p+1}^{m}(c_j - c_{j-1}) \left| \sum_{i=j}^{m} \frac{x_i}{c_i} \right|$$

$$\le \frac{1}{c_m} \left| \sum_{j=1}^{p}(c_j - c_{j-1}) \sum_{i=j}^{m} \frac{x_i}{c_i} \right| + \frac{c_m - c_p}{c_m} \varepsilon.$$

Now when $m \to \infty$, the first term goes to 0, so we obtain

$$\limsup_{m \to \infty} \left| \frac{1}{c_m} \sum_{i=1}^{m} x_i \right| \le \varepsilon,$$

which proves the existence of the required limit. ∎

(6.1.7) Proposition. *Let $p \ge 1$ be a fixed number. Let (X_n) be a nonnegative submartingale, and let (a_i) be a positive increasing predictable sequence with $a_i \to \infty$. Suppose*

$$(6.1.7a) \qquad \sum_{i=1}^{\infty} E_1 \left(\frac{X_{i+1}^p - X_i^p}{a_{i+1}^p} \right) < \infty.$$

Then $\lim_{n \to \infty} X_n/a_n = 0$ a.e.

Proof. First, by Kronecker's lemma (6.1.6) applied to (6.1.7a), we have

$$(6.1.7b) \qquad E_1 \left[\frac{X_{m+1}^p}{a_{m+1}^p} \right] - E_1 \left[\frac{X_1^p}{a_{m+1}^p} \right] = \sum_{i=1}^{m} E_1 \left[\frac{X_{i+1}^p - X_i^p}{a_{m+1}^p} \right] \to 0.$$

By Jensen's inequality (2.3.10), the process (X_n^p) is also a submartingale. Let $m < n$; apply (6.1.4a) to the finite submartingale $(X_m^p, X_{m+1}^p, \cdots, X_n^p)$, replacing a_i by a_i^p and λ by λ^p; apply the conditional expectation E_1; then take the limit as $n \to \infty$. The result is:

$$\lambda^p \, \mathbf{P}_1 \left\{ \sup_{i \geq m} \frac{X_i^p}{a_i^p} \geq \lambda^p \right\} \leq E_1 \frac{X_m^p}{a_m^p} + \sum_{i=m}^{\infty} E_1 \left[\frac{X_{i+1}^p - X_i^p}{a_{i+1}^p} \right].$$

The first term on the right converges to 0 by (6.1.7b). The second term converges to 0 by (6.1.7a). Therefore $\lim X_n/a_n = 0$ a.e. $\blacksquare$

The following theorem is due to Paul Lévy if $p = 2$. The particular case where the Y_i are independent and with zero expectation is one of the two classical Kolmogorov laws of large numbers. (The second one is proved below: see the case $p = 1$ of (6.1.15).)

(6.1.8) Theorem (Y. S. Chow). *Let p be a fixed number, $1 < p \leq 2$. Let $M_n = \sum_{i=1}^{n} Y_i$ be a martingale, and let a_i be a positive increasing predictable sequence. Suppose*

$$(6.1.8a) \qquad \sum_{i=1}^{\infty} E_1 \left(\frac{|Y_i|^p}{a_i^p} \right) < \infty.$$

Then $\lim_n M_n/a_n = 0$ a.e.

Proof. We take first the easy case $p = 2$. Since a_i is predictable, and $E_i Y_{i+1} = 0$, we have

$$E_1 \left[\frac{M_{i+1}^2 - M_i^2}{a_{i+1}^2} \right] = E_1 E_i \left[\frac{Y_{i+1}(2M_i + Y_{i+1})}{a_{i+1}^2} \right]$$

$$= E_1 \left[\frac{2M_i}{a_{i+1}^2} E_i Y_{i+1} \right] + E_1 E_i \left[\frac{Y_{i+1}^2}{a_{i+1}^2} \right]$$

$$= E_1 \left[\frac{Y_{i+1}^2}{a_{i+1}^2} \right].$$

Now apply Proposition (6.1.7).

For the case $p < 2$, we will prove that

$$(6.1.8b) \qquad E_1 \left(|M_{i+1}|^p - |M_i|^p \right) \leq 2 E_1 \left(|Y_{i+1}|^p \right).$$

This may be used in the same way as the martingale equality was used in the case $p = 2$ to finish the proof. We have

$$E_i \left(|M_i|^p \right) = |M_i|^p = |E_i M_i|^p$$

$$= \left| E_i(2M_i - M_{i+1}) \right|^p$$

$$\leq E_i \left(|2M_i - M_{i+1}|^p \right),$$

by Jensen's inequality (2.3.10). Therefore

$$
\begin{aligned}
E_i\big(|M_{i+1}|^p - |M_i|^p\big) & \\
\leq E_i\big(|M_{i+1}|^p + |2M_i - M_{i+1}|^p\big) &- 2E_i\big(|M_i|^p\big) \\
= E_i\big(|M_i + Y_{i+1}|^p + |M_i - Y_{i+1}|^p\big) &- 2E_i\big(|M_i|^p\big).
\end{aligned}
$$

(6.1.8c)

We will need the elementary inequality

(6.1.8d)
$$
|a + b|^p + |a - b|^p \leq 2\big(|a|^p + |b|^p\big),
$$

for $1 < p < 2$. To prove it, observe that we may assume $|a| \geq |b|$ by interchanging a and b. Then, writing $x = b/a$, it suffices to prove $(1+x)^p + (1-x)^p \leq 2(1 + |x|^p)$ for $-1 \leq x \leq 1$. Replacing x by $-x$ if necessary, it is enough to prove this for $0 \leq x \leq 1$. Also, the inequality is clear for $x = 0$. Let $F(x) = (1+x)^p + (1-x)^p - 2(1+x^p)$. If $G(x) = F(x)/x^p$, then we must show $G(x) \leq 0$ for $0 < x \leq 1$. But $G(1) = 2^p - 4 \leq 0$, so it is enough to show that the derivative $G'(x) > 0$. Now $G'(x) = pH(x)/x^{p+1}$, where $H(x) = 2 - (1+x)^{p-1} - (1-x)^{p-1}$, so we must show $H(x) > 0$ for $0 \leq x \leq 1$. But $H(0) = 0$ and $H'(x) = (p-1)\big((1-x)^{p-2} - (1+x)^{p-2}\big) > 0$ since $p - 2 < 0$ and $1 + x > 1 - x$. This completes the proof of the inequality (6.1.8d).

Applying (6.1.8d) to (6.1.8c), we obtain

$$
\begin{aligned}
E_i\big(|M_{i+1}|^p - |M_i|^p\big) &\leq 2E_i\big(|M_i|^p + |Y_{i+1}|^p\big) - 2E_i\big(|M_i|^p\big) \\
&= 2E_i\big(|Y_{i+1}|^p\big).
\end{aligned}
$$

Then apply E_1 to obtain (6.1.8b). ∎

Much more difficult is the case $p > 2$. The proof is based on a fundamental inequality of D. L. Burkholder, proved below (Theorem (6.3.6)). For simplicity we will take $a_i = i$. The exponent in the denominator must be $1 + p/2 < p$, so the condition is more difficult to satisfy than the corresponding condition with exponent p. The independent case of the following is due to Brunk [1948] and Chung [1947, 1951].

(6.1.9) Theorem (Y. S. Chow 1967b). *Let p be a fixed number with $2 < p < \infty$. Let $M_n = \sum_{i=1}^{n} Y_i$ be a martingale such that*

(6.1.9a)
$$
\sum_{i=1}^{\infty} E_1\left[\frac{|Y_i|^p}{i^{1+p/2}}\right] < \infty.
$$

Then $\lim M_n/n = 0$ a.e.

Proof. By Hölder's inequality (with exponents $p/(p-2)$ and $p/2$), we have

$$
\sum_{i=1}^{n} Y_i^2 \leq n^{1-2/p}\left(\sum_{i=1}^{n} |Y_i|^p\right)^{2/p}.
$$

Now raise both sides to the power $p/2$ and apply E_1:

$$E_1\left[\left|\sum_{i=1}^n Y_i^2\right|^{p/2}\right] \leq n^{p/2-1} E_1\left[\sum_{i=1}^n |Y_i|^p\right].$$

Next we apply the right side of (6.3.6) with $S_n = \left(\sum_{i=1}^n Y_i^2\right)^{1/2}$. There is a constant $K < \infty$ such that

(6.1.9b) $$E_1\left[\left|\sum_{i=1}^n Y_i\right|^p\right] \leq K\, n^{p/2-1} E_1\left(\sum_{i=1}^n |Y_i|^p\right).$$

Now by Proposition (6.1.7), it suffices to prove

$$\sum_{n=2}^\infty \frac{E_1\left|\sum_{i=1}^n Y_i\right|^p - E_1\left|\sum_{i=1}^{n-1} Y_i\right|^p}{n^p} < \infty.$$

Therefore it suffices to prove both that $\lim E_1 |\sum_{i=1}^n Y_i|^p/n^p = 0$ and that

$$\sum_{n=1}^\infty \left[\frac{1}{n^p} - \frac{1}{(n+1)^p}\right] E_1 \left|\sum_{i=1}^n Y_i\right|^p < \infty.$$

But by (6.1.9b),

$$E_1\frac{\left|\sum_{i=1}^n Y_i\right|^p}{n^p} \leq Kn^{-p/2-1}\sum_{i=1}^n E_1\left(|Y_i|^p\right),$$

which converges to 0 by (6.1.9a) and Kronecker's lemma. Again using (6.1.9b), and since $1/n^p - 1/(n+1)^p \leq (n^p + pn^{p-1} - n^p)/n^p(n+1)^p \leq p/n^{p+1}$, we have

$$\sum_{n=1}^\infty \left[\frac{1}{n^p} - \frac{1}{(n+1)^p}\right] E_1\left(\left|\sum_{i=1}^n Y_i\right|^p\right)$$

$$\leq \sum_{n=1}^\infty \left[\frac{1}{n^p} - \frac{1}{(n+1)^p}\right] Kn^{p/2-1} E_1\left(\sum_{i=1}^n |Y_i|^p\right)$$

$$\leq KpE_1\left(\sum_{n=1}^\infty n^{-p/2-2}\left(\sum_{i=1}^n |Y_i|^p\right)\right)$$

$$\leq KpE_1\left(\sum_{i=1}^\infty |Y_i|^p \sum_{n=i}^\infty n^{-p/2-2}\right)$$

$$\leq \frac{2Kp}{p+2} E_1\left(\sum_{i=1}^\infty |Y_i|^p i^{-p/2-1}\right) < \infty.$$

■

A generalization of independence: star-mixing

We restrict attention now to a probability space $(\Omega, \mathcal{F}, \mathbf{P})$.

Let $(\mathcal{A}_n)_{n \geq 1}$ be a family of σ-algebras; typically $\mathcal{A}_n$ is generated by a single random variable Y_n. Let $\mathcal{F}_n = \sigma(\bigcup_{i \leq n} \mathcal{A}_i)$; typically $\mathcal{F}_n$ is the σ-algebra generated by the random variables $Y_1, Y_2, \cdots, Y_n$. "Mixing" means intuitively that events separated by a large time span (i.e., by high powers of the "shift operator" on the process) are nearly independent. To obtain almost everywhere convergence results, a degree of uniformity in mixing is required. An early notion is *star-mixing*, introduced by Blum, Hanson, and Koopmans [1963]; their law of large numbers, proved under Kolmogorov's condition involving second moments, was reduced to the law of large numbers [in the case $p = 2$] for martingale differences by P. Bártfai (see Révész [1968], p. 140).

The family $(\mathcal{A}_n)$ is called *star-mixing* iff there exists a positive integer N and a function $f(n)$ defined for $n \geq N$, such that $f(n) \downarrow 0$ and if $n \geq N$, $A \in \mathcal{F}_m$, $B \in \mathcal{A}_{m+n}$, then

$$|\mathbf{P}(A \cap B) - \mathbf{P}(A)\mathbf{P}(B)| \leq f(n)\mathbf{P}(A)\mathbf{P}(B).$$

(6.1.10) Lemma. *Assume that $(\mathcal{A}_n)$ is star-mixing with constant N and function $f(n)$. Then for any σ-algebra $\mathcal{C} \subseteq \mathcal{F}_m$, any $n \geq N$, and any integrable random variable X measurable with respect to $\mathcal{A}_{m+n}$,*

$$\left| \mathbf{E}^{\mathcal{C}}[X] - \mathbf{E}[X] \right| \leq f(n)\mathbf{E}\left[|X|\right].$$

Proof. Let Z_k be simple and $\mathcal{A}_{m+n}$-measurable so that $0 \leq Z_k \uparrow X^+$. Fix k. Then $Z_k = \sum_{i=1}^{j} b_i \mathbf{1}_{B_i}$, where $b_i \geq 0$ and $B_i \in \mathcal{A}_{m+n}$. Choose $A \in \mathcal{C}$ such that $\mathbf{P}(A) > 0$. Then

$$\left| \mathbf{E}[Z_k \mid A] - \mathbf{E}[Z_k] \right| \leq \left| \sum_{i=1}^{j} b_j \frac{\mathbf{P}(B_i \cap A)}{\mathbf{P}(A)} - \sum_{i=1}^{j} b_i \mathbf{P}(B_i) \right|$$

$$\leq \sum_{i=1}^{j} b_i \left| \frac{\mathbf{P}(B_i \cap A)}{\mathbf{P}(A)} - \mathbf{P}(B_i) \right|$$

$$\leq f(n) \sum_{i=1}^{j} b_i \mathbf{P}(B_i) = f(n)\mathbf{E}[Z_k].$$

Then if $k \to \infty$, we obtain

$$\left| \mathbf{E}[X^+ \mid A] - \mathbf{E}[X^+] \right| \leq f(n)\mathbf{E}[X^+].$$

This remains true if A is replaced by $\mathcal{C}$ (integrate over sets $A \in \mathcal{C}$), and also if X^+ is replaced by X^-. The lemma follows. ∎

If (Y_n) is a stochastic process, we will say it is *star-mixing* iff the σ-algebras $\mathcal{A}_n = \sigma(Y_n)$ are star-mixing.

(6.1.11) Theorem. *Let (Y_n) be star-mixing, mean 0 ($\mathbf{E}[Y_n] = 0$), and L_1-bounded ($\|Y_n\|_1 \leq K < \infty$). Let p be such that $1 \leq p < \infty$. Assume that*

(6.1.11a) $\sum_{i=1}^{\infty} \mathbf{E}\left[|Y_i|^p\right]/i^p < \infty$, *if* $p \leq 2$, *or*

(6.1.11b) $\sum_{i=1}^{\infty} \mathbf{E}\left[|Y_i|^p\right]/i^{1+p/2} < \infty$, *if* $p > 2$.

Then

(6.1.11c)
$$\frac{1}{n}\sum_{i=1}^{n} Y_i \to 0 \quad a.s.$$

Proof. Fix an integer k with $0 \leq k < N$, and apply Lemma (6.1.10) to the process $X_m = Y_{mN+k}$. If $\mathcal{H}_m = \sigma(Y_{mN+k}, Y_{(m-1)N+k}, \cdots, Y_{N+k})$ for $m \geq 1$ and $\mathcal{H}_0 = \{\Omega, \emptyset\}$, then

$$\left|\mathbf{E}^{\mathcal{H}_{m-1}}[Y_{mN+k}] - \mathbf{E}[Y_{mN+k}]\right| \leq f(N)\mathbf{E}\left[|Y_{mN+k}|\right].$$

Now $\mathbf{E}[Y_{mN+k}] = 0$ and $\mathbf{E}\left[|Y_{mN+k}|\right] \leq K$, so given ε we may choose N so large that

(6.1.11d)
$$\left|\mathbf{E}^{\mathcal{H}_{m-1}}[Y_{mN+k}]\right| \leq \varepsilon K$$

for $m \geq 1$ and $0 \leq k < N$. Fix k and N. Let $T_m = Y_{mN+k}$ for $m \geq 1$. Then $U_m = T_m - \mathbf{E}^{\mathcal{H}_{m-1}}[T_m]$ defines a martingale difference sequence. The sequence (T_m) satisfies (6.1.11a) or (6.1.11b), as appropriate. Now "centering," that is replacing T_m by U_m, does not change this, because

$$|U_m| \leq |T_m| + \mathbf{E}^{\mathcal{H}_{m-1}}\left[|T_m|\right]$$
$$\leq 2\max\left\{|T_m|, \mathbf{E}^{\mathcal{H}_{m-1}}\left[|T_m|\right]\right\},$$

so by Jensen's inequality

$$|U_m|^p \leq 2^p \max\left\{|T_m|^p, \left(\mathbf{E}^{\mathcal{H}_{m-1}}\left[|T_m|\right]\right)^p\right\}$$
$$\leq 2^p \max\left\{|T_m|^p, \mathbf{E}^{\mathcal{H}_{m-1}}\left[|T_m|^p\right]\right\}$$
$$\leq 2^p \left(|T_m|^p + \mathbf{E}^{\mathcal{H}_{m-1}}\left[|T_m|^p\right]\right).$$

This implies

(6.1.11e) $\quad \mathbf{E}\left[|U_m|^p\right] \leq 2^p \left(\mathbf{E}\left[|T_m|^p\right] + \mathbf{E}\left[|T_m|^p\right]\right) = 2^{p+1}\mathbf{E}\left[|T_m|^p\right].$

Therefore the law of large numbers for martingale differences (Theorems (6.1.8) and (6.1.9)) implies that

$$\lim_{r} \frac{1}{r}\sum_{m=1}^{r} \left[Y_{mN+k} - \mathbf{E}^{\mathcal{H}_{m-1}}[Y_{mN+k}]\right] = 0 \quad a.s.$$

With (6.1.11d), this implies that

$$\left| \limsup_r \frac{1}{r} \sum_{m=1}^{r} Y_{mN+k} \right| \le \varepsilon K.$$

Since ε is arbitrary, the left side is 0. This holds for each k with $0 \le k < N$, which implies the theorem. ∎

Complements

(6.1.12) Inequality (6.1.11e) can be improved to $\mathbf{E}\left[|U_m|^p\right] \le 2^p \mathbf{E}\left[|T_m|^p\right]$. In the estimation of $|U_m|^p$, use the inequality $|x+y|^p \le 2^{p-1}\left(|x|^p + |y|^p\right)$, which is true since the function $|x|^p$ is convex.

(6.1.13) (Example of a star-mixing process.) A stationary ergodic Markov chain with countably many states and transition probabilities p_{ij} is star-mixing if and only if there is a number β with $0 < \beta < 1$ such that for all j,

$$\sup_i p_{ij} \le (1 + \beta) \inf_i p_{ij}$$

(Blum, Hanson, & Koopmans [1963]).

(6.1.14) (Qualitative star-mixing.) The family $(\mathcal{A}_n)$ is *Q-star-mixing* iff there exists a positive integer N and a constant $\alpha > 0$, such that if $n \ge N$, $A \in \mathcal{F}_m$, $B \in \mathcal{A}_{m+n}$, then

$$|\mathbf{P}(A \cap B) - \mathbf{P}(A)\mathbf{P}(B)| \le \alpha \mathbf{P}(A)\mathbf{P}(B).$$

Q-star-mixing is sufficient for a nonanticipating converse of the dominated ergodic theorem: If (Y_n) is a stationary Q-star-mixing positive process, $X_n = (1/n) \sum_{i=1}^{n} Y_i$, and $Y_1 \notin L \log L$, then there is a stopping time τ such that $\mathbf{E}\left[|X_\tau|\right] = \infty$ (Edgar, Millet, & Sucheston [1982]).

In Theorem (6.1.11), if star-mixing is replaced by Q-star-mixing, the conclusion (6.1.11c) should be replaced by

$$\limsup_n \frac{1}{n} \sum_{i=1}^{n} Y_n < \infty \quad \text{a.s.}$$

(6.1.15) (Laws of large numbers.) The Kolmogorov strong law of large numbers for independent random variables Y_i with the same distribution and finite expectation has a martingale proof. This has been hailed as a marked success of the theory. J. L. Doob observed (see Doob [1953], p. 343) that $X_{-n} = (1/n) \sum_{i=1}^{n} Y_i$ is a *reversed* martingale, hence converges a.e. In fact it suffices to assume that the Y_i's are *exchangeable*. This is extended here.

Let I be a set. A *permutation* on I is a bijection of the set I onto itself. The *support* of a permutation π is the set $\{i \in I : \pi(i) \ne i\}$. A family

$(Y_i)_{i \in I}$ of random variables is called *exchangeable* iff, for each permutation π with finite support, the family $(Y_i)_{i \in I}$ has the same distribution as the family $(Y_{\pi(i)})_{i \in I}$.

Suppose $(Y_n)_{n \in \mathbb{N}}$ is an exchangeable family of positive integrable random variables. For $m \in \mathbb{N}$, let

$$X_{-m} = \frac{(Y_1 + \cdots + Y_m)^p}{m},$$
$$\mathcal{F}_{-m} = \sigma \{ X_{-r} : r \geq m \}.$$

Then $(X_{-n})_{n \in \mathbb{N}}$ is a reversed submartingale if $0 < p \leq 1$; and a reversed supermartingale if $1 \leq p < \infty$. For $0 < p \leq 1$, the sequence X_{-n} converges a.s. The limit is 0 if $0 < p < 1$ (Edgar & Sucheston [1981]).

In order to prove this, we begin with a simple lemma.

Lemma. *Let P_m be the set of permutations of $\mathbb{N}$ with support in the set $\{1, \cdots, m\}$. Let $y_1, \cdots, y_m$ be nonnegative real numbers, let $1 \leq k \leq m$, and let $0 < p \leq 1$. Then*

$$\frac{1}{m!} \sum_{\pi \in P_m} \frac{(y_{\pi(1)} + \cdots + y_{\pi(k)})^p}{k} \geq \frac{(y_1 + \cdots + y_m)^p}{m}.$$

The reverse inequality holds if $1 \leq p < \infty$.

Proof. Indeed, for all integers $k \leq m$, we have

$$(m-1)!k(y_1 + \cdots + y_m) = \sum_{\pi \in P_m} (y_{\pi(1)} + \cdots + y_{\pi(k)})$$
$$\leq \sum_{\pi \in P_m} (y_{\pi(1)} + \cdots + y_{\pi(k)})^p (y_1 + \cdots + y_m)^{1-p}.$$

Hence

$$(m-1)!k(y_1 + \cdots y_m)^p \leq \sum_{\pi \in P_m} (y_{\pi(1)} + \cdots + y_{\pi(k)})^p.$$

∎

Now assume that $(Y_i)_{i \in \mathbb{N}}$ are exchangeable, and $Y_i > 0$. Let $S_m = Y_1 + \cdots Y_m$ and $X_{-m} = S_m^p/m$ for $m \in \mathbb{N}$. For all $\pi \in P_m$, the family

$$(Y_1, \cdots, Y_m, S_m, S_{m+1}, \cdots)$$

has the same joint distribution as the family

$$(Y_{\pi(1)}, \cdots, Y_{\pi(m)}, S_m, S_{m+1}, \cdots).$$

Hence, for all $k \leq m$,

$$\mathbf{E}^{\mathcal{F}_{-m}} \left[(Y_1 + \cdots + Y_k)^p \right] = \mathbf{E}^{\mathcal{F}_{-m}} \left[(Y_{\pi(1)} + \cdots + Y_{\pi(k)})^p \right].$$

Therefore, by the lemma, if $0 < p \leq 1$,

$$\mathbf{E}^{\mathcal{F}_{-m}}\left[X_{-k}\right] = \mathbf{E}^{\mathcal{F}_{-m}}\left[\frac{(Y_1 + \cdots + Y_k)^p}{k}\right]$$

$$= \mathbf{E}^{\mathcal{F}_{-m}}\left[\frac{1}{m!}\sum_{\pi \in P_m}\frac{(Y_{\pi(1)} + \cdots + Y_{\pi(k)})^p}{k}\right]$$

$$\geq \mathbf{E}^{\mathcal{F}_{-m}}\left[\frac{(Y_1 + \cdots Y_m)^p}{m}\right] = X_{-m}.$$

If $1 \leq p < \infty$, the reverse inequality holds.

Now we prove convergence in the case $0 < p \leq 1$. Then X_{-m} is dominated by $\mathbf{E}^{\mathcal{F}_{-m}}\left[Y_1^p\right]$ and converges a.e. by (1.2.8), say to X. Assume $p < 1$. If $X > 0$ on a nonnull set A, then $\lim X_{-2m} = X$ implies that on A we have $(S_{2m}/S_m)^p \to 2$, if $S_m = \sum_{i=1}^m Y_i$, hence $(Y_{m+1}+\cdots+Y_{2m})/(Y_1+\cdots Y_m) \to 2^{1/p} - 1 > 1$. This is a contradiction because X is determined by the exchangeable sequence (Y_n) so that the random variable X is invariant under the permutation $(1, \cdots, m) \leftrightarrow (m+1, \cdots, 2m)$.

The convergence of X_{-m} to zero when $p < 1$ can be proved in many ways; see for example Neveu [1965a], p. 153; and Bru, Heinich, & Lootgieter [1981]. For an ergodic approach, see (8.6.18), below.

It is easy to see that X_{-m} diverges for $p > 1$: compare with the case $p = 1$ (see (6.1.16), below). Yet the information that the process is a reversed supermartingale can be useful: For finite stretches, the "direct" and reverse processes agree, so supermartingale results proved above, for example (6.1.2), are applicable.

(6.1.16) For $p = 1$, the limit X in (6.1.15) can be identified as $\mathbf{E}^{\mathcal{S}}\left[Y_1\right]$, where $\mathcal{S}$ is the σ-algebra of "exchangeable" events: that is, events depending on the sequence (Y_n) and invariant under permutations of finite support. It can be proved that $\mathcal{S}$ agrees (up to null sets) with the "tail" σ-algebra: see, for example, Meyer [1966], pp. 149–150.

Remarks

Another way to achieve the generality replacing the expectation by E_1 would involve using a regular conditional distribution with respect to $\mathcal{F}_1$ (see Billingsley [1979], pp. 390, 399). Then the case of general $\mathcal{F}_1$ can be deduced from the case of trivial $\mathcal{F}_1$ and finite μ.

For Proposition (6.1.4) of Hájek-Rényi-Chow, see Hájek & Rényi [1956], Frank [1966], Bauer [1981], and Chow [1960a].

That the process

$$Z_n = \frac{(Y_1 + Y_2 + \cdots Y_n)^p}{n}$$

of (6.1.15) is an *amart* was observed by A. Gut [1982]. He used the Marcinkiewicz theorem to observe that Z_n, and hence Z_τ, converges a.s., and since it is uniformly integrable, $\mathbf{E}\left[Z_\tau\right]$ converges. This motivated us to show that in fact the process is a reversed *submartingale*, which does not use the Marcinkiewicz theorem, but instead proved it.

6.2. Decompositions of submartingales

This section discusses Doob's decomposition and Krickeberg's decomposition; the quadratic variation of a process; and convergence of martingale transforms.

A probability space $(\Omega, \mathcal{F}, \mathbf{P})$ and a stochastic basis $(\mathcal{F}_n)_{n \in \mathbb{N}}$ will be fixed. If $(X_n)_{n \in \mathbb{N}}$ is an adapted process, we may refer to it using a bold letter:

$$\mathbf{X} = (X_n).$$

Then sets of equations, such as $X_n = M_n + A_n$ for all n will be abbreviated in the obvious way: $\mathbf{X} = \mathbf{M} + \mathbf{A}$.

Recall that a process $\mathbf{V} = (V_n)$ is called *predictable* iff V_{n+1} is $\mathcal{F}_n$-measurable for all n.

Doob's decomposition

We begin with a decomposition theorem for submartingales.

(6.2.1) Theorem. *A submartingale [supermartingale] $\mathbf{X}$ can be uniquely written as $\mathbf{X} = \mathbf{M} + \mathbf{A}$ [$\mathbf{X} = \mathbf{M} - \mathbf{A}$], where $\mathbf{M}$ is a martingale and $\mathbf{A}$ is a predictable increasing process with $A_1 = 0$. If $\mathbf{X}$ is L_1-bounded, then so is $\mathbf{M}$, and then the limit $A_\infty = \lim_n A_n$ is integrable.*

Proof. Let $\mathbf{X} = (X_n)$ be a submartingale. Let $M_1 = X_1$ and $A_1 = 0$, then define recursively $M_{n+1} = M_n + X_{n+1} - \mathbf{E}^{\mathcal{F}_n}[X_{n+1}]$ and $A_{n+1} = A_n + \mathbf{E}^{\mathcal{F}_n}[X_{n+1}] - X_n$. Then $\mathbf{A} = (A_n)$ is predictable, positive, and increasing, and $\mathbf{M} = (M_n)$ is a martingale.

To check the uniqueness, suppose that $X_n = M'_n + A'_n$ is another decomposition with the same properties. Then

$$\begin{aligned} A'_{n+1} - A'_n &= \mathbf{E}^{\mathcal{F}_n}\left[A'_{n+1} - A'_n\right] \\ &= \mathbf{E}^{\mathcal{F}_n}\left[X_{n+1} - X_n - (M'_{n+1} - M'_n)\right] \\ &= A_{n+1} - A_n - 0 = A_{n+1} - A_n. \end{aligned}$$

Since $A_1 = 0 = A'_1$, this means that $A_n = A'_n$ for all n. Thus also $M'_n = X_n - A'_n = X_n - A_n = M_n$.

Assume now that we have $\sup_n \mathbf{E}[|X_n|] < \infty$. Since $A_n \geq 0$, we have $\sup_n \mathbf{E}[M_n^+] < \sup_n \mathbf{E}[X_n^+] < \infty$. But M_n is a martingale, so $\sup_n \mathbf{E}[|M_n|] = \sup_n(2\mathbf{E}[M_n^+] - \mathbf{E}[M_1]) < \infty$. Then $\mathbf{A}$ is also L_1-bounded, so the integrability of the limit follows from the monotone convergence theorem.

The supermartingale case is obtained by applying the submartingale case to the process $(-X_n)$. ∎

Next is the Krickeberg decomposition.

(6.2.2) **Theorem.** *An L_1-bounded submartingale* $\mathbf{X}$ *can be written as* $\mathbf{X} = \mathbf{M} - \mathbf{R}$, *where* $\mathbf{M}$ *is a positive martingale, and* $\mathbf{R}$ *is a positive super-martingale. If* $\mathbf{X}$ *is a martingale, then also* $\mathbf{R}$ *is a positive martingale and* $\mathbf{E}[M_1] + \mathbf{E}[R_1] = \sup_n \mathbf{E}[|X_n|]$.

Proof. It is easy to see that $\mathbf{X}^+ = (X_n^+)$ is also a submartingale (see also the proof of Theorem (6.1.4)).

The sequence $\mathbf{M}$ will be obtained as the martingale part in the Riesz decomposition of X_n^+. Actually, the argument is easier than for general amarts (1.4.6), since for $n \geq p$

$$\mathbf{E}^{\mathcal{F}_n}\left[X_{p+1}^+\right] = \mathbf{E}^{\mathcal{F}_n}\left[\mathbf{E}^{\mathcal{F}_p}\left[X_{p+1}^+\right]\right] \geq \mathbf{E}^{\mathcal{F}_n}\left[X_p^+\right].$$

Set $M_n = \lim_p \uparrow \mathbf{E}^{\mathcal{F}_n}\left[X_p^+\right]$. The supermartingale $\mathbf{R} = \mathbf{M} - \mathbf{X}$ is positive.

Finally, if $\mathbf{X}$ is a martingale, then

$$\begin{aligned}
\mathbf{E}[R_1] &= \mathbf{E}[M_1] - \mathbf{E}[X_1] \\
&= \lim_p \uparrow \mathbf{E}\left[X_p^+\right] - \mathbf{E}[X_1] \\
&= \lim_p \uparrow \left(\mathbf{E}[X_p] + \mathbf{E}\left[X_p^- - \mathbf{E}[X_1]\right]\right) \\
&= \lim_p \uparrow \mathbf{E}\left[X_p^-\right].
\end{aligned}$$

Therefore $\mathbf{E}[M_1] + \mathbf{E}[R_1] = \lim \mathbf{E}\left[X_p^+\right] + \lim \mathbf{E}\left[X_p^-\right] = \sup_n \mathbf{E}[|X_n|]$. ∎

The Krickeberg decomposition is not unique; see (6.2.7).

The *quadratic variation* of a process $\mathbf{X}$ is the random variable Q defined as

$$Q = X_1^2 + \sum_{i=1}^{\infty}(X_{i+1} - X_i)^2.$$

It is a curious fact (Austin [1966]) that the quadratic variation of any positive martingale is finite. First we consider bounded positive supermartingales.

(6.2.3) **Theorem.** *Let* Q *be the quadratic variation of a positive supermartingale* (X_n) *bounded by a constant* c. *Then: (a)* Q *is integrable:* $\mathbf{E}[Q] \leq 2c\mathbf{E}[X_1]$. *(b) If* $X_n = M_n - A_n$ *is Doob's decomposition, then* $\mathbf{E}[A_\infty] \leq \mathbf{E}[X_1]$ *and* (M_n) *is* L_2-*bounded:* $\|M_n\|_2^2 \leq 2c\mathbf{E}[X_1]$.

Proof. The quadratic variation Q is $\lim Q_n$, where

$$Q_n = X_1^2 + \sum_{i=1}^{n-1}(X_{i+1} - X_i)^2 = X_n^2 + 2\sum_{i=1}^{n-1}X_i(X_i - X_{i+1}).$$

Doob's decomposition $\mathbf{X} = \mathbf{M} - \mathbf{A}$ yields

$$X_i - X_{i+1} = (M_i - M_{i+1}) + (A_{i+1} - A_i).$$

Therefore we have

$$\mathbf{E}\left[Q_n\right] = \mathbf{E}\left[X_n^2 + 2\sum_{i=1}^{n-1} \mathbf{E}^{\mathcal{F}_i}\left[X_i(X_i - X_{i+1})\right]\right]$$

$$= \mathbf{E}\left[X_n^2 + 2\sum_{i=1}^{n-1} X_i(A_{i+1} - A_i)\right]$$

$$\leq c\mathbf{E}\left[X_n + 2\sum_{i=1}^{n-1}(A_{i+1} - A_i)\right]$$

$$= c\mathbf{E}\left[X_n + 2A_n\right] \leq 2c\mathbf{E}\left[X_n + A_n\right]$$

$$= 2c\mathbf{E}\left[M_n\right] = 2c\mathbf{E}\left[M_1\right] = 2c\mathbf{E}\left[X_1\right].$$

This proves (a). Since $X_n = M_n - A_n$, we have $\mathbf{E}\left[A_n\right] = \mathbf{E}\left[A_n\right] - \mathbf{E}\left[A_1\right] = \mathbf{E}\left[X_1\right] - \mathbf{E}\left[X_n\right] \leq \mathbf{E}\left[X_1\right]$, so we have by monotone convergence, $\mathbf{E}\left[A_\infty\right] \leq \mathbf{E}\left[X_1\right]$. (This does not require $X_n \leq c$.)

For the last part of (b), we prove that $\|M_n\|_2^2 \leq \mathbf{E}\left[Q\right]$. The simple "centering" identity

$$\mathbf{E}\left[(X - \mathbf{E}\left[X\right])^2\right] = \mathbf{E}\left[X^2\right] - \mathbf{E}\left[X\right]^2$$

is also true with conditional expectation in place of expectation. This fact will be used twice. First, we have

$$\mathbf{E}^{\mathcal{F}_i}\left[(M_{i+1} - M_i)^2\right] = \mathbf{E}^{\mathcal{F}_i}\left[M_{i+1}^2\right] - M_i^2.$$

Applying the centering identity again, since

$$M_{i+1} - M_i = (X_{i+1} - X_i) - \mathbf{E}^{\mathcal{F}_i}\left[(X_{i+1} - X_i)\right],$$

we have

$$\mathbf{E}^{\mathcal{F}_i}\left[(M_{i+1} - M_i)^2\right] = \mathbf{E}^{\mathcal{F}_i}\left[(X_{i+1} - X_i)^2\right] - \left(\mathbf{E}^{\mathcal{F}_i}\left[X_{i+1} - X_i\right]\right)^2$$

$$\leq \mathbf{E}^{\mathcal{F}_i}\left[(X_{i+1} - X_i)^2\right].$$

Now we can compute:

$$\|M_n\|_2^2 = \mathbf{E}\left[M_n^2\right]$$

$$= \mathbf{E}\left[M_1^2 + \sum_{i=1}^{n-1} \mathbf{E}^{\mathcal{F}_i}\left[M_{i+1}^2 - M_i^2\right]\right]$$

$$= \mathbf{E}\left[M_1^2 + \sum_{i=1}^{n-1} \mathbf{E}^{\mathcal{F}_i}\left[(M_{i+1} - M_i)^2\right]\right]$$

$$\leq \mathbf{E}\left[X_1^2 + \sum_{i=1}^{n-1} \mathbf{E}^{\mathcal{F}_i}\left[(X_{i+1} - X_i)^2\right]\right] \leq \mathbf{E}\left[Q\right].$$

So by (a), we have $\|M_n\|_2^2 \leq 2c\mathbf{E}\left[X_1\right]$.   ∎

Next we prove a maximal inequality for the quadratic variation of an arbitrary positive supermartingale. It will be convenient to state it in terms of a homogeneous expression, namely the *square function* of **X**, defined by $S = \sqrt{Q}$.

(6.2.4) Theorem. *Let (X_n) be a positive supermartingale, and let $S = \sqrt{Q}$ be its square function. Then, for every $\lambda > 0$, we have*

$$\mathbf{P}\left\{\sup_n X_n \leq \lambda, S \geq \lambda\right\} \leq \frac{2}{\lambda}\mathbf{E}\left[X_1\right]$$

$$\mathbf{P}\{S \geq \lambda\} \leq \frac{3}{\lambda}\mathbf{E}\left[X_1\right].$$

Hence S and Q are finite a.s.

Proof. Let $c > 0$ be a constant. The process $(X_n \wedge c)$ is also a supermartingale. Write $Q^{(c)}$ for its quadratic variation. By Theorem (6.2.3(a)), we have $\mathbf{E}\left[Q^{(c)}\right] \leq 2c\mathbf{E}\left[X_1 \wedge c\right] \leq 2c\mathbf{E}\left[X_1\right]$. Now on the set $\{\sup_n X_n \leq c\}$ we have $Q = Q^{(c)}$, so

$$\begin{aligned}
\mathbf{P}\{\sup_n X_n \leq c, S \geq \lambda\} &= \mathbf{P}\{\sup_n X_n \leq c, Q \geq \lambda^2\} \\
&= \mathbf{P}\{\sup_n X_n \leq c, Q^{(c)} \geq \lambda^2\} \\
&\leq \mathbf{P}\{Q^{(c)} \geq \lambda^2\} \\
&\leq \frac{1}{\lambda^2}\mathbf{E}\left[Q^{(c)}\right] \leq \frac{2c}{\lambda^2}\mathbf{E}\left[X_1\right].
\end{aligned}$$

Now choose $c = \lambda$ to obtain the first inequality in the theorem.

By Proposition (6.1.2), we have $\mathbf{P}\{\sup_n X_n \geq \lambda\} \leq (1/\lambda)\mathbf{E}\left[X_1\right]$, so

$$\begin{aligned}
\mathbf{P}\{S \geq \lambda\} &= \mathbf{P}\{S \geq \lambda, \sup_n X_n \leq \lambda\} + \mathbf{P}\{S \geq \lambda, \sup_n X_n > \lambda\} \\
&\leq \frac{2}{\lambda}\mathbf{E}\left[X_1\right] + \frac{1}{\lambda}\mathbf{E}\left[X_1\right] = \frac{3}{\lambda}\mathbf{E}\left[X_1\right]. \qquad \blacksquare
\end{aligned}$$

(6.2.5) Proposition.

(i) *If X_n is a martingale, then its square function S satisfies*

$$\mathbf{P}\{S \geq \lambda\} \leq \frac{6}{\lambda}\sup_n \mathbf{E}\left[|X_n|\right].$$

(ii) *If X_n is a positive submartingale, then*

$$\mathbf{P}\{S \geq \lambda\} \leq \frac{12}{\lambda}\sup_n \mathbf{E}\left[X_n\right].$$

Proof. Since the inequalities are trivial if $\mathbf{X}$ is not L_1-bounded, we may assume that $\mathbf{X}$ is L_1-bounded, and apply the Krickeberg decomposition (6.2.2): $\mathbf{X} = \mathbf{M} - \mathbf{R}$, where $\mathbf{M}$ is a positive martingale and $\mathbf{R}$ is a positive supermartingale. The difference sequences satisfy

$$X_n - X_{n-1} = (M_n - M_{n-1}) - (R_n - R_{n-1}).$$

Let S be the square function of $\mathbf{X}$, let S' be the square function of $\mathbf{M}$, and let S'' be the square function of $\mathbf{R}$. By the triangle inequality in the sequence space $l_2(\mathbb{N})$, we have $S \leq S' + S''$. Therefore, by Proposition (6.2.4),

$$\mathbf{P}\{S \geq \lambda\} \leq \mathbf{P}\{S' \geq \lambda/2\} + \mathbf{P}\{S'' \geq \lambda/2\}$$
$$\leq \frac{6}{\lambda}\mathbf{E}[M_1] + \frac{6}{\lambda}\mathbf{E}[R_1].$$

Next, if $\mathbf{X}$ is a martingale, then by (6.2.2) $\sup_n \mathbf{E}\left[|X_n|\right] = \mathbf{E}[M_1] + \mathbf{E}[R_1]$, which proves (i). If $\mathbf{X}$ is a positive submartingale, then $\mathbf{E}[M_1] = \lim_p \uparrow \mathbf{E}[X_p] = \sup_n \mathbf{E}[X_n]$ and $R_1 = M_1 - X_1$ implies that $\mathbf{E}[R_1] \leq \sup_n \mathbf{E}[X_n]$, which proves (ii). ∎

Part (i) is proved by Burkholder [1973] with constant 3 rather than 6.

Martingale transforms

Let $\mathbf{X} = (X_n)$ be a sequence of random variables. Its *difference sequence* is the process $\mathbf{Y} = (Y_n)$ so that $X_n = \sum_{i=1}^{n} Y_i$. If $\mathbf{V} = (V_n)$ is another process, then the *transform* of $\mathbf{X}$ by $\mathbf{V}$ is the process $\mathbf{Z} = (Z_n)$ given by $Z_n = \sum_{i=1}^{n} V_i Y_i$. We will sometimes write $\mathbf{Z} = \mathbf{V} * \mathbf{X}$.

If X_n is the fortune of a gambler at time n, then the transform Z_n may be viewed as the result of controlling $\mathbf{X}$ by $\mathbf{V}$. If we assume that $\mathbf{V}$ is predictable, then multiplication of Y_i by V_i is equivalent to changing the stakes for the ith game on the basis of information available before the ith game.

If $\mathbf{X}$ is a martingale, and $\mathbf{V}$ is predictable, then the transform $\mathbf{Z} = \mathbf{V} * \mathbf{X}$ is also a martingale (see Section 3.2).

We will now prove that the transform of an L_1-bounded martingale by a bounded predictable process converges. This is a surprising result, since (see (6.2.8)) the transform need not be L_1-bounded.

(6.2.6) Theorem. *Let $\mathbf{X}$ be an L_1-bounded submartingale or an L_1-bounded supermartingale. If $\mathbf{V}$ is an L_∞-bounded predictable process, then the transform $\mathbf{Z} = \mathbf{V} * \mathbf{X}$ converges a.s.*

Proof. We may assume that $|V_n| \leq 1$ for all n by multiplying by a constant. If $\mathbf{X}$ is an L_1-bounded submartingale, and $\mathbf{X} = \mathbf{X}' - \mathbf{X}''$ is the Krickeberg decomposition (6.2.2), then $\mathbf{V} * \mathbf{X} = \mathbf{V} * \mathbf{X}' - \mathbf{V} * \mathbf{X}''$, and $\mathbf{X}', \mathbf{X}''$ are both positive supermartingales; therefore it is enough to prove the result when $\mathbf{X}$ is a positive supermartingale.

First suppose the positive supermartingale $\mathbf{X}$ is L_∞-bounded, say $X_n \leq c$ a.s. Let $\mathbf{X} = \mathbf{M} - \mathbf{A}$ be the Doob decomposition of $\mathbf{X}$. Since $\mathbf{V} * \mathbf{X} = \mathbf{V} * \mathbf{M} - \mathbf{V} * \mathbf{A}$, it suffices to show that both of these parts converge a.s.

Convergence of $\mathbf{V} * \mathbf{A}$ follows from the monotone convergence of A_n to A_∞, since

$$|(\mathbf{V} * \mathbf{A})_n - (\mathbf{V} * \mathbf{A})_{n-1}| \leq |V_n| |A_n - A_{n-1}|$$
$$\leq A_n - A_{n-1},$$

and, using (6.2.3(b)), $\mathbf{E}[A_\infty] \leq \mathbf{E}[X_1] \leq c$.

Next, we consider convergence of $\mathbf{V} * \mathbf{M}$. By Theorem (6.2.3(b)), $\mathbf{M}$ is L_2-bounded. This implies that $\mathbf{V} * \mathbf{M}$ is also L_2-bounded, since

$$(\mathbf{V} * \mathbf{M})_1^2 = V_1^2 M_1^2 \leq M_1^2,$$
$$\left[(\mathbf{V} * \mathbf{M})_{n+1} - (\mathbf{V} * \mathbf{M})_n \right]^2 = V_{n+1}^2 (M_{n+1} - M_n)^2$$
$$\leq (M_{n+1} - M_n)^2,$$

and

$$\mathbf{E}\left[(\mathbf{V} * \mathbf{M})_n^2 \right] = \mathbf{E}\left[(\mathbf{V} * \mathbf{M})_1^2 + \sum_{k=1}^{n-1} \left[(\mathbf{V} * \mathbf{M})_{k+1} - (\mathbf{V} * \mathbf{M})_k \right]^2 \right]$$
$$\leq \mathbf{E}\left[M_1^2 + \sum_{k=1}^{n-1} (M_{k+1} - M_k)^2 \right]$$
$$= \mathbf{E}\left[M_n^2 \right] \leq 2c\mathbf{E}[X_1].$$

Now $\mathbf{V} * \mathbf{M}$ is L_2-bounded, hence L_1-bounded, and therefore it converges a.s.

For a general positive supermartingale $\mathbf{X}$, fix a positive constant c and consider the process $\mathbf{X}'$ defined by $X_n' = X_n \wedge c$. The process $\mathbf{X}'$ is a bounded positive supermartingale. So by the previous part of the proof, the process $\mathbf{V} * \mathbf{X}'$ converges a.s. The transform $\mathbf{V} * \mathbf{X}'$ agrees with $\mathbf{V} * \mathbf{X}$ on the set $\Omega_c = \{\sup_n |X_n| \leq c\}$. Therefore $\mathbf{V} * \mathbf{X}$ converges a.s. on Ω_c. By the maximal inequality (6.1.2a), the space Ω is a.s. a countable union of sets $\Omega_m = \{\sup_n |X_n| \leq m\}$, so $\mathbf{V} * \mathbf{X}$ converges a.s. Finally, if $\mathbf{X}$ is an L_1-bounded supermartingale, then $-\mathbf{X}$ is an L_1-bounded submartingale, so $\mathbf{V} * \mathbf{X} = -\left[\mathbf{V} * (-\mathbf{X}) \right]$ converges a.s. ∎

Complements

(6.2.7) The Krickeberg decomposition $\mathbf{X} = \mathbf{M} - \mathbf{R}$ of Theorem (6.2.2) is not unique, since any positive martingale can be added to each term. If M_n is defined as $\lim_p \mathbf{E}^{\mathscr{F}_n}\left[X_p^+ \right]$, as in the proof of Theorem (6.2.2), then the decomposition is minimal, in the sense that there is no positive martingale dominated by both $\mathbf{M}$ and $\mathbf{R}$.

(6.2.8) (Example of unbounded transform.) Let $\Omega = \{1, 2, 3, \cdots\}$ with $\mathbf{P}\{k\} = 1/k(k+1)$, so that $\mathbf{P}\{k, k+1, \cdots\} = 1/k$. Define the process $\mathbf{X}$ by

$$X_n(k) = \begin{cases} -1 & \text{if } k \leq n \\ n & \text{if } k > n. \end{cases}$$

Then $\|X_n\|_1 \leq 2$ for all n. The process is adapted to the stochastic basis $(\mathcal{F}_n)$, where $\mathcal{F}_n$ is the finite σ-algebra with the $n+1$ atoms

$$\{1\}, \{2\}, \cdots, \{n\}, \{n+1, n+2, \cdots\}.$$

Let $Y_n = X_n - X_{n-1}$ be the difference sequence. To prove that (X_n) is a martingale, we will show that $\mathbf{E}^{\mathcal{F}_n}[Y_{n+1}] = 0$. Now

$$Y_n(k) = \begin{cases} 0 & \text{if } k < n; \\ -n & \text{if } k = n; \\ 1 & \text{if } k > n. \end{cases}$$

The only atom of $\mathcal{F}_n$ on which Y_{n+1} does not vanish is $A = \{n+1, n+2, \cdots\}$. Thus we must show $\mathbf{E}[1_A Y_{n+1}] = 0$. Indeed,

$$\mathbf{E}[1_A Y_{n+1}] = -(n+1)\mathbf{P}\{n+1\} + 1 \cdot \mathbf{P}\{n+2, n+3, \cdots\}$$
$$= -(n+1)\frac{1}{(n+1)(n+2)} + \frac{1}{n+2} = 0.$$

We will transform by the process $V_i = (-1)^{i+1}$. Let $\mathbf{Z} = \mathbf{V} * \mathbf{X}$. Then we have $Z_n(2k-1) = -(2k-1)$ for $n \geq 2k-1$ and $Z_n(2k) = 2k+1$ for $n \geq 2k$. Thus

$$\lim_n \|Z_n\|_1 \geq \sum_{k=1}^{\infty} \left((2k-1)\mathbf{P}\{2k-1\} + (2k+1)\mathbf{P}\{2k\}\right) = \infty.$$

Hence $\mathbf{Z}$ is not L_1-bounded and the limit $Z_\infty = \lim_{n \to \infty} Z_n$ (which exists a.s. by Theorem (6.2.6)) is not integrable.

Remarks

Theorem (6.2.6) on the convergence of martingale transforms is due to Burkholder [1966]. For sharp L_p inequalities for martingale transforms, see Burkholder [1984], [1988], and [1989].

6.3. The norm of the square function of a martingale

In this section we will study the relationship between a norm of a martingale and the corresponding norm of the square function of the martingale.

Burkholder's inequalities (Theorem (6.3.6)) show that the L_p norms are "equivalent" in a strong sense for $1 < p < \infty$.

Suppose $\mathbf{X} = (X_n)$ is a stochastic process adapted to the stochastic basis $(\mathcal{F}_n)$. We will use the generic notations: set $X_0 = 0$; the *difference process*: $\Delta X_n = X_n - X_{n-1}$; *maximal functions*: $\mathbf{X}^* = \sup_{k \in \mathbb{N}} |X_k|$ and $X_n^* = \max_{1 \le k \le n} |X_k|$; *square functions*: $S(\mathbf{X}) = \left(\sum_{k=1}^{\infty} (\Delta X_k)^2 \right)^{1/2}$ and $S_n(\mathbf{X}) = \left(\sum_{k=1}^{n} (\Delta X_k)^2 \right)^{1/2}$; the L_p *norm of the process*: $\|\mathbf{X}\|_p = \sup_k \|X_k\|_p$.

In this section, we will consider a process $\mathbf{X} = (X_n)$, its difference process $Y_n = \Delta X_n$, and its square function process $S_n = S_n(\mathbf{X})$. By convention $X_0 = Y_0 = 0$. We will also sometimes use the convention $X_\infty = \limsup X_n$, so that X_τ is defined for stopping times τ that take the value ∞.

(6.3.1) Lemma. *Let $\mathbf{X}$ be an L_1-bounded martingale or positive submartingale. Let $\lambda > 0$. Define $\tau = \inf \{ n : |X_n| > \lambda \}$ (finite or infinite). Then*

$$(6.3.1a) \qquad \|S_{\tau-1}\|_2^2 + \|X_{\tau-1}\|_2^2 \le 2\mathbf{E}\left[X_{\tau-1} X_\tau \right] \le 2\lambda \|\mathbf{X}\|_1.$$

In the martingale case, the first inequality is equality.

Proof. Since $|X_{\tau-1}| \le \lambda$, we have $\mathbf{E}\left[|X_{\tau-1} X_\tau| \right] \le \lambda \mathbf{E}\left[|X_\tau| \right] \le \lambda \|\mathbf{X}\|_1$. Hence it suffices to prove the first inequality. By elementary algebra,

$$
\begin{aligned}
S_{n-1}^2 + X_{n-1}^2 &= 2 \sum_{1 \le j \le k \le n-1} Y_j Y_k \\
&= 2 \sum_{1 \le j \le n-1} Y_j (X_{n-1} - X_{j-1}) \\
&= 2 X_{n-1}^2 - 2 \sum_{1 \le j \le n-1} X_{j-1} Y_j \\
&= 2 \left[X_n X_{n-1} - \sum_{1 \le j \le n} X_{j-1} Y_j \right].
\end{aligned}
$$

Now note that $\mathbf{E}\left[X_{j-1} Y_j \right] = \mathbf{E}\left[\mathbf{E}^{\mathcal{F}_{j-1}} \left[X_{j-1} Y_j \right] \right] = \mathbf{E}\left[X_{j-1} \mathbf{E}^{\mathcal{F}_{j-1}} \left[Y_j \right] \right] \ge 0$ (or $= 0$ in the martingale case). Hence applying the expectation, we have

$$\|S_{n-1}\|_2^2 + \|X_{n-1}\|_2^2 \le 2\mathbf{E}\left[X_{n-1} X_n \right]$$

with equality in the martingale case. Now by the optional sampling theorem (1.4.29), the process $(X_{\tau \wedge n})$ is also a martingale (or positive submartingale). So this inequality remains true:

$$\|S_{\tau \wedge (n-1)}\|_2^2 + \|X_{\tau \wedge (n-1)}\|_2^2 \le 2\mathbf{E}\left[X_{\tau \wedge (n-1)} X_{\tau \wedge n} \right]$$

with equality in the martingale case. Now we have $|X_{\tau \wedge n}| \le |X_\tau|$ except on the set $\{\tau = \infty\}$, where $|X_{\tau \wedge n}| \le \lambda$. Hence the sequence $\big(|X_{\tau \wedge (n-1)} X_{\tau \wedge n}|\big)$ is dominated by the integrable random variable $\lambda(|X_\tau| + \lambda)$. Now $X_{\tau \wedge n} \to X_\tau$ as $n \to \infty$, even on $\{\tau = \infty\}$, by the convergence theorem (1.2.5). So we obtain the first inequality of (6.3.1a) with the dominated convergence theorem. ∎

We now prove the submartingale analog of Theorem (6.1.3). Recall that this theorem was proved for positive supermartingales (Lemma (6.2.4)) using Krickeberg's decomposition. The estimates given here for positive submartingales are sharper than those in Proposition (6.2.5(ii)).

(6.3.2) Lemma. *Let* $\mathbf{X}$ *be a martingale or a positive submartingale; let* $\mathbf{X}^*$ *be its maximal function and* S *its square function. Then for every* $\lambda > 0$

(6.3.2a) $$\mathbf{P}\{\mathbf{X}^* \le \lambda, S > \lambda\} \le \frac{2}{\lambda}\|\mathbf{X}\|_1,$$

(6.3.2b) $$\mathbf{P}\{S > \lambda\} \le \frac{3}{\lambda}\|\mathbf{X}\|_1.$$

Proof. By Proposition (6.1.5), we have $\lambda\mathbf{P}\{\mathbf{X}^* > \lambda\} \le \|\mathbf{X}\|_1$, so (6.3.2a) implies (6.3.2b). So it suffices to prove (6.3.2a). Let $\tau = \inf\{n : |X_n| > \lambda\}$. On the set $A = \{\tau = \infty\}$, we have $\mathbf{X}^* \le \lambda$. Hence by Lemma (6.3.1),

$$\mathbf{P}\{\mathbf{X}^* \le \lambda, S > \lambda\} \le \mathbf{P}\{S_\tau > \lambda\}$$
$$\le \frac{1}{\lambda^2}\mathbf{E}\left[S_\tau^2\right]$$
$$\le \frac{2}{\lambda}\|\mathbf{X}\|_1.$$
∎

Next we will consider square functions of sequences of random variables other than $\mathbf{X}$. We use the the notations $S(\mathbf{X})$ and $S_n(\mathbf{X})$ for square functions as defined above.

We begin with a process $\mathbf{X} = (X_n)$, fix a positive number θ, and consider the sequence $T_n = S_n(\theta\mathbf{X}) \vee X_n^*$. We first obtain a weak maximal inequality for T_n, and then apply standard arguments to obtain the desired strong L_p inequality.

(6.3.3) Lemma. *Let* $\mathbf{X}$ *be a positive submartingale, let* θ *be a positive number and let* $\beta = \sqrt{1 + 2\theta^2}$. *Then for every positive* λ,

$$\lambda\mathbf{P}\{T_n > \beta\lambda\} \le 3\,\mathbf{E}\left[X_n\,\mathbf{1}_{\{T_n > \lambda\}}\right].$$

Proof. Let $Z_n = X_n\,\mathbf{1}_{\{S_n(\theta\mathbf{X}) > \lambda\}}$. Let $\tau = \inf\{n : S_n(\theta\mathbf{X}) > \lambda\}$. On the set $B = \{S_n(\theta\mathbf{X}) > \beta\lambda, X_n^* \le \lambda\}$, we have $S_n(\theta\mathbf{X}) > \lambda$ since $\beta > 1$, hence

$\tau \leq n$ and $Z_n^* \leq \lambda$ since $Z_n \leq X_n$. Also on B, we have $|Y_\tau| = |X_\tau - X_{\tau-1}| \leq X_\tau \vee X_{\tau-1} \leq X_n^* \leq \lambda$, so that, since $S_{\tau-1}(\theta \mathbf{X}) \leq \lambda$, we have

$$
\begin{aligned}
(1+2\theta^2)\lambda^2 &= \beta^2 \lambda^2 \\
&\leq \big(S_n(\theta \mathbf{X})\big)^2 \\
&= \big(S_{\tau-1}(\theta \mathbf{X})\big)^2 + \theta^2 Y_\tau^2 + \theta^2 \sum_{j=\tau+1}^{n} Y_j^2 \\
&\leq \lambda^2 + \theta^2 \lambda^2 + \theta^2 \sum_{j=\tau+1}^{n} (Z_j - Z_{j-1})^2 \\
&\leq (1+\theta^2)\lambda^2 + \theta^2 \big(S_n(\mathbf{Z})\big)^2.
\end{aligned}
$$

This implies that $S_n(\mathbf{Z}) > \lambda$ holds on B. Now $\mathbf{Z}$ is a submartingale, since the product of an increasing adapted sequence and a positive submartingale is a submartingale and the process $\mathbf{1}_{\{S_n(\theta \mathbf{X})>\lambda\}}$ is increasing. Applying Lemma (6.3.2) to the submartingale

$$(Z_1, Z_2, \cdots, Z_n, Z_n, \cdots),$$

we obtain

$$
\begin{aligned}
\lambda \mathbf{P}\{X_n^* \leq \beta\lambda, S_n(\theta \mathbf{X}) > \lambda\} &\leq \lambda \mathbf{P}\{Z_n^* \leq \lambda, S_n(\mathbf{Z}) > \lambda\} \\
&\leq 2\|\mathbf{Z}\|_1 = 2\mathbf{E}\big[X_n \mathbf{1}_{\{S_n(\theta \mathbf{X})>\lambda\}}\big].
\end{aligned}
$$

Now applying the classical weak Doob's inequality (1.4.18), we have

$$\lambda \mathbf{P}\{X_n^* > \lambda\} \leq \mathbf{E}\big[X_n \mathbf{1}_{\{X_n^*>\lambda\}}\big] \leq \mathbf{E}\big[X_n \mathbf{1}_{\{T_n>\lambda\}}\big].$$

Combining,

$$
\begin{aligned}
\lambda \mathbf{P}\{T_n > \beta\lambda\} &\leq \lambda \mathbf{P}\{X_n^* > \lambda\} + \lambda \mathbf{P}\{S_n(\theta \mathbf{X}) > \beta\lambda, X_n^* \leq \lambda\} \\
&\leq 3\mathbf{E}\big[X_n \mathbf{1}_{\{T_n>\lambda\}}\big]. \qquad \blacksquare
\end{aligned}
$$

Now that we have a weak inequality, we may use the three-function inequality to pass to a strong inequality between the process $\mathbf{X}$ and its square function S.

(6.3.4) Proposition. *Let $\mathbf{X} = (X_n)$ be a positive submartingale. Let $1 < p < \infty$. Then*

(6.3.4a)
$$\|S_n\|_p \leq \frac{3\sqrt{2e}\,p^{3/2}}{p-1}\|X_n\|_p$$

(6.3.4b)
$$\|S\|_p \le \frac{3\sqrt{2e}\, p^{3/2}}{p-1} \|\mathbf{X}\|_p.$$

Proof. It suffices to prove (6.3.4a), since (6.3.4b) follows when $n \to \infty$. Applying (3.1.17) and (6.3.3), we obtain

$$\theta\|S_n\|_p \le \|T_n\|_p \le 3\beta^p q \|X_n\|_p,$$

where $q = p/(p-1)$. The best choice of θ is the one that minimizes $\beta^p/\theta = (1+2\theta^2)^{p/2}/\theta$. So θ is the positive solution of the equation $2\theta^2 - 2p\theta^2 + 1 = 0$, namely $\theta = 1/\sqrt{2(p-1)}$. Then

$$\frac{3\beta^p q}{\theta} = \frac{3\sqrt{2}p^{3/2}}{p-1}\left(1 + \frac{1}{p-1}\right)^{(p-1)/2}.$$

The last factor increases to $e^{1/2}$ as $p \to \infty$, so inequality (6.3.4a) is proved. ∎

A duality argument will now give an inverse inequality.

(6.3.5) **Proposition.** *Let* $\mathbf{X} = (X_n)$ *be a positive martingale. Let* $1 < p < \infty$ *and set* $q = p/(p-1)$. *Then*

(6.3.5a)
$$\|X_n\|_p \le \frac{3\sqrt{2e}\, q^{3/2}}{\sqrt{p-1}} \|S_n\|_p$$

(6.3.5b)
$$\|\mathbf{X}\|_p \le \frac{3\sqrt{2e}\, q^{3/2}}{\sqrt{p-1}} \|S\|_p.$$

Proof. It suffices to prove (6.3.5a). Fix the integer n. Without loss of generality, we may assume that $\|S_n\|_p < \infty$. Write $Y_j = \Delta X_j$ and $Y_n^* = \sup_{1 \le j \le n} |Y_j|$. Since $X_n \le nY_n^*$ and $Y_n^* \le S_n$, we may conclude that $X_n \in L_p^+$. Set $R_n = X_n^{p-1}$ and $R_j = \mathbf{E}^{\mathcal{F}_j}[R_n]$ for $1 \le j \le n$. Let $q = p/(p-1)$ be the conjugate exponent. Then $R_n \in L_q$, since $X_n \in L_p$; also $\|R_n\|_q = \|X_n\|_p^{p-1}$ and $\mathbf{E}[X_n R_n] = \|X_n\|_p^p$. Now $(R_1, R_2, \cdots, R_n)$ is a positive martingale. Therefore (6.3.4a) applies to it (with q replacing p); also using orthogonality and the inequalities of Schwartz and Hölder, we have

$$\|X_n\|_p^p = \mathbf{E}\,[X_n R_n] = \mathbf{E}\,[(X_{n-1} + \Delta X_n)(R_{n-1} + \Delta R_n)]$$
$$= \mathbf{E}\,[X_{n-1}R_{n-1} + \Delta X_n \Delta R_n]$$
$$= \mathbf{E}\,[X_{n-2}R_{n-2} + \Delta X_{n-1}\Delta R_{n-1} + \Delta X_n \Delta R_n]$$
$$= \cdots = \mathbf{E}\left[\sum_{j=1}^{n} \Delta X_j \Delta R_j\right]$$
$$\leq \mathbf{E}\,[S_n(\mathbf{X})S_n(\mathbf{R})]$$
$$\leq \|S_n(\mathbf{X})\|_p\, \|S_n(\mathbf{R})\|_q$$
$$\leq \|S_n(\mathbf{X})\|_p\, \frac{3\sqrt{2e}\,q^{3/2}}{q-1}\,\|R_n\|_q$$
$$= \frac{3\sqrt{2e}\,q^{3/2}}{\sqrt{q-1}}\,\|X_n\|_p^{p-1}\,\|S_n(\mathbf{X})\|_p.$$

Finally, divide by $\|X_n\|_p^{p-1}$ to obtain (6.3.5a).   ∎

(6.3.6) Theorem (Burkholder). *Let* $1 < p < \infty$ *and* $q = p/(p-1)$. *There are constants* $c_p = (p-1)/(6\sqrt{2e}\,p^{3/2})$ *and* $C_p = 6\sqrt{2e}\,q^{3/2}/\sqrt{q-1}$ *such that for every martingale* (X_n),

$$c_p\|S_n\|_p \leq \|X_n\|_p \leq C_p\|S_n\|_p,$$
$$c_p\|S\|_p \leq \|X\|_p \leq C_p\|S\|_p.$$

Proof. As usual, it suffices to prove the first pair of inequalities. Fix n. We apply the Krickeberg decomposition (6.2.2) to the martingale

$$(X_1, X_2, \cdots, X_n, X_n, \cdots).$$

In fact the proof of the decomposition is easy in the present case; the general case is only invoked to motivate the choice of $\mathbf{M}$ and $\mathbf{R}$: Set

$$M_n = \lim_p \uparrow\ \mathbf{E}^{\mathcal{F}_n}\left[X_p^+\right] = X_n^+,$$
$$M_j = \lim_p \uparrow\ \mathbf{E}^{\mathcal{F}_j}\left[X_p^+\right] = \mathbf{E}^{\mathcal{F}_j}\left[X_n^+\right] \qquad \text{for } 1 \leq j \leq n-1.$$

Now if $R_j = M_j - X_j$, then $\mathbf{M} = (M_j)$ and $\mathbf{R} = (R_j)$ are both positive martingales, with $\mathbf{X} = \mathbf{M} - \mathbf{R}$. Note that $M_n = X_n^+$ and $R_n = X_n^-$; so $\|M_n\|_p \leq \|X_n\|_p$ and $\|R_n\|_p \leq \|X_n\|_p$, hence $\|M_n\|_p + \|R_n\|_p \leq 2\|X_n\|_p$.

Now the difference sequences satisfy $\Delta X_j = \Delta M_j - \Delta R_j$. Thus by the triangle inequality in n-dimensional Euclidean space, we have $S_n(\mathbf{X}) \leq S_n(\mathbf{M}) + S_n(\mathbf{R})$. Therefore, by (6.3.4), we have

$$\|S_n(\mathbf{X})\|_p \leq \|S_n(\mathbf{M})\|_p + \|S_n(\mathbf{R})\|_p$$
$$\leq \frac{3\sqrt{2e}\,p^{3/2}}{p-1}\left(\|M_n\|_p + \|R_n\|_p\right)$$
$$\leq \frac{6\sqrt{2e}\,p^{3/2}}{p-1}\,\|X_n\|_p = c_p^{-1}\|X_n\|_p,$$

which proves the left-hand inequality.

For the right-hand inequality, we again use duality. Since X_n need not be positive, this may be done using the *duality map* $J: L_p \to L_q$ defined by:

$$(Jf)(\omega) = \begin{cases} |f(\omega)|^p / f(\omega) & \text{if } f(\omega) \neq 0 \\ 0 & \text{otherwise.} \end{cases}$$

(In the case of complex random variables, division by $f(\omega)$ is replaced with division by the complex conjugate $\overline{f(\omega)}$.) If $f \in L_p$, then $Jf \in L_q$, $\|Jf\|_q = \|f\|_p^{p-1}$, and $\mathbf{E}\left[f \cdot Jf\right] = \|f\|_p^p$ as before.

Now proceed as in the proof of Proposition (6.3.5) using JX_n. ∎

Complements

(6.3.7) Theorem. *Let Φ be an Orlicz function, let Ψ be its conjugate and let φ be its derivative. Suppose $\xi(u) = \Psi(\varphi(u))$ is an Orlicz function satisfying the (Δ_2) condition. Suppose $X \in L_\Phi$ and T are positive functions satisfying*

$$(6.3.7a) \qquad \lambda \mathbf{P}\{T > \beta \lambda\} \leq \alpha \mathbf{E}\left[X \, \mathbf{1}_{\{T > \lambda\}}\right]$$

for constants α, β. Then, if c is the constant in the (Δ_2) condition, i.e. $\xi(\beta u) \leq c\xi(u)$ for all u, and $d > 0$, then

$$dM_\xi\left(\frac{T}{\alpha\beta(c+d)\|X\|_\Phi}\right) \leq 1 \qquad \text{and} \qquad \|T\|_\xi \leq \alpha\beta(c+1)\|X\|_\Phi.$$

Proof. Verify the hypothesis of the three-function inequality (3.1.2) with $g = T/(\alpha\beta\|X\|_\Phi)$, $f = X/\|X\|_\Phi$, $h = T/(\alpha\|X\|_\Phi)$, $a = 1$, and $b = c + d$. Therefore

$$(c+d)M_\xi\left(\frac{T}{\alpha\beta(c+d)\|X\|_\Phi}\right) - M_\xi\left(\frac{T}{\alpha(c+d)\|X\|_\Phi}\right) \leq M_\Phi(f) = 1.$$

(By Corollary (3.1.9), this difference of M_ξ's is not $\infty - \infty$.) By (Δ_2),

$$M_\xi\left(\frac{T}{\alpha(c+d)\|X\|_\Phi}\right) \leq cM_\xi\left(\frac{T}{\alpha\beta(c+d)\|X\|_\Phi}\right),$$

so

$$dM_\xi\left(\frac{T}{\alpha\beta(c+d)\|X\|_\Phi}\right) \leq 1.$$

Choosing $d = 1$ gives $\|T\|_\xi \leq \alpha\beta(c+1)\|X\|_\Phi$. ∎

(6.3.8) Corollary. *Let* $\mathbf{X}$ *be a positive submartingale,* $\theta > 0$, *and* $\beta = \sqrt{1 + 2\theta^2}$. *Then*

$$\|S_n\|_\xi \leq \frac{3\beta(c+1)}{\theta} \, \|X_n\|_\Phi.$$

Proof. Apply the preceding Theorem, using Lemma (6.3.3), where $\alpha = 3$. ∎

(6.3.9) (Square function and maximal function.) Since the control of the square function S by $\mathbf{X}$ (6.3.8) is similar to control of the maximal function X^* by $\mathbf{X}$ (3.1.13), one may expect that S may be equivalent to X^*. This is indeed the case:

Let Φ be an Orlicz function such that $\Phi(u)/u \to \infty$, satisfying the (Δ_2) condition $\Phi(2u) \leq c\Phi(u)$ for some constant $c > 0$. Then there exist constants b_c and B_c such that for every martingale $\mathbf{X}$, we have $b_c M_\Phi[S(\mathbf{X})] \leq M_\Phi[X^*] \leq B_c M_\Phi[S(\mathbf{X})]$ (Burkholder, Davis, & Gundy [1972]).

An L_1 version of this result is also true:

There exists constants b, B such that for every L_1-bounded martingale $\mathbf{X}$, we have $b\mathbf{E}[S(\mathbf{X})] \leq \mathbf{E}[X^*] \leq B\mathbf{E}[S(\mathbf{X})]$ (B. Davis [1969]).

There is further a recent extension to more abstract spaces due to Johnson & Schechtman [1988].

(6.3.10) (Converse false.) Corollary (6.3.8) shows that if $\sup \|X_n\|_\Phi < \infty$, then $\sup \|S_n\|_\xi < \infty$. The converse is false: choose $X_n = X_1$ not in $L \log^k L$; then $S_n = 0$ is in $L \log^{k-1} L$.

(6.3.11) (Best constants.) The inequalities in (6.3.6) can be improved. Let $1 < p < \infty$, $1/p + 1/q = 1$, and $p^* = \max\{p, q\}$. Then the constant c_p can be replaced by $(p^* - 1)^{-1}$, and C_p by $p^* - 1$. The left-hand side inequality is optimal if $1 < p \leq 2$; the right-hand side if $2 \leq p < \infty$. The best constants in other cases are not known. The results extend to martingales taking values in a Hilbert space. (See D. L. Burkholder [1991], Theorem 3.3, or Burkholder [1988].)

Remarks

The first version of the inequality (6.3.6) is in Burkholder [1966].

6.4. Lifting

This section will discuss liftings for measure spaces. A measure defines an equivalence relation on the measurable sets: two sets are equivalent if they agree almost everywhere. A *lifting* is a choice of one set from each equivalence class in such a way as to preserve as much of the structure as possible. (The technical definition is given below.) The existence of liftings may be established using the martingale convergence theorem, so the topic is appropriate for discussion here. Liftings have also been used to choose a good basis for differentiation of integrals.

Throughout Section 6.4, we will write $(\Omega, \mathcal{F}, \mu)$ for a nonzero complete σ-finite measure space.

We will write $A \equiv B$ to mean $\mu(A \bigtriangleup B) = 0$. Similarly, we will write $f \equiv g$ to mean $\{\, \omega \in \Omega : f(\omega) \neq g(\omega) \,\}$ has measure 0.

In some places, it will be important to distinguish between an equivalence class of measurable functions and an actual function. In order to do this, we will use the following notational conventions. For $0 < p < \infty$, the notation $\mathcal{L}_p(\Omega, \mathcal{F}, \mu) = \mathcal{L}_p(\mu) = \mathcal{L}_p$ will denote the set of all real-valued $\mathcal{F}$-measurable functions f on Ω with $\int |f|^p \, d\mu < \infty$. The notation $\mathcal{L}_\infty(\Omega, \mathcal{F}, \mu)$ will denote the set of all bounded real-valued $\mathcal{F}$-measurable functions on Ω. The notation $L_p(\Omega, \mathcal{F}, \mu)$ will denote the quotient of $\mathcal{L}_p(\Omega, \mathcal{F}, \mu)$ obtained by identifying functions that agree almost everywhere. If $f \in \mathcal{L}_p$, the corresponding equivalence class will be written with a dot, $f^\bullet$, when it is important to distinguish between the two. For $f \in \mathcal{L}_\infty$, we will write $\|f\|_u$ for the uniform norm:

$$\|f\|_u = \sup\{\, |f(\omega)| : \omega \in \Omega \,\},$$

and $\|f\|_\infty$ for the usual $\mathcal{L}_\infty$ seminorm:

$$\|f\|_\infty = \inf\{\, M \in \mathbb{R} : \mu\{\, \omega : |f(\omega)| > M \,\} = 0 \,\}.$$

We also write $\|f^\bullet\|_\infty = \|f\|_\infty$ for the quotient norm on L_∞.

Existence of liftings

There are two common formulations of the basic definitions concerning liftings. We use here the two terms *lifting* and *density* to distinguish between them. (Some authors use "lifting" to mean what is here called "density," and others use "density" for what is here called "lower density.")

Roughly speaking, a "lifting" chooses one element of each equivalence class of L_∞ in a nice way.

(6.4.1) Definition. A *lifting* for the measure space $(\Omega, \mathcal{F}, \mu)$ is a function $\rho \colon \mathcal{L}_\infty(\Omega, \mathcal{F}, \mu) \to \mathcal{L}_\infty(\Omega, \mathcal{F}, \mu)$ satisfying

   (i) $\rho(f) \equiv f$;
   (ii) if $f \equiv g$, then $\rho(f) = \rho(g)$;
   (iii) $\rho(1) = 1$;
   (iv) if $f \geq 0$, then $\rho(f) \geq 0$;
   (v) $\rho(f + g) = \rho(f) + \rho(g)$; $\rho(af) = a\rho(f)$, $a \in \mathbb{R}$;
   (vi) $\rho(fg) = \rho(f)\rho(g)$.

(In fact, condition (iv) follows from the others.)

(6.4.2) Definition. A *density* for $(\Omega, \mathcal{F}, \mu)$ is a function $\delta \colon \mathcal{F} \to \mathcal{F}$ satisfying

   (a) $\delta(A) \equiv A$;
   (b) if $A \equiv B$, then $\delta(A) = \delta(B)$;
   (c) $\delta(\Omega) = \Omega$; $\delta(\emptyset) = \emptyset$;
   (d) $\delta(A \cap B) = \delta(A) \cap \delta(B)$;
   (e) $\delta(A \cup B) = \delta(A) \cup \delta(B)$.

A *lower density* is a function δ satisfying (a), (b), (c), (d). An *upper density* is a function δ satisfying (a), (b), (c), (e).

If $\mathcal{G}$ is an algebra of subsets of Ω that meets each μ-equivalence class of $\mathcal{F}$ exactly once, then $\mathcal{G}$ is the range of a unique density, namely $\delta(A)$ is the element of $\mathcal{G}$ in the equivalence class of A. Conversely, if δ is a density for $(\Omega, \mathcal{F}, \mu)$, then the range of δ is an algebra of sets that meets each equivalence class exactly once.

If δ is a lower density, then the function δ' defined by $\delta'(A) = \Omega \setminus \delta(\Omega \setminus A)$ is an upper density, and $\delta'(A) \supseteq \delta(A)$. We will say that δ' is the upper density *corresponding* to δ. If δ is a density, then $\delta' = \delta$.

It will take some effort to produce a nontrivial example of a lifting or a density. But upper and lower densities can be easily exhibited. Consider the measure space $\mathbb{R}$ with Lebesgue measure λ. The Lebesgue density theorem (7.1.12) can be used to show that

$$\delta(A) = \left\{ x : \lim_{h \downarrow 0} \frac{\lambda(A \cap (x - h, x + h))}{2h} = 1 \right\}$$

defines a lower density (called *Lebesgue lower density*),

$$\delta'(A) = \left\{ x : \limsup_{h \downarrow 0} \frac{\lambda(A \cap (x - h, x + h))}{2h} > 0 \right\}$$

defines an upper density, and $\delta(A) \subseteq \delta'(A)$.

Liftings and densities are closely connected.

(6.4.3) Proposition. *(1) Let ρ be a lifting. If $A \in \mathcal{F}$, then the function $\rho(\mathbf{1}_A)$ is an indicator function. The equation $\rho(\mathbf{1}_A) = \mathbf{1}_{\delta(A)}$ defines a density δ.*

(2) If δ is a density, then there is a unique lifting ρ such that $\rho(\mathbf{1}_A) = \mathbf{1}_{\delta(A)}$ for $A \in \mathcal{F}$.

Proof. (1) Let ρ be a lifting. If $A \in \mathcal{F}$, then by condition (vi) of Definition (6.4.1), we have $\rho(\mathbf{1}_A)^2 = \rho(\mathbf{1}_A^2) = \rho(\mathbf{1}_A)$. So for all $\omega \in \Omega$, either $\rho(\mathbf{1}_A)(\omega) = 0$ or $\rho(\mathbf{1}_A)(\omega) = 1$. Therefore, $\rho(\mathbf{1}_A)$ is an indicator function, say $\mathbf{1}_{\delta(A)}$. Properties (a), (b), (c) of Definition (6.4.2) for δ follow from properties (i), (ii), (iii) of Definition (6.4.1) for ρ. Next, note that $\rho(\mathbf{1}_{\Omega \setminus A}) = \rho(1 - \mathbf{1}_A) = 1 - \rho(\mathbf{1}_A)$, so $\delta(\Omega \setminus A) = \Omega \setminus \delta(A)$. Also $\mathbf{1}_{\delta(A \cap B)} = \rho(\mathbf{1}_{A \cap B}) = \rho(\mathbf{1}_A \mathbf{1}_B) = \rho(\mathbf{1}_A)\rho(\mathbf{1}_B) = \mathbf{1}_{\delta(A)} \mathbf{1}_{\delta(B)} = \mathbf{1}_{\delta(A) \cap \delta(B)}$, so $\delta(A \cap B) = \delta(A) \cap \delta(B)$. Finally, by complementation, we get $\delta(A \cup B) = \delta(A) \cup \delta(B)$.

(2) Write $\mathcal{S}$ for the set of $\mathcal{F}$-simple functions. If $f = \sum c_i \mathbf{1}_{A_i} \in \mathcal{S}$, define $\rho(f) = \sum c_i \mathbf{1}_{\delta(A_i)}$. It can be checked that ρ is well defined and $\|\rho(f)\|_u = \|f\|_\infty$. The conditions (i) – (vi) hold for functions in $\mathcal{S}$. The closure of $\mathcal{S}$ in the $\mathcal{L}_\infty$-norm is $\mathcal{L}_\infty(\Omega, \mathcal{F}, \mu)$, so ρ can be extended uniquely to $\mathcal{L}_\infty$ in such a way that $\|\rho(f)\|_u = \|f\|_\infty$ holds for all $f \in \mathcal{L}_\infty$. Now it can

be easily checked that (i) – (vi) hold for functions in $\mathcal{L}_\infty$. Suppose that ρ' is another lifting for $(\Omega, \mathcal{F}, \mu)$ that satisfies $\rho'(\mathbf{1}_A) = \mathbf{1}_{\delta(A)}$ for $A \in \mathcal{F}$. Clearly $\rho(f) = \rho'(f)$ for $f \in \mathcal{S}$. But if $\|f\|_\infty = M$, then $-M \le f \le M$ almost everywhere, so $-M \le \rho'(f) \le M$ everywhere, and thus $\|\rho'(f)\|_u \le \|f\|_\infty$. Thus ρ and ρ' are both continuous, so they coincide on $\mathcal{L}_\infty$. ∎

This proposition shows that the problem of existence of a lifting is equivalent to the problem of existence of a density. The next few results have the goal of proving such existence. The first one has a standard sort of Zorn's lemma proof, but checking the details takes some space.

(6.4.4) Proposition. *Let δ be a lower density for the nonzero complete σ-finite measure space $(\Omega, \mathcal{F}, \mu)$, and let δ' be the corresponding upper density. Then there is a density δ^* for $(\Omega, \mathcal{F}, \mu)$ satisfying $\delta(A) \le \delta^*(A) \le \delta'(A)$ for all $A \in \mathcal{F}$.*

Proof. Consider the collection $\mathcal{R}$ of all $\mathcal{G}$ satisfying:

$\mathcal{G}$ is an algebra of subsets of Ω,

$\mathcal{G}$ meets each μ-equivalence class at most once,

$\delta(A) \subseteq A \subseteq \delta'(A)$ for all $A \in \mathcal{G}$.

Then $\mathcal{R}$ is partially ordered by inclusion, and not empty, since $\{\emptyset, \Omega\} \in \mathcal{R}$. The union of a chain of elements of $\mathcal{R}$ is again an element. By Zorn's lemma, $\mathcal{R}$ has a maximal element $\mathcal{G}^*$. It remains only to show that $\mathcal{G}^*$ meets each equivalence class at least once.

So let $A_0 \in \mathcal{F}$, and define

$$A_0' = \bigcup_{C \in \mathcal{G}^*} (\delta(C \cup A_0) \setminus C).$$

We will show that $A_0 \equiv A_0'$ and $A_0' \in \mathcal{G}^*$.

First, if $C, D \in \mathcal{G}^*$, then

$$\delta(C \cup A_0) \cap \delta(D \cup (\Omega \setminus A_0)) \setminus (C \cup D) \subseteq \delta(C \cup D) \setminus (C \cup D) = \emptyset,$$

so $\left(\delta(C \cup A_0) \setminus C\right) \cap \left(\delta(D \cup (\Omega \setminus A_0)) \setminus D\right) = \emptyset$. Therefore, if $D \in \mathcal{G}^*$, we have

$$A_0' \cap \left(\delta(D \cup (\Omega \setminus A_0)) \setminus D\right) = \emptyset$$

by the definition of A_0', we get $\delta(A_0) \subseteq A_0'$. By completeness, we have $A_0' \in \mathcal{F}$ and $A_0' \equiv A_0$.

Next, if $D \in \mathcal{G}^*$, we have

$$\delta((\Omega \setminus D) \cup (\Omega \setminus A_0')) = \delta((\Omega \setminus D) \cup (\Omega \setminus A_0)) \subseteq (\Omega \setminus D) \cup (\Omega \setminus A_0');$$

so $D \cap A_0' \subseteq \delta'(D \cap A_0')$. Now if $E \in \mathcal{G}^*$, then

$$\delta((\Omega \setminus E) \cup A_0') = \delta((\Omega \setminus E) \cup A_0) \subseteq (\Omega \setminus E) \cup A_0,$$

so $E \setminus A_0' \subseteq \delta'(E \setminus A_0')$.

Let $\mathcal{G}'$ be the algebra generated by $\mathcal{G}^*$ and A_0', so

$$\mathcal{G}' = \{\, C \cup (D \cap A_0') \cup (E \setminus A_0') : C, D, E \in \mathcal{G}^* \,\}.$$

Then $\mathcal{G}' \supseteq \mathcal{G}^*$, and we claim $\mathcal{G}' \in \mathcal{R}$. If $B \in \mathcal{G}'$, say $B = C \cup (D \cap A_0') \cup (E \setminus A_0')$, then (since δ' is an upper density), $B \subseteq \delta'(C) \cup \delta'(D \cap A_0') \cup \delta'(E \setminus A_0') = \delta'(C \cup (D \cap A_0') \cup (E \setminus A_0')) = \delta'(B)$. Also $\Omega \setminus B \in \mathcal{G}'$, so $\Omega \setminus B \subseteq \delta'(\Omega \setminus B)$ or $\delta(B) \subseteq B$. To show that $\mathcal{G}'$ meets each equivalence class at most once, suppose $B_1, B_2 \in \mathcal{G}'$ and $B_1 \equiv B_2$. Then $\mu(B_1 \triangle B_2) = 0$, so $B_1 \triangle B_2 \subseteq \delta'(B_1 \triangle B_2) = \emptyset$, so $B_1 = B_2$.

Therefore $\mathcal{G}' \in \mathcal{R}$. By the maximality of $\mathcal{G}^*$, we have $\mathcal{G}' = \mathcal{G}^*$. Therefore $A_0' \in \mathcal{G}^*$. This completes the proof that $\mathcal{G}^*$ meets each equivalence class. Now $\mathcal{G}^*$ is the range of the required density δ^*.   ∎

(6.4.5) Corollary. *Lebesgue measure on the real line $\mathbb{R}$ admits a lifting.*

Proof. Apply Proposition (6.4.4) to the Lebesgue lower density.   ∎

Note that if $(\Omega, \mathcal{F}, \mu)$ is a complete measure space, and if $\mathcal{G} \subseteq \mathcal{F}$ is a σ-algebra containing all of the μ-null sets in $\mathcal{F}$, then $(\Omega, \mathcal{G}, \mu|\mathcal{G})$ is also a complete measure space. The collection of all such $\mathcal{G}$ is partially ordered by inclusion. In the next few steps we will be preparing for an application of Zorn's lemma in this situation.

(6.4.6) Lemma. *Let $(\Omega, \mathcal{F}, \mu)$ be a complete finite measure space, let $\mathcal{G} \subseteq \mathcal{F}$ be a σ-algebra containing all of the null sets of $\mathcal{F}$, and let δ be a density on $\mathcal{G}$. Let $A \in \mathcal{F}$. Then there is a density δ^* on the σ-algebra $\mathcal{G}'$ generated by $\mathcal{G}$ and A that extends δ.*

Proof. Consider

$$\mathcal{C}_1 = \{\, C \in \mathcal{G} : \mu(C \cap A) = 0 \,\}$$
$$\mathcal{C}_2 = \{\, C \in \mathcal{G} : \mu(C \setminus A) = 0 \,\}.$$

Both collections are closed under countable unions, and μ is finite, so there exist sets $C_1 \in \mathcal{C}_1$ and $C_2 \in \mathcal{C}_2$ with $\mu(C \setminus C_1) = 0$ for all $C \in \mathcal{C}_1$ and $\mu(C \setminus C_2) = 0$ for all $C \in \mathcal{C}_2$. Now

$$\mu(C_1 \cap C_2) = \mu(C_1 \cap C_2 \cap A) + \mu((C_1 \cap C_2) \setminus A) = 0,$$

so $\delta(C_1) \cap \delta(C_2) = \emptyset$. Define $A^* = (A \setminus \delta(C_1)) \cup \delta(C_2)$. Then $A^* \equiv A$. Define δ^* on $\mathcal{G}'$ by:

$$\delta^*((B_1 \cap A) \cup (B_2 \setminus A)) = (\delta(B_1) \cap A^*) \cup (\delta(B_2) \setminus A^*)$$

for $B_1, B_2 \in \mathcal{G}$.

If $(B_1 \cap A) \cup (B_2 \setminus A) \equiv (B_3 \cap A) \cup (B_4 \setminus A)$, then $\mu((B_1 \triangle B_3) \cap A) = 0$, so $\delta(B_1) \triangle \delta(B_3) \subseteq \delta(C_1)$. Similarly $\delta(B_2) \triangle \delta(B_4) \subseteq \delta(C_2)$. Since $\delta(C_1) \cap \delta(C_2) = \emptyset$, we see that $(\delta(B_1) \triangle \delta(B_3)) \cap A^* = \emptyset$ and $\delta(B_2) \setminus A^* = \delta(B_4) \setminus A^*$. This shows that δ^* is well defined and constant on equivalence classes.

The remaining parts of the verification that δ^* is a density that extends δ are straightforward.   ∎

(6.4.7) Lemma. *Let $(\Omega, \mathcal{F}, \mu)$ be a complete probability space. Suppose $\mathcal{F}_1 \subseteq \mathcal{F}_2 \subseteq \cdots$ are sub-σ-algebras of $\mathcal{F}$ and $\mathcal{F}_1$ contains all of the null sets. Suppose δ_n is a density on $\mathcal{F}_n$ $(n = 1, 2, \cdots)$ and δ_{n+1} extends δ_n for all n. Then there is a density δ^* on the σ-algebra $\mathcal{G}$ generated by $\bigcup_{n=1}^{\infty} \mathcal{F}_n$ such that δ^* extends each δ_n.*

Proof. It suffices to show that there is a lower density δ on $\mathcal{G}$ that extends the δ_n. Let ρ_n be the lifting for $(\Omega, \mathcal{F}_n, \mu)$ corresponding to δ_n as in Proposition (6.4.3). If $A \in \mathcal{F}$, note that the expression $\rho_n\left(\mathbf{E}^{\mathcal{F}_n}\left[\mathbf{1}_A\right]\right)$ is a well defined element of $\mathcal{L}_\infty$, even though the conditional expectation $\mathbf{E}^{\mathcal{F}_n}\left[\mathbf{1}_A\right]$ is only defined up to null sets.

Define, for $A \in \mathcal{G}$,

$$\delta(A) = \left\{ \omega : \lim_{n \to \infty} \rho_n\left(\mathbf{E}^{\mathcal{F}_n}\left[\mathbf{1}_A\right]\right)(\omega) = 1 \right\}.$$

We will verify that δ is a lower density on $\mathcal{G}$ that extends all δ_n. By the martingale convergence theorem (1.4.7), $\mathbf{E}^{\mathcal{F}_n}\left[\mathbf{1}_A\right]$ converges a.e. to $\mathbf{E}^{\mathcal{G}}\left[\mathbf{1}_A\right] = \mathbf{1}_A$, so $\delta(A) \equiv A$. Also, if $A \in \mathcal{F}_m$, then for $n \geq m$,

$$\rho_n\left(\mathbf{E}^{\mathcal{F}_n}\left[\mathbf{1}_A\right]\right) = \rho_n(\mathbf{1}_A) = \mathbf{1}_{\delta_n(A)} = \mathbf{1}_{\delta_m(A)},$$

so $\delta(A) = \delta_m(A)$. Thus δ extends all δ_n. In particular, $\delta(\Omega) = \Omega$ and $\delta(\emptyset) = \emptyset$. It remains only to verify that $\delta(A \cap B) = \delta(A) \cap \delta(B)$.

Let $A, B \in \mathcal{G}$. Then $\mathbf{1}_{A \cap B} \leq \mathbf{1}_A$, so

$$\mathbf{E}^{\mathcal{F}_n}\left[\mathbf{1}_{A \cap B}\right] \leq \mathbf{E}^{\mathcal{F}_n}\left[\mathbf{1}_A\right] \quad \text{a.e.,}$$

so $\rho_n(\mathbf{E}^{\mathcal{F}_n}\left[\mathbf{1}_{A \cap B}\right]) \leq \rho_n(\mathbf{E}^{\mathcal{F}_n}\left[\mathbf{1}_A\right])$, and therefore we have $\delta(A \cap B) \subseteq \delta(A)$. Similarly $\delta(A \cap B) \subseteq \delta(B)$. Therefore $\delta(A \cap B) \subseteq \delta(A) \cap \delta(B)$.

On the other hand, $\mathbf{1}_A \leq \mathbf{1}_{A \cup B}$, so for $\omega \in \delta(A)$,

$$1 = \lim \rho_n\left(\mathbf{E}^{\mathcal{F}_n}\left[\mathbf{1}_A\right]\right)(\omega) \leq \liminf \rho_n\left(\mathbf{E}^{\mathcal{F}_n}\left[\mathbf{1}_{A \cup B}\right]\right)(\omega)$$
$$\leq \limsup \rho_n\left(\mathbf{E}^{\mathcal{F}_n}\left[\mathbf{1}_{A \cup B}\right]\right)(\omega) \leq 1,$$

and similarly for $\omega \in \delta(B)$. If $\omega \in \delta(A) \cap \delta(B)$, then

$$\lim \rho_n\left(\mathbf{E}^{\mathcal{F}_n}\left[\mathbf{1}_{A \cap B}\right]\right)(\omega)$$
$$= \lim \left[\rho_n\left(\mathbf{E}^{\mathcal{F}_n}\left[\mathbf{1}_A\right]\right)(\omega) + \rho_n\left(\mathbf{E}^{\mathcal{F}_n}\left[\mathbf{1}_B\right]\right)(\omega) - \rho_n\left(\mathbf{E}^{\mathcal{F}_n}\left[\mathbf{1}_{A \cup B}\right]\right)(\omega) \right]$$
$$= 1 + 1 - 1 = 1.$$

Hence $\delta(A) \cap \delta(B) \subseteq \delta(A \cap B)$. This completes the proof that δ is a lower density. ∎

(6.4.8) Maharam's lifting theorem. *Let $(\Omega, \mathcal{F}, \mu)$ be a finite complete measure space. There is a lifting for $(\Omega, \mathcal{F}, \mu)$.*

Proof. By Proposition (6.4.3), it suffices to show that there is a density. We may assume $\mu(\Omega) = 1$ by multiplying μ by a constant. Consider the collection $\mathcal{R}$ of all ordered pairs $(\mathcal{G}, \delta)$, where $\mathcal{G} \subseteq \mathcal{F}$ is a σ-algebra containing all of the null sets in $\mathcal{F}$, and δ is a density for $(\Omega, \mathcal{G}, \mu)$. Then $\mathcal{R}$ is partially ordered by the definition: $(\mathcal{G}_1, \delta_1) \leq (\mathcal{G}, \delta_2)$ iff $\mathcal{G}_1 \subseteq \mathcal{G}_2$ and δ_2 extends δ_1. Also, $\mathcal{R}$ is not empty: let $\mathcal{G}$ consist only of the null sets and their complements, and $\delta(A) = 0$ for $\mu(A) = 0$; $\delta(A) = \Omega$ for $\mu(\Omega \setminus A) = 0$. Let $\mathcal{C} = \{(\mathcal{G}_i, \delta_i) : i \in I\}$ be a chain in $\mathcal{R}$. We must show that $\mathcal{C}$ has an upper bound in $\mathcal{R}$. If $\mathcal{C}$ has a largest element, it is clear. If there is an increasing sequence $i_1 \leq i_2 \leq \cdots$ such that

$$\bigcup_n \mathcal{G}_{i_n} = \bigcup_{i \in I} \mathcal{G}_i,$$

then the upper bound exists by Lemma (6.4.7). If there is no such sequence, then $\bigcup_{i \in I} \mathcal{G}_i$ is a σ-algebra, and the unique extension of all the δ_i is a density on it. So by Zorn's lemma, there is a maximal element $(\mathcal{G}^*, \delta^*)$ of $\mathcal{R}$. By Lemma (6.4.6), we must have $\mathcal{G}^* = \mathcal{F}$. So δ^* is a density for $(\Omega, \mathcal{F}, \mu)$. ∎

The existence of a lifting for a complete σ-finite measure space follows easily from this.

Properties of liftings

In this section the letter ρ will denote both a lifting for a measure space $(\Omega, \mathcal{F}, \mu)$ and the corresponding density: so $\rho(\mathbf{1}_A) = \mathbf{1}_{\rho(A)}$.

We begin with certainly the most important property for applications of liftings. As we know, if $\{h_i : i \in I\}$ is a bounded set in $\mathcal{L}_\infty$, then it has an essential supremum (4.1.1). But if the family is uncountable, the essential supremum is often not equal to the "upper envelope" or pointwise supremum:

$$h(\omega) = \sup_{i \in I} h_i(\omega) \qquad \omega \in \Omega.$$

Indeed, the function h defined in this way need not even be a measurable function. If the functions h_i are representatives according to a lifting ρ, that is $\rho(h_i) = h_i$, then we get the surprising conclusion that the upper envelope is the essential supremum.

(6.4.9) Measurability of the upper envelope. *Let ρ be a lifting for a complete measure space $(\Omega, \mathcal{F}, \mu)$. Suppose $\{h_i : i \in I\}$ is a family of bounded measurable functions, and $h_i \leq \rho(h_i)$ for all i. Then*

(1) *the pointwise supremum $h = \sup h_i$ is measurable;*
(2) *if h' is measurable and $h' \geq h_i$ a.e. for all i, then $h' \geq h$ a.e.;*
(3) *if h is bounded, then $h \leq \rho(h)$.*

Proof. We consider first the special case where $\mu(\Omega) < \infty$, $|h_i| \leq 1$ for all i, and $\{\, h_i : i \in I \,\}$ is directed upward. Choose a sequence $h_{i_1} \leq h_{i_2} \leq \cdots$ with $\lim_k \int h_{i_k} \, d\mu = \sup_i \int h_i \, d\mu$. Let $g = \sup_k h_{i_k}$, so that $g^\bullet$ is the supremum of $\{\, h_i^\bullet : i \in I \,\}$, or g is the essential supremum of $\{\, h_i : i \in I \,\}$. Then $g^\bullet \geq h_i^\bullet$ for all i, so $\rho(g) \geq \rho(h_i) \geq h_i$ everywhere. Thus $\rho(g) \geq \sup_i h_i = h \geq g$, so h is measurable, $h \equiv g$, and h is the essential supremum of $\{\, h_i : i \in I \,\}$. Also, $\rho(h) = \rho(g) \geq h$.

For the next case, suppose $\mu(\Omega) < \infty$ and $|h_i| \leq 1$ for all i, but $\{\, h_i : i \in I \,\}$ is not directed upward. The first case may be applied to the collection of finite suprema $h_{i_1} \vee h_{i_2} \vee \cdots \vee h_{i_n}$, which is directed upward. We must check that if $\rho(f) \geq f$ and $\rho(g) \geq g$, then $\rho(f \vee g) \geq f \vee g$. But $f \vee g \geq f$, so $\rho(f \vee g) \geq \rho(f) \geq f$; similarly $\rho(f \vee g) \geq g$; so $\rho(f \vee g) \geq f \vee g$.

Next suppose $\mu(\Omega) = \infty$ and $|h_i| \leq 1$ for all i. Write Ω as a disjoint union $\bigcup_{n=1}^\infty \Omega_n$ of sets of finite measure, and apply the previous case on each of the sets $\rho(\Omega_n)$.

Finally, in order to drop the condition $|h_i| \leq 1$, for a fixed positive integer n, apply the previous case to $h \wedge n = \sup(h_i \wedge n)$. But this is measurable for all n, so h itself is measurable. (Possibly $h(\omega) = \infty$ for some ω, and possibly $h \notin \mathcal{L}_\infty$.) ∎

The preceding theorem can be used to formulate uncountable versions of the standard limit theorems for the Lebesgue integral (Fatou's lemma, monotone convergence, dominated convergence, and so on) for uncountable sets $\{\, h_i : i \in I \,\}$ of functions with $\rho(h_i) = h_i$.

(6.4.10) Corollary. *Let $\{\, C_i : i \in I \,\} \subseteq \mathcal{F}$, and suppose $\rho(C_i) \supseteq C_i$ for all i. Then $C = \bigcup_{i \in I} C_i \in \mathcal{F}$ and $\rho(C) \supseteq C$. If $\{\, C_i : i \in I \,\}$ is directed upward, then $\mu(C) = \sup \mu(C_i)$.*

Proof. Apply Theorem (6.4.9) to the measurable functions $h_i = 1_{C_i}$. ∎

(6.4.11) Proposition. *Suppose $f \in \mathcal{L}_\infty (\Omega, \mathcal{F}, \mu)$ and $h \colon \mathbb{R} \to \mathbb{R}$ is bounded and continuous. Then $\rho(h \circ f) = h \circ \rho(f)$.*

Proof. The function f is bounded, say $|f| \leq a$. Let

$$\mathcal{A} = \{\, h \in C([-a, a]) : \rho(h \circ f) = h \circ \rho(f) \,\}.$$

Then $\mathcal{A}$ contains the constants, $\mathcal{A}$ contains the identity function $h(x) = x$, and $\mathcal{A}$ is closed under sums and products. Also $\mathcal{A}$ is closed under limits of uniformly convergent sequences. By the Weierstrass approximation theorem, $\mathcal{A} = C([-a, a])$. ∎

(6.4.12) Corollary. *Suppose* $f \in \mathcal{L}_\infty(\Omega, \mathcal{F}, \mu)$ *satisfies* $\rho(f) = f$. *If* $F \subseteq \mathbb{R}$ *is a closed set, then* $f^{-1}(F) \in \mathcal{F}$ *and* $\rho(f^{-1}(F)) \subseteq f^{-1}(F)$. *If* $G \subseteq \mathbb{R}$ *is an open set, then* $f^{-1}(G) \in \mathcal{F}$ *and* $\rho(f^{-1}(G)) \supseteq f^{-1}(G)$.

Proof. Suppose G is open. The family

$$\mathcal{H} = \{\, h \in C_b(\mathbb{R}) : 0 \le h \le \mathbf{1}_G \,\}$$

is directed upward and $\sup \mathcal{H} = \mathbf{1}_G$. Now $\rho(h \circ f) = h \circ \rho(f) = h \circ f$ for all $h \in \mathcal{H}$. So by Theorem (6.4.9), we have $\mathbf{1}_G \circ f = \sup_{h \in \mathcal{H}}(h \circ f)$ is measurable and $\rho(\mathbf{1}_G \circ f) \ge \mathbf{1}_G \circ f$. But $\mathbf{1}_G \circ f = \mathbf{1}_{f^{-1}(G)}$, so this means $f^{-1}(G) \in \mathcal{F}$ and $\rho(f^{-1}(G)) \supseteq f^{-1}(G)$.   ■

A similar argument will show that if $f \in \mathcal{L}_\infty$, $\rho(f) = f$, and $h \colon \mathbb{R} \to \mathbb{R}$ is lower semicontinuous, then $\rho(h \circ f) \ge h \circ f$.

The last application concerns "separable modifications" for stochastic processes. The result is formulated for a stochastic process with values in an interval of the line, but similar proofs can be used in other situations as well.

(6.4.13) Definition. Let $(\Omega, \mathcal{F}, \mathbf{P})$ be a complete probability space, and $[a, b] \subseteq \mathbb{R}$ a compact interval. A *stochastic process* in $[a, b]$ is a family $(X_t)_{t \in \mathbb{R}}$ of random variables with $a \le X_t \le b$. The stochastic process $(Y_t)_{t \in \mathbb{R}}$ is a *modification* of (X_t) iff for all $t \in \mathbb{R}$, we have $X_t = Y_t$ a.s. (the exceptional set may depend on t). A stochastic process (X_t) is *separable* iff there is a countable set $I \subseteq \mathbb{R}$ and $\Omega_0 \in \mathcal{F}$ such that $\mathbf{P}(\Omega_0) = 1$, and for all $\omega \in \Omega_0$, the graph

$$\{\, (t, X_t(\omega)) \in \mathbb{R} \times [a, b] : t \in \mathbb{R} \,\}$$

is contained in the closure of

$$\{\, (t, X_t(\omega)) : t \in I \,\}.$$

(6.4.14) Theorem. *Let* $(X_t)_{t \in \mathbb{R}}$ *be a stochastic process in* $[a, b]$. *Let* ρ *be a lifting for* $(\Omega, \mathcal{F}, \mathbf{P})$. *Then* $Y_t = \rho(X_t)$ *is a separable modification of* (X_t).

Proof. For $A \subseteq \mathbb{R}$ and $F \subseteq [a, b]$, define

$$V(A, F) = \{\, \omega \in \Omega : Y_t(\omega) \in F \text{ for all } t \in A \,\}.$$

If F is closed, then by Corollary (6.4.12), we have $\rho(Y_t^{-1}(F)) \subseteq Y_t^{-1}(F)$ for all t. Now for fixed $F \subseteq [a, b]$ closed, and $G \subseteq \mathbb{R}$ open, the family

$$\{\, V(A, F) : A \text{ finite}, A \subseteq G \,\}$$

is directed downward, so $\bigcap V(A,F) = V(G,F)$ is measurable and

$$\mathbf{P}\big(V(G,F)\big) = \inf\big\{\, \mathbf{P}\big(V(A,F)\big) : A \text{ finite}, A \subseteq G \,\big\}.$$

So there is a countable set $I_{G,F} \subseteq G$ with $\mathbf{P}(V(G,F)) = \mathbf{P}(V(I_{G,F},F))$. Now as G runs through a countable base for the open sets in $\mathbb{R}$, and F runs through a countable base for the closed sets in $[a,b]$, let I be the union of the sets $I_{G,F}$. So I is countable. Let Ω_0 be the complement of the union of the exceptional sets

$$V(I_{G,F},F) \setminus V(G,F).$$

The set Ω_0 has probability 1, since there are only countably many such exceptional sets, each of probability 0. Now fix $\omega_0 \in \Omega_0$ and $t_0 \in \mathbb{R}$. If G is an open neighborhood of t_0 and F is the complement of an open neighborhood of $Y_{t_0}(\omega_0)$, we see that $\omega_0 \notin V(G,F)$, so $\omega_0 \notin V(I_{G,F},F)$, so there exists $t \in I \cap G$ with $Y_t(\omega_0) \notin F$. As G decreases and F increases, this shows that the point $(t_0, Y_{t_0}(\omega_0))$ is in the closure of the points $(t, Y_t(\omega_0))$ with $t \in I$. This shows that (Y_t) is separable. ∎

Complements

The axiom of choice is used in the proofs of Proposition (6.4.4) and Theorem (6.4.8). It is known that it cannot be avoided. If ρ is a lifting for $\mathbb{R}$, then

$$f^\bullet \mapsto \rho(f)(0)$$

is a linear functional on $L_\infty(\mathbb{R})$ that is not induced by an element of $L_1(\mathbb{R})$, so its restriction to the unit ball B of L_∞ does not have the property of Baire for the weak-star topology on B (see Christensen [1974], p. 99). The map

$$A \mapsto \rho(\mathbf{1}_A)(0)$$

is a finitely additive measure on a σ-algebra (say the Borel sets in $\mathbb{R}$) that is not σ-additive. It is not possible to prove the existence of any of these in ZF set theory without choice. R. Solovay [1970] constructed a model of ZF (with the principle of dependent choices) in which they are impossible.

A lifting chooses representatives for bounded functions. It was observed by von Neumann [1931] that a similar construction is not possible for unbounded real-valued functions. In fact, he proved more. A mapping $\rho\colon \mathcal{L}_p(\Omega,\mathcal{F},\mu) \to \mathcal{L}_p(\Omega,\mathcal{F},\mu)$ is called a *linear lifting* iff:

(1) $\rho(f) \equiv f$;
(2) if $f \equiv g$, then $\rho(f) = \rho(g)$;
(3) if $f \geq 0$ then $\rho(f) \geq 0$;
(4) $\rho(f+g) = \rho(f) + \rho(g)$; $\rho(af) = a\rho(f)$.

(6.4.15) Proposition. *Let $1 \le p < \infty$. Suppose $\mathcal{L}_p(\Omega, \mathcal{F}, \mu)$ admits a linear lifting. Then $(\Omega, \mathcal{F}, \mu)$ is purely atomic.*

Proof. Let ρ be a linear lifting. If $(\Omega, \mathcal{F}, \mu)$ is not purely atomic, then there is $E \in \mathcal{F}$ such that $0 < \mu(E) < \infty$ and E contains no atoms. For each $\omega \in E$, the map $f^\bullet \mapsto \rho(f)(\omega)$ is a positive linear functional, hence continuous; call it u_ω. Now for each positive integer n, there exist disjoint sets $E_1^{(n)}, E_2^{(n)}, \cdots, E_n^{(n)} \subseteq E$ with $\mu(E_k^{(n)}) = \mu(E)/n$. Let

$$F^{(n)} = \bigcup_{k=1}^{n} \left\{ \omega : \rho\left(1_{E_k^{(n)}}\right)(\omega) = 1 \right\}.$$

Then $F^{(n)} \equiv E$, for all n, so

$$\mu\left(E \cap \bigcap_{n=1}^{\infty} F^{(n)}\right) = \mu(E) > 0.$$

Choose $\omega \in E \cap \bigcap_{n=1}^{\infty} F^{(n)}$. For each n, there exists k, $1 \le k \le n$, such that $\omega \in E_k^{(n)}$, so

$$\left\langle 1_{E_k^{(n)}}, u_\omega \right\rangle = \rho\left(1_{E_k^{(n)}}\right)(\omega) = 1.$$

Then

$$\|u_\omega\| \ge \frac{1}{\left\|1_{E_k^{(n)}}\right\|_p} = n^{1/p}\mu(E)^{-1/p},$$

which goes to ∞ as $n \to \infty$. This contradicts the continuity of u_ω. So in fact, $(\Omega, \mathcal{F}, \mu)$ is purely atomic. ∎

Remarks

The problem of whether a lifting exists for Lebesgue measure on the real line was proposed by A. Haar. This was solved by J. von Neumann [1931]. The existence of a lifting for a general σ-finite complete measure space was proved by D. Maharam [1958]. Her proof relied on the classification of measure algebras. A more direct proof was given by A. & C. Ionescu Tulcea [1961], based on the martingale convergence theorem. The proof given here is a variant of that proof.

The standard reference on liftings is the comprehensive book by A. & C. Ionescu Tulcea [1969]. It contains several applications not discussed here.

7

Derivation

In this chapter we will consider the topic of *derivation*. We begin with the classical derivation theorems in $\mathbb{R}$ and $\mathbb{R}^d$. This can be done by considering an appropriate stochastic basis indexed by a directed set, and applying the martingale convergence theorems of Chapter 4.

The theory of derivation bases in general is considered next. There are many parallels here with the martingale limit theorems on directed sets, but we do not derive the derivation basis material from the stochastic basis material.

Finally, we consider the special derivation bases known as D-bases and Busemann-Feller bases. Characterization of the bases that differentiate the indefinite integrals of all functions in a given Orlicz space L_Ψ occupies most of the space here.

7.1. Derivation in $\mathbb{R}$

This section discusses derivation of functions defined on the space $\mathbb{R}$. This is the classical theory of derivation for the Lebesgue integral. We will carry out the discussion using martingale proofs in this simple setting as an example to be followed in the later, more complicated, settings.

As always, $\mathbb{R}$ denotes the set of real numbers. We will write $\mathcal{F}$ for the σ-algebra of Lebesgue measurable subsets of $\mathbb{R}$, and λ for Lebesgue measure on $\mathcal{F}$. The Lebesgue *outer measure* of a set Q will be denoted $\lambda^*(Q)$, and the *inner measure* by $\lambda_*(Q)$. The outer measure may be defined as:

$$\lambda^*(Q) = \inf\{\lambda(A) : A \in \mathcal{F}, A \supseteq Q\}.$$

Similarly, the inner measure is:

$$\lambda_*(Q) = \sup\{\lambda(A) : A \in \mathcal{F}, A \subseteq Q\}.$$

It is known that these values may also be written as:

$$\lambda^*(Q) = \inf\{\lambda(U) : U \text{ open}, U \supseteq Q\}$$
$$\lambda_*(Q) = \sup\{\lambda(K) : K \text{ compact}, K \subseteq Q\}.$$

Stochastic bases

In order to apply the results from Chapter 4, we will consider some stochastic bases on $\mathbb{R}$.

Let $a < b$ be real numbers. A *finite subdivision* of $[a, b]$ is a finite subset of $[a, b]$, including a and b. We write $\mathcal{D}[a, b]$ or $\mathcal{D}[a, b)$ for the set of all subdivisions of $[a, b]$. If $t = \{a = x_0 < x_1 < \cdots < x_n = b\}$ is a subdivision, the *induced partition* is $\pi_t = \{[x_{i-1}, x_i) : i = 1, 2, \cdots, n\}$; and the *induced σ-algebra* is the σ-algebra $\mathcal{F}_t$ on $\Omega = [a, b)$ with atoms $[x_{i-1}, x_i)$, $i = 1, 2, \cdots, n$. Note that $\mathcal{D}[a, b)$ is a directed set when ordered by inclusion. If $t_1 \subseteq t_2$, we will write $t_1 \leq t_2$ and say that t_2 *refines* t_1. Note that $t_1 \leq t_2$ implies $\mathcal{F}_{t_1} \subseteq \mathcal{F}_{t_2}$. Thus,

$$\left(\mathcal{F}_t\right)_{t \in \mathcal{D}[a,b)}$$

is a stochastic basis on $\Omega = [a, b)$.

Suppose S is a subset of $[a, b]$, including a and b (such as a countable dense subset). We write $\mathcal{D}_S[a, b)$ for the subset of $\mathcal{D}[a, b)$ consisting of those subdivisions involving only elements of S. Then

$$\left(\mathcal{F}_t\right)_{t \in \mathcal{D}_S[a,b)}$$

is again a stochastic basis on $\Omega = [a, b)$.

There is an easy way to generate martingales adapted to these stochastic bases. Note that we may begin with an *arbitrary* function F, not necessarily continuous, or even measurable.

(7.1.1) Proposition (Difference-Quotient Martingale). *Let*

$$F \colon [a, b] \to \mathbb{R}$$

be a function. Define simple functions f_t as follows: If

$$t = \{a = x_0 < x_1 < \cdots < x_n = b\}$$

is a subdivision of $[a, b)$, then

(7.1.1a) $$f_t(x) = \sum_{i=1}^{n} \frac{F(x_i) - F(x_{i-1})}{x_i - x_{i-1}} \mathbf{1}_{[x_{i-1}, x_i)}.$$

Let S be any subset of $[a, b)$. Then

$$\left(f_t\right)_{t \in \mathcal{D}_S[a,b)}$$

is a martingale adapted to the stochastic basis

$$\left(\mathcal{F}_t\right)_{t \in \mathcal{D}_S[a,b)}.$$

Proof. We must prove: if $s, t \in \mathcal{D}_S[a, b)$ and $s \leq t$, then $\mathbf{E}_\lambda \left[f_t \mid \mathcal{F}_s \right] = f_s$. Since s and t are finite, by induction it is enough to prove this when t contains exactly one more point than s. Suppose

$$s = \{a = x_0 < x_1 < \cdots < x_n = b\},$$
$$t = \{a = x_0 < x_1 < \cdots < x_{i-1} < y < x_i < \cdots < x_n = b\}.$$

Now any element A of $\mathcal{F}_s$ is a finite union of intervals $[x_{j-1}, x_j)$, so the equation

$$\int_A f_s \, d\lambda = \int_A f_t \, d\lambda$$

must be checked only for $A = [x_{j-1}, x_j)$, $j = 1, 2, \cdots, n$. This is trivial except for $j = i$. For $j = i$, it reduces to the identity

$$\frac{F(x_i) - F(y)}{x_i - y} (x_i - y) + \frac{F(y) - F(x_{i-1})}{y - x_{i-1}} (y - x_{i-1}) = F(x_i) - F(x_{i-1}).$$

Therefore, (f_t) is a martingale. ∎

It is probably not hard to believe that the convergence of a difference-quotient martingale defined by a function F is related to the differentiability of the function F. This will be considered more fully below.

The Vitali covering theorem

The technical lemma for derivation in ℝ is a theorem of Vitali. But it is much more than a technicality. Variants of the theorem will be seen below.

The theorem will use some terminology for intervals in ℝ. It is spelled out here so that the similar ideas below will be more easily recognized. An interval E in ℝ is *nondegenerate* iff E has positive length. If E is a bounded nondegenerate interval, then it has a midpoint (or center) $x \in \mathbb{R}$ and a radius $r > 0$, so that E has one of the forms $(x-r, x+r)$, $[x-r, x+r)$, $(x-r, x+r]$, $[x-r, x+r]$. If $\alpha > 1$, then the α-*halo* of such an interval E is the closed interval with the same center and α times the radius: $[x - \alpha r, x + \alpha r]$. If $\mathcal{V}$ is a (finite or infinite) collection of nondegenerate bounded intervals, then the α-*halo* of $\mathcal{V}$ is the union of the α-halos of the intervals in $\mathcal{V}$. If $Q \subseteq \mathbb{R}$ and $\mathcal{V}$ is a collection of nondegenerate intervals, then we say that $\mathcal{V}$ is a *Vitali cover* of Q iff, for each $\varepsilon > 0$ and each $x \in Q$, there is $E \in \mathcal{V}$ with radius $< \varepsilon$ and $x \in E$. Thus if $x \in Q \cap U$, where U is an open set, then there exists $E \in \mathcal{V}$ with $x \in E \subseteq U$.

(7.1.2) Vitali covering theorem. *Let Q be a subset of ℝ, and suppose $\mathcal{V}$ is a Vitali cover of Q by nondegenerate intervals. Then*

(a) *There exists a pairwise disjoint countable family $\{E_n\} \subseteq \mathcal{V}$ (finite or infinite) such that*

$$\lambda^* \left(Q \setminus \bigcup_n E_n \right) = 0.$$

(b) If $\lambda^*(Q) < \infty$, then, for each $\varepsilon > 0$, there exists a pairwise disjoint finite family $\{E_1, E_2, \cdots, E_p\} \subseteq \mathcal{V}$ such that

$$\lambda^*\left(Q \setminus \bigcup_{n=1}^{p} E_n\right) < \varepsilon.$$

Proof (Banach). Note that $\lambda(\overline{E} \setminus E) = 0$ for any interval E, and that if $\overline{E} \cap \overline{F} = \emptyset$ then $E \cap F = \emptyset$. Therefore we may assume that the intervals in $\mathcal{V}$ are closed intervals. We may assume also that $Q \neq \emptyset$.

First assume $\lambda^*(Q) < \infty$. Choose an open set $V \supseteq Q$ with $\lambda(V) < \infty$. Then

$$\mathcal{V}_0 = \{ E \in \mathcal{V} : E \subseteq V \}$$

is also a Vitali cover of Q. The disjoint sequence $\{E_n\}$ of intervals will be constructed recursively. Since Q is nonempty, $\mathcal{V}_0$ is also nonempty. Choose $E_1 \in \mathcal{V}_0$. If $Q \subseteq E_1$, we are done; so suppose not, that is, $Q \setminus E_1 \neq \emptyset$. Then $U_1 = V \setminus E_1$ is open and $U_1 \cap Q \neq \emptyset$. Let

$$\delta_1 = \sup \{ \lambda(E) : E \in \mathcal{V}_0, E \subseteq U_1 \}.$$

Choose $E_2 \in \mathcal{V}_0$ such that $E_2 \subseteq U_1$ and $\lambda(E_2) > \delta_1/2$. Now suppose we have chosen $E_1, E_2, \cdots, E_n$. If $Q \subseteq E_1 \cup E_2 \cup \cdots \cup E_n$, we are done; suppose not, that is, $Q \setminus (E_1 \cup \cdots \cup E_n) \neq \emptyset$. Then $U_n = V \setminus (E_1 \cup \cdots \cup E_n)$ is open and $U_n \cap Q \neq \emptyset$. Let

$$\delta_n = \{ \lambda(E) : E \in \mathcal{V}, E \subseteq U_n \}.$$

Choose $E_{n+1} \in \mathcal{V}_0$ such that $E_{n+1} \subseteq U_n$ and $\lambda(E_{n+1}) > \delta_n/2$. Note that we have $E_{n+1} \cap E_k = \emptyset$ for $k < n+1$.

If this recursive process ends, we are done. Assume it does not. Then we get an infinite sequence $\{E_1, E_2, \cdots\} \subseteq \mathcal{V}_0$ of pairwise disjoint intervals. Write $B = Q \setminus \bigcup E_n$. We claim that $\lambda^*(B) = 0$. Let H_n be the 5-halo of E_n for each n. Then for each positive integer p,

$$\lambda\left(\bigcup_{n=p}^{\infty} H_n\right) \leq \sum_{n=p}^{\infty} \lambda(H_n) \leq 5 \sum_{n=p}^{\infty} \lambda(E_n) = 5\lambda\left(\bigcup_{n=p}^{\infty} E_n\right) \leq 5\lambda(V) < \infty.$$

Thus $\sum_{n=1}^{\infty} \lambda(H_n) < \infty$, and hence $\lambda\left(\bigcup_{n=p}^{\infty} H_n\right) \to 0$ as $p \to \infty$. We will show that $B \subseteq \bigcup_{n=p}^{\infty} H_n$ for all p. Fix p. Let

$$x \in B = Q \setminus \bigcup_{n=1}^{\infty} E_n \subseteq Q \setminus \bigcup_{n=1}^{p} E_n \subseteq U_p.$$

Since $\mathcal{V}_0$ is a Vitali cover of Q, there is $E \in \mathcal{V}_0$ with $x \in E \subseteq U_p$. Now $\lambda(E_n) \to 0$ and $\delta_n < \lambda(E_{n+1})$, so $\delta_n \to 0$. Thus, there is n with $\delta_n < \lambda(E)$, so that $E \not\subseteq U_n$. Let n be the smallest integer with $E \not\subseteq U_n$. Clearly $p < n$. Thus

$$E \cap (E_1 \cup \cdots \cup E_n) \neq \emptyset,$$
$$E \cap (E_1 \cup \cdots \cup E_{n-1}) = \emptyset,$$

so $E \cap E_n \neq \emptyset$. But $E \subseteq U_{n-1}$, so $\lambda(E) \leq \delta_{n-1} < 2\lambda(E_n)$. Therefore [since $E \cap E_n \neq \emptyset$, $\lambda(E) < 2\lambda(E_n)$, $\lambda(H_n) = 5\lambda(E_n)$] we have $E \subseteq H_n$. Thus $x \in E \subseteq \bigcup_{n=p}^{\infty} H_n$. This shows $B \subseteq \bigcup_{n=p}^{\infty} H_n$ for all p, and thus $\lambda^*(B) = 0$.

If $\varepsilon > 0$ is given, choose p so that $\sum_{n=p+1}^{\infty} \lambda(H_n) < \varepsilon$. Thus

$$\lambda^*\left(Q \setminus \bigcup_{n=1}^{p} H_n\right) < \varepsilon.$$

For the general case $\lambda^*(Q) = \infty$, apply the above, replacing Q by $Q \cap (n, n+1)$ and let $V = (n, n+1)$ to get a countable disjoint family $\mathcal{A}_n \subseteq \mathcal{V}$ of subintervals of $(n, n+1)$ covering almost all of $Q \cap (n, n+1)$. Since the set of integers has Lebesgue measure zero, the union

$$\bigcup_{n=-\infty}^{\infty} \mathcal{A}_n$$

is the required countable pairwise disjoint family. ∎

The use that we will make of the Vitali covering theorem is to deduce the Vitali condition (V).

(7.1.3) Corollary. *Let $a < b$ be real numbers, and let S be a dense subset of $[a, b)$. Then the stochastic basis*

$$\left(\mathcal{F}_t\right)_{t \in \mathcal{D}_S[a,b)}$$

satisfies the covering condition (V).

Proof. Let $A_t \subseteq [a, b)$ be given for each $t \in \mathcal{D}_S[a, b)$, $A_t \in \mathcal{F}_t$, and let $A^* = \text{e} \lim \sup A_t$. Fix $t_0 \in \mathcal{D}_S[a, b)$ and $\varepsilon > 0$. For each positive integer m there is (since S is dense) a partition t_m with mesh $< 1/m$ and $t_m \geq t_0$. Then we have $A^* \subseteq \text{ess sup}_{t \geq t_m} A_t$; choose $t_{m1}, t_{m2}, \cdots \geq t_m$ with $A^* \subseteq \bigcup_{i=1}^{\infty} A_{t_{mi}}$ a.e. Let

$$B = A^* \cap \bigcap_{m=1}^{\infty} \bigcup_{i=1}^{\infty} A_{t_{mi}}.$$

Then $B \subseteq \bigcup_{i=1}^{\infty} A_{t_{mi}}$ for all m and $B = A^*$ a.e.

Now $A_{t_{mi}} \in \mathcal{F}_{t_{mi}}$, so it has the form of a finite union

$$A_{t_{mi}} = \bigcup_k [a_{mik}, b_{mik}),$$

with $b_{mik} - a_{mik} < 1/m$. Therefore, the collection obtained by combining all of the intervals, $\mathcal{V} = \{[a_{mik}, b_{mik})\}_{m,i,k}$, is a Vitali cover of B. Thus by the Vitali covering theorem, there exist a finite disjoint collection $\{E_1, E_2, \cdots, E_p\}$ of intervals from $\mathcal{V}$ with $\lambda(B \setminus (E_1 \cup E_2 \cup \cdots \cup E_p)) < \varepsilon$. But $E_1 \cup E_2 \cup \cdots \cup E_p$ is of the form $A(\tau)$ for some incomplete stopping time τ. Indeed, for each E_j, choose a pair m, i with $E_j \subseteq A_{t_{mi}}$ and $E_j \in \mathcal{F}_{t_{mi}}$; then define $\tau = t_{mi}$ on E_j.   ∎

Derivation

It is now a routine (but interesting) matter to apply the martingale convergence results of Chapter 4 to prove the classical derivation theorems on $\mathbb{R}$. We need *pointwise* a.e. convergence, not just essential convergence, so we use a countable directed set.

(7.1.4) Theorem. *Let $a < b$ be real numbers, and let $F: [a, b] \to \mathbb{R}$ be monotone. Then the derivative $F'(x)$ exists and is finite for almost all $x \in (a, b)$.*

Proof. Assume F is nondecreasing. Then F is continuous except at a countable number of points. Let S be a countable set, dense in $[a, b]$, containing a, b, and all points of discontinuity of F. Consider the difference-quotient martingale

$$(f_t)_{t \in \mathcal{D}_S[a,b]}$$

as in Proposition (7.1.1). This martingale is L_1-bounded. Indeed, if

$$t = \{a = x_0 < x_1 < \cdots < x_n = b\},$$

then

$$\int |f_t| \, d\lambda = \int f_t \, d\lambda$$
$$= \sum_{i=1}^{n} \frac{F(x_i) - F(x_{i-1})}{x_i - x_{i-1}} (x_i - x_{i-1})$$
$$= F(b) - F(a) < \infty.$$

Therefore, by (4.2.11), the martingale converges essentially, and (since $\mathcal{D}_S[a, b]$ is countable) almost everywhere. Say $f_t \to f$ a.e. We claim that if x is a point of $(a, b) \setminus S$ where $f_t(x) \to f(x)$, then F is differentiable at x and $F'(x) = f(x)$. [This will show that F' exists and is finite a.e.] Indeed, let $\varepsilon > 0$ be given. Choose $t_0 \in \mathcal{D}_S[a, b]$ such that, for all $t \geq t_0$, we have $|f_t(x) - f(x)| \leq \varepsilon$. Write $t_0 = \{a = x_0 < \cdots < x_n = b\}$. Now $x \notin S$, so we have $x_{i-1} < x < x_i$ for some i. If $y, z \in S$ with $x_{i-1} < y < x < z < x_i$, then the subdivision

$$t = \{a = x_0 < x_1 < \cdots < x_{i-1} < y < z < x_i < \cdots < x_n = b\}$$

refines t_0. Thus $|f_t(x) - f(x)| \leq \varepsilon$, so

(7.1.4a)
$$\left| \frac{F(z) - F(y)}{z - y} - f(x) \right| \leq \varepsilon.$$

Now it follows that (7.1.4a) holds for *all* y, z satisfying $x_{i-1} < y < x < z < x_i$, since points not belonging to S are points of continuity of F. This shows that

$$\limsup_{y \to x} \frac{F(x) - F(y)}{x - y} \leq f(x) + \varepsilon,$$

$$\liminf_{y \to x} \frac{F(x) - F(y)}{x - y} \geq f(x) - \varepsilon.$$

Since ε was arbitrary, we have

$$\lim_{y \to x} \frac{F(x) - F(y)}{x - y} = f(x).$$
∎

An examination of the preceding proof might naturally lead to a generalization to functions of bounded variation.

(7.1.5) Definition. Let $F \colon [a, b] \to \mathbb{R}$ be a function. The *variation* of F on a subdivision $t = \{a = x_0 < \cdots < x_n = b\}$ is

$$V_t(F) = \sum_{i=1}^{n} |F(x_i) - F(x_{i-1})|.$$

The *variation* of F on a subset S of $[a, b]$ is

$$V_S(F) = \sup \{ V_t(F) : t \in \mathcal{D}_S[a, b] \}.$$

The function F is said to have *bounded variation* on $[a, b]$ (or *finite variation*) iff $V_{[a,b]}(F) < \infty$.

It is easily seen that for $a < b < c$, we have $V_{[a,c]}(F) = V_{[a,b]}(F) + V_{[b,c]}(F)$.

(7.1.6) Proposition. Let $F \colon [a, b] \to \mathbb{R}$ have bounded variation. Then F is the difference of two nondecreasing functions.

Proof. Let $F_1(x) = V_{[a,x]}(F)$ and $F_2(x) = F_1(x) - F(x)$. Then $F = F_1 - F_2$. Clearly F_1 is nondecreasing. To see that F_2 is also nondecreasing, for $x < y$ compute

$$F_2(y) - F_2(x) = V_{[a,y]}(F) - F(y) - V_{[a,x]} + F(x)$$
$$= V_{[x,y]} - (F(y) - F(x)) \geq 0.$$
∎

(7.1.7) Theorem. *Let $F\colon [a,b] \to \mathbb{R}$ be a function of bounded variation. Then F' exists and is finite almost everywhere in (a,b).*

Proof. Since F is the difference of two nondecreasing functions, this follows from Theorem (7.1.4).   ∎

An alternative proof of this fact would use the difference-quotient martingale (f_t) associated with F, and note that the martingale is L_1-bounded, since the L_1 bound of $(f_t)_{t \in \mathcal{D}_S[a,b]}$ is exactly equal to the variation $V_S(F)$. Then the proof would proceed exactly as in (7.1.4). This method would require the result that a function of bounded variation is continuous except at countably many points, which is easy to prove, but no easier than Proposition (7.1.6). These remarks also lead to another corollary.

(7.1.8) Corollary. *Let $F\colon [a,b] \to \mathbb{R}$ have bounded variation. Let S be a countable dense subset of $[a,b]$, including a, b, and all points of discontinuity of F. Then the difference-quotient martingale $(f_t)_{t \in \mathcal{D}_S[a,b]}$, defined as in (7.1.1), converges a.e. to F'.*

In Chapter 4, a special role was played by the closed martingales; that is, martingales of the form $f_t = \mathbf{E}^{\mathcal{F}_t}[f]$, where $f \in L_1$. When we translate into the present language, we obtain the following.

(7.1.9) Theorem. *Let $f \in L_1([a,b])$, and let F be the indefinite integral*

$$F(x) = \int_a^x f(y)\, dy.$$

Then $F'(x) = f(x)$ for almost all $x \in (a,b)$.

Proof. Let (f_t) be the difference-quotient martingale associated to F as in (7.1.1). Let $t = \{a = x_0 < \cdots < x_n = b\}$ be a subdivision of $[a,b]$. It is only a matter of applying the definition to see that

$$\mathbf{E}_\lambda[f \mid \mathcal{F}_t] = f_t.$$

If S is a dense set as in the proof of (7.1.4), then the σ-algebra $\mathcal{F}_\infty$ generated by

$$\bigcup_{t \in \mathcal{D}_S[a,b)} \mathcal{F}_t$$

is the algebra of all Borel sets in $[a,b)$. But (f_t) converges to $\mathbf{E}_\lambda[f \mid \mathcal{F}_\infty]$, which is (a.e.) f. Therefore $F' = f$ a.e.   ∎

We know that a martingale is of the form $\mathbf{E}_\lambda[f \mid \mathcal{F}_t]$ if and only if it is uniformly integrable.

(7.1.10) Proposition. *Let $F \colon [a, b] \to \mathbb{R}$ be a function, and let*

$$(f_t)_{t \in \mathcal{D}[a,b)}$$

be the difference-quotient martingale associated with F. Then the following are equivalent.

(a) *(f_t) is uniformly integrable.*
(b) *F is absolutely continuous: that is: for every $\varepsilon > 0$ there is $\delta > 0$ such that $\sum_{i=1}^{n} |F(b_i) - F(a_i)| < \varepsilon$ for every finite pairwise disjoint family $\{(a_1, b_1), (a_2, b_2), \cdots, (a_n, b_n)\}$ of open subintervals of (a, b) with $\sum_{i=1}^{n}(b_i - a_i) < \delta$.*

Proof. Lebesgue measure is atomless, so uniform integrability is characterized as in (2.3.2(3)). Suppose that (f_t) is uniformly integrable. Let $\varepsilon > 0$. Then there is $\delta > 0$ such that $\int_A |f_t| \, d\lambda < \varepsilon$ for any $t \in \mathcal{D}[a, b)$ and any set A with $\lambda(A) < \delta$. If $\{(a_1, b_1), (a_2, b_2), \cdots, (a_n, b_n)\}$ is a pairwise disjoint family of subintervals of (a, b) with $\sum_{i=1}^{n}(b_i - a_i) < \delta$, let t consist of a, b, and all of the endpoints a_i, b_i. Then $A = \bigcup_{i=1}^{n}(a_i, b_i)$ belongs to $\mathcal{F}_t$ and $\lambda(A) < \delta$. Therefore

$$\sum_{i=1}^{n} |F(b_i) - F(a_i)| = \int_A |f_t| \, d\lambda < \varepsilon.$$

Conversely, suppose that (b) is satisfied. Let $\varepsilon > 0$ be given. Then there exists δ as in (b). Let A be any measurable set with $\lambda(A) < \delta$. We claim that $\int_A |f_t| \, d\lambda \leq \varepsilon$ for all t. There is an open set $U \supseteq A$ with $\lambda(U) < \delta$. It is enough to show that $\int_U |f_t| \, d\lambda \leq \varepsilon$, since $\int_A |f_t| \, d\lambda \leq \int_U |f_t| \, d\lambda$. Now U is a countable disjoint union of open intervals, so by the monotone convergence theorem, it is enough to consider the case where U is a finite disjoint union of open intervals. Choose $t_1 \in \mathcal{D}[a, b)$ containing all the endpoints of intervals in U and all the points of t. If $t_1 = \{a = x_0 < \cdots < x_n = b\}$, then

$$\int_U |f_t| \, d\lambda \leq \int_U |f_{t_1}| \, d\lambda$$
$$= \sum |F(x_i) - F(x_{i-1})|,$$

where the last sum is over a set of indices i such that $\sum(x_i - x_{i-1}) = \lambda(U) < \delta$, so that the result is at most ε. ∎

(7.1.11) Corollary. *A function $F \colon [a, b] \to \mathbb{R}$ is absolutely continuous if and only if it has the form*

$$F(x) = c + \int_a^x f \, d\lambda \qquad x \in [a, b]$$

for some function $f \in L_1([a, b])$.

Complements

(7.1.12) (Lebesgue density theorem.) Deduce from the results of this section the Lebesgue density theorem: Let $A \subseteq \mathbb{R}$ be a measurable set with $\lambda(A) > 0$. Then almost every point $x \in A$ is a *point of density* of A in the sense that for every $\varepsilon > 0$, there is $\delta > 0$ so that

$$\frac{\lambda\big(A \cap (x - \delta, x + \delta)\big)}{2\delta} > 1 - \varepsilon.$$

Remarks

The Vitali covering theorem dates from 1905. The proof given here is due to S. Banach [1924]. This result is, as we have seen, closely connected to the covering condition (V) for stochastic bases. Because of this connection, the covering conditions considered in Chapter 4 are often known generically as "Vitali conditions," and condition (V) is sometimes known simply as "the Vitali condition."

7.2. Derivation in $\mathbb{R}^d$

The next topic is derivation in Euclidean space $\mathbb{R}^d$, for $d = 2, 3, \cdots$. Examples of derivation bases to be considered are squares, disks, and intervals. We will write λ for d-dimensional Lebesgue measure on $\mathbb{R}^d$ and λ^* for d-dimensional Lebesgue outer measure.

Substantial sets

Let $\mathcal{C}$ be a collection of nonempty bounded open sets in $\mathbb{R}^d$. Suppose $\mathcal{C}$ is *substantial*: that is, there is a constant M such that, for every $C \in \mathcal{C}$, there is an open ball B with $C \subseteq B$ and $\lambda(B) < M\lambda(C)$. We will also assume that the sets $C \in \mathcal{C}$ have boundary of Lebesgue measure 0: $\lambda(\partial C) = 0$. [Recall that a point a is a *boundary point* of a set $C \subseteq \mathbb{R}^d$ iff every neighborhood of a meets both C and $\mathbb{R}^d \setminus C$.] For example, when C is a convex set, we do have $\lambda(\partial C) = 0$. If B is an open ball, we write $r(B)$ for its radius.

Let $\mathcal{V}$ be a subcollection of $\mathcal{C}$ and let $Q \subseteq \mathbb{R}^d$. We say that $\mathcal{V}$ is a *Vitali cover* of Q iff for every $x \in Q$ and every $\varepsilon > 0$, there is $E \in \mathcal{V}$ with diam $E < \varepsilon$ and $x \in E$.

(7.2.1) Theorem. *Let $\mathcal{C}$ be a substantial collection of nonempty open bounded subsets of $\mathbb{R}^d$ with boundaries of measure 0. Let Q be a set in $\mathbb{R}^d$, and suppose $\mathcal{V}$ is a Vitali cover of Q by sets of $\mathcal{C}$. Then there exists a pairwise disjoint countable family $\{E_n\} \subseteq \mathcal{V}$ (finite or infinite) such that $\lambda^*\big(Q \setminus \bigcup_{n=1}^{\infty} E_n\big) = 0$. If $\lambda^*(Q) < \infty$, then for each $\varepsilon > 0$, there exists a pairwise disjoint finite family $\{E_1, E_2, \cdots, E_p\} \subseteq \mathcal{V}$ such that $\lambda^*\big(Q \setminus \bigcup_{n=1}^{p} E_n\big) < \varepsilon$.*

Proof. First suppose Q is a bounded set.

We apply (4.2.6) repeatedly. Let $\varepsilon > 0$ be given. Fix an open set $W \supseteq Q$ with $\lambda(W) < (1 + \varepsilon)\lambda^*(Q)$. Then

$$W_1 = \bigcup \{ C \in \mathcal{C} : C \subseteq W \}$$

is an open set. Apply 4.2.6: there exist disjoint $E_1, E_2, \cdots, E_{p_1} \in \mathcal{C}$ with $E_i \subseteq W$ and $\sum_{i=1}^{p_1} \lambda(E_i) > 2^{-1} M^{-1} 3^{-d} \lambda(W_1)$. Now $W \setminus \bigcup_{i=1}^{p_1} \overline{E_i}$ is an open set; let

$$W_2 = \bigcup \{ C \in \mathcal{C} : C \subseteq W \setminus \textstyle\bigcup_{i=1}^{p_1} \overline{E_i} \}.$$

Repeat: there exist $E_{p_1+1}, \cdots, E_{p_2} \in \mathcal{C}$ with $E_i \subseteq W \setminus \bigcup_{i=1}^{p_1} \overline{E_i}$ and

$$\sum_{i=p_1+1}^{p_2} \lambda(E_i) > 2^{-1} M^{-1} 3^{-d} \lambda(W_2).$$

Continue in this way.

Thus, for each k,

$$\sum_{i=1}^{p_k} \lambda(E_i) > \left[1 - \left(1 - 2^{-1} M^{-1} 3^{-d} \right)^k \right] \lambda^*(Q).$$

Take $p = p_k$ for large enough k to obtain $\lambda^* \left(Q \setminus \bigcup_{n=1}^{p} E_n \right) < \varepsilon$, and allow an infinite sequence of E_i to obtain $\lambda^* \left(Q \setminus \bigcup_{n=1}^{\infty} E_n \right) = 0$.

The case of unbounded Q may be done as before: consider $Q \cap S$ for a disjoint family of open cubes S that cover $\mathbb{R}^d$ (except for a set of measure zero).   ∎

A $\mathcal{C}$-*partition* of an open set U is a countable disjoint collection $\{C_i\} \subseteq \mathcal{C}$, contained in U, such that

$$\lambda \left(U \setminus \bigcup_{i=1}^{\infty} C_i \right) = 0.$$

We write $\mathcal{D}(U)$ for the set of all $\mathcal{C}$-partitions of U. The ordering is refinement: if

$$t = \{ C_i : i \in \mathbb{N} \}$$
$$t' = \{ C'_j : j \in \mathbb{N} \},$$

then we say t' refines t iff each C'_j is contained in some C_i. (And therefore each C_i is, except for a set of measure zero, the union of some collection of the C'_j.)

We write $\mathcal{F}_t$ for the σ-algebra on U generated by the partition t. Theorem (7.2.1) implies that the stochastic basis $(\mathcal{F}_t)_{t \in \mathcal{D}(U)}$ satisfies the covering condition (V):

(7.2.2) Proposition. *Let U be a bounded open set. The collection $\mathcal{D}(U)$ of C-partitions of U is directed by refinement, and satisfies condition* (V).

Proof. $\mathcal{D}(U)$ is directed by refinement: Let $t_1, t_2 \in \mathcal{D}(U)$. For each $C_1 \in t_1$ and $C_2 \in t_2$, consider the intersection $W = C_1 \cap C_2$. The collection of all $C \in \mathcal{C}$ contained in W is a Vitali cover of W, so by (7.2.1), W is (up to null sets) the disjoint union of sets $C \in \mathcal{C}$.

(V) is done as in (7.1.3). ∎

As usual, (V) implies convergence.

(7.2.3) Theorem. *Let f be locally integrable on $\mathbb{R}^d$. Then*

$$\lim_{\substack{\text{diam } E \to 0 \\ E \in \mathcal{C} \\ x \in E}} \int_E f \, d\lambda = f(x)$$

for almost all $x \in \mathbb{R}^d$.

Proof. Consider the restriction of f to a large disk R. If

$$t = \{\, C_i : i \in \mathbb{N} \,\}$$

is a C-partition of R, define

$$f_t = \sum_{i \in \mathbb{N}} \left(\frac{1}{\lambda(C_i)} \int_{C_i} f \, d\lambda \right) \mathbf{1}_{C_i}.$$

It is easy to check that $(f_t)_{t \in \mathcal{D}(R)}$ is the closed martingale $\mathbf{E}_\lambda[f \,|\, \mathcal{F}_t]$. We may therefore conclude by (4.2.11) that (f_t) converges essentially to f.

Now $L_1(\lambda, R)$ is separable, so there is a countable collection $\mathcal{C}_1$ that is L_1-dense in

$$\{\, \mathbf{1}_C : C \in \mathcal{C}, C \subseteq R \,\}.$$

For each positive integer n, choose a C-partition $t_n \in \mathcal{D}(R)$ such that

(1) $t_1 \leq t_2 \leq \cdots$;
(2) diam $E < 1/n$ for all $E \in t_n$;
(3) $\lambda\{\text{ess sup}_{t \geq t_n} f_t > f + 1/n\} < 2^{-n}$;
(4) $\lambda\{\text{ess inf}_{t \geq t_n} f_t < f - 1/n\} < 2^{-n}$.

Then, for each n, choose a countable family $\{t_{ni}\}_{i=0}^{\infty}$ such that

(1) $t_{n0} = t_n$;
(2) $t_{ni} \geq t_n$;
(3) if $E \in t_n$, and $F \subseteq E$, $F \in \mathcal{C}_1$, then $F \in t_{ni}$ for some i.

Now if

$$C_n = \{\sup_i f_{t_{ni}} > f + 1/n\} \cup \{\inf_i f_{t_{ni}} < f - 1/n\},$$

then $\lambda(C_n) < 2^{-n+1}$, so almost every $x \in R$ belongs to only finitely many C_n. Let $A_{ni} = \bigcup_{E \in t_{ni}} E$. Then $\lambda(R \setminus \bigcap_{n,i} A_{ni}) = 0$. So almost all $x \in R$ belong to $\bigcap_{n,i} A_{ni}$.

Fix a point x with $x \in \bigcap_{n,i} A_{ni}$ and $x \notin C_n$ for $n \geq n_0$. Fix $n \geq n_0$. Then there is $E_0 \in t_n$ such that $x \in E_0^\circ$; let r_n be half the distance from x to the boundary of E_0. Then we claim: for all $E \in C$ with $x \in E$ and diam $E < r_n$, we have

$$\left| \frac{1}{\lambda(E)} \int_E f \, d\lambda - f(x) \right| \leq \frac{1}{n}.$$

Indeed, if $E \in C_1$, then $E \in t_{ni}$ for some i, so

$$\frac{1}{\lambda(E)} \int_E f \, d\lambda = f_{t_{ni}}(x) \leq f(x) + \frac{1}{n},$$

and similarly

$$\frac{1}{\lambda(E)} \int_E f \, d\lambda \geq f(x) - \frac{1}{n}.$$

Now by the dominated convergence theorem, the real number

$$\frac{1}{\lambda(E)} \int_E f \, d\lambda$$

is a continuous function of $E \in C$ (as long as $E \subseteq E_0$), so the inequalities hold for all $E \in C$. This shows that

$$\sup_{\substack{E \in C \\ x \in E \\ \text{diam } E < r_n}} \left| \frac{1}{\lambda(E)} \int_E f \, d\lambda - f(x) \right| \leq \frac{1}{n}.$$

Now as $n \to \infty$, we have $r_n \to 0$, so

(7.2.3b)
$$\lim_{\substack{E \in C \\ \text{diam } E \to 0 \\ x \in E}} \frac{1}{\lambda(E)} \int_E f \, d\lambda = f(x).$$

The point x was chosen from a set in R with complement of measure zero, so (7.2.3b) holds almost everywhere in R. Finally, it follows that (7.2.3b) holds almost everywhere in $\mathbb{R}^2$. ∎

Note that the condition $\lambda(\partial C) = 0$ is required to show that $\mathcal{D}(C)$ is a directed set. The derivation theorem (7.2.3) in fact holds more generally: see (7.4.3) and (7.4.11).

Disks and cubes

As a simple example, consider derivation using disks. (In $\mathbb{R}^3$ the analogous theory uses balls, and in higher numbers of dimensions, the appropriate analogs of balls.)

An (open) *disk* is a set of the form

$$B(x,r) = \{\, y \in \mathbb{R}^2 : |x - y| < r \,\},$$

where $x \in \mathbb{R}^2$ and $r > 0$. We say it has *center* x and *radius* r. To simplify the notation, we will sometimes write $r(E)$ for the radius of a disk E. If f is a locally integrable function, we will be interested in the limit

$$\lim_{r \to 0} \frac{1}{\pi r^2} \int_{B(x,r)} f \, d\lambda.$$

The conclusion will be that this converges to $f(x)$ for almost every x.

(7.2.4) Corollary. *Let f be a locally integrable function on $\mathbb{R}^2$. Then*

$$(7.2.4a) \qquad\qquad \lim_{r \to 0} \frac{1}{\pi r^2} \int_{B(x,r)} f \, d\lambda = f(x)$$

for almost all $x \in \mathbb{R}^2$; and

$$(7.2.4b) \qquad\qquad \lim_{\substack{E \text{ disk} \\ r(E) \to 0 \\ x \in E}} \frac{1}{\lambda(E)} \int_E f \, d\lambda = f(x)$$

for almost all $x \in \mathbb{R}^2$.

Another way in which derivation in $\mathbb{R}^d$ can be done uses cubes as the basic cells. (For $\mathbb{R}^2$, we call them squares.) What we want are squares with sides parallel to the coordinate axes.

A *square* is a set in $\mathbb{R}^2$ of the form

$$\{\, (x,y) \in \mathbb{R}^2 : a < x < a + r, b < y < b + r \,\}$$

where $(a,b) \in \mathbb{R}^2$ is the *initial vertex* and $r > 0$ is the *edge length*. If E is a square, we will write $e(E)$ for the edge length of E. So (7.2.3) yields the following.

(7.2.5) Corollary. *Let f be a locally integrable function on $\mathbb{R}^2$. Then*

$$\lim_{\substack{e(E) \to 0 \\ x \in E \\ E \text{ square}}} \frac{1}{\lambda(E)} \int_E f \, d\lambda = f(x)$$

for almost all $x \in \mathbb{R}^2$.

Intervals

We consider next another natural example of a stochastic basis. It occurs in $\mathbb{R}^n$. For simplicity consider the case $\mathbb{R}^2$. This is a situation where the sets are not substantial in the sense described above.

We define an *interval* to be a set $[a, b) \times [c, d)$. An *interval partition* of the square $V = [0, 1) \times [0, 1)$ is a finite disjoint set t of intervals with union V. The corresponding σ-algebra $\mathcal{F}_t$ has the intervals of t as atoms. The set of all interval partitions of V is ordered by refinement.

There is a mode of convergence that corresponds to this stochastic basis. Let $f \in L_1(V)$. We are interested in the convergence of the indefinite integral of f to $f(x)$ in this sense: For every $\varepsilon > 0$, there is $\delta > 0$ such that for all intervals $E \subseteq V$ with $x \in E$ and diam $E < \delta$, we have

$$\left| \frac{1}{\lambda(E)} \int_E f \, d\lambda - f(x) \right| < \varepsilon.$$

The *diameter* of a rectangle is the length of the diagonal. An equivalent way to view this convergence requires both dimensions of the rectangle to approach zero, but postulates no relation between them.

Now this convergence is true for functions $f \in L_\infty$ (and even for functions $f \in L \log L$), but it fails in general for $f \in L_1$. (This is discussed below in Section 7.4.) We exhibit here an L_1 function that is not equal a.e. to the derivative (in the sense of the interval basis) of its indefinite integral.

(7.2.6) Lemma. *Let $N > 1$ be an integer, and let S be an interval. Then there exist intervals $E_1, E_2, \cdots, E_N$ contained in S such that the sets $I = \bigcap_{k=1}^{N} E_k$ and $U = \bigcup_{k=1}^{N} E_k$ satisfy $\lambda(I) = \lambda(E_k)/N$; $\lambda(E_k) = \lambda(S)/N$; $\lambda(U) \geq (\log N)\lambda(E_k)$.*

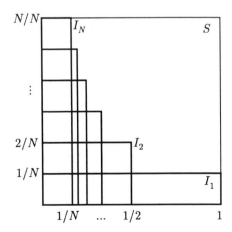

Figure (7.2.6). Illustration for Lemma 7.2.6.

Proof. If $S = [a, b) \times [c, d)$, let

$$E_k = \left[a + \frac{b-a}{k}, c + \frac{(d-c)k}{N} \right).$$

Thus $\lambda(E_k) = \lambda(S)/N$, $\lambda(I) = \lambda(S)/N^2 = \lambda(E_k)/N$, and

$$\lambda(U) = \left[\frac{\lambda(S)}{N} \right] \left[1 + \frac{1}{2} + \frac{1}{3} + \cdots + \frac{1}{N} \right] > (\log N)\lambda(E_k).$$ ∎

Rephrased, this shows:

(7.2.7) Lemma. *Let* $N > 1$ *be an integer and* S *an interval. There exist sets* $I \subseteq U \subseteq S$, *finite unions of intervals, such that*

$$\lambda(U) \geq \frac{\log N}{N} \lambda(S),$$

and, for every $x_0 \in U$, *there is an interval* $E_0 \subseteq S$ *with* $x_0 \in E_0$ *and*

$$\lambda(E_0 \cap I) = \frac{1}{N} \lambda(E_0).$$

(7.2.8) Lemma. *Let* $N > 1$ *be an integer and* S *an interval. There is a set* $B \subseteq S$, *a finite union of intervals, such that* $\lambda(B) \leq \big(2/(N \log N)\big)\lambda(S)$ *and for almost every* $x_0 \in S$, *there is an interval* $E_0 \subseteq S$ *with* $\lambda(E_0 \cap B) \geq (1/N)\lambda(E_0)$.

Proof. First, apply Lemma (7.2.7) in S. We get sets I^1 and U^1 with $\lambda(U^1) > (\log N/N)\lambda(S)$. Then subdivide the remainder $S \setminus U^1$ into intervals, and apply Lemma (7.2.7) to each of them. Let I^2 and U^2 be the unions of the corresponding results. We have $\lambda(U^2) > (\log N/N)\lambda(S \setminus U^1)$. Continue with the remainder. Stop when enough of the set is covered:

$$\lambda\left(U^1 \cup \cdots \cup U^m\right) > \left(1 - \frac{1}{N \log N}\right)\lambda(S).$$

Let $U^{m+1} = I^{m+1} = S \setminus (U^1 \cup \cdots \cup U^m)$. Let the set B be $I^1 \cup I^2 \cup \cdots \cup I^m \cup I^{m+1}$. Then (almost) every $x_0 \in S$ belongs to some U^k, hence to some interval E_0 with $\lambda(E_0 \cap B) \geq (1/N)\lambda(E_0)$ (equality for $k \leq m$). We have

$$\lambda(B) = \sum_{k=1}^{m+1} \lambda(I^k)$$

$$\leq \sum_{k=1}^{m} \frac{1}{N \log N} \lambda(U^k) + \lambda(U^{m+1})$$

$$\leq \frac{1}{N \log N} \lambda(S) + \lambda(U^{m+1})$$

$$\leq \frac{2}{N \log N} \lambda(S).$$ ∎

A slight variant:

(7.2.9) Lemma. *Let $N > 1$ be an integer and S an interval. Then there is a function w defined on S with only values 0 and N, such that*

$$\int_S w \log w \, d\lambda \le 2\lambda(S),$$

with the convention $0 \log 0 = 0$; and for almost every $x_0 \in S$, there is an interval $E_0 \subseteq S$ with $x_0 \in E_0$ and

$$\int_{E_0} w \, d\lambda \ge \lambda(E_0).$$

Proof. Let B be as in (7.2.8), and let $w = N \, \mathbf{1}_B$. ∎

Finally we come to the counterexample. It is a nonnegative integrable function for which the upper derivate (for the interval base) is infinite almost everywhere.

(7.2.10) Theorem. *There is a nonnegative function $f \in L_1\big([0,1) \times [0,1)\big)$ with infinite upper derivate a.e.*

Proof. Let N_m be a sequence of integers with $N_m > e^{e^m}$. Write $V = [0,1) \times [0,1)$. For each m, subdivide V into intervals with diameter $< 1/m$, then on each interval apply Lemma (7.2.9) with integer N_m. Thus we get a function $w_m \colon V \to [0, \infty)$, with values 0 and N_m, such that $\int w_m \log w_m \, d\lambda \le 2$ and for almost all $x_0 \in V$, there is an interval E_0 with diam $E_0 < 1/m$ and $\int_{E_0} w_m \, d\lambda \ge \lambda(E_0)$. Since w_m has only values 0 and N_m, with the usual convention $0 \log 0 = 0$ we have $w_m \log w_m = w_m \log N_m$.

Now let $f = \sum_{m=1}^{\infty} e^{m/2} w_m$. Then $f \ge 0$. Also,

$$\int_V f \, d\lambda = \sum \int e^{m/2} w_m \, d\lambda$$

$$= \sum \int e^{-m/2} e^m w_m \, d\lambda$$

$$\le \sum e^{-m/2} \int (\log N_m) w_m \, d\lambda$$

$$= \sum e^{-m/2} \int w_m \log w_m \, d\lambda$$

$$\le \sum e^{-m/2} 2 < \infty.$$

Thus $f \in L_1$. For almost all x_0 and any m, there is an interval E_m with diameter $< 1/m$ such that

$$\int_{E_m} f \, d\lambda \ge e^{m/2} \int_{E_m} w_m \, d\lambda \ge e^{m/2} \lambda(E_m).$$

Thus the upper derivate of f at x_0 is $\geq e^{m/2}$. This is true for all m, so the upper derivate is infinite.   ∎

Complements

(7.2.11) (A generalized derivation theorem of Besicovitch.) A measure ν defined on the Borel sets of $\mathbb{R}^d$ is *locally finite* iff $\nu(B(0,R)) < \infty$ for all $R > 0$. Let us say that a collection $\mathcal{V}$ of closed balls in $\mathbb{R}^d$ is a *Vitali cover* of a set $Q \subseteq \mathbb{R}^d$ iff, for every $x \in Q$, the collection $\mathcal{V}$ contains balls of arbitrarily small radius centered at x.

Suppose ν is a locally finite measure on $\mathbb{R}^d$, Q is a subset of $\mathbb{R}^d$, and $\mathcal{V}$ is a Vitali cover of Q by closed balls. Then $\mathcal{V}$ has a countable, pairwise disjoint, subcollection $\{E_1, E_2, \cdots\}$ such that

$$\nu^*\left(Q \setminus \bigcup_n E_n\right) = 0.$$

(Besicovitch [1945], [1946], [1947]). This Vitali theorem may be used as in this section for derivation theorems with ν replacing Lebesgue measure λ.

(7.2.12) (Intervals with edges of the same size.) Let k, m be integers, $1 \leq k \leq m$, and let $\mathcal{C}$ be the collection of intervals in $[0,1]^m$ whose edges have no more than k different sizes. If f is in $L \log^{k-1} L$, then for almost every x,

$$\lim_{\substack{\text{diam } E \to 0 \\ E \in \mathcal{C} \\ x \in E}} \int_E f \, d\lambda = f(x).$$

Remarks

The counterexample concerning the interval basis is due to Bohr. It can be adapted to show that the space $L \log L$ is exactly the correct space for differentiability in the interval basis; see Hayes & Pauc [1970], page 98. Similarly, in $\mathbb{R}^d$, Zygmund [1934] showed that $L \log^{d-1} L$ is exactly the correct space. This is also true in other settings: see (9.2.3) and the introduction to Section 9.4. Complement 7.2.12, due to Zygmund [1967], was extended to the setting of substantial sets by Frangos & Sucheston [1985], Theorem 6.2.

7.3. Abstract derivation

We have seen above that, in some special cases, derivation theorems can be viewed as consequences of martingale convergence theorems. But it is not convenient to view them that way in general. It is also possible, in some special cases, to view martingale convergence theorems as consequences of derivation theorems (7.3.23). But again, it is not convenient to view them that way in general. In this section we will discuss derivation in the general abstract setting. In the next section we will specialize somewhat in order to obtain more precise results.

Nonmeasurable sets

In the most general setting for derivation (defined below), we must deal with nonmeasurable sets. We will specify our notation and terminology concerning them here. Let $(\Omega, \mathcal{F}, \mu)$ be a σ-finite measure space. Let $Q \subseteq \Omega$ be a set. The *outer measure* of Q is

$$\mu^*(Q) = \inf \{ \mu(A) : A \in \mathcal{F}, A \supseteq Q \},$$

and the *inner measure* of Q is

$$\mu_*(Q) = \sup \{ \mu(A) : A \in \mathcal{F}, A \subseteq Q \}.$$

There exists a measurable set $C \supseteq Q$ with $\mu(C) = \mu^*(Q)$. Such a set C is unique up to null sets, and is called an *outer envelope* of Q. For example, C could be the essential infimum of all measurable sets A with $A \supseteq Q$, and therefore is the intersection of a countable family of such sets A. We may sometimes write $\overline{Q}$ for an outer envelope of Q. A measurable set $\underline{Q} \subseteq Q$ with $\mu(\underline{Q}) = \mu_*(Q)$ similarly exists, and is called an *inner envelope* of Q.

Derivation bases

The abstract setting for derivation theory is the "derivation basis." It corresponds (very roughly) to the "stochastic basis" in the convergence theory of Chapter 4. The idea is to assign to every point $\omega \in \Omega$ a net $(E_t)_{t \in J}$ of sets (or even a family of such nets). We are interested in relations such as

$$\lim_{t \in J} \frac{1}{\mu(E_t)} \int_{E_t} f \, d\mu = f(\omega).$$

The directed set J may or may not be the same for all of the nets (E_t) belonging to a given derivation basis.

Let $(\Omega, \mathcal{F}, \mu)$ be a σ-finite measure space. (It is possible to study derivation without the assumption of σ-finiteness, but we do not pursue it here. For example, derivation with respect to Hausdorff [fractal] measures requires such a theory.) A *deriving net* $(\omega, J, (E_t))$ consists of a point $\omega \in \Omega$, a directed set J, and a net $(E_t)_{t \in J}$, where $E_t \in \mathcal{F}$, $0 < \mu(E_t) < \infty$.

Note that $\omega \in E_t$ is not specified in the definition, although this is the case in all the examples we shall study. The abstract definition allows, for example, a case like this: Ω is the plane $\mathbb{R}^2$, and for $\omega \in \Omega$, $r > 0$, the set E_r is an annulus $\{ y \in \mathbb{R}^2 : r < |y - \omega| < 2r \}$ centered at ω.

(7.3.1) Definition. A *derivation basis* on $(\Omega, \mathcal{F}, \mu)$ is a collection **B** of deriving nets satisfying:

    (1) for every $\omega \in \Omega$, there is at least one deriving net $(\omega, J, (E_t)) \in$ **B**;
    (2) if we have $(\omega, J, (E_t)_{t \in J}) \in$ **B** and $J' \subseteq J$ is cofinal in J, then we also have $(\omega, J', (E_t)_{t \in J'}) \in$ **B**.

If $(\omega, J, (E_t)) \in \mathbf{B}$, we will say that (E_t) *converges to* ω according to $\mathbf{B}$. When the derivation basis $\mathbf{B}$ is understood, we will write

$$E_t \Rightarrow \omega.$$

The sets E_t that occur in a derivation basis $\mathbf{B}$ are known as the *constituents* of $\mathbf{B}$. The collection of all the constituents of $\mathbf{B}$ is the *constituency* of $\mathbf{B}$ (or the *spread* of $\mathbf{B}$).

Normally, throughout this chapter, a measure space $(\Omega, \mathcal{F}, \mu)$ and a derivation basis $\mathbf{B}$ will be fixed.

Vitali covers and derivation

(7.3.2) Definition. Let $Q \subseteq \Omega$ be a (possibly nonmeasurable) set. A collection $\mathcal{V}$ of constituents is a $\mathbf{B}$-*fine cover* of Q iff, for every $\omega \in Q$, there exists $E_t \Rightarrow \omega$ with $E_t \in \mathcal{V}$ for all t. Let $C \in \mathcal{F}$ be a measurable set. The collection $\mathcal{V}$ of constituents is a $\mathbf{B}$-*fine almost-cover* of C iff there is a set Q such that C is an outer envelope of Q and $\mathcal{V}$ is a $\mathbf{B}$-fine cover of Q.

A collection $\mathcal{U}$ of constituents is a *full* $\mathbf{B}$-*fine cover* of Q iff, for every $\omega \in Q$ and every deriving net $E_t \Rightarrow \omega$, there is an index t_0 such that $E_t \in \mathcal{U}$ for all $t \geq t_0$. Observe that the intersection of two full $\mathbf{B}$-fine covers of a set Q is again a full $\mathbf{B}$-fine cover of Q; and the intersection of a full $\mathbf{B}$-fine cover of Q with a $\mathbf{B}$-fine cover of Q is again a $\mathbf{B}$-fine cover of Q.

Let γ be a real-valued function defined on the constituency of $\mathbf{B}$. The *upper derivate* of γ (with respect to μ) is

$$D^{\star}\gamma(\omega) = \sup_{E_t \Rightarrow \omega} \limsup_{t} \frac{\gamma(E_t)}{\mu(E_t)}.$$

The *lower derivate* of γ (with respect to μ) is

$$D_{\star}\gamma(\omega) = \inf_{E_t \Rightarrow \omega} \liminf_{t} \frac{\gamma(E_t)}{\mu(E_t)}.$$

If $D^{\star}\gamma(\omega) = D_{\star}\gamma(\omega)$, then we say γ is *differentiable* at ω, and write $D\gamma(\omega)$ for the common value. If γ is differentiable at almost every $\omega \in \Omega$, then we say that γ is *differentiable*.

One family of examples of set functions γ is the family of integrals. An *integral* is a set function γ of the form

$$\gamma(E) = \int_E f \, d\mu,$$

where f is a measurable function such that the integral exists for all constituents E. Then we say that $\mathbf{B}$ *differentiates the integral* of f iff $D\gamma = f$ a.e.

Some examples of derivation bases were used in the first two sections of this chapter. In the measure space $\mathbb{R}$, for the *interval basis*, we postulate that $E_n \Rightarrow \omega$ iff E_n is a sequence of closed intervals, containing the point ω, with positive lengths converging to 0. Then Theorem (7.1.9) can be interpreted to say that the interval basis in $\mathbb{R}$ differentiates all L_1 integrals.

In $\mathbb{R}^2$, the *centered disk basis* is described by saying that $E_n \Rightarrow \omega$ iff E_n is a sequence of disks, centered at ω, with radius converging to 0. The *disk basis* is similar: $E_n \Rightarrow \omega$ iff E_n is a sequence of disks, containing the point ω, with radius converging to 0. Corollary (7.2.4) states that both of these bases differentiate all L_1 integrals.

Description of the bases corresponding to (7.2.3) and (7.2.5) is left to the reader.

For the *interval basis* on $\mathbb{R}^2$, we postulate that $E_n \Rightarrow \omega$ iff E_n is a sequence of intervals, containing the point ω, such that diam $E_n \to 0$. Theorem (7.2.10) shows that the interval basis fails to differentiate some L_1 integrals.

Many derivation bases (including all of those considered above) have a useful approximation property. If Ω is a metric space, and μ is a σ-finite Borel measure on Ω, then μ is a *Radon measure* iff, for every Borel set B, we have

$$\mu(B) = \sup \{ \mu(K) : K \text{ compact}, K \subseteq B \},$$
$$\mu(B) = \inf \{ \mu(U) : U \text{ open}, U \supseteq B \}.$$

Lebesgue measure on $\mathbb{R}^d$ is a Radon measure. If Ω is a complete separable metric space, then every finite Borel measure on Ω is a Radon measure (for example, Halmos [1950], (10), p. 40).

Now if $\mathbf{B}$ is any of the bases considered above, and if $C \in \mathcal{F}, 0 < \lambda(C) < \infty$, and $\varepsilon > 0$, then there is an open set $U \supseteq C$ with $\lambda(U) < \lambda(C) + \varepsilon$. However,

$$\mathcal{U} = \{ E \text{ constituent} : E \subseteq U \}$$

is a full $\mathbf{B}$-fine cover of C.

For abstract derivation bases we will use the following definition. A derivation basis $\mathbf{B}$ has *small overflow* iff for every $C \in \mathcal{F}$ with $\mu(C) < \infty$ and every $\varepsilon > 0$, there exist a set $C_0 \subseteq C$ with $\mu(C \setminus C_0) = 0$ and a full $\mathbf{B}$-fine cover $\mathcal{U}$ of C_0 such that for any $A_1, A_2, \cdots, A_n \in \mathcal{U}$, we have $\mu(\bigcup_{i=1}^n A_i \setminus C) < \varepsilon$. We will prove below (Proposition (7.4.2)) that many of the commonly used derivation bases have small overflow.

The strong Vitali property

We have seen that the Vitali covering theorem is an important tool for the classical derivation theorems considered above. Derivation bases for which corresponding properties hold will have useful derivation properties. We begin with a property for derivation bases that roughly corresponds to the covering property (V) for stochastic bases.

(7.3.3) Definition. A derivation basis **B** satisfies the *strong Vitali property* iff, for every $C \in \mathcal{F}$ with $0 < \mu(C) < \infty$, every **B**-fine almost-cover $\mathcal{V}$ of C, and every $\varepsilon > 0$, there exist finitely many pairwise disjoint constituents $A_1, A_2, \cdots, A_n \in \mathcal{V}$ with

$$\mu\left(C \setminus \bigcup_{i=1}^{n} A_i\right) < \varepsilon,$$

$$\mu\left(\bigcup_{i=1}^{n} A_i \setminus C\right) < \varepsilon.$$

We will prove that a derivation basis **B** with the strong Vitali property differentiates all L_1 integrals. First, we prove a basic lemma for such integrals.

(7.3.4) Lemma. Let $f \in L_1$, and define $\gamma(A) = \int_A f \, d\mu$ for $A \in \mathcal{F}$. For any $\varepsilon > 0$ there exists $\delta > 0$ so that if $A_1, A_2, \cdots, A_n$ are disjoint constituents and $C \in \mathcal{F}$ such that $\mu(C \setminus \bigcup A_i) < \delta$ and $\mu(\bigcup A_i \setminus C) < \delta$, then

$$\left| \sum_{i=1}^{n} \gamma(A_i) - \gamma(C) \right| \leq \varepsilon.$$

Proof. Since $f \in L_1$, given $\varepsilon > 0$ there is $\delta > 0$ so that if $G \in \mathcal{F}$ with $\mu(G) < \delta$, then $\int_G |f| \, d\mu < \varepsilon/2$. Now let $A_1, A_2, \cdots, A_n$ be disjoint constituents and $C \in \mathcal{F}$ with $\mu(C \setminus \bigcup A_i) < \delta$ and $\mu(\bigcup A_i \setminus C) < \delta$. Then

$$\left| \sum \gamma(A_i) - \gamma(C) \right| = \left| \sum \int_{A_i} f \, d\mu - \int_C f \, d\mu \right|$$

$$= \left| \int_{\bigcup A_i} f \, d\mu - \int_C f \, d\mu \, d\mu \right|$$

$$\leq \int_{\bigcup A_i \setminus C} |f| \, d\mu + \int_{C \setminus \bigcup A_i} |f| \, d\mu$$

$$\leq \frac{\varepsilon}{2} + \frac{\varepsilon}{2} = \varepsilon.$$

∎

(7.3.5) Theorem. Let **B** *be a derivation basis satisfying the strong Vitali property. Then* **B** *differentiates all L_1 integrals.*

Proof. Let γ be an L_1 integral; say $\gamma(A) = \int_A f \, d\mu$, with $f \in L_1$. We claim first that $D^\star \gamma \leq f$ a.e. If not, there exist $a < b$ such that $\mu^*\{f < a < b < D^\star \gamma\} > 0$. Hence there is a set $Q \subseteq \{f < a < b < D^\star \gamma\}$ with $0 < \mu^*(Q) < \infty$. Let C be an outer envelope of Q. Now

$$\mathcal{V} = \left\{ E \text{ constituent} : \frac{\gamma(E)}{\mu(E)} > b \right\}$$

is a **B**-fine cover of Q, and therefore a **B**-fine almost-cover of C. Let $\varepsilon > 0$. By the strong Vitali property and Lemma (7.3.4), there exist disjoint constituents $A_1, A_2, \cdots, A_n \in \mathcal{V}$ with $\mu(C \setminus \bigcup A_i) < \varepsilon$, $\mu(\bigcup A_i \setminus C) < \varepsilon$, $|\sum \gamma(A_i) - \gamma(C)| < \varepsilon$. Thus

$$\gamma(C) + \varepsilon \geq \sum \gamma(A_i) \geq b \sum \mu(A_i)$$
$$= b\mu(\bigcup A_i) \geq b\mu(C) - \varepsilon|b|.$$

This is true for all $\varepsilon > 0$, so $\gamma(C) \geq b\mu(C)$. But $f < a$ on C, so $\gamma(C) = \int_C f \, d\mu < a\mu(C) < b\mu(C)$, a contradiction. Therefore $D^\star \gamma \leq f$ a.e.

Similarly, $D_\star \gamma \geq f$ a.e. Clearly $D_\star \gamma \leq D^\star \gamma$, so we have $D_\star \gamma = D^\star \gamma = f$ a.e. ∎

The weak Vitali property

Next we will consider a condition analogous to condition (V_1) for stochastic bases.

(7.3.6) Definition. The derivation basis **B** has the *weak Vitali property* iff, for every $C \in \mathcal{F}$ with $0 < \mu(C) < \infty$, every **B**-fine almost-cover $\mathcal{V}$ of C, and every $\varepsilon > 0$, there exist finitely many constituents $A_1, A_2, \cdots, A_n \in \mathcal{V}$ with

(a) $\mu \left(C \setminus \bigcup_{i=1}^n A_i \right) < \varepsilon$,
(b) $\mu \left(\bigcup_{i=1}^n A_i \setminus C \right) < \varepsilon$,
(c) $\sum_{i=1}^n \mu(A_i) - \mu \left(\bigcup_{i=1}^n A_i \right) < \varepsilon$.

(The left-hand side of (a) is called the "deficit"; the left-hand side of (b) is called the "overflow"; the left-hand side of (c) is called the "overlap.")

There are some useful alternative formulations of the definition.

(7.3.7) Proposition. *Let* **B** *be a derivation basis. The following are equivalent*

(1) *The weak Vitali property: For every $C \in \mathcal{F}$ with $0 < \mu(C) < \infty$, every* **B**-*fine almost-cover $\mathcal{V}$ of C, and every $\varepsilon > 0$, there exist finitely many constituents $A_1, A_2, \cdots, A_n \in \mathcal{V}$ with $\mu \left(C \setminus \bigcup A_i \right) < \varepsilon$, $\mu \left(\bigcup A_i \setminus C \right) < \varepsilon$, $\sum \mu(A_i) - \mu \left(\bigcup A_i \right) < \varepsilon$.*

(2) *For every $C \in \mathcal{F}$ with $0 < \mu(C) < \infty$, every* **B**-*fine almost-cover $\mathcal{V}$ of C, and every $\varepsilon > 0$, there exist finitely many constituents $A_1, A_2, \cdots, A_n \in \mathcal{V}$ with $\left\| \sum \mathbf{1}_{A_i} - \mathbf{1}_C \right\|_1 < \varepsilon$.*

(3) *For every $C \in \mathcal{F}$ with $0 < \mu(C) < \infty$, every* **B**-*fine almost-cover $\mathcal{V}$ of C, and every $\varepsilon > 0$, there exist countably many constituents $A_1, A_2, \cdots \in \mathcal{V}$ with $\mu \left(C \setminus \bigcup A_i \right) = 0$, $\mu \left(\bigcup A_i \setminus C \right) < \varepsilon$, $\sum \mu(A_i) - \mu \left(\bigcup A_i \right) < \varepsilon$.*

(4) *For every $C \in \mathcal{F}$ with $0 < \mu(C) < \infty$, every* **B**-*fine almost-cover $\mathcal{V}$ of C, and every $\varepsilon > 0$, there exist countably many constituents $A_1, A_2, \cdots \in \mathcal{V}$ with $\left\| \sum \mathbf{1}_{A_i} - \mathbf{1}_C \right\|_1 < \varepsilon$.*

Proof. (1) $\implies$ (3). Let C, $\mathcal{V}$, ε be given. Write $\varepsilon' = \varepsilon/3$. Apply (a) to obtain finitely many constituents $A_{11}, A_{21}, \cdots \in \mathcal{V}$ with $\mu(C \setminus \bigcup A_{i1}) < \varepsilon'/2$, $\mu(\bigcup A_{i1} \setminus C) < \varepsilon'/2$, $\sum \mu(A_{i1}) - \mu(\bigcup A_{i1}) < \varepsilon'/2$. Then apply (a) to the set $C_1 = C \setminus \bigcup A_{i1}$ to obtain finitely many constituents $A_{12}, A_{22}, \cdots \in \mathcal{V}$ with $\mu(C_1 \setminus \bigcup A_{i2}) < \varepsilon'/4$, $\mu(\bigcup A_{i2} \setminus C_1) < \varepsilon'/4$, $\sum \mu(A_{i2}) - \mu(\bigcup A_{i2}) < \varepsilon'/4$. Continue in the same way: $C_k = C_{k-1} \setminus \bigcup_i A_{ik}$, $\mu(C_{k-1} \setminus \bigcup A_{ik}) < \varepsilon'/2^k$, $\mu(\bigcup_i A_{ik} \setminus C_{k-1}) < \varepsilon'/2^k$, $\sum_i \mu(A_{ik}) - \mu(\bigcup_i A_{ik}) < \varepsilon'/2^k$. Then the countable set $\{A_{ik}\}$ satisfies the conditions of (3):

$$\mu\left(C \setminus \bigcup_{i,k} A_{ik}\right) = \mu\left(\bigcap_k C_k\right) = \lim_k \mu(C_k) = 0;$$

$$\mu\left(\bigcup_{i,k} A_{ik} \setminus C\right) = \sum_k \mu\left(\bigcup_i A_{ik} \setminus C_{k-1}\right) < \varepsilon' < \varepsilon;$$

$$\sum_{i,k} \mu(A_{ik}) = \sum_k \sum_i \mu(A_{ik})$$

$$\leq \sum_k \mu\left(\bigcup_i A_{ik}\right) + \varepsilon'$$

$$\leq \sum_k \left[\mu(C_{k-1} \setminus C_k) + \mu\left(\bigcup_i A_{ik} \setminus C_{k-1}\right)\right] + \varepsilon'$$

$$\leq \mu(C) + 2\varepsilon'$$

$$\leq \mu\left(\bigcup_{i,k} A_{ik}\right) + 3\varepsilon'$$

$$= \mu\left(\bigcup_{i,k} A_{ik}\right) + \varepsilon.$$

(3) $\implies$ (4). Suppose (3) is satisfied. Then

$$\left\|\sum_i \mathbf{1}_{A_i} - \mathbf{1}_C\right\|_1 \leq \left\|\sum \mathbf{1}_{A_1} - \mathbf{1}_{\cup A_i}\right\|_1 + \left\|\mathbf{1}_{\cup A_i} - \mathbf{1}_C\right\|_1$$

$$= \sum \mu(A_i) - \mu(\bigcup A_i) + \mu(\bigcup A_i \setminus C) + \mu(C \setminus \bigcup A_i)$$

$$\leq \varepsilon + \varepsilon + 0 = 2\varepsilon.$$

(4) $\implies$ (2). Suppose the conclusion is true with a countably infinite collection of sets $A_1, A_2, \cdots$. By monotone convergence,

$$\lim_n \left\|\sum_{i=1}^n \mathbf{1}_{A_i} - \mathbf{1}_C\right\|_1 = \left\|\sum_{i=1}^\infty \mathbf{1}_{A_i} - \mathbf{1}_C\right\|_1 < \varepsilon.$$

Thus, for some n, we have $\left\|\sum_{i=1}^n \mathbf{1}_{A_i} - \mathbf{1}_C\right\|_1 < \varepsilon$.

(2) $\Longrightarrow$ (1). Suppose $\|\sum \mathbf{1}_{A_i} - \mathbf{1}_C\|_1 < \varepsilon$. Then $\mathbf{1}_{\bigcup A_i \setminus C} \leq |\sum \mathbf{1}_{A_i} - \mathbf{1}_C|$, so $\mu(\bigcup A_i \setminus C) = \|\bigcup A_i \setminus C\|_1 < \varepsilon$. Next, $\mathbf{1}_{C \setminus \bigcup A_i} \leq |\sum \mathbf{1}_{A_i} - \mathbf{1}_C|$, so $\mu(C \setminus \bigcup A_i) < \varepsilon$. Finally,

$$\sum \mu(A_i) - \mu\left(\bigcup A_i\right) = \left\|\sum \mathbf{1}_{A_i} - \mathbf{1}_{\bigcup A_i}\right\|_1$$
$$\leq \left\|\sum \mathbf{1}_{A_i} - \mathbf{1}_C\right\|_1 + \left\|\mathbf{1}_C - \mathbf{1}_{\bigcup A_i}\right\|_1 \leq 3\varepsilon. \qquad \blacksquare$$

By analogy with the stochastic basis version, we may expect that the weak Vitali property is necessary and sufficient for derivation of L_∞ integrals. (See below, Theorem (7.3.11).) First, the appropriate lemma.

(7.3.8) Lemma. *Let $f \in L_\infty$, and define $\gamma(A) = \int_A f \, d\mu$ for $A \in \mathcal{F}$, $\mu(A) < \infty$. Let $A_1, A_2, \cdots, A_n$ be constituents and let $C \in \mathcal{F}$. Then if $\|\sum \mathbf{1}_{A_i} - \mathbf{1}_C\|_1 < \varepsilon$, we have*

$$\left|\sum_{i=1}^{n} \gamma(A_i) - \gamma(C)\right| \leq \varepsilon \|f\|_\infty.$$

Proof.

$$\left|\sum \gamma(A_i) - \gamma(C)\right| = \left|\int \left(\sum \mathbf{1}_{A_i} - \mathbf{1}_C\right) f \, d\mu\right|$$
$$\leq \left\|\sum \mathbf{1}_{A_i} - \mathbf{1}_C\right\|_1 \|f\|_\infty \leq \varepsilon \|f\|_\infty. \qquad \blacksquare$$

Derivation of L_∞ integrals is closely related to the Lebesgue density theorem, and its abstract analogs.

(7.3.9) Definition. Let A be a measurable set. A point ω is a *point of density* of A iff for every $\varepsilon > 0$ there is $E_t \Rightarrow \omega$ with $\limsup_t \mu(A \cap E_t)/\mu(E_t) \geq 1 - \varepsilon$. (Equivalently, the set function $\gamma(E) = \mu(A \cap E)$ has $D^\star\gamma(\omega) = 1$.) A derivation basis has the *density property* iff almost every point of every measurable set A is a point of density of the set A.

Of course, it is enough to check sets of finite measure:

(7.3.10) Proposition. *Suppose, for every measurable set C of finite measure, almost every point of C is a point of density of C. Then $\mathbf{B}$ has the density property.*

Proof. Let A be a measurable set of infinite measure. Then (since μ is σ-finite), there is an increasing sequence C_n of sets of finite measure with $\bigcup C_n = A$. Now we know that almost every point of C_n is a point of density of C_n. Hence for almost every point ω of A, there is n such that ω is a point of density of C_n. But for any $E_t \Rightarrow \omega$,

$$\limsup_t \frac{\mu(E_t \cap A)}{\mu(E_t)} \geq \limsup_t \frac{\mu(E_t \cap C_n)}{\mu(E_t)},$$

so ω is a point of density of A as well. $\qquad \blacksquare$

Now we prove that the density property is characterized by the weak Vitali property.

(7.3.11) Theorem. *Let* **B** *be a derivation basis. The following are equivalent:*

(a) **B** *has the density property.*
(b) *If* $C \in \mathcal{F}$ *then* **B** *differentiates the integral of* $\mathbf{1}_C$.
(c) **B** *differentiates all* L_∞ *integrals.*
(d) **B** *has the weak Vitali property.*

Proof. (c) $\Longrightarrow$ (b) and (b) $\Longrightarrow$ (a) are easy.

(d) $\Longrightarrow$ (c). Let γ be an L_∞ integral; say $\gamma(A) = \int_A f\, d\mu$, with $f \in L_\infty$. We claim first that $D^\star\gamma \le f$ a.e. If not, there exist $a < b$ such that $\mu^*\{f < a < b < D^\star\gamma\} > 0$. Hence there is a set $Q \subseteq \{f < a < b < D^\star\gamma\}$ with $0 < \mu^*(Q) < \infty$. Let C be an outer envelope of Q. Now

$$\mathcal{V} = \left\{ E \text{ constituent} : \frac{\gamma(E)}{\mu(E)} > b \right\}$$

is a **B**-fine almost-cover of C. Let $\varepsilon > 0$. By the weak Vitali property and Lemma (7.3.8), there exist constituents $A_1, A_2, \cdots, A_n \in \mathcal{V}$ with $\| \sum \mathbf{1}_{A_i} - \mathbf{1}_C \|_1 < \varepsilon$ and $|\sum \gamma(A_i) - \gamma(C)| < \varepsilon$. Thus

$$\gamma(C) + \varepsilon \ge \sum \gamma(A_i) \ge b \sum \mu(A_i)$$
$$\ge b\mu(C) - \varepsilon|b|.$$

This is true for all $\varepsilon > 0$, so $\gamma(C) \ge b\mu(C)$. But $f < a$ on C, so $\gamma(C) = \int_C f\, d\mu < a\mu(C) < b\mu(C)$, a contradiction. Therefore $D^\star\gamma \le f$.

Similarly, $D_\star\gamma \ge f$.

(a) $\Longrightarrow$ (d). Suppose the density property holds. Let $\mathcal{V}$ be a **B**-fine cover of Q and C an outer envelope of Q; suppose $0 < \mu(C) < \infty$; let $\varepsilon > 0$. Choose α with $0 < \alpha < 1$ such that

$$0 < \left(\frac{1}{\alpha} - 1\right) \mu(C) < \varepsilon.$$

Now if $Y \subseteq Q$ and $\mu^*(Y) > 0$, let

$$\mathcal{V}(Y, \alpha) = \{ E \in \mathcal{V} : \mu^*(Y \cap E) > \alpha\, \mu(E) \},$$
$$r_Y = \sup \{ \mu(E) : E \in \mathcal{V}(Y, \alpha) \}.$$

From the density property applied to an outer envelope $\overline{Y}$ of Y, we see that some point of Y is a point of density of $\overline{Y}$, so there exists $\omega \in Y$ and $E_t \Rightarrow \omega$, with $E_t \in \mathcal{V}$, so $\mu(\overline{Y} \cap E_t)/\mu(E_t) \to 1 > \alpha$. Thus $\mathcal{V}(Y, \alpha) \ne \emptyset$, and therefore $r_Y > 0$. If $Y \subseteq Q$ and $\mu^*(Y) = 0$, write $r_Y = 0$.

Now fix β with $0 < \beta < 1$. Let $X_1 = Q$. Then $\mu^*(X_1) > 0$, so $r_{X_1} > 0$. There exists $A_1 \in \mathcal{V}$ with $\mu(A_1) > \beta r_{X_1}$ and $\mu(C \cap A_1) > \alpha\mu(A_1)$. Let

$X_2 = X_1 \setminus A_1$. Continue recursively: Suppose $A_1, A_2, \cdots, A_n \in \mathcal{V}$ have been defined such that

$$\mu(\overline{X_i} \cap A_i) > \alpha\mu(A_i), \qquad \mu(A_i) > \beta r_{X_i},$$

where $X_{i+1} = X_i \setminus \bigcup_{j=1}^{i} A_j$ and $\overline{X_i}$ is an outer envelope of X_i. If we have $\mu^*(X_{n+1}) = 0$, then the recursive construction stops. Otherwise, let $A_{n+1} \in \mathcal{V}$ satisfy $\mu(A_{n+1}) > \beta r_{X_{n+1}}$ and $\mu(\overline{X_{n+1}} \cap A_{n+1}) > \alpha\mu(A_{n+1})$. So we get a (finite or infinite) sequence of sets $A_1, A_2, \cdots \in \mathcal{V}$ such that $A_i \cap X_i$ are disjoint subsets of Q,

$$\mu(C) \geq \mu(C \cap \bigcup A_i) \geq \mu\big(\bigcup(A_i \cap \overline{X_i})\big)$$
$$= \sum \mu(\overline{X_i} \cap A_i) > \alpha \sum \mu(A_i).$$

Thus $\sum \mu(A_i) < (1/\alpha)\mu(C \cap \bigcup A_i) < \infty$.

Now we claim that $\mu(C \setminus \bigcup A_i) = 0$. If the sequence A_i is finite, then $\mu^*(X_{N+1}) = 0$ for some N, so $\mu^*\big(Q \setminus \bigcup_{i=1}^{N} A_i\big) = 0$ as claimed. So suppose the sequence A_i is infinite. Then $\beta \sum_i r_{X_i} < \sum \mu(A_i) < \infty$, so $r_{X_i} \to 0$. Let $X_\infty = Q \setminus \bigcup A_i$. Then $X_\infty \subseteq X_n$ for all n. Thus $\mathcal{V}(X_\infty, \alpha) \subseteq \mathcal{V}(X_n, \alpha)$, so $r_{X_\infty} \leq r_{X_n}$ for all n, so $r_{X_\infty} = 0$. Therefore $\mu^*(X_\infty) = 0$, or $\mu(C \setminus \bigcup A_i) = 0$ as claimed.

Now we have $\mu(C) = \mu(C \cap \bigcup A_i)$. Then

$$\left\| \sum \mathbf{1}_{A_i} - \mathbf{1}_C \right\|_1 \leq \sum \mu(A_i) - \mu(C) < \left(\frac{1}{\alpha} - 1\right)\mu(C) < \varepsilon.$$

This shows that the weak Vitali property holds. ∎

Orlicz functions

Next we come to the analogs for differentiation bases of the covering conditions (V_Φ) for stochastic bases. We will retain the same terminology.

(7.3.12) Definition. Let Φ be an Orlicz function. The derivation basis **B** has *property* (V_Φ) iff, for every $C \in \mathcal{F}$ with $0 < \mu(C) < \infty$, every **B**-fine almost-cover $\mathcal{V}$ of C, and every $\varepsilon > 0$, there exist finitely many constituents $A_1, A_2, \cdots, A_n \in \mathcal{V}$ with

$$\left\| \sum_{i=1}^{n} \mathbf{1}_{A_i} - \mathbf{1}_C \right\|_\Phi < \varepsilon.$$

Note in particular that the weak Vitali property is exactly property (V_Φ) in the case $L_\Phi = L_1$.

We will need to recall two facts about Orlicz functions:

(2.1.22) If g_n has integer values, and $\|g_n\|_\Phi \to 0$, then $\|g_n\|_1 \to 0$.

(2.1.20) If Φ is finite, then $\mu(A_n) \to 0$ if and only if $\|\mathbf{1}_{A_n}\|_\Phi \to 0$.

The typical case where Φ is not finite is $L_\Phi = L_\infty$. When Φ is finite, condition (V_Φ) may be reformulated.

(7.3.13) Proposition. *Suppose* Φ *is a finite Orlicz function and* **B** *is a derivation basis. The following are equivalent.*

(1) (V_Φ): *For every* $C \in \mathcal{F}$ *with* $0 < \mu(C) < \infty$, *every* **B**-*fine almost-cover* $\mathcal{V}$ *of* C, *and every* $\varepsilon > 0$, *there exist finitely many constituents* $A_1, A_2, \cdots, A_n \in \mathcal{V}$ *with* $\|\sum 1_{A_i} - 1_C\|_\Phi < \varepsilon$.

(2) *For every* $C \in \mathcal{F}$ *with* $0 < \mu(C) < \infty$, *every* **B**-*fine almost-cover* $\mathcal{V}$ *of* C, *and every* $\varepsilon > 0$, *there exist finitely many constituents* $A_1, A_2, \cdots, A_n \in \mathcal{V}$ *with* $\mu(C \setminus \bigcup A_i) < \varepsilon$, $\mu(\bigcup A_i \setminus C) < \varepsilon$, *and* $\|\sum 1_{A_i} - 1_{\bigcup A_i}\|_\Phi < \varepsilon$.

Proof. (1) $\Longrightarrow$ (2). Let $\varepsilon > 0$. Use (2.1.20) to choose $\varepsilon' > 0$ so that $\varepsilon' < \varepsilon$ and $\mu(D) < 2\varepsilon'$ implies $\|1_D\|_\Phi < \varepsilon/2$. Then by (2.1.22) choose $\delta < \varepsilon/2$ so that if g has integer values and $\|g\|_\Phi < \delta$, then $\|g\|_1 < \varepsilon'$.

Now suppose C and $\mathcal{V}$ are given. Then by (1) there exist $A_1, \cdots, A_n \in \mathcal{V}$ with $\|\sum 1_{A_i} - 1_C\|_\Phi < \delta$. Now $1_{\bigcup A_i \setminus C} \le |\sum 1_{A_i} - 1_C|$, so $\|1_{\bigcup A_i \setminus C}\|_\Phi < \delta$, and thus $\mu(\bigcup A_i \setminus C) < \varepsilon' < \varepsilon$. Similarly, $1_{C \setminus \bigcup A_i} \le |\sum 1_{A_i} - 1_C|$, so $\mu(C \setminus \bigcup A_i) < \varepsilon' < \varepsilon$. Finally, $\mu(C \triangle \bigcup A_i) \le 2\varepsilon'$, so $\|1_{C \triangle \bigcup A_i}\|_\Phi < \varepsilon/2$, and thus

$$\left\|\sum 1_{A_i} - 1_{\bigcup A_i}\right\|_\Phi \le \left\|\sum 1_{A_i} - 1_C\right\|_\Phi + \left\|1_C - 1_{\bigcup A_i}\right\|_\Phi \le \delta + \frac{\varepsilon}{2} < \varepsilon.$$

(2) $\Longrightarrow$ (1). Let $C, \mathcal{V}, \varepsilon$ be given. By (2.1.20) choose $\delta > 0$ so that $\delta < \varepsilon/3$ and $\mu(D) < \delta$ implies $\|1_D\|_\Phi < \varepsilon/3$. There exist $A_1, A_2, \cdots, A_n \in \mathcal{V}$ with $\mu(C \setminus \bigcup A_i) < \delta$, $\mu(\bigcup A_i \setminus C) < \delta$, $\|\sum 1_{A_i} - 1_{\bigcup A_i}\|_\Phi < \delta$. Therefore $\|1_{C \setminus \bigcup A_i}\|_\Phi < \varepsilon/3$ and $\|1_{\bigcup A_i \setminus C}\|_\Phi < \varepsilon/3$ so $\|\sum 1_{A_i} - 1_C\|_\Phi < \varepsilon$. ∎

(7.3.14) Proposition. *If property* (V_Φ) *holds, then the weak Vitali property holds.*

Proof. Apply (2.1.22) to (7.3.7(2)). ∎

Thus we see that property (V_Φ) implies that all L_∞ integrals are differentiable. But of course many more integrals may also be differentiable.

(7.3.15) Lemma. *Let* Φ *and* Ψ *be conjugate Orlicz functions,* Φ *finite. Let* $f \in L_\Psi$, *and define* $\gamma(A) = \int_A f \, d\mu$ *for* $A \in \mathcal{F}$, $\mu(A) < \infty$. *Let* $A_1, A_2, \cdots, A_n$ *be constituents and let* $C \in \mathcal{F}$. *If* $\|\sum 1_{A_i} - 1_C\|_\Phi < \varepsilon$, *then*

$$\left|\sum_{i=1}^n \gamma(A_i) - \gamma(C)\right| \le 2\varepsilon \|f\|_\Psi.$$

Proof.

$$|\gamma(A_i) - \gamma(C)| = \left|\int \left(\sum 1_{A_i} - 1_C\right) f \, d\mu\right|$$

$$\le \int \left|\sum 1_{A_i} - 1_C\right| |f| \, d\mu$$

$$\le 2\left\|\sum 1_{A_i} - 1_C\right\|_\Phi \|f\|_\Psi \le 2\varepsilon \|f\|_\Psi.$$ ∎

The proof of the following result is omitted, since it is almost identical to those of Theorems (7.3.5) and (7.3.11) (d) $\Longrightarrow$ (c); this time use Lemma (7.3.15).

(7.3.16) Theorem. *Let Φ and Ψ be conjugate Orlicz functions, Φ finite. Let $\mathbf{B}$ be a derivation basis satisfying property (V_Φ). Then $\mathbf{B}$ differentiates all L_Ψ integrals.*

The converse, however, is more involved. As in (4.3.11), we prove it under the assumption of (Δ_2) at ∞.

(7.3.17) Theorem. *Let Φ and Ψ be conjugate Orlicz functions. Assume Φ is finite and satisfies condition (Δ_2) at ∞. Let $\mathbf{B}$ be a derivation basis. Suppose $\mathbf{B}$ differentiates all L_Ψ integrals. Then $\mathbf{B}$ has property (V_Φ).*

Proof. (I) We begin with some simplifications. First, we may assume that $\Phi(1) > 0$: Indeed, let $a > 0$ be such that $\Phi(a) > 0$. The Orlicz function Φ_0 defined by $\Phi_0(u) = \Phi(au)$ satisfies $\|f\|_{\Phi_0} = a\|f\|_\Phi$ and $\Phi_0(1) > 0$. The conjugate is $\Psi_0(v) = (1/a)\Psi(v)$, so $L_{\Psi_0} = L_\Psi$.

Next, by (Δ_2), there exist M and u_0 such that $\Phi(2u) \leq M\Phi(u)$ for $u \geq u_0$. Since $\Phi(1) > 0$, we may assume (by replacing the constant M) that $\Phi(2u) \leq M\Phi(u)$ for all $u \geq 1$. Also $\Phi(0) = 0$, so this means $\Phi(2n) \leq M\Phi(n)$ for all nonnegative integers n.

If $A \in \mathcal{F}$ and $\mu(A) < \infty$, then $1_A \in L_\Psi$, so $\mathbf{B}$ differentiates the integral of 1_A. Therefore $\mathbf{B}$ differentiates the integral of the complement $1_{\Omega\setminus A}$.

(II) Let $\mathcal{A} = \{A_1, A_2, \cdots\}$ be a finite or countably infinite collection of constituents. Write $\bigcup\mathcal{A} = \bigcup_i A_i$; $n_\mathcal{A} = \sum 1_{A_i}$; $e_\mathcal{A} = n_\mathcal{A} - 1_{\cup\mathcal{A}}$. Note if $\mathcal{A} = \emptyset$, then $n_\mathcal{A} = e_\mathcal{A} = 0$. We claim that

$$(7.3.17a) \qquad \int \Phi(n_\mathcal{A})\, d\mu \leq M \int \Phi(e_\mathcal{A})\, d\mu + \Phi(1)\mu(\textstyle\bigcup\mathcal{A}).$$

Indeed, on the set $\{n_\mathcal{A} \leq 1\}$ we have $\Phi(n_\mathcal{A}) = \Phi(1)1_{\cup\mathcal{A}}$; and on the set $\{n_\mathcal{A} \geq 2\}$ we have

$$\Phi(n_\mathcal{A}) = \Phi(e_\mathcal{A} + 1) \leq \Phi(2e_\mathcal{A}) \leq M\Phi(e_\mathcal{A}).$$

(III) Next claim: Suppose $\alpha, \beta > 0$, $C \in \mathcal{F}$, $0 < \mu(C) < \infty$, $\mathcal{V}$ is a $\mathbf{B}$-fine almost-cover of C, $h \in L_\Psi$, $h \geq 0$, $\mu\{h > 0\} < \infty$, and $\mu(C\setminus\{h > 0\}) > 0$. Then there is $B \in \mathcal{V}$ with $\mu(B) > 0$ and

$$(7.3.17b) \qquad \int_B h\, d\mu + \alpha\mu(B \setminus C) \leq \beta\mu(B).$$

To see this, observe that $\mathbf{B}$ differentiates the integrals of h and $1_{\Omega\setminus C}$, and therefore $\mathbf{B}$ differentiates the integral of $h + \alpha 1_{\Omega\setminus C}$. This derivative is 0 a.e. on $C \setminus \{h > 0\}$, which has positive measure, so there is $B \in \mathcal{V}$ with

$$\frac{\int_B h\, d\mu + \alpha\mu(B \setminus C)}{\mu(B)} < \beta.$$

This proves the claim.

(IV) Next claim: Let $\eta > 0$, $\mu(C) < \infty$, $\mathcal{V}$ a **B**-fine almost-cover of C, and $\mathcal{A}$ a finite or countably infinite set of constituents. Let $\beta > 0$ satisfy

$$\beta \left(1 - \frac{\beta}{\varphi(1)}\right)^{-1} < \eta \quad \text{and} \quad (1 - \eta)\left(1 - \frac{\beta}{\varphi(1)}\right)^{-1} < 1.$$

Suppose $\mathcal{A}$ satisfies

(i) $\int \Phi(e_\mathcal{A})\,d\mu \leq \eta\mu(C \cap \bigcup\mathcal{A})$,

(ii) $(1 - \eta)\int n_\mathcal{A}\,d\mu \leq \mu(C \cap \bigcup\mathcal{A})$,

(iii) $\mu(C \setminus \bigcup\mathcal{A}) > 0$,

(iv) $\varphi(n_\mathcal{A}) \in L_\Psi$.

Then there exists $B \in \mathcal{V}$ with $\mu(B) > 0$ and

(v) $\int_B \varphi(n_\mathcal{A})\,d\mu + \varphi(1)\mu(B \setminus C) \leq \beta\mu(B)$.

Furthermore, if $B \in \mathcal{V}$ satisfies (v), then the collection $\mathcal{B} = \mathcal{A} \cup \{B\}$ satisfies

(vi) $\int \Phi(e_\mathcal{B})\,d\mu \leq \eta\mu(C \cap \bigcup\mathcal{B})$,

(vii) $(1 - \eta)\int n_\mathcal{B}\,d\mu \leq \mu(C \cap \bigcup\mathcal{B})$.

To prove this, we will apply claim (III) with $\alpha = \varphi(1)$ and $h = \varphi(n_\mathcal{A})$. Of course $\varphi(1) < \infty$ since Φ is finite, and $\varphi(1) > 0$ since $\Phi(1) > 0$. The condition $\mu(C \setminus \{h > 0\}) > 0$ follows from (iii); and $\mu\{h > 0\} < \infty$ follows from (ii). Thus from (7.3.17b) we get (v). Now suppose B satisfies (v). Then

$$\mu\big(B \setminus (C \setminus \bigcup\mathcal{A})\big) \leq \mu(B \setminus C) + \mu(B \cap \bigcup\mathcal{A})$$

$$\leq \mu(B \setminus C) + \frac{1}{\varphi(1)}\int_B \varphi(n_\mathcal{A})\,d\mu$$

$$\leq \frac{\beta}{\varphi(1)}\mu(B).$$

Thus

$$\mu\big(B \cap (C \setminus \bigcup\mathcal{A})\big) = \mu(B) - \mu\big(B \setminus (C \setminus \bigcup\mathcal{A})\big) \geq \left(1 - \frac{\beta}{\varphi(1)}\right)\mu(B).$$

Therefore

$$\int_B \varphi(n_\mathcal{A})\,d\mu \leq \beta\mu(B) \leq \beta\left(1 - \frac{\beta}{\varphi(1)}\right)^{-1}\mu\big(B \cap (C \setminus \bigcup\mathcal{A})\big)$$

$$\leq \eta\mu\big(B \cap (C \setminus \bigcup\mathcal{A})\big).$$

But now

$$\int \Phi(e_\mathcal{B})\,d\mu = \int_{\bigcup\mathcal{A}\setminus B} \Phi(e_\mathcal{B})\,d\mu + \int_{B \cap \bigcup\mathcal{A}} \Phi(e_\mathcal{B})\,d\mu$$

$$= \int_{\bigcup\mathcal{A}\setminus B} \Phi(e_\mathcal{A})\,d\mu + \int_{B \cap \bigcup\mathcal{A}} \Phi(n_\mathcal{A})\,d\mu$$

$$= \int_{\bigcup\mathcal{A}} \Phi(e_\mathcal{A})\,d\mu + \int_{B \cap \bigcup\mathcal{A}} \big(\Phi(n_\mathcal{A}) - \Phi(n_\mathcal{A} - 1)\big)\,d\mu$$

$$\leq \int \Phi(e_\mathcal{A})\,d\mu + \int_B \varphi(n_\mathcal{A})\,d\mu.$$

Here we used the inequality $\Phi(n) - \Phi(n-1) \leq \varphi(n)$ (2.1.19c). Now

$$\int \Phi(e_B) \, d\mu \leq \int \Phi(e_A) \, d\mu + \int_B \varphi(n_A) \, d\mu$$
$$\leq \mu(C \cap \textstyle\bigcup\mathcal{A}) + \eta\mu(B \cap (C \setminus \textstyle\bigcup\mathcal{A}))$$
$$= \eta\mu(C \cap \textstyle\bigcup\mathcal{B}).$$

This proves (vi). For (vii), compute

$$(1 - \eta) \int n_B \, d\mu = (1 - \eta) \int n_A \, d\mu + (1 - \eta)\mu(B)$$
$$\leq \mu(C \cap \textstyle\bigcup\mathcal{A}) + (1 - \eta) \left(1 - \frac{\beta}{\varphi(1)}\right)^{-1} \mu(B \cap (C \setminus \textstyle\bigcup\mathcal{A}))$$
$$\leq \mu(C \cap \textstyle\bigcup\mathcal{A}) + \mu(B \cap (C \setminus \textstyle\bigcup\mathcal{A}))$$
$$= \mu(C \cap \textstyle\bigcup\mathcal{B}).$$

This proves (vii).

(V) Now we are ready to establish (V_Φ). Let $C \in \mathcal{F}$ with $0 < \mu(C) < \infty$, let $\mathcal{V}$ be a **B**-fine almost-cover of C, and let $\varepsilon > 0$. We may assume that $\varepsilon < 1/4$. Choose $\varepsilon_1 > 0$ so that $\int \Phi(f) \, d\mu \leq \varepsilon_1$ implies $\|f\|_\Phi < \varepsilon$ [by Corollary (2.1.18(C))]. Then choose $\eta > 0$ so that

$$\eta\mu(C) \leq \varepsilon_1 \quad \text{and} \quad \frac{\eta}{1 - \eta}\mu(C) < \varepsilon.$$

We will apply the assertion of (IV) recursively.

Begin with $\mathcal{A}_0 = \emptyset$. Then $\mathcal{A}_0$ satisfies (i)–(iv). (Even if $\varphi(0) > 0$, still $\Psi(\varphi(0)) = 0$.) Thus $\mathcal{D}_1 \neq \emptyset$, where

$$\mathcal{D}_1 = \left\{ B \in \mathcal{V} : \int_B \varphi(0) \, d\mu + \varphi(1)\mu(B \setminus C) \leq \beta\mu(B) \right\}.$$

Choose $B_1 \in \mathcal{D}_1$, $\mu(B_1) \geq (1/2) \sup \{ \mu(B) : B \in \mathcal{D}_1 \}$. Then we write $\mathcal{A}_1 = \{B_1\}$ so that $\mathcal{A}_1$ satisfies (i), (ii), and (iv). If $\mathcal{A}_1$ does not satisfy (iii), the recursive construction stops. Otherwise continue. Suppose $\mathcal{A}_k = \{B_1, B_2, \cdots, B_k\}$ is defined, satisfying (i)–(iv). Then $\mathcal{D}_{k+1} \neq \emptyset$, where

$$\mathcal{D}_{k+1} = \left\{ B \in \mathcal{V} : \int_B \varphi(n_{\mathcal{A}_k}) \, d\mu + \varphi(1)\mu(B \setminus C) \leq \beta\mu(B) \right\}.$$

Choose $B_{k+1} \in \mathcal{D}_{k+1}$ with $\mu(B_{k+1}) \geq (1/2) \sup \{ \mu(B) : B \in \mathcal{D}_{k+1} \}$. Then let $\mathcal{A}_{k+1} = \mathcal{A}_k \cup \{B_{k+1}\}$. Thus $\mathcal{A}_{k+1}$ satisfies (i), (ii), and (iv). If it does not satisfy (iii), stop. Otherwise continue.

We therefore get $\mathcal{A} = \bigcup_k \mathcal{A}_k = \{B_1, B_2, \cdots\}$, a finite or countably infinite collection of constituents. We claim that $\mathcal{A}$ has the properties required by (V_Φ), namely: $\mu(C \setminus \bigcup \mathcal{A}) = 0$, $\mu(\bigcup \mathcal{A} \setminus C) < \varepsilon$, $\|e_\mathcal{A}\|_\Phi < \varepsilon$. Each $\mathcal{A}_k$ satisfies (i), so (by monotone convergence) so does $\mathcal{A}$:

$$\int \Phi(e_\mathcal{A}) \, d\mu \leq \eta \mu(C \cap \bigcup \mathcal{A}) \leq \eta \mu(C) \leq \varepsilon_1.$$

By the choice of ε_1, we have $\|e_\mathcal{A}\|_\Phi < \varepsilon$. Each $\mathcal{A}_k$ satisfies (ii), so $\mathcal{A}$ does also. Therefore

$$(1 - \eta) \, \mu(\bigcup \mathcal{A}) \leq (1 - \eta) \int n_\mathcal{A} \, d\mu \leq \mu(C \cap \bigcup \mathcal{A}) < \infty.$$

Also,

$$
\begin{aligned}
\mu(\bigcup \mathcal{A} \setminus C) &\leq \mu(\bigcup \mathcal{A}) - \mu(C \cap \bigcup \mathcal{A}) \\
&\leq \left(1 - (1 - \eta)\right) \mu(\bigcup \mathcal{A}) = \eta \mu(\bigcup \mathcal{A}) \\
&\leq \frac{\eta}{1 - \eta} \mu(C) < \varepsilon.
\end{aligned}
$$

Note $\|2e_\mathcal{A}\|_\Phi < 2\varepsilon < 1/2$, so by (3.1.20) we have $\varphi(2e_\mathcal{A}) \in L_\Psi$. If $A = \{n_\mathcal{A} = 1\}$, then $\mu(A) \leq \mu(\bigcup \mathcal{A}) < \infty$, so $\varphi(1_A) \in L_\Psi$. Now $n_\mathcal{A} \leq 1_A + 2e_\mathcal{A}$ with disjoint supports, so $\varphi(n_\mathcal{A}) \leq \varphi(1_A) + \varphi(2e_\mathcal{A})$. Thus $\varphi(e_\mathcal{A}) \in L_\Psi$. That is, $\mathcal{A}$ satisfies (iv).

Now we claim $\mu(C \setminus \bigcup \mathcal{A}) = 0$. If not, then $\mathcal{A}$ satisfies (i)–(iv). There is B as in (IV), so $B \in \mathcal{D}_k$ for all k and $\mu(B) > 0$. Thus $\mu(B) \leq 2\mu(B_k)$ for all k. But $\sum \mu(B_k) < \infty$, a contradiction, so $\mu(C \setminus \bigcup \mathcal{A}) = 0$. Therefore (V_Φ) is verified.   ∎

Property (FV_Φ)

Next we come to a Vitali type of property, with the advantage that condition (Δ_2) is not required for the converse.

Let Φ be an Orlicz function. A derivation basis has *property* (FV_Φ) iff, for every $\varepsilon > 0$, every $C \in \mathcal{F}$ with $0 < \mu(C) < \infty$, and every **B**-fine almost-cover $\mathcal{V}$ of C, there exist constituents $A_1, A_2, \cdots, A_n \in \mathcal{V}$ and nonnegative scalars $a_1, a_2, \cdots, a_n$ such that

$$\left\| \sum_{i=1}^n a_i \, 1_{A_i} - 1_C \right\|_\Phi < \varepsilon,$$

$$\left| \sum_{i=1}^n a_i \mu(A_i) - \mu(C) \right| < \varepsilon.$$

It is easy to see that (V_Φ) implies (FV_Φ) with $a_i = 1$.

(7.3.18) Proposition. *Suppose $L_\Phi \subseteq L_1$. Then* **B** *has property* (FV$_\Phi$) *if and only if for every $\varepsilon > 0$, every $C \in \mathcal{F}$ with $0 < \mu(C) < \infty$, and every* **B**-*fine almost-cover $\mathcal{V}$ of C, there exist $A_1, A_2, \cdots, A_n \in \mathcal{V}$, and $a_1, a_2, \cdots, a_n \geq 0$ such that $\|\sum a_i \mathbf{1}_{A_i} - \mathbf{1}_C\|_\Phi < \varepsilon$.*

Proof. If **B** has property (FV$_\Phi$), then clearly the other condition holds. For the converse, suppose the condition stated is satisfied. For given $\varepsilon > 0$, there is $\delta > 0$ so that $\|f\|_\Phi < \delta$ implies $\|f\|_1 < \varepsilon$. We may also assume $\delta < \varepsilon$. Then given C and $\mathcal{V}$, there exist $A_1, A_2, \cdots, A_n \in \mathcal{V}$, and $a_1, a_2, \cdots, a_n \geq 0$ such that $\|\sum a_i \mathbf{1}_{A_i} - \mathbf{1}_C\|_\Phi < \delta < \varepsilon$. Then we know that $\|\sum a_i \mathbf{1}_{A_i} - \mathbf{1}_C\|_1 < \varepsilon$, so

$$\left| \sum a_i \mu(A_i) - \mu(C) \right| = \left| \int \left(\sum a_i \mathbf{1}_{A_i} - \mathbf{1}_C \right) d\mu \right| < \varepsilon. \quad \blacksquare$$

The usual lemma will be useful in the proof of convergence. The proof (similar to that of Lemma (7.3.15)) is left to the reader.

(7.3.19) Lemma. *Let Φ and Ψ be conjugate Orlicz functions, Φ finite. Suppose* **B** *has property* (FV$_\Phi$). *Let $f \in L_\Psi$, and define $\gamma(A) = \int_A f \, d\mu$. If $A_1, A_2, \cdots, A_n$ are constituents and $a_1, a_2, \cdots, a_n \geq 0$ with $\|\sum a_i \mathbf{1}_{A_i} - \mathbf{1}_C\|_\Phi \leq \varepsilon$, then*

$$\left| \sum a_i \gamma(A_i) - \gamma(C) \right| \leq 2\varepsilon \|f\|_\Psi.$$

(7.3.20) Theorem. *Let Φ and Ψ be conjugate Orlicz functions, Φ finite. Suppose* **B** *has property* (FV$_\Phi$). *Then* **B** *differentiates all L_Ψ integrals.*

Proof. Let $f \in L_\Psi$ and $\gamma(A) = \int_A f \, d\mu$. We will show that $D^\star \gamma \leq f$. Suppose not: then there exist $a < b$ with $\mu^*\{f < a < b < D^\star \gamma\} > 0$. Thus there is $Q \subseteq \{f < a < b < D^\star \gamma\}$, $0 < \mu^*(Q) < \infty$. Let C be an outer envelope of Q. Then

$$\left\{ E \text{ constituent} : \frac{\gamma(E)}{\mu(E)} > b \right\}$$

is a **B**-fine almost-cover of C. Let $\varepsilon > 0$ be given. By (FV$_\Phi$) and Lemma (7.3.19), there exist $A_1, A_2, \cdots, A_n \in \mathcal{V}$, and $a_1, a_2, \cdots, a_n \geq 0$ with

$$\left| \sum a_i \mu(A_i) - \mu(C) \right| < \varepsilon,$$

$$\left| \sum a_i \gamma(A_i) - \gamma(C) \right| < \varepsilon.$$

Then

$$\gamma(C) + \varepsilon \geq \sum a_i \gamma(A_i) \geq b \sum a_i \mu(A_i)$$
$$\geq b\mu(C) - |b|\varepsilon.$$

This is true for all $\varepsilon > 0$, so $\gamma(C) \geq b\mu(C)$. But $f < a$ on C, so

$$\gamma(C) = \int_C f \, d\mu \leq a\mu(C) < b\mu(C) \leq \gamma(C),$$

a contradiction. Therefore $D^\star \gamma \leq f$.

Similarly, $D_\star \gamma \geq f$. $\quad \blacksquare$

For the converse, we first take the case $L_\Phi \subseteq L_1$.

(7.3.21) Theorem. *Let Φ and Ψ be conjugate Orlicz functions with Φ finite and $L_\Phi \subseteq L_1$. Suppose $\mathbf{B}$ differentiates all L_Ψ integrals. Then $\mathbf{B}$ has property* (FV$_\Phi$).

Proof. Suppose (FV$_\Phi$) fails. There is $\varepsilon > 0$, $Q \subseteq \Omega$, and a $\mathbf{B}$-fine cover $\mathcal{V}$ of Q such that, if C is an outer envelope of Q, $a_i \geq 0$ and $A_i \in \mathcal{V}$ satisfy $|\sum a_i \mu(A_i) - \mu(C)| < \varepsilon$, then

$$(7.3.21\text{a}) \qquad \left\| \sum a_i \, \mathbf{1}_{A_i} - \mathbf{1}_C \right\|_\Phi \geq \varepsilon.$$

We assume also that $\varepsilon < \mu(C)$.

By (2.1.23), there is $\varepsilon' > 0$ so that $|f| \leq 1$, $\|f\|_1 < \varepsilon'$ imply $\|f\|_\Phi < \varepsilon/2$. Assume also that $\varepsilon' < \varepsilon$. Since $L_\Phi \subseteq L_1$, there is $\varepsilon'' > 0$ so that $\|f\|_\Phi < \varepsilon''$ implies $\|f\|_1 < \varepsilon'/2$. Assume also that $\varepsilon'' < \varepsilon/2$.

We begin with an application of the Hahn-Banach theorem (5.1.2). Consider three subsets of L_Φ:

$$\mathcal{C}_1 = \left\{ \sum a_i \, \mathbf{1}_{A_i} : a_i \geq 0, \, A_i \in \mathcal{V}, \, |\sum a_i \mu(A_i) - \mu(C)| < \varepsilon'/2 \right\},$$
$$\mathcal{C}_2 = \{ \xi \in L_\Phi : \xi \leq \mathbf{1}_C \},$$
$$\mathcal{C}_3 = \{ \xi \in L_\Phi : \|\xi\|_\Phi < \varepsilon'' \}.$$

All three of the sets are convex. $\mathcal{C}_3$ is open, so $\mathcal{C}_2 + \mathcal{C}_3$ is convex and open. We claim that $\mathcal{C}_1 \cap (\mathcal{C}_2 + \mathcal{C}_3) = \emptyset$. Suppose $\xi = \sum a_i \, \mathbf{1}_{A_i} \in \mathcal{C}_1$ and $\xi = \xi_2 + \xi_3$ with $\xi_2 \in \mathcal{C}_2$, $\xi_3 \in \mathcal{C}_3$. Then $\|\xi_3\|_1 < \varepsilon'/2$. Let $B = \{\xi > 1\}$. Now

$$-\frac{\varepsilon'}{2} < \sum a_i \mu(A_i) - \mu(C) = \int (\xi - \mathbf{1}_C) \, d\mu$$
$$= \int_{B \cap C} (\xi - 1) \, d\mu + \int_{\Omega \backslash C} \xi \, d\mu - \int_{C \backslash B} (1 - \xi) \, d\mu$$
$$\leq \int_{B \cap C} |\xi_3| \, d\mu + \int_{\Omega \backslash C} |\xi_3| \, d\mu - \int_{C \backslash B} (1 - \xi) \, d\mu$$
$$\leq \frac{\varepsilon'}{2} - \int_{C \backslash B} (1 - \xi) \, d\mu.$$

Thus $\int_{C \backslash B} (1 - \xi) \, d\mu < \varepsilon'$ and $\|(1 - \xi) \mathbf{1}_{C \backslash B}\|_\Phi < \varepsilon/2$. But then

$$|\xi - \mathbf{1}_C| = (\xi - 1) \mathbf{1}_{B \cap C} + (1 - \xi) \mathbf{1}_{C \backslash B} + |\xi| \mathbf{1}_{\Omega \backslash C} \leq |\xi_3| + (1 - \xi) \mathbf{1}_{C \backslash B},$$

so

$$\|\xi - \mathbf{1}_C\|_\Phi \leq \|\xi_3\|_\Phi + \|(1 - \xi) \mathbf{1}_{C \backslash B}\|_\Phi < \frac{\varepsilon}{2} + \frac{\varepsilon}{2} = \varepsilon.$$

This contradicts (7.3.21a). In fact $\mathcal{C}_1 \cap (\mathcal{C}_2 + \mathcal{C}_3) = \emptyset$. Then by the Hahn-Banach theorem, there is a functional $x^* \in L_\Phi^*$ with $x^*(\xi) \geq 1$ for all $\xi \in \mathcal{C}_1$

and $x^*(\xi) < 1$ for all $\xi \in C_2 + C_3$. By (2.2.24), since $x^*(\xi) < 1$ for all $\xi \in C_2$ and C_3, the functional is of the form $x^*(\xi) = \int \xi f \, d\mu$ for some $f \in L_\Psi$.

Now we claim that **B** does not differentiate the integral γ of f. If $E \in \mathcal{V}$, then

$$\xi = \frac{\mu(C) - \varepsilon'/4}{\mu(E)} \, 1_E \in C_1,$$

so $x^*(\xi) \geq 1$, which means for $\gamma(E) = \int_E f \, d\mu$

$$\frac{\gamma(E)}{\mu(E)} \geq \frac{1}{\mu(C) - \varepsilon'/4}.$$

Since $\mathcal{V}$ is a **B**-fine cover of Q, we conclude that $D^\star \gamma(\omega) \geq 1/(\mu(C) - \varepsilon'/4)$ for all $\omega \in Q$. Now if $D^\star \gamma = f$ a.e., we would have $f(\omega) \geq 1/(\mu(C) - \varepsilon'/4)$ a.e. on C. But $1_C \in C_2$, so we would have

$$1 > \gamma(C) = \int_C f \, d\mu = \frac{\mu(C)}{\mu(C) - \varepsilon'/4} > 1,$$

a contradiction. Thus $D^\star \gamma = f$ fails on a set of positive measure. Therefore **B** does not differentiate the integral of the L_Ψ function f. ∎

(7.3.22) Next we should consider the case $L_\Phi \not\subseteq L_1$. Let Φ and Ψ be conjugate Orlicz functions. Suppose **B** differentiates all L_Ψ integrals. If $A \in \mathcal{F}$, $\mu(A) < \infty$, then $1_A \in L_\Psi$. So **B** has the density property, and therefore **B** differentiates all L_∞ integrals. Thus **B** differentiates integrals of functions in $L_\Psi + L_\infty$.

We have seen in Proposition (2.2.13) that $L_\Psi + L_\infty$ is itself an Orlicz space L_{Ψ_s}, where Ψ_s is the shifted Orlicz function given by

(7.3.22a) $$\Psi_s(v) = \begin{cases} 0, & v \leq 1, \\ \Psi(v - 1), & v > 1. \end{cases}$$

A calculation shows that the conjugate Orlicz function is then

(7.3.22b) $$\Phi_s(u) = \Phi(u) + u,$$

and $L_{\Phi_s} = L_\Phi \cap L_1$. Since $L_{\Phi_s} \subseteq L_1$, the case of Φ_s and Ψ_s is covered by the previous material. (Φ is finite if and only if the derivative φ is finite, if and only if ψ is unbounded, if and only if $\lim \Psi(v)/v = \infty$.) Thus we have proved:

Theorem. Let Ψ be a finite Orlicz function with $\lim \Psi(v)/v = \infty$. Let Ψ_s and Φ_s be defined by (7.3.22a) and (7.3.22b). Then the following are equivalent:

(1) **B** differentiates all L_Ψ integrals;
(2) **B** differentiates all L_{Ψ_s} integrals;
(3) **B** has property (FV_{Φ_s}).

The important case not yet completed is $L_\Psi = L_1$, $L_\Phi = L_\infty$. We will discuss this in the next section.

Complements

(7.3.23) (Stochastic basis.) Let $(\Omega, \mathcal{F}, \mathbf{P})$ be a probability space, let J be a countable directed set, and let $(\mathcal{F}_t)_{t \in J}$ be a stochastic basis consisting of *finite* σ-algebras. For each $t \in J$ and $\omega \in \Omega$, let $E_t(\omega)$ be the atom of $\mathcal{F}_t$ containing ω. We may define a derivation basis $\mathbf{B}$ on Ω by postulating

$$\left(E_t(\omega) \right)_{t \in J'} \Rightarrow \omega$$

for $J' \subseteq J$ cofinal. Now if $X \in L_1$, then there correspond a martingale

$$X_t = \mathbf{E}^{\mathcal{F}_t}[X]$$

and an integral

$$\gamma(A) = \int_A X \, d\mathbf{P}.$$

Now of course

$$\frac{\gamma\left(E_t(\omega) \right)}{\mathbf{P}\left(E_t(\omega) \right)} = X_t(\omega),$$

so $D\gamma = X$ a.e. if and only if $X_t \to X$ a.s.

The analogies between the stochastic basis covering properties of Chapter 4 and the derivation basis Vitali properties of this chapter can be exhibited in this setting. For example, $(\mathcal{F}_t)$ satisfies the covering condition (V) if and only if $\mathbf{B}$ has the strong Vitali property.

(7.3.24) (A "demiconvergence" result.) Suppose a derivation basis $\mathbf{B}$ has the weak Vitali property. Then $\mathbf{B}$ "lower differentiates" all nonnegative L_1 integrals. That is: if $f \geq 0$, $f \in L_1$, and $\gamma(A) = \int_A f \, d\mu$ for all $A \in \mathcal{F}$, then $D_* \gamma \geq f$ a.e. (If $f \geq 0$, we may dispense with Lemma (7.3.8) in the proof of the first part of (d) $\Longrightarrow$ (c) in Theorem (7.3.11). See Hayes & Pauc [1970], Proposition 2.1, page 19.)

(7.3.25) (Overflow.) Talagrand's definition for property (V_Φ) (Talagrand[1986]) is as in Proposition (7.3.12) part (2), except that the overflow condition $\mu(\bigcup A_i \setminus C) < \varepsilon$ is omitted. Of course, if $\mathbf{B}$ has small overflow, then the condition is not needed. But for completely general $\mathbf{B}$ it is needed. For example, suppose we have the trivial derivation basis defined on $[0,1]$ so that $E_n \Rightarrow \omega$ iff $E_n = [0,1]$ for all n. It is easy to arrange zero deficit and overlap, but the integral of no nonconstant $f \in L_\Psi$ has derivative f.

(7.3.26) $((\text{FV}_\Phi).)$ Talagrand's definition for (FV_Φ) (Talagrand [1986]) is: for every $\varepsilon > 0$, every $C \in \mathcal{F}$ with $0 < \mu(C) < \infty$, and every $\mathbf{B}$-fine almost-cover $\mathcal{V}$ of C, there exist constituents $A_1, A_2, \cdots, A_n \in \mathcal{V}$ and nonnegative scalars $a_1, a_2, \cdots, a_n$ such that

$$\left\| \sum a_i \mathbf{1}_{A_i} - 1 \wedge \sum a_i \mathbf{1}_{A_i} \right\|_\Phi < \varepsilon,$$

$$\left| \sum a_i \mu(A_i) - \mu(C) \right| < \varepsilon.$$

For general derivation bases, this is in fact not sufficient for differentiation of L_Ψ integrals. For example: Let $(\Omega, \mathcal{F}, \mu)$ be $[0,1]$ with Lebesgue measure. Enumerate the rationals in $(0,1)$ as r_n. Let $E_n = [0, r_n]$, $n \in \mathbb{N}$, and postulate $E_n \Rightarrow \omega$ for all $\omega \in [0,1]$. Then certainly $\mathbf{B}$ differentiates the integral of no nonconstant $f \in L_\Psi$. But if $\mathcal{V}$ is a $\mathbf{B}$-fine almost-cover of any set C, and $\varepsilon > 0$, then there exist arbitrarily large n with $|r_n - \mu(C)| < \varepsilon$, so $E_n \in \mathcal{V}$. Now let $A_1 = E_n$ and $a_1 = 1$, so $\|a_1 \mathbf{1}_{A_1} - 1 \wedge a_1 \mathbf{1}_{A_1}\|_\Phi = 0$ and $|a_1 \mu(A_1) - \mu(C)| < \varepsilon$.

(7.3.27) (Both inequalities are needed in (FV_Φ).) Let $\xi = \sum a_i \mathbf{1}_{A_i}$. Then $\|\xi - \mathbf{1}_C\|_2$ small does not imply $|\sum a_i \mu(A_i) - \mu(C)|$ is small. Let $(\Omega, \mathcal{F}, \mu)$ be $\mathbb{R}$ with Lebesgue measure. Take $C = [0,1]$, $A_i = [0,1] \cup [i, i+1]$ and $a_i = 1/n$ for $i = 1, 2, \cdots n$. Then $\|\sum a_i \mathbf{1}_{A_i} - \mathbf{1}_C\|_2 = 1/\sqrt{n}$ but $|\sum a_i \mu(A_i) - \mu(C)| = 1$.

Remarks

A reference on derivation theory is Hayes & Pauc [1970]. Much of the material in this section follows their treatment. Theorem (7.3.17) on the necessity of (V_Φ) is due to C. A. Hayes [1976]. We have used his proof. Property (FV_Φ) is due to Talagrand [1986] (note (7.3.26)).

7.4. D-bases

The most often used derivation bases are of special kinds. In this section we will consider the D-bases and a still more special case, the Busemann-Feller bases.

D-bases

Let $(\Omega, \mathcal{F}, \mu)$ be a σ-finite measure space. A *D-basis* on $(\Omega, \mathcal{F}, \mu)$ is a pair $(\mathcal{E}, \delta)$, where $\mathcal{E}$ is a family of measurable sets E with $0 < \mu(E) < \infty$, and δ is a function $\delta \colon \mathcal{E} \to (0, \infty)$, such that, for every $\omega \in \Omega$ and every $\varepsilon > 0$, there exists $E \in \mathcal{E}$ with $\omega \in E$ and $\delta(E) < \varepsilon$. Given a D-basis $(\mathcal{E}, \delta)$, we may define a derivation basis by specifying $E_n \Rightarrow \omega$ iff $\omega \in E_n \in \mathcal{E}$ and $\delta(E_n) \to 0$. A derivation basis $\mathbf{B}$ that can be specified in this way will also be called a D-basis. We may even write $\mathbf{B} = (\mathcal{E}, \delta)$.

For example, Ω may be a metric space, with metric ρ, and $\delta(E)$ may be the *diameter* of E:

$$\delta(E) = \operatorname{diam} E = \sup \{\, \rho(x, y) : x, y \in E \,\}.$$

Some of the derivation bases in Sections 7.1 and 7.2 are D-bases with δ as diameter: the interval basis in $\mathbb{R}^d$; the disk basis in $\mathbb{R}^2$ (but not the centered disk basis).

We have seen a correspondence between stochastic bases and derivation bases. The condition of a countable cofinal subset for stochastic bases corresponds roughly to being a D-basis for derivation bases.

The measurability of the upper and lower derivates is true for certain D-bases.

(7.4.1) Proposition. *Let Ω be a metric space, let μ be a Radon measure on μ, let $(\mathcal{E}, \delta)$ be a D-basis where $\mathcal{E}$ is a collection of open sets, and δ is diameter. If $\gamma\colon \mathcal{E} \to \mathbb{R}$ is any set function, then $D^\star\gamma$ and $D_\star\gamma$ are measurable functions.*

Proof. Fix a real number t. For positive integers m and k, let

$$P_{mk} = \bigcup \left\{ E \in \mathcal{E} : \delta(E) < \frac{1}{k},\ \frac{\gamma(E)}{\mu(E)} > t + \frac{1}{m} \right\}.$$

Then P_{mk} is an open set. Now

$$\{D^\star\gamma > t\} = \bigcup_{m=1}^{\infty} \bigcap_{k=1}^{\infty} P_{mk}$$

is a measurable set (a $G_{\delta\sigma}$ set). This is true for all t, so $D^\star\gamma$ is a measurable function (a function of the second Baire class). Similarly $D_\star\gamma$ is measurable. ∎

The next result shows that the most common derivation bases have small overflow.

(7.4.2) Proposition. *Let Ω be a metric space, let μ be a Radon measure, and let $(\mathcal{E}, \delta)$ be a D-basis, where δ is diameter. Then $(\mathcal{E}, \delta)$ has small overflow.*

Proof. Let $C \in \mathcal{F}$, $0 < \mu(C) < \infty$. Let $\varepsilon > 0$ be given. There is an open set $U \supseteq C$ with $\mu(U) < \mu(C) + \varepsilon$. The family

$$\mathcal{U} = \{ E \in \mathcal{E} : E \subseteq U \}$$

is a full **B**-fine cover of C, and if $A_1, A_2, \cdots, A_n \in \mathcal{U}$, then $\mu(\bigcup A_i \setminus C) \leq \mu(U \setminus C) < \varepsilon$. ∎

Properties (A) and (C)

The derivation basis **B** has *property* (A) iff for every $C \in \mathcal{F}$ with $0 < \mu(C) < \infty$ there is a constant M such that for every **B**-fine almost-cover $\mathcal{V}$ of C and every $\varepsilon > 0$, there exist $A_1, A_2, \cdots, A_n \in \mathcal{V}$ with $\mu(\bigcup A_i \setminus C) < \varepsilon$, and

$$\left\| \sum 1_{A_i} \right\|_\infty < M \sum \mu(A_i).$$

The differentiation theorem follows the usual outline.

(7.4.3) Theorem. *Suppose* **B** *has property* (A). *Then* **B** *differentiates all L_1 integrals.*

Proof. Let $f \in L_1$, and define $\gamma(A) = \int_A f \, d\mu$. We claim $f \geq D^\star\gamma$. If not, then $\mu^*\{f < D^\star\gamma\} > 0$, so there exist $a < b$ with $\mu^*\{f < a < b < D^\star\gamma\} > 0$. Let $Q \subseteq \{f < a < b < D^\star\gamma\}$ with $0 < \mu^*(Q) < \infty$, and let C be an outer envelope of Q. Let M be the constant of property (A). Now $f \in L_1$, so there is $\varepsilon > 0$ such that $\mu(G) < \varepsilon$ implies $\int_G |f - a| \, d\mu < (b - a)/(2M)$. Now

$$\mathcal{V} = \left\{ E \text{ constituent} : \frac{\gamma(E)}{\mu(E)} > b \right\}$$

is a **B**-fine almost-cover of C. There exist constituents $A_1, A_2, \cdots, A_n \in \mathcal{V}$ with $\mu(\bigcup A_i \setminus C) < \varepsilon$ and $\|\sum \mathbf{1}_{A_i}\|_\infty < M \sum \mu(A_i)$. Now for each i,

$$\int_{A_i \setminus C} (f - a) \, d\mu = \int_{A_i} (f - a) \, d\mu - \int_{A_i \cap C} (f - a) \, d\mu$$
$$> (b - a)\mu(A_i) - 0.$$

Now if $G = \bigcup A_i \setminus C$, then $\mu(G) < \varepsilon$, and thus

$$(b - a) \sum \mu(A_i) \leq \sum \int_{A_i \setminus C} (f - a) \, d\mu = \int_G \left(\sum \mathbf{1}_{A_i} \right) (f - a) \, d\mu$$
$$\leq \left\| \sum \mathbf{1}_{A_i} \right\|_\infty \int_G |f - a| \, d\mu < M \sum \mu(A_i) \frac{b - a}{2M}$$
$$= \frac{b - a}{2} \sum \mu(A_i),$$

a contradiction. Thus $f \geq D^\star\gamma$.

Similarly $f \leq D_\star\gamma$. ∎

For the converse, we use the Hahn-Banach theorem, but only on a finite-dimensional space.

(7.4.4) Theorem. *Suppose the D-basis* $\mathbf{B} = (\mathcal{E}, \delta)$ *differentiates all L_1 integrals. Then* **B** *has property* (A).

Proof. First, **B** differentiates all L_1 integrals, so **B** has the density property, and therefore the weak Vitali property. In particular, **B** has small overflow.

Let $C \in \mathcal{F}$ with $0 < \mu(C) < \infty$. Suppose property (A) fails. Choose $M_k \uparrow \infty$ with

$$\sum_{k=1}^\infty \frac{1}{M_k} < \frac{\mu(C)}{2}.$$

Then for each k, there exists a **B**-fine almost-cover $\mathcal{V}$ of C such that $A_1, A_2, \cdots, A_n \in \mathcal{V}_k$ implies $\|\sum \mathbf{1}_{A_i}\|_\infty > M_k \sum \mu(A_i)$. We may assume $\delta(E) < 1/k$ for all $E \in \mathcal{V}_k$.

Fix k. The following "functional" version of the inequality is also true: if $A_1, A_2, \cdots, A_n \in \mathcal{V}_k$ and $a_1, a_2, \cdots, a_n \geq 0$, then $\|\sum a_i \mathbf{1}_{A_i}\|_\infty > M_k \sum a_i \mu(A_i)$. This may be seen by approximating the scalars a_i with rational numbers, multiplying through by a common denominator, and observing that repetitions are allowed in the list $A_1, A_2, \cdots A_n$. By the weak Vitali property, there exist finitely many sets $A_{1k}, A_{2k}, \cdots \in \mathcal{V}_k$ with $\mu(\bigcup_i A_{ik} \setminus C) < \mu(C)/2$ and $\mu(C \setminus \bigcup_i A_{ik}) < \mu(C)/2$. Let $\mathcal{A}_k$ be the finite algebra on Ω generated by the sets $A_{1k}, A_{2k}, \cdots$. Consider two subsets of $L_\infty(\Omega, \mathcal{A}_k, \mu)$:

$$\mathcal{C}_1 = \left\{ \sum_i a_i \mathbf{1}_{A_{ik}} : a_i \geq 0, \sum_i a_i \mu(A_{ik}) = 1 \right\},$$

$$\mathcal{C}_2 = \{ \xi \in L_\infty(\mathcal{A}_k) : \xi \leq M_k \}.$$

These sets are disjoint, convex, and closed. $\mathcal{C}_1$ is compact. Thus there is $f_k \in L_1(\Omega, \mathcal{A}_k, \mu)$ with $\int \xi f_k \, d\mu > 1$ for all $\xi \in \mathcal{C}_1$ and $\int \xi f_k \, d\mu \leq 1$ for all $\xi \in \mathcal{C}_2$. Because of $\mathcal{C}_2$, we have $f_k \geq 0$ and $\|f_k\|_1 \leq 1/M_k$. Now

$$\frac{1}{\mu(A_{ik})} \mathbf{1}_{A_{ik}} \in \mathcal{C}_1,$$

so $\int_{A_{ik}} f_k \, d\mu > \mu(A_{ik})$.

We may therefore construct such a function f_k for each k. Let $f = \sum_k f_k$. Then $f \geq 0$, and $\|f\|_1 \leq \sum 1/M_k < \infty$, so $f \in L_1$. But we claim $\mathbf{B}$ does not differentiate the integral γ of f. Let

$$B = \bigcap_k \bigcup_i A_{ik}.$$

Then $\mu(C \setminus B) \leq \sum_k \mu(C \setminus \bigcup_i A_{ik}) \leq \mu(C)/2$, so $\mu(C \cap B) \geq \mu(C)/2$. For every $\omega \in B$ and every k, there is $E \in \mathcal{V}_k \subseteq \mathcal{E}$ with $\omega \in E$, $\delta(E) < 1/k$ and $\int_E f \, d\mu \geq \int_E f_k \, d\mu > \mu(E)$. Therefore $D^\star \gamma \geq 1$ on B. But if $f \geq 1$ on B, then

$$\frac{\mu(C)}{2} \leq \mu(B) \leq \int_B f \, d\mu \leq \int f \, d\mu \leq \sum \frac{1}{M_k} < \frac{\mu(C)}{2},$$

a contradiction. Therefore $D^\star \gamma > f$ on a set of positive measure. ∎

There is another property closely related to property (A). The derivation basis $\mathbf{B}$ has *property* (C) iff, for every $\varepsilon > 0$, there exists a constant M such that if $C \in \mathcal{F}$, $\varepsilon < \mu(C) < \infty$, and $\mathcal{V}$ is a $\mathbf{B}$-fine almost-cover of C, then for every $\eta > 0$ there exist $A_1, A_2, \cdots, A_n \in \mathcal{V}$ with $\|\sum \mathbf{1}_{A_i}\|_\infty < M\mu(\bigcup A_i)$ and $\mu(\bigcup A_i \setminus C) < \eta$.

If $\mathbf{B}$ has small overflow, the definition may be simplified. Recall that $\mathbf{B}$ has *small overflow* iff, for every $C \in \mathcal{F}$ with $\mu(C) < \infty$ and every $\eta > 0$, there exists a set $C_0 \subseteq C$ with $\mu(C \setminus C_0) = 0$ and a full $\mathbf{B}$-fine cover $\mathcal{U}$ of C_0 such that, for any $A_1, A_2, \cdots, A_n \in \mathcal{U}$, we have $\mu(\bigcup A_i \setminus C) < \eta$.

Now if **B** has small overflow, property (C) may be stated: For every $\varepsilon > 0$, there exists a constant M such that if $C \in \mathcal{F}$, $\varepsilon < \mu(C) < \infty$, and $\mathcal{V}$ is a **B**-fine almost-cover of C, then there exists $A_1, \cdots, A_n \in \mathcal{V}$ with $\|\sum \mathbf{1}_{A_i}\|_\infty < M\mu(\bigcup A_i)$. Indeed, if this is true and $\eta > 0$ is given, choose C_0 and $\mathcal{U}$ as in the definition of small overflow; then $\varepsilon < \mu(C_0) < \infty$ and $\mathcal{V} \cap \mathcal{U}$ is a **B**-fine almost-cover of C_0, so that there exist $A_1, \cdots, A_n \in \mathcal{V} \cap \mathcal{U}$ with $\|\sum \mathbf{1}_{A_i}\|_\infty < M\mu(\bigcup A_i)$ and $\mu(\bigcup A_i \setminus C) < \eta$.

Property (A) may be similarly simplified.

Properties (A) and (C) are equivalent for a D-basis with small overflow:

(7.4.5) Theorem. *Let* $\mathbf{B} = (\mathcal{E}, \delta)$ *be a D-basis with small overflow. The following are equivalent:*

(1) *Property (A): for every* $C \in \mathcal{F}$, $\mu(C) > 0$, *there is* M *such that for every* **B**-*fine almost-cover* $\mathcal{V}$ *of* C, *there exist* $A_1, A_2, \cdots, A_n \in \mathcal{V}$ *with* $\|\sum \mathbf{1}_{A_i}\|_\infty < M \sum \mu(A_i)$.

(2) *For every* $C \in \mathcal{F}$, $\mu(C) > 0$, *there is* M *such that for every* **B**-*fine almost-cover* $\mathcal{V}$ *of* C, *there exist* $A_1, A_2, \cdots, A_n \in \mathcal{V}$ *with* $\|\sum \mathbf{1}_{A_i}\|_\infty < M\mu(\bigcup A_i)$.

(3) *Property (C): for every* $\varepsilon > 0$ *there is* M *such that if* $C \in \mathcal{F}$, $\mu(C) > \varepsilon$, *and* $\mathcal{V}$ *is a* **B**-*fine almost-cover of* C, *then there exist* $A_1, A_2, \cdots, A_n \in \mathcal{V}$ *with* $\|\sum \mathbf{1}_{A_i}\|_\infty < M\mu(\bigcup A_i)$.

Proof. (2) $\Longrightarrow$ (1) and (3) $\Longrightarrow$ (2) are easy.

(1) $\Longrightarrow$ (2). Since (A) holds, **B** differentiates all L_1 integrals, so the density property holds: that is, the weak Vitali property holds.

But suppose (2) fails. There is $C \in \mathcal{F}$, $\mu(C) > 0$, such that there exist **B**-fine almost-covers $\mathcal{V}_k$ of C for which $A_1, A_2, \cdots, A_n \in \mathcal{V}_k$ implies $\|\sum \mathbf{1}_{A_i}\|_\infty \geq 2^k \mu(\bigcup A_i)$. Now for each k, use the weak Vitali property to choose finite sets $\mathcal{B}_k = \{B_{1k}, B_{2k}, \cdots\} \subseteq \mathcal{V}_k$ with $\delta(B_{ik}) < 1/k$, $\sum_i \mu(B_{ik}) - \mu(\bigcup_i B_{ik}) < 2^{-k}$, $\mu(\bigcup_i B_{ik} \setminus C) < 2^{-k}$, and $\mu(C \setminus \bigcup_i B_{ik}) < 2^{-k}$. Note that $\sum \mu(A_j) - \mu(\bigcup A_j) < 2^{-k}$ also holds for any subfamily $\{A_j\}$ of $\mathcal{B}_k$. Now $\widetilde{C} = \limsup_k \bigcup_i B_{ik}$ satisfies $\mu(C \setminus \widetilde{C}) = 0$, and $\bigcup_k \mathcal{B}_k$ is a **B**-fine cover of $\widetilde{C}$.

We may apply (1) to the set $\widetilde{C}$, to obtain a corresponding constant M. Choose j so large that $2^{j-2} > M$. Now $\mathcal{V} = \bigcup_{k=j}^\infty \mathcal{B}_k$ is a **B**-fine cover of $\widetilde{C}$, so there exist $A_1, A_2, \cdots, A_n \in \mathcal{V}$ with $\|\sum \mathbf{1}_{A_i}\|_\infty < M \sum \mu(A_i)$. Each set A_i belongs to some $\mathcal{B}_k$ with $k \geq j$; call it $\mathcal{B}_{k(i)}$. Now

$$\sum_{i=1}^n \mu(A_i) = \sum_{k=j}^\infty \sum_{k(i)=k} \mu(A_i)$$

$$\leq \sum_{k=j}^\infty \left[\mu\left(\bigcup_{k(i)=k} A_i \right) + 2^{-k} \right]$$

$$\leq \sum_{k=j}^{\infty} \left[2^{-k} \left\| \sum_{k(i)=k} \mathbf{1}_{A_i} \right\|_\infty + 2^{-k} \right]$$

$$\leq \sum_{k=j}^{\infty} 2^{-k+1} \left\| \sum_{i=1}^{n} \mathbf{1}_{A_i} \right\|_\infty$$

$$\leq 2^{-j+2} \left\| \sum_{i=1}^{n} \mathbf{1}_{A_i} \right\|_\infty,$$

a contradiction. Hence (2) holds.

(2) $\Longrightarrow$ (3). Suppose (2) holds. Then (A) holds, so again the weak Vitali property holds.

Suppose (3) fails. Then there exist $\varepsilon > 0$, $C_m \in \mathcal{F}$ with $\mu(C_m) > \varepsilon$, and **B**-fine almost-covers $\mathcal{V}_m$ of C_m such that $A_1, A_2, \cdots, A_n \in \mathcal{V}_m$ implies $\| \sum \mathbf{1}_{A_i} \|_\infty \geq 2^m \mu(\bigcup A_i)$. For each m, choose by the weak Vitali property a finite set $\mathcal{B}_m = \{B_{1m}, B_{2m}, \cdots\} \subseteq \mathcal{V}_m$ with $\delta(B_{im}) < 1/m$, $\mu(C_m \setminus \bigcup_i B_{im}) < 2^{-m}$, and $\mu(\bigcup_i B_{im} \setminus C_m) < 2^{-m}$. Now let $C = \limsup_m \bigcup_i B_{im}$. Then $\bigcup_{m=1}^{\infty} \mathcal{B}_m$ is a **B**-fine cover of C and $\mu(C) \geq \varepsilon$. Apply (2) to C to obtain a constant M. Choose j so that $2^{j-1} > M$. Then $\mathcal{V} = \bigcup_{m=j}^{\infty} \mathcal{B}_m$ is a **B**-fine cover of C. Thus there exist $A_1, A_2, \cdots, A_n \in \mathcal{V}$ with $\| \sum \mathbf{1}_{A_i} \|_\infty < M \mu(\bigcup A_i)$. Each set A_i is in some $\mathcal{B}_{m(i)}$. Now

$$\mu \left(\bigcup_{i=1}^{n} A_n \right) \leq \sum_{m=j}^{\infty} \mu \left(\bigcup_{m(i)=m} A_i \right)$$

$$\leq \sum_{j=m}^{\infty} 2^{-m} \left\| \sum_{m(i)=m} \mathbf{1}_{A_i} \right\|_\infty$$

$$\leq 2^{-j+1} \left\| \sum_{i=1}^{n} \mathbf{1}_{A_i} \right\|_\infty,$$

a contradiction. Hence (3) holds.   ∎

Halo theorems

The classical derivation theorems, such as the theorems in Sections 7.1 and 7.2, use the Vitali covering theorem. Banach's proof of this theorem uses a simple geometric fact about "halos" of sets in $\mathbb{R}^d$. It is useful to generalize this idea.

We will use the "essential supremum" or "essential union" of a family of sets (see Section 4.1). In many cases, this is the same as the ordinary union:

(7.4.6) **Proposition.** *Let Ω be a metric space and let μ be a Radon measure on Ω. Let $(A_i)_{i\in I}$ be a family of open subsets of Ω. Then*

$$A = \bigcup_{i\in I} A_i$$

is an essential supremum of the family.

Proof. Certainly A is open (hence measurable) and $A \supseteq A_i$ for all i. Suppose also B is measurable and $B \supseteq A_i$ a.e. for all i. We claim that $B \supseteq A$ a.e. Suppose not. Then $\mu(A \setminus B) > 0$. Thus there is a compact set $K \subseteq A\setminus B$, with $\mu(K) > 0$. Now $(A_i)_{i\in I}$ is an open cover of K, so there is a finite subcover, $A_{i_1}\cup A_{i_2}\cup\cdots\cup A_{i_n} \supseteq K$. Then $K \subseteq (A_{i_1}\cup A_{i_2}\cup\cdots\cup A_{i_n})\setminus B$. In fact $\mu(A_i \setminus B) > 0$ for some i, which contradicts $B \supseteq A_i$ a.e. ∎

Let $\mathbf{B} = (\mathcal{E},\delta)$ be a D-basis, let $A \in \mathcal{E}$, and let $\alpha > 0$. The α-*halo* of A is

$$H(\alpha, A) = \text{ess sup } \{ E \in \mathcal{E} : E\cap A \neq \emptyset,\ \delta(E) \leq \alpha\delta(A) \}.$$

For example, in the cube basis of $\mathbb{R}^2$, given a square A, the halo $H(2, A)$ is the (open) square with the same center as A and 5 times the side of A. The key element in Banach's proof of the Vitali covering theorem is the existence of a constant M with $\mu(H(2, A)) \leq M\mu(A)$ for all squares A. This idea can be generalized.

(7.4.7) **Theorem.** *Let Ω be a metric space, and let $\mathbf{B} = (\mathcal{E},\delta)$ be a D-basis with small overflow, where δ is diameter. Suppose all $E \in \mathcal{E}$ are closed sets. Suppose $\alpha > 1$ and $M < \infty$ exist such that $\mu(H(\alpha, A)) \leq M\mu(A)$ for all $A \in \mathcal{E}$. Then $\mathbf{B}$ has the strong Vitali property.*

Proof. Let $Q \subseteq \Omega$, $0 < \mu^*(Q) < \infty$, let $\mathcal{V}$ be a $\mathbf{B}$-fine cover of Q, and let C be an outer envelope of Q. Suppose $\varepsilon > 0$ is given. By small overflow, there is a $\mathbf{B}$-fine cover $\mathcal{V}_1 \subseteq \mathcal{V}$ of Q with $\delta(E) \leq 1$ for all $E \in \mathcal{V}_1$ and

$$\mu\left(\text{ess sup } E \setminus C\right) < \varepsilon.$$
$$\quad\ _{E\in\mathcal{V}_1}$$

We will proceed by transfinite induction. To begin, let

$$\beta_1 = \sup\{\delta(E) : E \in \mathcal{V}_1\},$$

and choose $A_1 \in \mathcal{V}_1$ with $\delta(A_1) \geq \beta_1/\alpha$. Let

$$\mathcal{V}_2 = \{E \in \mathcal{V}_1 : E\cap A_1 = \emptyset\}.$$

If $\mathcal{V}_2 = \emptyset$, the recursive construction stops; otherwise continue: define

$$\beta_2 = \sup\{\delta(E) : E \in \mathcal{V}_2\},$$

and choose $A_2 \in \mathcal{V}_2$ with $\delta(A_2) \geq \beta_2/\alpha$. Suppose τ is an ordinal, and pairwise disjoint sets $(A_\gamma)_{\gamma < \tau}$ have been chosen. Let

$$\mathcal{V}_\tau = \{ E \in \mathcal{V}_1 : E \cap A_\gamma = \emptyset \text{ for all } \gamma < \tau \}.$$

If $\mathcal{V}_\tau = \emptyset$, the recursive construction stops; otherwise continue: define

$$\beta_\tau = \sup \{ \delta(E) : E \in \mathcal{V}_\tau \},$$

and choose $A_\tau \in \mathcal{V}_\tau$ with $\delta(A_\tau) \geq \beta_\tau/\alpha$. Now the sets A_γ are disjoint, and their union has outer measure at most $\mu(C) + \varepsilon < \infty$, so there are only countably many of them. (That is, the construction stops at some countable ordinal.) If $E \in \mathcal{V}_1$, then $E \cap A_\gamma \neq \emptyset$ for some γ. The series $\sum \mu(A_\gamma)$ converges. Thus we may choose a finite set I of ordinals so that

$$\sum_{\gamma \notin I} \mu(A_\gamma) < \frac{\varepsilon}{M}$$

and $\mu(\bigcup_{\gamma \in I} A_\gamma \setminus C) < \varepsilon$. We claim that $\mu(C \setminus \bigcup_{\gamma \in I} A_\gamma) < \varepsilon$, or equivalently $\mu^*(Q \setminus \bigcup_{\gamma \in I} A_\gamma) < \varepsilon$. If $\omega \in Q \setminus \bigcup_{\gamma \in I} A_\gamma$, then there is a net $E_t \Rightarrow \omega$, $E_t \in \mathcal{V}_1$. But the finite union $\bigcup_{\gamma \in I} A_\gamma$ is closed, so there is t with $E_t \cap \bigcup_{\gamma \in I} A_\gamma = \emptyset$. Fix such a t. Now if γ is the least ordinal with $E_t \cap A_\gamma \neq \emptyset$, then $\gamma \notin I$ and $E_t \in \mathcal{V}_\gamma$, so $\delta(E_\gamma) \leq \beta_\gamma \leq \alpha\delta(A_\gamma)$, and thus $\omega \in E_t \subseteq H(\alpha, A_\gamma)$. This shows that

$$Q \setminus \bigcup_{\gamma \in I} A_\gamma \subseteq \bigcup_{\gamma \notin I} H(\alpha, A_\gamma).$$

Therefore

$$\mu^* \left(Q \setminus \bigcup_{\gamma \in I} A_\gamma \right) \leq \sum_{\gamma \notin I} \mu(H(\alpha, A_\gamma)) \leq M \sum_{\gamma \notin I} \mu(A_\gamma) < \varepsilon. \qquad \blacksquare$$

Since $H(\alpha, A)$ is an essential supremum, the assertion

$$\mu\big(H(\alpha, A)\big) \leq M\mu(A)$$

is the same as: for any finite collection $E_1, E_2, \cdots, E_n \in \mathcal{E}$ with $E_i \cap A \neq \emptyset$ and $\delta(E_i) \leq \alpha\delta(A)$, we have

$$\mu \left(\bigcup_{i=1}^n E_i \right) \leq M\mu(A).$$

Dropping the requirement that the constituents be closed, we may still prove the weak Vitali property.

(7.4.8) Theorem. *Let Ω be a metric space, let μ be a Radon measure, and let $\mathbf{B} = (\mathcal{E}, \delta)$ be a D-basis, where δ is diameter. Suppose there exist $\alpha > 1$ and $M < \infty$ such that $\mu(H(\alpha, A)) \leq M\mu(A)$ for all $A \in \mathcal{E}$. Then $\mathbf{B}$ has the weak Vitali property.*

Proof. Let $0 < r < 1$. Define

$$\mathcal{E}_r = \{ K : K \text{ compact}, \ K \subseteq E, \ \mu(K) \geq r\mu(E) \text{ for some } E \in \mathcal{E} \}.$$

Then $(\mathcal{E}_r, \delta)$ is also a D-basis, where δ is diameter. Its constituents are closed sets. Let $H_r(\alpha, K)$ be the α-halo for the basis $\mathcal{E}_r$ and $H(\alpha, E)$ be the α-halo for the basis $\mathcal{E}$. If $K \subseteq E$, $\mu(K) \geq r\mu(E)$, then $H_r(\alpha, K) \subseteq H(\alpha, E)$, so

$$\mu(H_r(\alpha, K)) \leq \mu(H(\alpha, E)) \leq M\mu(E) \leq \frac{M}{r}\mu(K).$$

Therefore by the previous theorem, $\mathcal{E}_r$ has the strong Vitali property.

Now let $C \in \mathcal{F}$, $0 < \mu(C) < \infty$, $\mathcal{V}$ a $\mathbf{B}$-fine almost-cover of C, and $\varepsilon > 0$. Choose $r < 1$ so that

$$\left(\frac{1}{r} - 1 \right) (\mu(C) + \varepsilon) \leq \varepsilon.$$

Then the set of all compact K such that $K \subseteq E$ and $\mu(K) \geq r\mu(E)$ for some E in $\mathcal{V}$ is a $(\mathcal{E}_r, \delta)$-fine almost-cover of C. By the strong Vitali property, there exist disjoint $K_1, K_2, \cdots, K_n$ with $\mu(\bigcup K_i \setminus C) < \varepsilon$ and $\mu(C \setminus \bigcup K_i) < \varepsilon$. It is easily checked that the corresponding sets $E_i \in \mathcal{V}$ satisfy $\mu(C \setminus \bigcup E_i) < \varepsilon$, $\mu(\bigcup E_i \setminus C) < 2\varepsilon$, and $\sum \mu(E_i) - \mu(\bigcup E_i) < 3\varepsilon$. ∎

This may be used to prove a covering theorem that may be of independent interest. A *homothety* of $\mathbb{R}^d$ is a function $\theta \colon \mathbb{R}^d \to \mathbb{R}^d$ of the form $\theta(x) = rx + a$, where $r > 0$ and $a \in \mathbb{R}^d$. Recall the notation λ for d-dimensional Lebesgue measure on $\mathbb{R}^d$. If $\theta(x) = rx + a$, then for every measurable set $C \subseteq \mathbb{R}^d$, we have $\lambda(\theta(C)) = r^d \lambda(C)$.

(7.4.9) Homothety filling theorem. *Let $U \subseteq \mathbb{R}^d$ be an open set of finite measure, let $C \subseteq \mathbb{R}^d$ be a bounded measurable set of positive measure, and let $\varepsilon > 0$. Then there exist homotheties $\theta_1, \theta_2, \cdots, \theta_n$ such that*

(1) $\bigcup_i \theta_i(C) \subseteq U$,
(2) $\lambda(U \setminus \bigcup \theta_i(C)) < \varepsilon$,
(3) $\sum \lambda(\theta_i(C)) < \lambda(U) + \varepsilon$,
(4) $\operatorname{diam} \theta_i(C) < \varepsilon$.

Proof. The family

$$\mathcal{E} = \{ \theta(C) : \theta \text{ homothety} \}$$

together with $\delta = $ diameter, defines a D-basis on $\mathbb{R}^d$. Since C is bounded, it is contained in an interval I of $\mathbb{R}^d$. Let J be the interval with the same center as I, but 5 times the side. Then clearly $H(2, C) \subseteq J$. Thus we have $\lambda(H(2, C)) \leq M\lambda(C)$, where $M = \lambda(J)/\lambda(C)$. But then for any homothety θ, also $\lambda(H(2, \theta(C))) \leq M\lambda(\theta(C))$. Thus by Theorem (7.4.8), the derivation basis $\mathbf{B} = (\mathcal{E}, \delta)$ has the weak Vitali property. Now

$$\mathcal{V} = \{\, \theta(C) : \theta \text{ homothety}, \ \theta(C) \subseteq U, \delta(\theta(C)) < \varepsilon \,\}$$

is a $\mathbf{B}$-fine cover of U, so the result follows from the weak Vitali property.

∎

As usual (Proposition (7.3.7)), the same result is true if we allow countably many sets $\theta_i(C)$ and change (2) to: $\lambda(U \setminus \bigcup \theta_i(C)) = 0$.

Weak halo

A related sort of halo will be discussed next. Under the right conditions, it leads to a necessary and sufficient condition for the density property. Let $(\mathcal{E}, \delta)$ be a D-basis on $(\Omega, \mathcal{F}, \mu)$. Let $\alpha, \eta > 0$, and let $C \in \mathcal{F}$. The *weak η-halo* of C is:

$$S(\alpha, \eta, C) = \text{ess sup } \{\, E \in \mathcal{E} : \mu(C \cap E) > \alpha\mu(E), \delta(E) < \eta \,\}.$$

The *weak halo* of C is the essential union over all $\eta > 0$, or:

$$S(\alpha, C) = \text{ess sup } \{\, E \in \mathcal{E} : \mu(C \cap E) > \alpha\mu(E) \,\}.$$

It is not hard to guess that if $\gamma(E) = \mu(C \cap E)$, then $S(\alpha, \eta, C)$ and $S(\alpha, C)$ are related to the set $\{D_*\gamma > \alpha\}$. The next result will make this precise.

A D-basis $(\mathcal{E}, \delta)$ has the *weak halo evanescence property* iff: for every α, $0 < \alpha < 1$, every decreasing sequence of measurable sets of finite measure $C_n \downarrow \emptyset$, and every decreasing sequence of positive numbers $\eta_n \downarrow 0$, we have $\lim_n \mu(S(\alpha, \eta_n, C_n)) = 0$.

(7.4.10) Theorem. *Let Ω be a metric space, let μ be a Radon measure on Ω, let $\mathbf{B} = (\mathcal{E}, \delta)$ be a D-basis, where $\mathcal{E}$ is a collection of open sets of finite measure, and δ is diameter. Then $\mathbf{B}$ has the weak halo evanescence property if and only if $\mathbf{B}$ has the density property.*

Proof. First suppose that $\mathbf{B}$ has the density property. Since $\mathcal{E}$ consists of open sets and μ is a Radon measure, the essential supremum in the definition of the weak halo is an actual union. Let $C \in \mathcal{F}$, $0 < \mu(C) < \infty$. Then by Theorem (7.3.11) $\mathbf{B}$ differentiates the integral of $\mathbf{1}_C$, so if $\gamma(E) = \mu(E \cap C)$, then for $0 < \alpha < 1$ we have

$$\bigcap_{\eta > 0} S(\alpha, \eta, C) = \{D\gamma > \alpha\} = C.$$

Suppose $C_n \downarrow \emptyset$ and $\eta_n \downarrow 0$. For each n, we have therefore

$$\lim_k \mu(S(\alpha, \eta_k, C_n)) = \mu(C_n).$$

Given $\varepsilon > 0$, choose n_0 so that $\mu(C_{n_0}) < \varepsilon/2$; then choose k_0 so that $\mu(S(\alpha, \eta_{k_0}, C_{n_0})) < \varepsilon$. If $n > \max\{n_0, k_0\}$, then $\mu(S(\alpha, \eta_n, C_n)) < \varepsilon$. This shows that $\lim_n \mu(S(\alpha, \eta_n, C_n)) = 0$. Thus **B** has the weak halo evanescence property.

Conversely, suppose that **B** has the weak halo evanescence property. Recall (Proposition (7.4.1)) that the upper and lower derivates are measurable functions under the conditions assumed. Let $C \in \mathcal{F}$, $0 < \mu(C) < \infty$. We want to show that almost every point of C is a point of density of C. Let γ be the integral of $\mathbf{1}_C$: we want to show that $D_\star \gamma \geq 1$ a.e. on C. Suppose not. There exists $\alpha > 0$ such that $\mu\{\omega \in C : D_\star \gamma(\omega) < 1 - \alpha\} > 0$. Thus there is a compact set $K \subseteq C \cap \{D_\star \gamma < 1 - \alpha\}$ with $\mu(K) > 0$. Now

$$\mathcal{V}_1 = \left\{ E \in \mathcal{E} : \frac{\mu(K \cap E)}{\mu(E)} < 1 - \alpha, \ \delta(E) < 1 \right\}$$

is a (**B**-fine) open cover of K. There are finitely many sets $A_{11}, A_{21}, \cdots \in \mathcal{V}_1$ with $K \subseteq \bigcup_i A_{i1}$, an open set. Next,

$$\mathcal{V}_2 = \left\{ E \in \mathcal{E} : \frac{\mu(K \cap E)}{\mu(E)} < 1 - \alpha, \ \delta(E) < \frac{1}{2}, \ E \subseteq \bigcup_i A_{i1} \right\}$$

is a (**B**-fine) open cover of K. There are finitely many sets $A_{12}, A_{22}, \cdots \in \mathcal{V}_2$ with $K \subseteq \bigcup_i A_{i2}$. Continuing in this way, we obtain sets $A_{ik} \in \mathcal{E}$ such that

$$\frac{\mu(K \cap A_{ik})}{\mu(A_{ik})} < 1 - \alpha,$$

$$\delta(A_{ik}) < \frac{1}{k},$$

$$K \subseteq \bigcup_i A_{ik},$$

$$\bigcup_i A_{i,k+1} \subseteq \bigcup_i A_{ik}.$$

Now K is closed, so $K = \bigcap_k \bigcup_i A_{ik}$. Thus the sets defined by $C_k = \bigcup_i A_{ik} \setminus K$ satisfy $C_k \downarrow \emptyset$. But

$$\mu(A_{ik}) = \mu(A_{ik} \cap C_k) + \mu(A_{ik} \cap K) < \mu(A_{ik} \cap C_k) + (1 - \alpha)\mu(A_{ik}),$$

so $\mu(A_{ik} \cap C_k) > \alpha\mu(A_{ik})$. Thus $A_{ik} \subseteq S(\alpha, 1/k, C_k)$. Now for every k, we have $K \subseteq \bigcup_i A_{ik} \subseteq S(\alpha, 1/k, C_k)$, so by the weak halo evanescence property $\mu(K) = 0$. This contradiction shows that in fact $D\gamma = 1$ a.e. on C. ∎

Busemann-Feller bases

A collection $\mathcal{A}$ of subsets of $\mathbb{R}^d$ is said to be *closed under homotheties* iff $A \in \mathcal{A}$ implies $\theta(A) \in \mathcal{A}$ for all homotheties θ.

A D-basis $(\mathcal{E}, \delta)$ on $(\Omega, \mathcal{F}, \mu)$ is a *Busemann-Feller basis* iff:

(1) Ω is $\mathbb{R}^d$ for some $d \geq 1$,
(2) μ is d-dimensional Lebesgue measure λ,
(3) $\mathcal{E}$ is a collection of bounded open sets,
(4) $\mathcal{E}$ is closed under homotheties,
(5) δ is diameter.

Many of the examples of derivation bases in Sections 7.1 and 7.2 are essentially Busemann-Feller bases; in order to make them Busemann-Feller bases we must remove the boundaries from the constituents. This makes no essential difference in these cases, since the boundaries are sets of measure zero. Since the constituents are open sets, and Lebesgue measure λ on $\mathbb{R}^d$ is a Radon measure, the essential suprema appearing in the general theory are unions in this case (as in (7.4.10)). For example, the weak halo is

$$S(\alpha, C) = \bigcup \{ E \in \mathcal{E} : \mu(C \cap E) > \alpha\mu(E) \}.$$

Weak halo conditions are particularly useful for Busemann-Feller bases. The weak halo evanescence property may be simplified in the following way:

The derivation basis **B** has *property* (WH) iff, for every α, $0 < \alpha < 1$, there exists $M < \infty$ such that for every $C \in \mathcal{F}$, we have $\mu(S(\alpha, C)) \leq M\mu(C)$.

(7.4.11) Theorem. *Let* **B** *be a Busemann-Feller basis. Then* **B** *has property* (WH) *if and only if* **B** *has the density property.*

Proof. Clearly (WH) implies the weak halo evanescence property, hence the density property.

For the converse, suppose (WH) fails. Then there exist α, $0 < \alpha < 1$, and $B_m \in \mathcal{F}$ with $\mu(S(\alpha, B_m)) > 4^m\mu(B_m)$. Let J be the unit interval in $\mathbb{R}^d$. Fix m. There exists a finite union S_m of constituents E with $\mu(E \cap B_m) > \alpha\mu(E)$ such that $\mu(S_m) > 4^m\mu(B_m)$. By Theorem (7.4.9), there exist homotheties $\theta_1, \theta_2, \cdots$ such that $\bigcup_i \theta_i(S_m) \subseteq J$, $\lambda(J \setminus \bigcup_i \theta_i(S_m)) = 0$, $\sum_i \lambda(\theta_i(S_m)) < 2$, and $\delta(\theta_i(S_m)) < 1/m$. Let $C_m = \bigcup_i \theta_i(B_m)$. Then: for almost every $\omega \in J$, there exists a constituent E with $\omega \in E$, $\delta(E) < 1/m$, and $\lambda(E \cap C_m) > \alpha\mu(E)$. Also, $\lambda(C_m) \leq \sum_i \lambda(\theta_i(B_m)) < 4^{-m} \sum_i \mu(\theta_i(S_m)) < 2 \cdot 4^{-m}$.

Now let $C = \bigcup_{m=1}^{\infty} C_m$. Thus $\lambda(C) \leq \sum \lambda(C_m) \leq \sum 2 \cdot 4^{-m} \leq 2/3$. We claim that **B** does not differentiate the integral γ of $\mathbf{1}_C$. For almost every $\omega \in J$ and every m, there is a constituent E with $\omega \in E$, $\delta(E) < 1/m$, and $\lambda(E \cap C) \geq \lambda(E \cap C_m) > \alpha\lambda(E)$. Therefore $D^{\star}\gamma(\omega) \geq \alpha$, so $D^{\star}\gamma = \mathbf{1}_C$ fails on the set $J \setminus C$, which has positive measure.   ∎

For derivation of L_1 integrals, we should consider a refinement of (WH). We postulate not only for each α there is a constant M, but more precisely how M depends on α as $\alpha \to 0$.

The derivation basis $\mathbf{B}$ has property (WH$_1$) iff there is a constant K such that, for every α, $0 < \alpha < 1$, and every $C \in \mathcal{F}$, we have

$$\mu\big(S(\alpha, C)\big) \leq \frac{K}{\alpha}\,\mu(C).$$

More generally, let Ψ be an Orlicz function. The derivation basis $\mathbf{B}$ has property (WH$_\Psi$) iff there is a constant K such that, for every α, $0 < \alpha < 1$, and every $C \in \mathcal{F}$, we have

$$\mu\big(S(\alpha, C)\big) \leq K\Psi\left(\frac{1}{\alpha}\right)\mu(C).$$

The proof of the following is like the previous proof. (Note that the conjugate function Φ is not being used.)

(7.4.12) Theorem. *Let $\mathbf{B}$ be a Busemann-Feller basis, and let Ψ be an Orlicz function with $0 < \Psi(v) < \infty$ for all $v > 0$. Suppose $\mathbf{B}$ differentiates all L_Ψ integrals. Then $\mathbf{B}$ has property (WH$_\Psi$).*

Proof. Suppose (WH$_\Psi$) fails. Choose numbers $K_m \uparrow \infty$ so that

$$(7.4.12a) \qquad \sum_{m=1}^{\infty} \frac{1}{K_m} < \Psi(1).$$

There exist $B_m \in \mathcal{F}$ and $\alpha_m < 1$ with $\lambda(S(\alpha_m, B_m)) > K_m \Psi(1/\alpha_m)\mu(B_m)$. Proceeding as in (7.4.11), we obtain sets C_m contained in the unit interval J of $\mathbb{R}^d$ such that $\lambda(C_m) < 1/\big(K_m \Psi(1/\alpha_m)\big)$, and for almost every $\omega \in J$, there is a constituent E with $\omega \in E$, $\delta(E) < 1/m$ and $\lambda(E \cap C_m) > \alpha_m \lambda(E)$. Now consider the nonnegative function

$$f = \sup_{m} \frac{1}{\alpha_m}\mathbf{1}_{C_m},$$

and its integral $\gamma(E) = \int_E f\,d\lambda$. We have

$$\int \Psi(f)\,d\lambda \leq \sum \int \Psi\left(\frac{1}{\alpha_m}\mathbf{1}_{C_m}\right)d\lambda$$

$$(7.4.12b) \qquad = \sum \lambda(C_m)\Psi\left(\frac{1}{\alpha_m}\right) < \sum \frac{1}{K_m}$$

so $f \in L_\Psi$. But we claim that $D^\star\gamma = f$ fails on a set of positive measure. For almost every $\omega \in J$ and every m, there is a constituent E with $\omega \in E$, $\delta(E) < 1/m$, and $\lambda(E \cap C_m) > \alpha_m \lambda(E)$. Thus

$$\gamma(E) = \int_E f\,d\lambda \geq \int_E \frac{1}{\alpha_m}\mathbf{1}_{C_m}\,d\lambda = \frac{1}{\alpha_m}\lambda(C_m \cap E) > \lambda(E).$$

Thus $D^\star\gamma(\omega) \geq 1$ on J. But if $f \geq 1$ on J, then we would have $\int \Psi(f)\,d\lambda \geq \int_J \Psi(1)\,d\lambda = \Psi(1)$, which contradicts (7.4.12a) and (7.4.12b). Therefore $\lambda(f < D^\star\gamma) > 0$.   ∎

For a converse result, we will consider a "functional" variant of condition (WH_Ψ). The derivation basis $\mathbf{B}$ has property (FH_Ψ) iff there is a constant K such that, for all disjoint bounded sets $C_1, C_2, \cdots, C_n \in \mathcal{F}$ and all nonnegative scalars $c_1, c_2, \cdots, c_n$, we have

$$\mu\left(\operatorname{ess\,sup}\left\{E \text{ constituent} : \sum c_i \mu(C_i \cap E) > \mu(E)\right\}\right)$$
$$\leq K \sum \Psi(c_i)\mu(C_i).$$

Note that (FH_Ψ) implies (WH_Ψ) by taking a single set C_1 and scalar $c_1 = 1/\alpha$. On the other hand, if (FH_Ψ) holds, then the inequality remains true even for infinite lists $C_1, C_2, \cdots$ and $c_1, c_2, \cdots$. Condition (FH_Ψ) is more complex than (WH_Ψ), but it is necessary and sufficient for differentiation of L_Ψ integrals.

Note an alternative formulation: If we write $g = \sum c_i \mathbf{1}_{C_i}$ (a nonnegative simple function), then the inequality $\sum c_i \mu(C_i \cap E) > \mu(E)$ becomes $\int_E g\,d\mu > \mu(E)$, and $\sum \Psi(c_i)\mu(C_i)$ becomes $\int \Psi(g)\,d\mu$.

(7.4.13) Theorem. *Let $\mathbf{B}$ be a Busemann-Feller basis, and let Ψ be an Orlicz function satisfying (Δ_2) with $0 < \Psi(v) < \infty$ for all $v > 0$. Then $\mathbf{B}$ differentiates all L_Ψ integrals if and only if $\mathbf{B}$ has property (FH_Ψ).*

Proof. Assume (FH_Ψ). Property (FH_Ψ) implies (WH_Ψ), which implies (WH). Thus (7.4.11) $\mathbf{B}$ differentiates the integrals of all L_∞ functions.

Let $f \geq 0$ belong to L_Ψ, and let $\gamma(E) = \int_E f\,d\lambda$ for constituents E. We claim that $D\gamma = f$ a.e. Write $C_i = \{i-1 \leq f < i\}$; the sets C_i are disjoint. For $m \in \mathbb{N}$, let $f_m = f\,\mathbf{1}_{\{f<m\}}$, so $f_m \in L_\infty$. Therefore $\gamma_m(E) = \int_E f_m\,d\lambda$ satisfies $D\gamma_m = f_m$ a.e.

Fix $\varepsilon > 0$. Now $f \in L_\Psi$, where Ψ satisfies (Δ_2), so $\int \Phi(2f/\varepsilon)\,d\lambda < \infty$. On the set C_i, we have (for $i \geq 2$) $i/\varepsilon \leq 2(i-1)/\varepsilon \leq 2f/\varepsilon$, so

$$\sum_{i=2}^{\infty} \Psi\left(\frac{i}{\varepsilon}\right)\mu(C_i) \leq \int \Psi\left(\frac{2f}{\varepsilon}\right)d\lambda < \infty.$$

Now define

$$S_m = \bigcup\left\{E \text{ constituent} : \sum_{i=m+1}^{\infty} i\,\lambda(C_i \cap E) > \varepsilon\lambda(E)\right\}.$$

The sets decrease: $S_1 \supseteq S_2 \supseteq \cdots$. By condition (FH_Ψ),

$$\lambda(S_m) \leq K \sum_{i=m+1}^{\infty} \Psi\left(\frac{i}{\varepsilon}\right)\lambda(C_i);$$

so $\lambda(S_m) \to 0$. Thus $\lambda\left(\bigcap S_m\right) = 0$.

Now let ω be such that $D\gamma_m(\omega) = f_m(\omega)$ for all m and $\omega \notin \bigcap S_m$. Almost every ω satisfies these conditions. Let $E_n \Rightarrow \omega$. Then choose m so that $m > f(\omega)$ and $\omega \notin S_m$. Now

$$\frac{\gamma(E_n)}{\lambda(E_n)} = \sum_{i=1}^{m} \frac{\gamma(E_n \cap C_i)}{\lambda(E_n)} + \sum_{i=m+1}^{\infty} \frac{\gamma(E_n \cap C_i)}{\lambda(E_n)}.$$

By the definition of S_m,

$$\sum_{i=m+1}^{\infty} \frac{\gamma(E_n \cap C_i)}{\lambda(E_n)} \leq \sum_{i=m+1}^{\infty} \frac{i\,\lambda(E_n \cap C_i)}{\lambda(E_n)} \leq \varepsilon.$$

Also, $m > f(\omega)$, so $f(\omega) = f_m(\omega)$ and

$$\sum_{i=1}^{m} \frac{\gamma(E_n \cap C_i)}{\lambda(E_n)} = \frac{\gamma_m(E_n)}{\lambda(E_n)} \to f_m(\omega).$$

Thus

$$f(\omega) - \varepsilon \leq \liminf \frac{\gamma(E_n)}{\lambda(E_n)} \leq \limsup \frac{\gamma(E_n)}{\lambda(E_n)} \leq f(\omega) + \varepsilon.$$

This is true for all $E_n \Rightarrow \omega$, so

$$f(\omega) - \varepsilon \leq D_\star\gamma(\omega) \leq D^\star\gamma(\omega) \leq f(\omega) + \varepsilon.$$

Finally, since ε was arbitrary, we have $f(\omega) = D\gamma(\omega)$ a.e.

For the converse, suppose (FH$_\Psi$) fails. Choose $K_m \uparrow \infty$ with

(7.4.13a)
$$\sum_{m=1}^{\infty} \frac{2}{K_m} < \Psi(1).$$

There exist nonnegative simple functions g_m, with bounded support, such that

$$\lambda\left(\bigcup\left\{E : \int_E g_m \, d\lambda > \lambda(E)\right\}\right) > K_m \int \Psi(g_m) \, d\lambda.$$

For each m, there is a finite union S_m of constituents E with $\int_E g_m \, d\lambda > \lambda(E)$ such that $\lambda(S_m) > K_m \int \Psi(g_m) \, d\lambda$. Let J be the unit interval in $\mathbb{R}^d$. There exist homotheties $\theta_1, \theta_2, \cdots$ such that $\bigcup_i \theta_i(S_m) \subseteq J$, $\delta(\theta_i(S_m)) < 1/m$, $\lambda(J \setminus \bigcup_i \theta_i(S_m)) = 0$, and $\sum_i \lambda(\theta_i(S_m)) < 2$. Thus if $f_m = \sup_i g_m \circ \theta_i$, we have

$$\int \Psi(f_m) \, d\lambda \leq \sum_i \int \Psi(g_m \circ \theta_i) \, d\lambda \leq \frac{1}{K_m} \sum_i \lambda(\theta_i(S_m)) < \frac{2}{K_m},$$

and for almost every $\omega \in J$ there is a constituent E with $\omega \in E$, $\delta(E) < 1/m$ and $\int_E f_m \, d\lambda > \lambda(E)$.

Then consider $f = \sup_m f_m$. Now

$$(7.4.13b) \qquad \int \Psi(f) \, d\lambda \leq \sum \int \Psi(f_m) \, d\lambda < \sum \frac{2}{K_m},$$

so $f \in L_\Psi$. Let γ be the integral of f. We claim that $D^\star \gamma = f$ fails on a set of positive measure. For almost every $\omega \in J$ and every m, there is E with $\omega \in E$, $\delta(E) < 1/m$ and $\int f \, d\lambda > \lambda(E)$, so $D^\star \gamma \geq 1$ a.e. on J. But if $f \geq 1$ a.e. on J, then we would have $\int \Psi(f) \, d\lambda \geq \int_J \Psi(1) \, d\lambda = \Psi(1)$, which contradicts (7.4.13a) and (7.4.13b). ∎

The particular case of (FH_1) should be noted. The Busemann-Feller basis $\mathbf{B}$ differentiates all L_1 integrals if and only if: there is a constant K such that, for all disjoint bounded sets $C_1, C_2, \cdots, C_n \in \mathcal{F}$ and all nonnegative scalars $c_1, c_2, \cdots, c_n$, we have

$$\lambda \left(\bigcup \left\{ E \text{ constituent} : \sum c_i \lambda(C_i \cap E) > \lambda(E) \right\} \right) \leq K \sum c_i \lambda(C_i).$$

Complements

(7.4.14) (Interval basis.) Let $\mathbf{B}$ be the interval basis in $\mathbb{R}^2$. The weak halo $S(\alpha, J)$ of the unit square J of $\mathbb{R}^2$ has area $1 + (4/\alpha) \log(1/\alpha)$ (see Figure (7.4.14)). This is a Busemann-Feller basis, so we know that $\mathbf{B}$ does not differentiate all L_Ψ integrals unless $\Psi(v) \succ_\infty v \log v$. In particular, $\mathbf{B}$ does not differentiate all L_1 integrals. This is essentially the fact that is used in the counterexample (7.2.10).

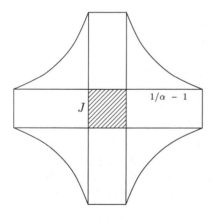

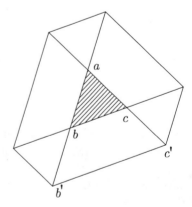

Figure (7.4.14).
$\lambda\big(S(\alpha, J)\big) = 1 + (4/\alpha) \log(1/\alpha)$.

Figure (7.4.15a). Weak halo.

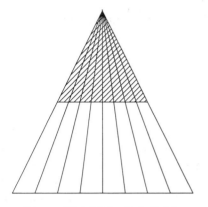
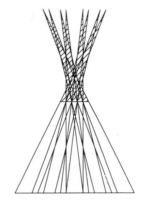

Figure (7.4.15b). Subdivide.   *Figure (7.4.15c).* Translate.

(7.4.15) (Rectangle basis.) In $\mathbb{R}^2$, the *rectangle basis* is the Busemann-Feller basis consisting of the (open) rectangles, with sides not necessarily parallel to the axes. This basis fails the density property. This can be seen using Theorem (7.4.11) as follows. The weak halo $S(1/2, C)$ of a triangle C with vertices a, b, c (shown in Figure (7.4.15a)) contains a triangle with vertices a, b', c', where b is the midpoint of the line segment from a to b' and c is the midpoint of the line segment from a to c'. Consider a triangle (shaded in Figure (7.4.15b)) and the corresponding larger triangle (white). Subdivide it into many smaller triangles as shown. Then translate the triangles as in Figure (7.4.15c). The weak halo $S(1/2, C)$ of the shaded portion C contains the white portion. By taking a large number of subdivisions and using appropriate translations, it may be arranged that the shaded area is as close to 0 as we like, while the white area remains large. This shows that the property (WH) fails. (Details of this construction are in Busemann & Feller [1934]. Or see Hayes & Pauc [1970], Section V.5, p. 104.)

Remarks

D-bases, Busemann-Feller bases, and many other variants are discussed by Hayes and Pauc [1970]. The reader may consult that volume for further information.

In the term "D-basis," the "D" is for Denjoy. See Denjoy [1951], Haupt [1953], Pauc [1953]. The general theory allows the function δ to be something other than the diameter. There is also the possibility of using another "disentanglement function" in the definition of the halo.

Property (A) is the derivation version of a condition of Astbury [1981b]. The equivalence of property (A) and property (C) (for special D-bases) is proved in Millet & Sucheston [1980c]. They also prove the sufficiency of property (C) for differentiation of L_1 integrals. Necessity of property (C) for differentiation of L_1 integrals is stated in Talagrand [1986].

Busemann-Feller bases come from a paper of Busemann and Feller [1934].

8

Pointwise ergodic theorems

In this chapter, we will prove some of the pointwise convergence theorems from ergodic theory. The main result of this chapter is the superadditive ratio ergodic theorem. It implies the Chacon-Ornstein theorem, the Kingman subadditive ergodic theorem, and, for positive operators, the Dunford-Schwartz theorem and Chacon's ergodic theorem involving "admissible sequences." We consider positive linear contractions T of L_1.

Our plan is as follows. We first prove weak maximal inequalities, from which we obtain the Hopf decomposition of the space Ω into the conservative part C and the dissipative part D (8.3.1). Assuming T conservative (that is, $\Omega = C$), we prove the Chacon-Ornstein theorem (8.5.4), i.e., the convergence to a finite limit of the ratio of sums of iterates of T applied to functions f and g. The limit is identified in terms of f, g, and the σ-algebra $\mathcal{C}$ of absorbing sets. The superadditive operator ergodic theorem is proved in the conservative case (8.4.6). It is then observed that the total contribution of Ω to C is a superadditive process with respect to the conservative operator T_C induced by T on C . Since the behavior of the ergodic ratio on the dissipative part D is obvious, the Chacon-Ornstein theorem (8.6.10) and, more generally, the superadditive ratio theorem (8.6.7) will follow on Ω. This affords considerable economy of argument, since the direct study of the contribution of D to C is not obvious even for additive processes.

The superadditive theory (or, equivalently, the subadditive theory) is mostly known for its applications, but in fact the notion of a superadditive process is shown to shed light on the earlier additive theory of L_1 operators.

Throughout this chapter, the term "operator" means a bounded linear transformation. An operator T defined on (equivalence classes of) real-valued measurable functions on Ω is called *positive* iff $f \geq 0$ a.e. implies $Tf \geq 0$ a.e. The operator T is a *contraction* on L_p iff $\|Tf\|_p \leq \|f\|_p$ for each $f \in L_p$. A positive operator T is called *sub-Markovian* iff it is a contraction on L_1. A positive operator T is called *Markovian* iff it preserves the integral, $\int Tf \, d\mu = \int f \, d\mu$ for $f \in L_1^+$, so that T is an isometry on L_1^+.

If f and g are extended real-valued functions, we write $f \vee g$ for the pointwise maximum, $f \wedge g$ for the pointwise minimum, $f^+ = f \vee 0$, and $f^- = (-f) \vee 0$.

Throughout Chapter 8, we will let $(\Omega, \mathcal{F}, \mu)$ be a σ-finite measure space, and T a positive contraction on $L_1(\Omega, \mathcal{F}, \mu)$. Observe that a positive linear operator on L_1 can be extended to act on the lattice of all extended real-valued functions f such that the negative part f^- is integrable. (See (8.4.9).)

8.1. Preliminaries

Given a norm-bounded sequence of elements f_n in a Banach space E, it is often important to find in E an element φ in some sense asymptotically close to a subsequence of f_n. Bounded subsets of reflexive Banach spaces are weakly sequentially compact (in fact, this is a characterization of reflexivity) so in $L_p(\Omega, \mathcal{F}, \mu)$ for $1 < p < \infty$ the weak limit of a subsequence of f_n will do for φ. Weak sequential compactness is not available in $L_1(\Omega, \mathcal{F}, \mu)$ without extra assumptions (uniform integrability; see Section 2.3), so we have to settle for less than a weak limit.

One procedure (8.1.4) is to consider elements of L_1^+ as members of L_1^{**}, i.e., finitely additive measures. Then by a theorem of Alaoglu, a subsequence of f_n converges weak-star to an element η in L_1^{**}, and the maximal countably additive measure dominated by η will do for φ. A related method consists in taking Banach limits; see for example Krengel [1985], p. 135. First we discuss the truncated limit, a more elementary and transparent method of constructing φ.

Truncated limits

A *weak unit* in L_1 is an element $u \in L_1^+$ such that, for each $f \in L_1^+$, if $u \wedge f = 0$, then $f = 0$. (This is the terminology used also in more general Banach lattices. See Remarks, below.) Any strictly positive integrable function u is a weak unit in L_1; such a function exists since $(\Omega, \mathcal{F}, \mu)$ is σ-finite. Let (f_n) be a sequence in L_1^+ with $\sup_n \|f_n\|_1 = M < \infty$. A function $\varphi \in L_1^+$ is called a *weak truncated limit* of (f_n) iff: for a weak unit u, the weak limit

$$\varphi_k = \operatorname*{w\,lim}_n f_n \wedge ku$$

exists for every $k \in \mathbb{N}$, and $\varphi_k \uparrow \varphi$. It is easy to see that in this definition the choice of the weak unit u is irrelevant. We will write

$$\varphi = \operatorname*{WTL}_n f_n.$$

If f_n is not positive, $\operatorname{WTL} f_n$ is defined as $\operatorname{WTL} f_n^+ - \operatorname{WTL} f_n^-$, assuming these expressions exist.

There is a compactness for weak truncated limits:

(8.1.1) Proposition. *Let (f_n) be a sequence in L_1 with $\sup_n \|f_n\|_1 = M < \infty$. Then there is a subsequence of (f_n) that has a weak truncated limit φ, and $\|\varphi\|_1 \le M$.*

Proof. It suffices to consider positive f_n. Let u be a weak unit. For each $k \in \mathbb{N}$, the sequence $(f_n \wedge ku)_{n \in \mathbb{N}}$ is a sequence bounded by ku. Thus $(f_n \wedge ku)$ is uniformly integrable, so it has a subsequence that converges weakly, say to φ_k. By the diagonal procedure, we obtain one subsequence (f_{n_j}) such that, for each k, the sequence $f_{n_j} \wedge ku$ converges weakly to φ_k as $j \to \infty$. Now the sequence φ_k is increasing and bounded in norm by M, so its pointwise limit $\varphi = \lim_k \varphi_k$ belongs to L_1^+. Thus $\text{WTL}_j\, f_{n_j} = \varphi$ and $\|\varphi\|_1 \leq M$. ∎

Next we prove a few of the basic properties of weak truncated limits. The operator WTL is additive and has the "Fatou property." The arguments will use elementary properties of operations in Banach lattices of functions, for example, $(f + h) \wedge h \leq (f \wedge h) + (g \wedge h)$ for nonnegative functions. This (and other similar properties) can be verified by considering various possible pointwise relations between f, g, and h. We will also use this fact (8.4.10): if an operator T is *positive* in the sense that $Tf \geq 0$ whenever $f \geq 0$, then T is continuous with respect to the norm topology (and therefore also with respect to the weak topology).

(8.1.2) Lemma. *Let* $f_n \geq 0$ *and* $g_n \geq 0$. *If* $\text{WTL}\, f_n = \varphi$, $\text{WTL}\, g_n = \gamma$, *and* $\text{WTL}(f_n + g_n) = \eta$, *then* $\varphi + \gamma = \eta$.

Proof. First,

$$(f_n + g_n) \wedge ku \leq (f_n \wedge ku) + (g_n \wedge ku),$$

hence $\eta \leq \varphi + \gamma$. For the other direction,

$$(f_n + g_n) \wedge 2ku \geq (f_n \wedge ku) + (g_n \wedge ku),$$

which implies that $\eta \geq \varphi + \gamma$. ∎

(8.1.3) Fatou's lemma. *Let* T *be a positive operator on* L_1, *and let* $f_n \geq 0$. *If* $\text{WTL}\, f_n = \varphi$ *and* $\text{WTL}\, Tf_n = \psi$, *then* $T\varphi \leq \psi$.

Proof. Let $\varphi_k = \text{w}\lim_n f_n \wedge ku$ and $\psi_k = \text{w}\lim_n Tf_n \wedge ku$, so that $\varphi_k \uparrow \varphi$ and $\psi_k \uparrow \psi$. Given $k \in \mathbb{N}$ and $\varepsilon > 0$, there exists m so large that $T(ku) \leq mu + g$, where $g \geq 0$ and $\|g\|_1 \leq \varepsilon$. Then

$$
\begin{aligned}
T(\varphi_k) &= T\big(\text{w}\lim_n (f_n \wedge ku)\big) \\
&= \text{w}\lim_n T(f_n \wedge ku) \\
&\leq \text{w}\lim \big(Tf_n \wedge T(ku)\big) \\
&\leq \text{w}\lim_n \big(Tf_n \wedge (mu + g)\big) \\
&\leq \text{w}\lim_n \big(Tf_n \wedge (mu)\big) + g \\
&\leq \psi_m + g \leq \psi + g.
\end{aligned}
$$

Since $\|g\|_1 < \varepsilon$, and ε is arbitrarily small, it follows that $T\varphi_k \leq \psi$. Letting $k \to \infty$, we conclude that $T\varphi \leq \psi$. ∎

The preceding result may be abbreviated: $T\ \text{WTL} \leq \text{WTL}\ T$.

These results are sufficient for our needs in this chapter. A few other results are stated as Complements below.

Decomposition of set-functions

Let $\mathcal{A}$ be an algebra of subsets of a set Ω. Sets introduced below will normally be assumed to be in $\mathcal{A}$. A finite, nonnegative, finitely-additive set function ψ on $\mathcal{A}$ is called a *charge*. The set function ψ is a *supercharge* if instead of finite additivity only finite superadditivity is required: $\psi(A \cup B) \geq \psi(A) + \psi(B)$ for each pair of disjoint sets A, B. Every supercharge ψ is *monotone*: $A \subseteq B$ implies $\psi(A) \leq \psi(B)$. Also note by induction that if the sets A_i are pairwise disjoint, then

$$\psi\left(\bigcup_{i=1}^{n} A_i\right) \geq \sum_{i=1}^{n} \psi(A_i),$$

and therefore (let $n \to \infty$)

$$\psi\left(\bigcup_{i=1}^{\infty} A_i\right) \geq \sum_{i=1}^{\infty} \psi(A_i).$$

We say that ψ is *countably superadditive*.

A charge ψ is called a *pure charge* iff it does not dominate any nontrivial measure on $\mathcal{A}$: if μ is a measure and $\mu \leq \psi$ on $\mathcal{A}$, then μ vanishes on $\mathcal{A}$. Similarly a *pure supercharge* is one that does not dominate any nontrivial charge. A *partition* of a set A is a collection of disjoint sets (in $\mathcal{A}$) with union A.

(8.1.4) Theorem. *Let ψ be a supercharge defined on an algebra $\mathcal{A}$ of subsets of Ω. Then:*

(1) ψ *admits a unique decomposition*

(8.1.4a) $$\psi = \psi_{\mathrm{m}} + \psi_{\mathrm{c}} + \psi_{\mathrm{s}},$$

  where ψ_{m} is a measure, ψ_{c} is a pure charge, and ψ_{s} is a pure supercharge.

(2) *The measure ψ_{m} is given by*

(8.1.4b) $$\psi_{\mathrm{m}}(A) = \inf_{\mathrm{c}} \sum_{i} \psi(A_i)$$

  where $\inf_{\mathrm{c}}$ denotes the inf over all countable partitions $\{A_1, A_2, \cdots\}$ of A. The charge ψ_{c} is given by

(8.1.4c) $$\psi_{\mathrm{c}}(A) = \inf_{\mathrm{f}} \sum_{i} (\psi - \psi_{\mathrm{m}})(A_i),$$

  where $\inf_{\mathrm{f}}$ denotes the inf over all finite partitions $\{A_1, A_2, \cdots, A_n\}$ of A.

(3) *If ψ is a charge, then ψ_{s} is 0. If ψ dominates a measure φ, then ψ_{m} also dominates φ.*

Proof. Let ψ be a supercharge. We first show that ψ_{m} defined by (8.1.4b) is a measure. Suppose $\{A_1, A_2, \cdots\}$ is a countable partition of a set A. Given $\varepsilon > 0$, choose for each i a countable partition $\{A_{i1}, A_{i2}, \cdots\}$ of A_i such that

$$\sum_k \psi(A_{ik}) \leq \psi_{\mathrm{m}}(A_i) + \varepsilon 2^{-i}.$$

Then $\{A_{ik} : i = 1, 2, \cdots ; k = 1, 2, \cdots\}$ is a countable partition of A. Therefore

$$\psi_{\mathrm{m}}(A) \leq \sum_{i,k} \psi(A_{ik}) \leq \sum_i \left[\psi_{\mathrm{m}}(A_i) + \varepsilon 2^{-i}\right] = \sum_i \psi_{\mathrm{m}}(A_i) + \varepsilon.$$

Since ε is arbitrary, we conclude

$$\psi_{\mathrm{m}}(A) \leq \sum_i \psi(A_i).$$

Next, we prove the reverse inequality. Given $\varepsilon > 0$, choose a countable partition $\{B_1, B_2, \cdots\}$ of A such that

(8.1.4d) $$\sum_k \psi(B_k) \leq \psi_{\mathrm{m}}(A) + \varepsilon.$$

Then

$$\sum_i \psi_{\mathrm{m}}(A_i) \leq \sum_i \left[\sum_k \psi(A_i \cap B_k)\right]$$
$$= \sum_k \left[\sum_i \psi(A_i \cap B_k)\right]$$
$$\leq \sum_k \psi(B_k),$$

since ψ is countably superadditive. Combining this with (8.1.4d), and taking into account the fact that ε is arbitrary, we conclude

$$\psi_{\mathrm{m}}(A) \geq \sum_i \psi_{\mathrm{m}}(A_i).$$

This completes the proof that ψ_{m} is a measure.

The proof that ψ_{c} defined by (8.1.4c) is finitely additive is similar: countable partitions are replaced by finite partitions. The details are omitted.

Next we show that ψ_{c} is a *pure* charge. Suppose μ is a measure on $\mathcal{A}$ and $\mu \leq \psi_{\mathrm{c}}$. Then, from (8.1.4c) we have $\mu \leq \psi - \psi_{\mathrm{m}}$. For each set $A \in \mathcal{A}$,

$$\psi_{\mathrm{m}}(A) = \inf_{\mathrm{c}} \sum_i \psi(A_i) \geq \inf_{\mathrm{c}} \sum_i \left[\psi_{\mathrm{m}}(A_i) + \mu(A_i)\right] = \psi_{\mathrm{m}}(A) + \mu(A),$$

and therefore $\mu(A) = 0$.

The proof that ψ_s is a pure supercharge is similar.

For the uniqueness, suppose that ψ also has the decomposition $\psi = \varphi_m + \varphi_c + \varphi_s$, where φ_m is a measure, φ_c is a pure charge, and φ_s is a pure supercharge. Note that the set function

$$\psi'(A) = \inf_f \sum_i \psi_s(A_i)$$

is a charge dominated by ψ_s, hence it vanishes. Similarly

$$\varphi'(A) = \inf_f \sum_i \varphi_s(A_i)$$

vanishes. Therefore, applying the operation $\inf_f$ to both sides of the equation $\psi_m + \psi_c + \psi_s = \varphi_m + \varphi_c + \varphi_s$, we obtain $\psi_m + \psi_c = \varphi_m + \varphi_c$. Similarly, applying the operation $\inf_c$ to this equation, we obtain $\psi_m = \varphi_m$. Therefore also $\psi_c = \varphi_c$ and $\psi_s = \varphi_s$.

If ψ is a charge, then $\psi_s = 0$ by the uniqueness.

Finally, suppose ψ dominates a measure φ. By the uniqueness, $\varphi_m = \varphi$. Applying the operation $\inf_c$ to the inequality $\psi \geq \varphi$, we have $\psi_m \geq \varphi_m = \varphi$. ∎

If ψ is a supercharge, we will write $M(\psi)$ for the maximal measure dominated by ψ, that is, the measure ψ_m of the theorem. Note that if the set functions ψ_1 and ψ_2 are defined on the same algebra or σ-algebra, then $M(\psi_1 + \psi_2) = M(\psi_1) + M(\psi_2)$. Indeed, if partitions close to the infima are obtained for each set function, then the common refinement of the partitions will be at least as close for each supercharge.

On the other hand, if the set functions are defined on different algebras or σ-algebras, then M is only superadditive. This is so since the infimum in the expression (8.1.4b) may correspond to different sequences of partitions for different supercharges.

(8.1.5) Corollary. *If $\psi_1, \cdots, \psi_n$ are supercharges defined on the same algebra, then $M(\sum_{i=1}^n \psi_i) = \sum_{i=1}^n M(\psi_i)$. Hence a sum of pure charges is a pure charge.*

If θ is a measure-preserving invertible point transformation and $\psi_i = \theta^{-i}\psi$, then the corollary may be applied to conclude $M(\sum_{i=1}^n \theta^{-i}\psi) = \sum_{i=1}^n M(\theta^{-i}\psi)$. Note that if θ is not invertible, the equality may fail, since the supercharges $\theta^{-i}\psi$ are defined on different algebras.

Complements

In complete analogy with weak truncated limits, we can define strong truncated limits, or simply *truncated limits*. That is: if $f_n \geq 0$, then we write $\varphi = \text{TL}\, f_n$ iff, for some weak unit u, $\lim_n \|\varphi_k - (f_n \wedge ku)\|_1 \to 0$ for all k, and $\varphi_k \uparrow \varphi$. For general f_n, write $\text{TL}\, f_n = \text{TL}\, f_n^+ - \text{TL}\, f_n^-$.

(8.1.6) (Weak and strong truncated limits 0 coincide for nonnegative functions.) Suppose $f_n \geq 0$. Then $\mathrm{WTL}\, f_n = 0$ if and only if $\mathrm{TL}\, f_n = 0$. To prove this, integrate the bounded sequence $f_n \wedge ku$.

A sequence (f_n) will be called *TL null* iff $\mathrm{TL}\,|f_n| = 0$. For this it suffices that $\mathrm{w\,lim}\,|f_n| \wedge u = 0$. In σ-finite measure spaces, *stochastic convergence* is defined as convergence in measure on sets of finite measure. Thus in particular, in probability spaces, stochastic convergence is convergence in probability.

(8.1.7) (TL convergence and stochastic convergence.) Let $f_n, \varphi \in L_1$. Then the following are equivalent:

(1) $\mathrm{TL}\, f_n = \varphi$.
(2) $f_n - \varphi$ is TL null.
(3) f_n converges to φ stochastically.

(8.1.8) (Sharpening of Proposition (8.1.1).) If $f_n \in L_1^+$, $\sup_n \|f_n\|_1 < \infty$, then there is a subsequence (f_{k_n}) of (f_n) such that $f_{k_n} = g_n + h_n$, where $g_n, h_n \in L_1^+$, the sequence (g_n) converges weakly, and the h_n have disjoint supports (hence $\mathrm{TL}\, h_n = 0$).

Remarks

The notion of truncated limits is particularly well suited for study of a large class of Banach lattices E, namely those satisfying the following two conditions:

(A) There is a *weak unit*; that is, an element $u \in E^+$ such that if $f \in E^+$ and $f \wedge u = 0$, then $f = 0$.

(B) Every norm bounded increasing sequence in E converges in norm.

Assumptions equivalent with (B) are:

(B$'$) E is weakly sequentially complete;

(B$''$) E contains no subspace isomorphic to c_0.

Condition (B) implies that E is order continuous, hence order intervals are compact and a weak unit exists if E is separable; thus the assumption (A) is not an important loss of generality (see Lindenstrauss & Tzafriri [1979]).

Truncated limits exist in Banach lattices satisfying (A) and (B); the proof is the same as the L_1 proof given here.

The method of truncated limits was developed by Akcoglu & Sucheston [1978], [1983]. The decomposition (8.1.8) of Dacunha-Castelle & Schreiber [1974] could be used instead in the real case, but this decomposition does not extend to Banach lattices satisfying (A) and (B); see Akcoglu & Sucheston [1984a].

Theorem (8.1.4), applied also in Chapter 4, is from Sucheston [1964]. For the purpose of the present chapter, the Yoshida & Hewitt [1953] decomposition in which ψ is assumed to be a charge is sufficient.

8.2. Weak maximal inequalities

There exist important ergodic results for operators acting on L_1. One such result is E. Hopf's maximal ergodic theorem. We will first give a version of Hopf's maximal theorem for lattices.

(8.2.1) Lemma. *Let L be a linear space of measurable functions with values in $(-\infty, \infty)$ such that L is a lattice under pointwise operations, let T be a positive linear operator on L, and let $h \in L$. For $N \in \mathbb{N}$, let*

$$h_N = \max_{1 \le n \le N} \sum_{i=0}^{n-1} T^i h$$

and $B_N = \{h_N > 0\}$. Then

$$h \, \mathbf{1}_{B_N} \ge h_N^+ - T(h_N^+).$$

Proof. Since T is positive, it satisfies $T(f \vee g) \ge Tf \vee Tg$. Thus

$$T(h_N^+) \ge T0 \vee Th_N$$
$$\ge 0 \vee (Th) \vee (Th + T^2 h) \vee \cdots \vee (Th + T^2 h + \cdots + T^N h)$$
$$= h_{N+1} - h \ge h_N - h.$$

For $\omega \in B_N$, we have $h_N(\omega) = h_N^+(\omega)$, so

$$h(\omega) \, \mathbf{1}_{B_N}(\omega) = h(\omega) \ge h_N(\omega) - T(h_N^+)(\omega) = h_N^+(\omega) - T(h_N^+)(\omega).$$

For $\omega \notin B_N$, we have $h_N^+(\omega) = 0$, so

$$h(\omega) \, \mathbf{1}_{B_N}(\omega) = 0 \ge 0 - T(h_N^+)(\omega) = h_N^+(\omega) - T(h_N^+)(\omega). \qquad \blacksquare$$

Let T be a positive operator on L_1. Obviously T extends to the positive cone of positive measurable functions by

$$Tf = \lim \uparrow Tf_n \le \infty$$

if $f_n \in L_1^+$ and $f_n \uparrow f$. (See (8.4.8).) More generally, Tf can be defined for any function such that either f^+ or f^- is integrable.

(8.2.2) Theorem (Hopf's maximal ergodic theorem). *Let T be a positive contraction on L_1, and let h be a measurable extended real-valued function such that h^+ is integrable. For $N \in \mathbb{N}$, let*

$$h_N = \max_{1 \le n \le N} \sum_{i=0}^{n-1} T^i h,$$

$B_N = \{h_N > 0\}$, and $B_\infty = \bigcup_{N=1}^{\infty} B_N$. Then

$$\int_{B_\infty} h \, d\mu \ge 0.$$

Proof. Since T is a positive contraction, we have

$$\int T(h_N^+) \, d\mu = \|T(h_N^+)\|_1 \le \|h_N^+\|_1 = \int h_N^+ \, d\mu.$$

The sequence h_N increases, so the sequence B_N increases. Applying Lemma (8.2.1) and the monotone convergence theorem, we have

$$\int_{B_\infty} h \, d\mu = \lim_{N \to \infty} \int_{B_N} h \, d\mu \ge \liminf_{N \to \infty} \int \left(h_N^+ - T(h_N^+) \right) d\mu \ge 0. \qquad \blacksquare$$

If f and g are functions, we will study the behavior of ratios of the form

$$R_n(f, g) = \frac{\sum_{i=0}^{n-1} T^i f}{\sum_{j=0}^{n-1} T^j g}.$$

For example, if $\mathbf{1}$ is the function identically 1, and $T\mathbf{1} = \mathbf{1}$, then $R_n(f, \mathbf{1})$ is the standard Cesàro mean

$$\frac{1}{n} \sum_{i=0}^{n-1} T^i f.$$

Recall that a σ-finite measure γ on $\mathcal{F}$ is said to be *equivalent* to μ iff $\mu(A) = 0$ holds if and only if $\gamma(A) = 0$. Then the Radon-Nikodým derivative $g = d\gamma/d\mu$ satisfies $0 < g < \infty$ a.e.

(8.2.3) Proposition. *Let f and g be measurable functions. Suppose f is integrable, and $0 < g < \infty$. Write $\gamma = g\mu$. Let $s = \sup_{n \in \mathbb{N}} R_n(f, g)$. Then, for all $\lambda > 0$, we have*

(8.2.3a) $$\gamma\{s > \lambda\} \le \frac{1}{\lambda} \int_{\{s > \lambda\}} \frac{f}{g} \, d\gamma.$$

Proof. Given $\lambda > 0$, let $h = f - \lambda g$. Then the set

$$\{s > \lambda\} = \bigcup_{n=1}^{\infty} \left\{ \sum_{i=0}^{n-1} T^i f > \lambda \sum_{j=0}^{n-1} T^j g \right\}$$

is the set B_∞ of Theorem (8.2.2), so we have

$$\int_{\{s > \lambda\}} (f - \lambda g) \, d\mu \ge 0,$$

and therefore

$$\int_{\{s > \lambda\}} \left(\frac{f}{g} - \lambda \right) d\gamma \ge 0.$$

This yields the required inequality. $\qquad \blacksquare$

The Cesàro averages of the powers of an operator will often be written as follows:

$$A_n = A_n(T) = \frac{1}{n} \sum_{i=0}^{n-1} T^i.$$

(8.2.4) Definition. An operator T is said to be L_p *mean bounded* iff

$$\sup_n \left\| A_n(T) \right\|_p < \infty.$$

Note that L_∞ mean boundedness for a positive operator T can be rewritten: $\sup_n \|A_n\|_\infty \le c$ if and only if $\sup_n A_n \mathbf{1} \le c$. If T satisfies this, and we use $g = \mathbf{1}$ in the proposition, we have $\gamma = \mu$ and

$$\{s > \lambda\} = \bigcup_{n=1}^\infty \left\{ \sum_{i=0}^{n-1} T^i f > \lambda \sum_{j=0}^{n-1} T^j \mathbf{1} \right\} \supseteq \bigcup_{n=1}^\infty \{A_n f > \lambda c\}.$$

Thus, given a nonnegative function f, the maximal function

$$f^* = \sup_n A_n f$$

satisfies the three-function inequality

$$(8.2.4a) \qquad \mu\{f^* > c\lambda\} \le \frac{1}{\lambda} \int_{\{s>\lambda\}} f \, d\mu.$$

(See Section 3.1 for sample consequences of this.)

We will discuss replacing s in this inequality by f. In the next lemma, we use a sequence T_n of positive operators, which will be chosen to be A_n for our application. Recall that R_0 is the heart of the largest Orlicz space $L_1 + L_\infty$. Equivalently, R_0 may be defined as the space of all measurable functions such that

$$f \mathbf{1}_{\{|f| \ge \lambda\}} \in L_1$$

for each $\lambda > 0$ (Proposition (2.2.14)).

(8.2.5) Lemma. *Let T_n be a sequence of positive linear operators on $L_1 + L_\infty$ such that $\sup_n \|T_n\|_\infty = c < \infty$. Assume for each $f \in L_1$ and $\lambda > 0$ that*

$$\mu\{\sup T_n f > c\lambda\} \le \frac{1}{\lambda} \|f\|_1.$$

Then for each $f \in R_0$ and $\lambda > 0$,

$$\mu\{\sup T_n f > 2c\lambda\} \le \frac{1}{\lambda} \int_{\{f>\lambda\}} |f| \, d\mu.$$

Proof. Let $f \in R_0$ be given. For $\lambda > 0$, write $f^\lambda = f \mathbf{1}_{\{|f|>\lambda\}}$. Then we have $|f| \le |f^\lambda| + \lambda$, so $\sup T_n f \le \sup T_n |f^\lambda| + c\lambda$. Therefore

$$\mu\{\sup T_n f > 2c\lambda\} \le \mu\{ \sup T_n(|f^\lambda|) > c\lambda\}$$

$$\le \frac{1}{\lambda} \|f^\lambda\|_1 = \frac{1}{\lambda} \int_{\{|f|>\lambda\}} |f| \, d\mu. \qquad \blacksquare$$

We may now obtain a maximal inequality by applying Lemma (8.2.5) to (8.2.4a).

(8.2.6) Theorem. *Let T be a positive linear operator on $L_1 + L_\infty$ such that $\|T\|_1 \leq 1$ and $\sup_n \|A_n(T)\|_\infty = c < \infty$. Then for each $f \in R_0^+$ and each $\lambda > 0$, the maximal function $f^* = \sup_n A_n f$ satisfies*

$$\mu\{f^* > 2c\lambda\} \leq \frac{1}{\lambda} \int_{\{f > \lambda\}} |f| \, d\mu.$$

The case when $c = 1$ deserves special mention. Then T is a contraction in L_∞ (as well as L_1). The following classical result is the maximal lemma usually used to obtain the Hopf-Dunford-Schwartz ergodic theorem (see, for example, Krengel [1985], p. 51).

(8.2.7) Theorem. *Let T be a positive operator on $L_1 + L_\infty$ that is a contraction on L_1 and on L_∞. Then for $f \in L_1$ and $\lambda > 0$, the maximal function $f^* = \sup_n A_n f$ satisfies*

$$\mu\{f^* > \lambda\} \leq \frac{1}{\lambda} \int_{\{f^* > \lambda\}} |f| \, d\mu.$$

Proof. Hopf's maximal ergodic theorem (8.2.2) can be applied to the function $h = f - \lambda$ since H^+ is integrable. Then $\int_{B_\infty} (f - \lambda) \, d\mu \geq 0$, where

$$B_\infty = \{\sup_N A_N(f - \lambda) > 0\}.$$

But $B_\infty \supseteq \{f^* > \lambda\} \supseteq \{f > \lambda\}$, so $f - \lambda \leq 0$ on $B_\infty \setminus \{f^* > \lambda\}$. Therefore

$$\int_{\{f^* > \lambda\}} (f - \lambda) \, d\mu \geq 0,$$

which yields the required inequality. ∎

Complements

(8.2.8) Theorem (8.2.2) remains true if the linearity of T is replaced by the assumptions

$$T0 = 0,$$
$$T(f + g) \geq Tf + Tg.$$

(8.2.9) (Point transformations.) A *measurable transformation* θ maps Ω to Ω and is such that $\theta^{-1}\mathcal{F} \subseteq \mathcal{F}$. If also $\mu(\theta^{-1}A) = \mu(A)$ for each $A \in \mathcal{F}$, then θ is called *measure preserving* or an *endomorphism*. An endomorphism θ defines an operator T by $Tf = f \circ \theta$. Such an operator is Markovian: It is positive and preserves the integral, $\int f \, d\mu = \int f \circ \theta \, d\mu$.

Remarks

Theorem (8.2.2) is due to E. Hopf [1954]. A great simplification of the proof is due to Garsia [1965].

An extension of Hopf's lemma to the non-linear setting was given by Lin & Wittmann [1991].

8.3. Hopf's decomposition

We will next consider Hopf's decomposition of the measure space Ω into "conservative" and "dissipative" parts.

Let T be a positive operator on L_1. The corresponding *potential* operator T_P is defined by

$$T_P f = \sum_{i=0}^{\infty} T^i f = \lim_{n \to \infty} \sum_{i=0}^{n-1} T^i f.$$

Pointwise convergence (possibly to ∞) of the series holds at least for functions $f \geq 0$. We will obtain the Hopf decomposition of the space Ω into the *conservative part* C and the *dissipative part* D, defined as follows. If $f \in L_1^+$, then $T_P f = 0$ or ∞ on C; and $T_P f < \infty$ on D. We say that T is a *conservative* operator iff $\Omega = C$, and T is a *dissipative* operator iff $\Omega = D$.

(8.3.1) **Theorem.** *Let T be a positive contraction on L_1. Then Ω uniquely decomposes as a disjoint union $C \cup D$ as above.*

Proof. For nonnegative integrable functions f, g, define

$$E(f,g) = \{T_P f = \infty, T_P g < \infty\}.$$

Then $T_P(f - \lambda g) = \infty$ on $E(f,g)$ for all $\lambda > 0$. Applying Hopf's maximal ergodic theorem (8.2.2) to the function $h = f - \lambda g$, we obtain (since $B_\infty \supseteq E(f,g)$):

$$0 \leq \int_{B_\infty} (f - \lambda g)\, d\mu \leq \int_\Omega f\, d\mu - \lambda \int_{E(f,g)} g\, d\mu.$$

Since this is true for all $\lambda > 0$, we have $\int_{E(f,g)} g\, d\mu = 0$.

Now $T_P g = \infty$ if and only if $T_P(T^n g) = \infty$, so $E(f,g) = E(f, T^n g)$. Hence we have $\int_{E(f,g)} T^n g\, d\mu = 0$ for all n. This shows that on $E(f,g)$, we have $T^n g = 0$ for all n. Thus, at each point ω with $T_P f(\omega) = \infty$, we have either $T_P g(\omega) = \infty$ or $T_P g(\omega) = 0$.

Since the measure μ is σ-finite, we may choose an integrable function $f_0 > 0$ a.e. Let $C = \{T_P f_0 = \infty\}$ and $D = \{T_P f_0 < \infty\}$. By the argument just given, with $f = f_0$, we see that C is as required for the conservative part. To see that D is as required for the dissipative part, let $h \geq 0$ be integrable, and apply the previous discussion with $f = h$ and $g = f_0$. At each point where $T_P h = \infty$, we have either $T_P f_0 = \infty$ or $T_P f_0 = 0$. But $T_P f_0 \geq f_0 > 0$, so $T_P h < \infty$ on D. ∎

Below, we will always write C for the conservative part and D for the dissipative part of Ω as given in the theorem.

Clearly, if there is a strictly positive integrable function p such that $Tp = p$, then T is conservative.

Complements

(8.3.2) (Point transformations.) Let T be the Markovian operator defined by an endomorphism θ of a probability space $(\Omega, \mathcal{F}, \mu)$. Then T is conservative, since $T\mathbf{1} = \mathbf{1}$.

Remarks

The decomposition Theorem (8.3.1) for invertible point transformations appears in E. Hopf [1937], and for positive contractions on L_1 in E. Hopf [1954].

8.4. The σ-algebra of absorbing sets

A subset A of Ω is called absorbing if (under T) no mass leaves A. Formally:

(8.4.1) Definition. If $A \subseteq \Omega$, write

$$L_1^+(A) = \{\, f \in L_1 : f \geq 0, f = 0 \text{ outside } A \,\}.$$

A set $A \in \mathcal{F}$ is *absorbing* iff $Tf \in L_1^+(A)$ for all $f \in L_1^+(A)$.

It will be proved that the conservative part C is absorbing. The absorbing subsets of C figure prominently in the identification of the limit in the ergodic theorem (see, for example, (8.5.3)). Therefore it will be useful to give several characterizations of this class. We will write $\mathcal{C}$ for the class of all absorbing subsets of the conservative part C.

The first characterization of $\mathcal{C}$ is in terms of the potential operator T_P.

(8.4.2) Lemma. *A subset A of C is absorbing if and only if it has the form $\{T_P f = \infty\}$ for some $f \in L_1^+$. The conservative part C is absorbing.*

Proof. For $f \in L_1^+$, write $C_f = \{T_P f = \infty\}$. Let A be an absorbing subset of C. Let f be an integrable function, 0 outside A and strictly positive inside A. Then $T_P f$ is 0 outside A, since A is absorbing, and $T_P f$ is positive inside A and hence ∞ since $A \subseteq C$. Therefore $A = C_f$.

Conversely, suppose $A = C_f$. Let $g \in L_1^+$ be 0 on A and strictly positive outside A. Since $T_P f$ is finite outside A, we can write $\Omega \setminus A$ as a disjoint union of sets $G_i = \{ig \leq T_P f < (i+1)g\}$ of finite measure. Now assume (for purposes of contradiction) that A is not absorbing. Then there is $h \in L_1^+(A)$ such that Th is not supported by A. Then for some i, the set $G = G_i$ satisfies $\|\mathbf{1}_G Th\|_1 = a > 0$. For each $k \in \mathbb{N}$, we have $\|\mathbf{1}_G T(kh)\|_1 = ka$. Let f_n be any sequence in L_1^+ such that f_n increases to ∞ on A. Then $\lim_n \|(kh \wedge f_n) - kh\|_1 = 0$, so that

$$\lim_n \left\| T\big((kh \wedge f_n) - kh\big) \right\|_1 = 0,$$

and hence $\lim_n \int_G T(kh \wedge f_n)\,d\mu - \int_G T(kh)\,d\mu = 0$, so $\lim_n \int_G T(kh \wedge f_n)\,d\mu = ka$ and therefore $\lim_n \int_G Tf_n\,d\mu \geq ka$. Thus $\lim_n \int_G Tf_n\,d\mu = \infty$.

We now obtain a contradiction by constructing a sequence f_n such that this relation fails. We know that $T_P f = \infty$ on A, so

$$f_n = \sum_{j=0}^{n-1} T^j f$$

increases to ∞ on A. Then $f_n \leq T_P f$ and $T f_n \leq T_P f$ so we have $\int_G T f_n \, d\mu \leq (i+1) \int_G g \, d\mu < \infty$. This completes the proof that A is absorbing.

Finally, $C = C_f$ for any strictly positive integrable function f, so C is absorbing. ∎

The next characterization of C will involve the adjoint T^* of the operator T. Since T is an operator on L_1, the adjoint T^* is an operator on L_∞ characterized by the *duality relation*

$$\int T f \cdot g \, d\mu = \int f \cdot T^* g \, d\mu \qquad f \in L_1, g \in L_\infty.$$

Since we are assuming that T is a positive operator, it follows that T^* is also a positive operator. We are also assuming that T is a contraction in L_1, so T^* is a contraction in L_∞. Thus $T^* 1 \leq 1$.

Now the operator T^* is monotonely continuous, that is, if $f_n \uparrow f \in L_\infty$, then $T^* f_n \uparrow T^* f$. (See (8.4.8).) Therefore, T^* (like T) extends uniquely to the convex cone of finite, nonnegative measurable functions, and the extension retains the monotone continuity (8.4.9). We will write T_P^* for the potential operator defined from the adjoint operator T^*.

(8.4.3) Lemma. *Let $A \in C$, and let $h \in L_\infty$ be nonnegative on A. Then $T_P^* h$ has only values 0 and ∞ on A. In particular, $T_P^* h = 0$ or ∞ on C for every function $h \in L_\infty$ such that $h \geq 0$ on C.*

Proof. Using the duality relation we see that if A is absorbing for T, then $\Omega \setminus A$ is absorbing for T^*. Therefore $1_A \cdot T^* h = 1_A \cdot T^*(h \cdot 1_A)$, and by induction it follows that $T^{*k} h \geq 0$ on A for all $k \in \mathbb{N}$. If the lemma fails, then there exists a positive number b and a set $B \subseteq A$ such that $0 < T_P^* h < b$ on B and $0 < \mu(B) < \infty$. Then for each k we have

$$\infty > b \cdot \mu(B) \geq \int \sum_{i=k}^{\infty} T^{*i} h \, 1_B \, d\mu = \int T^{*k} h \cdot \sum_{i=0}^{\infty} T^i 1_B \geq 0,$$

which implies that $T^{*k} h = 0$ on B. It follows that $T_P^* h = 0$ on B, which is a contradiction. ∎

(8.4.4) Lemma. *Let $A \in \mathcal{C}$ and let g be a finite, positive, measurable function. If $T^*g \le g$ on A or $T^*g \ge g$ on A, then $T^*g = g$ on A.*

Proof. Suppose $T^*g \le g$ on A, and assume first that $g \in L_\infty^+$. Set $h = g - T^*g$. Then $\sum_{i=0}^{n-1} T^{*i}h = g - T^{*n}g \le g < \infty$. By Lemma (8.4.3), we have $T_P^* h = 0$ on A, so $h = 0$ on A. That is, $g = T^*g$ on A.

Now suppose g is unbounded and $T^*g \le g$ on A. For a constant λ, let $g' = g \wedge \lambda$. Then $T^*g' \le T^*g \wedge T^*\lambda \le g \wedge \lambda = g'$ on A. By the previous argument, $T^*g' = g'$ on A. If we let $\lambda \uparrow \infty$, we will obtain by the monotone continuity of T^* that $T^*g = g$.

If $T^*g \ge g$ on A, proceed in a similar manner, using $h = T^*g' - g'$ and the fact that $T^{*n}g'$ is bounded in L_∞. ∎

The following theorem states the main results about $\mathcal{C}$.

(8.4.5) Theorem.

(1) *The class $\mathcal{C}$ of absorbing subsets of C is a σ-algebra of subsets of C.*

(2) *The class $\mathcal{C}$ is the class of all sets of the form $C_f = \{T_P f = \infty\}$, where $f \in L_1^+$.*

(3) *The class $\mathcal{C}$ is the class of all subsets A of C such that $T^*\mathbf{1}_A = \mathbf{1}_A$ on C.*

(4) *A nonnegative measurable function on C is $\mathcal{C}$-measurable if and only if $T^*h = h$ on C.*

(5) *A function $h \in L_\infty(C)$ is $\mathcal{C}$-measurable if and only if $T^*h = h$ on C.*

Proof. Let $\mathcal{I}$ denote the class of subsets A of C such that $T^*\mathbf{1}_A = \mathbf{1}_A$ on C.

First, $T^*\mathbf{1} \le \mathbf{1}$, so $T^*\mathbf{1}_C \le \mathbf{1}_C$ on C, hence by Lemma (8.4.4), we have $T^*\mathbf{1}_C = \mathbf{1}_C$ on C. More generally, the inequality $T^*\mathbf{1}_A \le \mathbf{1}_A$ holds on A, so if $A \in \mathcal{C}$, we have $T^*\mathbf{1}_A = \mathbf{1}_A$ on A.

Let $A \in \mathcal{I}$. If $B = C \setminus A$, then taking differences and using $T^*\mathbf{1}_C = \mathbf{1}_C$ on C, we obtain that $T^*\mathbf{1}_B = \mathbf{1}_B$ on A. But $T^*\mathbf{1}_B \le \mathbf{1}_B$ also holds on B, hence $T^*\mathbf{1}_B \le \mathbf{1}_B$ on C. Subtracting this from $T^*\mathbf{1}_C = \mathbf{1}_C$, we obtain $T^*\mathbf{1}_B = \mathbf{1}_B$ on C. Thus $\mathcal{I}$ is closed under complementation. Thus $\mathcal{C} \subseteq \mathcal{I}$.

If $A, B \in \mathcal{I}$, then on C we have

$$\mathbf{1}_A + \mathbf{1}_B = T^*(\mathbf{1}_A + \mathbf{1}_B) = T^*(\mathbf{1}_{A \cap B} + \mathbf{1}_{A \cup B})$$
$$= T^*\mathbf{1}_{A \cap B} + T^*\mathbf{1}_{A \cup B} \le \mathbf{1}_{A \cap B} + \mathbf{1}_{A \cup B} = \mathbf{1}_A + \mathbf{1}_B.$$

So we have equality, and therefore $T^*\mathbf{1}_{A \cap B} = \mathbf{1}_{A \cap B}$ and $T^*\mathbf{1}_{A \cup B} = \mathbf{1}_{A \cup B}$ on C. This shows that $\mathcal{I}$ is an algebra.

The operator T^* is monotonely continuous, so $\mathcal{I}$ is closed under increasing unions. Therefore $\mathcal{I}$ is a σ-algebra.

Suppose $A \in \mathcal{I}$. Then also $B = C \setminus A \in \mathcal{I}$. So $T^*\mathbf{1}_B = \mathbf{1}_B$. Now if $f \in L_1^+(A)$, then

$$\int_B Tf \, d\mu = \int T^*\mathbf{1}_B \cdot f \, d\mu = \int \mathbf{1}_B \cdot f \, d\mu = 0,$$

so Tf is supported in A. Therefore $A \in C$. We have shown that $\mathcal{I} \subseteq C$, and hence that $C = \mathcal{I}$.

Let H denote the convex cone of finite, positive, measurable functions on C such that $T^*h = h$ on C. Observe that H is closed under infimum (and supremum). Indeed, $T^*(h \wedge h') \leq T^*h \wedge T^*h' = h \wedge h'$ on C, so by Lemma (8.4.4), $T^*(h \wedge h') = h \wedge h'$ on C. Now we claim that a positive, finite, measurable function h belongs to H if and only if h is C-measurable. Indeed, if h is C-measurable, then it is the limit of an increasing sequence of linear combinations of indicator functions of sets in C, so $h \in H$ by the monotone continuity of T^*. Conversely, since $1_{\{h > a\}}$ is the limit of the increasing sequence

$$1 \wedge n(h - a)^+,$$

we see that if $h \in H$, then all sets of the form $\{h > a\}$ are in C, so h is C-measurable.

Finally, assume that h is in $L_\infty(C)$. If h is C-measurable, then so are h^+ and h^-, so $T^*h = h$ on C by the preceding part. Conversely, if $T^*h = h$ on C, then

$$T^*(h^+) = T^*h + T^*(h^-) \geq T^*h = h^+$$

on the support of h^+, hence $T^*(h^+) \geq h^+$ on C. Again by Lemma (8.4.4), it follows that $T^*(h^+) = h^+$ on C. Thus h^+ is C-measurable. Similarly, h^- is C-measurable. ∎

(8.4.6) Definition. Let T be a positive linear contraction on L_1, and let $A \in C$. The operator *induced* by A on T is the operator T_A on $L_1(A)$ defined by

$$T_A f = T(f \cdot 1_A).$$

The induced operator T_A is conservative and Markovian (that is, it preserves integrals). The powers of T_A are given by

$$T_A{}^k f = T^k(f \cdot 1_A).$$

The adjoint $T_A{}^*$ defined on $L_\infty(A)$ satisfies

$$T_A{}^{*k}(h) = T^{*k}(h \cdot 1_A)$$

on A. The σ-algebra of absorbing sets of T_A is $A \cap C$. All these statements follow from the preceding material, using the fact that A is absorbing.

The operator T acts like the conditional expectation in that C-measurable functions may be factored out. Thus the appearance of $\mathbf{E}^C$ in the limit theorems for T is not surprising.

(8.4.7) Proposition. *Let* $f \in L_1(C)$ *and let* $h \in L_\infty(C)$ *be* C-*measurable. Then* $T(h \cdot f) = h \cdot T(f)$.

Proof. Since T is monotonely continuous and linear, it is enough to consider the case $h = 1_A$, where $A \in C$. Since A is absorbing, $T(1_A \cdot f) = 1_A \cdot T(1_A \cdot f)$. Now $B = C \setminus A$ is also absorbing, so $0 = 1_A \cdot T(1_B \cdot f)$. Adding, we get $T(1_A \cdot f) = 1_A \cdot Tf$. ∎

Complements

(8.4.8) (Monotone continuity) We write $f_n \uparrow f$ if f_n is a sequence of functions, $f_n(\omega) \leq f_{n+1}(\omega)$ for all n and almost all ω, and $f(\omega) = \lim_n f_n(\omega)$ for almost all ω. A similar definition can be given for $f_n \downarrow f$. A positive operator T is *monotonely continuous* on the space E of functions iff: $f_n \uparrow f \in E$ implies $Tf_n \uparrow Tf$. By considering the functions $f - f_n$, we see that monotone continuity can by characterized by: $f_n \downarrow 0$ implies $Tf_n \downarrow 0$.

Every positive operator on L_p is monotonely continuous for $1 \leq p < \infty$, and every adjoint positive operator on L_∞ is monotonely continuous. The proofs are similar. Consider an adjoint operator T on L_∞, say $T = S^*$, where S is an operator on L_1. If $f_n \downarrow 0$ in L_∞, then Tf_n decreases to some nonnegative limit h. We must show $h = 0$ a.e. If $g \in L_1$, then

$$\int g \cdot h \, d\mu = \lim_n \int g \cdot Tf_n \, d\mu = \lim \int Sg \cdot f_n \, d\mu = 0.$$

This holds for all $g \in L_1$, so $h = 0$ a.e. (The relevant property of L_p is "order continuity." See Lemma (9.1.1).)

There is a converse in the case of L_∞: If $S \colon L_\infty \to L_\infty$ is monotonely continuous, then $S = T^*$ for some $T \colon L_1 \to L_1$.

(8.4.9) (Operator extension.) A positive operator defined on L_p can be extended, preserving monotone continuity, to the convex cone of all nonnegative measurable functions (see also Neveu [1965a], p. 188).

(8.4.10) (Continuity.) Every positive operator on a Banach lattice is automatically norm continuous. See (9.1.1), below (or Lindenstrauss & Tzafriri [1979], p. 2).

(8.4.11) (Point transformations.) Let T be the Markovian operator defined by an endomorphism θ of a probability space $(\Omega, \mathcal{F}, \mu)$. A set A is absorbing if and only if $A = \theta^{-1}A$ a.e. So in this case, C is the σ-algebra of *invariant sets*.

(8.4.12) (Monotone continuity.) In an order-continuous Banach lattice E, if $f_n \uparrow f \in E$, then $\|f_n - f\| \to 0$ (see Section 9.1 or Lindenstrauss & Tzafriri [1979], Proposition 1.a.8). Thus positive operators are monotonely continuous. For example, positive operators on L_p spaces $1 \leq p < \infty$ are monotonely continuous.

Remarks

Study of the σ-algebra C goes back to E. Hopf [1954].

8.5. The Chacon-Ornstein theorem (conservative case)

This famous theorem asserts the convergence of ratios of the form

$$R_n = R_n(f,g) = \frac{\sum_{i=0}^{n-1} T^i f}{\sum_{i=0}^{n-1} T^i g}$$

for $f, g \in L_1$, with $g > 0$. The case where T is conservative will be treated first. The nonconservative case will be done later (8.6.10) in the framework of superadditive processes.

As is customary in ergodic proofs, we establish convergence for larger and larger classes of functions. If f is of the form $s - Ts$, where $s \in L_1$, then the numerator "telescopes," and we have

$$R_n(f,g) = \frac{s}{\sum_{i=0}^{n-1} T^i g} - \frac{T^n s}{\sum_{i=0}^{n-1} T^i g}.$$

The first term clearly converges to 0 on C. We will show that the second term converges to 0 on Ω.

(8.5.1) Lemma. Let $s \in L_1^+$, $g \in L_1$, $g > 0$. Then

$$\frac{T^n s}{\sum_{i=0}^{n-1} T^i g}$$

converges to 0 a.e.

Proof. Fix $\varepsilon > 0$, let $h_n = T^n s - \varepsilon \sum_{i=0}^{n-1} T^i g$, $h_0 = s$, and $A_n = \{h_n > 0\}$. Now $\gamma = g\mu$ is a finite measure equivalent to μ. We claim that $\sum_n \gamma(A_n) < \infty$. To see this, first compute

$$\int_{A_{n+1}} T h_n \, d\mu \leq \int_{A_{n+1}} T(h_n^+) \, d\mu \leq \int_\Omega T(h_n^+) \, d\mu$$

$$\leq \int_\Omega h_n^+ \, d\mu = \int_{A_n} h_n \, d\mu.$$

Note that $\varepsilon g = T h_n - h_{n+1}$, so that

$$\varepsilon \int_{A_{n+1}} g \, d\mu = \int_{A_{n+1}} T h_n \, d\mu - \int_{A_{n+1}} h_{n+1} \, d\mu$$

$$\leq \int_{A_n} h_n \, d\mu - \int_{A_{n+1}} h_{n+1} \, d\mu.$$

The sum telescopes:

$$\sum_{n=1}^\infty \gamma(A_n) = \sum_{n=0}^\infty \int_{A_{n+1}} g \, d\mu \leq \frac{1}{\varepsilon} \int_{A_0} h_0 \, d\mu < \infty.$$

Therefore, by Borel-Cantelli we have $\gamma(\limsup A_n) = 0$, and therefore we have $\mu(\limsup A_n) = 0$. This shows that the ratio $T^n s / \sum_{i=0}^{n-1} T^i g$ converges to 0 a.e. ∎

We will need an approximation lemma. Functions that are close in L_1 norm have similar pointwise ratio behavior. This will allow us to derive the ratio theorem on L_1 from that on L_g. More precisely, we have:

(8.5.2) Lemma. *Let $g \in L_1$, $g > 0$ a.e., and let $e_k \in L_1$, $k \in \mathbb{N}$, satisfy $\sum_k \|e_k\|_1 < \infty$. Then $\lim_k \sup_n R_n(e_k, g) = 0$.*

Proof. Write, as usual, $\gamma = g\mu$. Fix $\lambda > 0$. Now apply Proposition (8.2.3). If $s_k = \sup_n R_n(|e_k|, g)$, then

$$\lambda \gamma\{s_k > \lambda\} \leq \int_{\{s_k > \lambda\}} \frac{|e_k|}{g} \, d\gamma \leq \int_\Omega |e_k| \, d\mu = \|e_k\|_1.$$

Therefore $\sum_k \gamma\{s_k > \lambda\} < \infty$. By Borel-Cantelli, we have $s_k \leq \lambda$ except for finitely many k, or $\limsup_k s_k \leq \lambda$. Since $\lambda > 0$ was arbitrary, we have $\lim_k s_k = 0$. Since $|R_n(e_k, g)| \leq R_n(|e_k|, g)$, we have the required result. ∎

We will now assume that the operator T is conservative ($C = \Omega$). Since T is sub-Markovian, we have $\|T^*\|_\infty \leq 1$, hence $T^*\mathbf{1} \leq \mathbf{1}$. But T is conservative, so this means that $T^*\mathbf{1} = \mathbf{1}$; thus T preserves the integral, that is, T is Markovian.

The σ-algebra C is used to identify the limit of $R_n(f, g)$. (See Section 2.3 for a discussion of conditional expectations on infinite measure spaces.)

(8.5.3) Proposition. *Let T be a conservative Markovian operator. Let $f, g \in L_1$ with $g > 0$ a.e. Let $\gamma = g\mu$ be a finite equivalent measure. Then the ratio $R_n(f, g)$ converges a.e. to the finite limit*

$$R(f, g) = \mathbf{E}_\gamma^C \left[\frac{f}{g} \right].$$

Proof. We may assume $f \geq 0$. If f is of the form $h \cdot g$, where $h \in L_\infty$ and $T^*h = h$, then by Proposition (8.4.7), we have $T^i(h \cdot g) = h \cdot T^i g$, so clearly $R_n(f, g) = h$. But also, since h is C-measurable, $\mathbf{E}_\gamma^C \left[fg^{-1} \right] = \mathbf{E}_\gamma^C [h] = h$.

If f is of the form $s - Ts$ where $s \in L_1$, then by Lemma (8.5.2), we have $R_n(f, g) \to 0$ a.e. But also for $A \in C$, we have $\int_A (f/g) \, d\gamma = \int \mathbf{1}_A \cdot f \, d\mu = \int \mathbf{1}_A \cdot (s - Ts) \, d\mu = \int (\mathbf{1}_A - T^*\mathbf{1}_A) \cdot s \, d\mu = 0$, so that $\mathbf{E}_\gamma^C \left[fg^{-1} \right] = 0$.

Thus the theorem is true for all f in the linear space

$$E_g = \{ h \cdot g + s - Ts : h \in L_\infty, T^*h = h, s \in L_1 \}.$$

We claim that E_g is dense in L_1. By the Hahn-Banach theorem (5.1.2), it suffices to prove: if $k \in L_\infty$ and $\int k \cdot f \, d\mu = 0$ for all $f \in E_g$, then $k = 0$ a.e. So suppose k is such a function. Then we have, for all $s \in L_1$,

$$\int (k - T^*k) \cdot s \, d\mu = \int k \cdot (s - Ts) \, d\mu = 0,$$

so that $T^*k = k$. Thus we have $k \cdot g \in E_g$, so $0 = \int k \cdot k \cdot g \, d\mu$, so that $k = 0$ since $g > 0$. This completes the proof that E_g is dense in L_1.

Finally, let f be a general element of L_1. Choose a sequence $f_k \in E_g$ with $\sum_{k=1}^{\infty} \|f - f_k\|_1 < \infty$. Write $e_k = f - f_k$ and $s_k = \sup_n R_n(|e_k|, g)$. Consider the inequality

$$|R_n(f, g) - R(f, g)| \leq |R_n(f_k, g) - R(f_k, g)| + s_k + \mathbf{E}_\gamma^{\mathcal{C}}\left[\frac{|e_k|}{g}\right].$$

The first term on the right converges to 0 for each fixed k as $n \to \infty$ because $f_k \in E_g$. The second term converges to 0 as $k \to \infty$ by Lemma (8.5.2). The third term converges to 0 as $k \to \infty$ since

$$\sum_k \int \mathbf{E}_\gamma^{\mathcal{C}}\left[\frac{|e_k|}{g}\right] d\gamma = \sum_k \|e_k\|_1 < \infty. \qquad \blacksquare$$

The formulas for the conditional expectation from Section 2.3 may be applied to this limit $R(f, g)$. For example, we have

$$R(f, g) = \frac{\mathbf{E}_\mu^{\mathcal{C}}[f]}{\mathbf{E}_\mu^{\mathcal{C}}[g]}$$

whenever the conditional expectations with respect to μ exist, in particular if μ is finite.

Let us consider next what can be done if $g \in L_1^+$, but $g = 0$ on a set of positive measure. Then $g\mu$ is no longer equivalent to μ. But there is still some finite measure equivalent to μ, say $\rho = r\mu$. The isomorphism between $L_1(\Omega, \mathcal{F}, \mu)$ and $L_1(\Omega, \mathcal{F}, \rho)$ maps T to the operator T_ρ defined by

$$T_\rho f = r^{-1} T(fr) \qquad f \in L_1(\rho).$$

Clearly we have $T_\rho^k f = r^{-1} T^k(fr)$, $R_n(T_\rho)(f, g) = R_n(T)(fr, gr)$, and T_ρ is conservative if T is. The σ-algebra of absorbing sets is the same for T and T_ρ. The spaces $L_\infty(\rho)$ and $L_\infty(\mu)$ are the same and the adjoint operators T_ρ^* and T^* are the same, because

$$\int f \cdot T_\rho^* g \, d\rho = \int (T_\rho f) \cdot g \, d\rho$$

$$= \int T(fr) \cdot g \, d\mu$$

$$= \int fr \cdot T^* g \, d\mu$$

$$= \int f \cdot T^* g \, d\rho.$$

Now if $f, g \in L_1$ and $g > 0$, then $f/r, g/r \in L_1(\rho)$, and the preceding results applied to the operator T_ρ give that

$$\lim_n R_n(f, g) = \frac{\mathbf{E}_\rho^{\mathcal{C}}[fr^{-1}]}{\mathbf{E}_\rho^{\mathcal{C}}[gr^{-1}]}.$$

It follows from the convergence of $R_n(f,g)$ (and can be checked directly) that the expression on the right does not depend on the finite equivalent measure ρ chosen.

The case when $g = 0$ on a set of finite measure can now be discussed. It is senseless to consider the limit $\lim_n R_n(f,g)$ if $R_n(f,g)$ is defined for no n because of 0 denominator. But the convergence can be proved on the set G where $R_n(f,g)$ is defined for sufficiently large n. Notation: if B is a measurable set, we will write $A \in B \cap C$ iff $A = B \cap C$ for some $C \in C$, or (equivalently) $A \subseteq B$ and $A \in C$.

(8.5.4) Proposition (conservative Chacon-Ornstein). *Let T be conservative, $f \in L_1$, $g \in L_1^+$. Then the ratio*

$$R_n(f,g) = \frac{\sum_{i=0}^{n-1} T^i f}{\sum_{i=0}^{n-1} T^i g}$$

converges to a finite limit $R(f,g)$ a.e. on the set $G = \{T_P g > 0\}$. The limit $R(f,g)$ is C-measurable, and satisfies $\int_A g \cdot R(f,g)\,d\mu = \int_A f\,d\mu$ for all $A \in G \cap C$. If $\rho = r\mu$ is any probability measure equivalent to μ, then

$$R(f,g) = \frac{\mathbf{E}_\rho^C\left[fr^{-1}\right]}{\mathbf{E}_\rho^C\left[gr^{-1}\right]}$$

a.e. on G.

Proof. We claim first that the support F of $\mathbf{E}_\rho^C[g]$ is $G = \{T_P g > 0\}$. By Theorem (8.4.5), $G \in C$. Since g is supported in G, so is $\mathbf{E}_\rho^C[g]$. Thus $F \subseteq G$. Conversely, since $G \setminus F \in C$, for each k,

$$0 = \int_{G \setminus F} \mathbf{E}_\rho^C[g]\,d\rho = \int_{G \setminus F} T^k g\,d\rho,$$

so $\rho(G \setminus F) = 0$.

By considering the induced operator T_G as in Definition (8.4.6), we may assume $G = \Omega$. Now

$$R_n(f,g) = \frac{R_n(f, \mathbf{E}_\rho^C[g])}{R_n(g, \mathbf{E}_\rho^C[g])}.$$

By Proposition (8.5.3), the numerator converges to $\mathbf{E}_\rho^C\left[fr^{-1}\right] / \mathbf{E}_\rho^C\left[gr^{-1}\right]$ and the denominator converges to 1. ∎

The ratio $R_n(f,g)$ for $f \in L_1$, $g \geq 0$, still converges to a finite limit on the set $\{T_P g > 0\}$, if T is not necessarily conservative and g is not necessarily integrable. These generalizations are treated in the framework of superadditive processes in Section 8.6.

Complements

(8.5.5) (Point transformations.) Theorem (8.5.3), in the special case of an endomorphism of a probability space, with $g = 1$, is called Birkhoff's theorem, and is essentially the oldest pointwise ergodic result (G. D. Birkhoff [1931]). The limit in Birkhoff's theorem is $\mathbf{E}^{\mathcal{C}}[f]$, where $\mathcal{C}$ is the σ-algebra of invariant sets.

Remarks

Theorem (8.5.4) is due to Chacon & Ornstein [1960]. The general (nonconservative) case will be proved below (8.6.10). The use of the Hahn-Banach theorem in the proof of the conservative case of the Chacon-Ornstein theorem originated with Neveu; see Neveu [1964] or Garsia [1970].

8.6. Superadditive processes

Let T be a sub-Markovian operator. The sequence (s_n) of measurable functions is called a *superadditive process* iff

$$(S_1) \qquad\qquad s_{k+n} \geq s_k + T^k s_n \qquad k, n \geq 0$$

and

$$(S_2) \qquad\qquad \gamma = \sup_n \frac{\int s_n \, d\mu}{n} < \infty.$$

The number γ in (S_2) is called the *time constant* of the process. A sequence (s_n) is called *extended superadditive* iff (S_1) is satisfied. A sequence (s_n) is *subadditive* iff $(-s_n)$ is superadditive; and *additive* iff it is both superadditive and subadditive. Note that (s_n) is additive if and only if it has the form

$$s_n = \sum_{i=0}^{n-1} T^i f$$

for some function f.

If T is Markovian, i.e., it preserves the integral, then $x_n = \int s_n \, d\mu$ is a numerical superadditive sequence, i.e., $x_{n+k} \geq x_k + x_n$, and the time constant $\gamma = \sup_{n \in \mathbb{N}} n^{-1} \int s_n \, d\mu$ is also given by $\gamma = \lim_{n \to \infty} n^{-1} \int s_n \, d\mu$ (see 8.6.13).

The theory of processes subadditive with respect to an operator induced by a measure preserving point transformation was initiated by J.F.C. Kingman in 1968, who gave important applications to probability. Here, the operator superadditive theory is mainly developed because of the light it sheds on operator ergodic theorems.

Mathematically, the superadditive and the subadditive cases are equivalent, but the first one is slightly simpler to treat for the following reason.

An arbitrary superadditive process (s_n) obviously dominates the additive process

$$\sum_{i=0}^{n-1} T^i s_1;$$

the difference is a positive superadditive processes which can be studied in the pleasing context of positive operators acting on positive functions. By the same argument, if the additive theorem is known, we can restrict our attention to the *positive* superadditive processes.

Dominants

We will compare a positive superadditive process to an additive process that dominates it but barely. A *dominant* of a positive superadditive process (s_n) is an L_1^+ function δ such $\sum_{i=0}^{n-1} T^i \delta \geq s_n$ for all n. An *exact dominant* of a positive superadditive process (s_n) with time constant γ is a dominant δ such that $\int \delta \, d\mu = \gamma$. We will show that such a δ exists in some cases .

We will use the functions

$$\varphi_m = \frac{1}{m} \sum_{i=1}^{m} (s_i - T s_{i-1}).$$

(8.6.1) Lemma. Let φ_m be as above. Then

$$\sum_{i=0}^{n-1} T^i \varphi_m \geq \left(1 - \frac{n-1}{m}\right) s_n$$

for $1 \leq n < m$.

Proof. First, $s_i \geq s_1 + T s_{i-1}$, so each term $s_i - T s_{i-1}$ is nonnegative. Next we have

$$\sum_{i=0}^{n-1} T^i (s_j - T s_{j-1}) = \sum_{i=0}^{n-1} T^i s_j - \sum_{i=1}^{n} T^i s_{j-1}$$

$$= (I - T^n) s_j + \sum_{i=1}^{n} T^i (s_j - s_{j-1}),$$

so that (remembering $s_0 = 0$)

$$m \sum_{i=0}^{n-1} T^i \varphi_m = (I - T^n) \sum_{j=1}^{m} s_j + \sum_{i=1}^{n} T^i s_m$$

$$= \sum_{i=0}^{n-1} s_i + s_n + \sum_{i=1}^{m-n} (s_{n+i} - T^n s_i) + \sum_{i=1}^{n-1} (T^i s_m - T^n s_{i+m-n}).$$

The first sum in the last line is nonnegative. Since $s_{n+i} - T^n s_i \geq s_n$, the second sum is $\geq (m-n)s_n$. Also,

$$T^i s_m - T^n s_{i+m-n} = T^i(s_m - T^{n-i}s_{i+m-n}) \geq T^i s_{n-i} \geq 0,$$

so the last sum is nonnegative. Therefore

$$m \sum_{i=0}^{n-1} T^i \varphi_m \geq [1 + (m-n)]s_n.$$

Now divide by m. ∎

(8.6.2) Proposition. *Suppose T is a Markovian operator, and (s_n) is a positive superadditive process. Then (s_n) admits an exact dominant.*

Proof. For this proof, we use the weak truncated limits (see Section 8.1). Define φ_m as before. Since T is Markovian,

$$\int \varphi_m \, d\mu = \frac{1}{m} \sum_{i=1}^{m} \left(\int s_i \, d\mu - \int s_{i-1} \, d\mu \right) = \frac{1}{m} \int s_m \, d\mu \leq \gamma.$$

Taking a subsequence of φ_m, still denoted by φ_m, and using a diagonal procedure, we may assume that the weak truncated limit $\lambda_i = \text{WTL}_m T^i \varphi_m$ exists for each i. Then by the additivity of weak truncated limits (8.1.2), $\text{WTL}_m \sum_{i=0}^{n-1} T^i \varphi_m$ exists and equals $\sum_{i=0}^{n-1} \lambda_i$. By Lemma 8.6.1,

(8.6.2a) $$\sum_{i=0}^{n-1} \lambda_i \geq s_n \qquad n \geq 1.$$

By the Fatou property of the weak truncated limits (T WTL $\leq$ WTL T, see 8.1.3), we have $T\lambda_{i-1} \leq \lambda_i$, and, more generally, $T^n \lambda_i \leq \lambda_{i+n}$. Hence we can write

$$\lambda_i = (\lambda_i - T\lambda_{i-1}) + T(\lambda_{i-1} - T\lambda_{i-2}) + \cdots + T^{i-1}(\lambda_1 - T\lambda_0) + T^i \lambda_0,$$

with all the summands positive. Therefore

$$\gamma \geq \int \lambda_i \, d\mu$$
$$= \int \left[(\lambda_i - T\lambda_{i-1}) + (\lambda_{i-1} - T\lambda_{i-2}) + \cdots + (\lambda_1 - T\lambda_0) + \lambda_0 \right] d\mu.$$

Define

(8.6.2b) $$\delta = \lambda_0 + \sum_{i=1}^{\infty} (\lambda_i - T\lambda_{i-1})$$
$$= \sum_{i=0}^{\infty} (\lambda_i - T\lambda_i).$$

Then $\int \delta\, d\mu \le \gamma$, and

$$\sum_{j=0}^{n-1} T^j \delta = (I - T^n) \sum_{i=0}^{\infty} \lambda_i$$

$$\ge \sum_{i=0}^{\infty} \lambda_i - \sum_{i=n}^{\infty} \lambda_i$$

$$= \sum_{i=0}^{n-1} \lambda_i \ge s_n$$

by (8.6.2a). This shows that $n \int \delta\, d\mu \ge \int s_n\, d\mu$ for each n, hence $\int \delta\, d\mu \ge \gamma$. Thus $\int \delta d\mu = \gamma$. This shows that δ is an exact dominant. ∎

A positive superadditive process (shifted) can be approximated also from below by additive processes. Note that a positive superadditive process is increasing: $s_{k+n} \ge s_k + T^k s_n \ge s_k$.

(8.6.3) Lemma. *Let s_k be a positive superadditive process, and let $a_k = s_k/k$. Then*

$$\sum_{i=0}^{n-1} T^i a_k \le s_{n+k-1}$$

for all $n \ge 0$.

Proof. We have

$$k \sum_{i=0}^{n-1} T^i a_k = \sum_{i=0}^{n-1} T^i s_k$$

$$\le \sum_{i=0}^{n-1} (s_{k+i} - s_i)$$

$$\le \sum_{i=0}^{n+k-1} s_i - \sum_{i=0}^{n-1} s_i \le k s_{n+k-1},$$

where the last inequality follows from the fact that s_n is increasing. ∎

Superadditive ratio theorem

(8.6.4)   Theorem. *Let T be a conservative operator, and let s_n be a positive superadditive process with an exact dominant δ. Let $C_\delta = \{\sum_{i=0}^{\infty} T^i \delta > 0\}$, and define $\sigma(A) = \int_A \delta\, d\mu$. Then*

(8.6.4a) $$\lim_n \frac{s_n}{\sum_{i=0}^{n-1} T^i \delta} = 1 \qquad \text{on } C_\delta$$

and

(8.6.4b) $\lim \dfrac{1}{n} \displaystyle\int_A s_n \, d\mu = \sup \dfrac{1}{n} \displaystyle\int_A s_n \, d\mu = \sigma(A)$ for $A \in C_\delta \cap C$.

Proof. Let $f, g \in L_1^+$. Set $C_g = \{\sum_{i=0}^\infty T^i g > 0\}$. From Proposition 8.5.4 we have that $\lim R_n(f, g) = R(f, g)$ where $R(f, g)$ is C-measurable and

(8.6.4c) $\displaystyle\int_A R(f, g) \cdot g \, d\mu = \int_A f \, d\mu$

for $A \in C_g \cap C$. Now $T^* 1_A = 1_A$, so the right-hand side does not change if f is replaced by $T^k f$. Therefore, for each k, we have $R(f, g) = R(T^k f, g)$. Furthermore, $C_{T^k g} = C_g$ and $R(f, g) = R(f, t^k g)$. So

$$R(f, g) = \lim_n \frac{\sum_{i=0}^{n-1} T^i f}{\sum_{i=k}^{n+k-1} T^i g} = \lim_n \frac{\sum_{i=0}^{n-1} T^i f}{\sum_{i=0}^{n+k-1} T^i g}.$$

So we have

(8.6.4d) $$R(f, g) = \lim_n \frac{\sum_{i=0}^{n-k} T^i f}{\sum_{i=0}^{n-1} T^i g}.$$

By Lemma 8.6.3 and the definition of a dominant, if $a_k = s_k/k$, then

(8.6.4e) $$\sum_{i=0}^{n-k} T^i a_k \le s_n \le \sum_{i=0}^{n-1} T^i \delta.$$

Write

$$\overline{R} = \limsup \frac{s_n}{\sum_{i=0}^{n-1} T^i \delta}$$

$$\underline{R} = \liminf \frac{s_n}{\sum_{i=0}^{n-1} T^i \delta}.$$

Then by (8.6.4d) and (8.6.4e) with $g = \delta$, we have on C_δ

$$R(a_k, \delta) \le \underline{R} \le \overline{R} \le R(\delta, \delta) = 1.$$

Again by (8.6.4c), for $A \in C_\delta \cap C$,

$$\int_A a_k \, d\mu = \int_A R(a_k, \delta)\delta \, d\mu \le \int_A \underline{R}\delta \, d\mu \le \int_A \overline{R}\delta \, d\mu \le \int_A \delta \, d\mu.$$

But $\lim \int_{C_\delta} a_k \, d\mu = \gamma = \int_{C_\delta} \delta \, d\mu$. Hence $\underline{R} = \overline{R}$ on C_δ. Also, $\left(\int_A s_k \, d\mu \right)$ is a superadditive sequence of numbers, so (8.6.16) it converges to

$$\sup_k \int_A a_k \, d\mu = \int_A \delta \, d\mu$$

since $1_A \delta$ is an exact dominant of $(1_A s_n)$. This implies (8.6.4b).   ∎

We now consider the general non-conservative case. For an extended superadditive process (s_n') it will be useful to consider the notion of an *extended exact dominant*.

(8.6.5) Definition. Let (s'_n) be an extended superadditive process. A measurable function δ' with values in $[-\infty, \infty]$ is an *extended exact dominant* for s'_n iff $\sum_{i=0}^{n-1} T^i \delta' \geq s'_n$ for all n and, for all $A \in \mathcal{C}$,

$$\int_A \delta' \, d\mu = \sigma'(A),$$

where $\sigma'(A)$ is the (finite or infinite) limit

$$\sigma'(A) = \lim_n \frac{1}{n} \int_A s'_n \, d\mu = \sup_n \frac{1}{n} \int_A s'_n \, d\mu.$$

The existence of extended exact dominants can be established by applying the results about exact dominants to processes of the form $1_A s'_n$ where $A \in \mathcal{C}$ and $\sigma'(A) < \infty$. (See also the proof of Theorem 8.6.7, below.)

We now show that an extended superadditive process divided by n can be integrated term by term. (Of course, this is false without superadditivity.)

(8.6.6) Lemma. *Let ψ_n be a sequence of measurable functions with ψ_n^- integrable. Suppose*

(8.6.6a) $\psi_k + \psi_n \leq \psi_{k+n}$ *for $k, n \geq 0$.*

If E is a conditional expectation operator, or an operator $\int_A$, then

$$-\infty < \lim_n \frac{1}{n} E \psi_n = E\left(\lim_n \frac{1}{n} \psi_n\right) \leq \infty.$$

Proof. We have by induction using (8.6.6a) that $\psi_1 \leq \psi_n/n$. Therefore ψ_n/n is bounded below by the integrable function $-\psi_1^-$, so that Fatou's lemma applies. Using (8.6.13), we have

$$-\infty < E\psi_1 \leq E\left(\lim \frac{1}{n}\psi_n\right)$$

$$\leq \lim E\left(\frac{1}{n}\psi_n\right) = E\left(\sup_n \frac{1}{n}\psi_n\right)$$

$$= E\left(\lim_n \frac{1}{n}\psi_n\right) \leq \infty.$$

∎

We will now consider the general non-conservative case. Here is the main theorem. Recall: C is the *conservative part*, D the *dissipative part*; T is *Markovian* iff it preserves the integral.

(8.6.7) **Theorem (Akcoglu-Sucheston).** *Let T be a sub-Markovian operator, s_n a positive superadditive process, and s'_n a positive extended superadditive process. Let $E = \{\sup_n s'_n > 0\}$. Then $\lim s_n/s'_n = h$ exists a.e. on $C \cap E$ and is C-measurable. Let δ [respectively, δ'] be an exact [extended exact] dominant of the process $s_n \mathbf{1}_C$ [$s'_n \mathbf{1}_C$] with respect to the Markovian operator T_C. Let ρ be a probability measure equivalent to μ, say $\rho = r\mu$. Set*

$$s = \sup_n \frac{1}{n} \mathbf{E}^{\mathcal{C}}_\rho [s_n \mathbf{1}_C/r]$$

$$s' = \sup_n \frac{1}{n} \mathbf{E}^{\mathcal{C}}_\rho [s'_n \mathbf{1}_C/r].$$

The following limits exist a.e. on C:

(8.6.7a)
$$\lim_n \frac{1}{n} \mathbf{E}^{\mathcal{C}}_\rho [s_n \mathbf{1}_C/r] = s = \mathbf{E}^{\mathcal{C}}_\rho [\delta/r] < \infty,$$
$$\lim_n \frac{1}{n} \mathbf{E}^{\mathcal{C}}_\rho [s'_n \mathbf{1}_C/r] = s' = \mathbf{E}^{\mathcal{C}}_\rho [\delta'/r] \leq \infty.$$

Also, $\lim s_n/s'_n = s/s'$ a.e. on $C \cap E$. If either T is Markovian or s_n is additive on D, then $\lim s_n/s'_n = \lim \uparrow s_n/\lim \uparrow s'_n$ a.e. on $D \cap E$.

Proof. To simplify the notation, we will assume that μ is a probability and $r = 1$. To obtain the general case, replace μ by ρ and $\mathbf{E}^{\mathcal{C}} [\cdot]$ by $\mathbf{E}^{\mathcal{C}}_\rho [\cdot \, r^{-1}]$. Since s_n and s'_n are increasing sequences, $\lim s_n/s'_n = \lim \uparrow s_n/\lim \uparrow s'_n$ exist a.e. on $D \cap E$ if $\lim s_n < \infty$ on D. This is true in the Markovian case because of the existence of a dominant for s_n (see Proposition 8.6.2). If s_n is additive on D, then it is of the form $\sum_{i=0}^{n-1} T^i \delta$ with δ integrable, hence $\lim s_n < \infty$ on D.

Now consider C. If (s_n) is (extended) superadditive, then so is $(s_n \mathbf{1}_A)$ for each absorbing set A, both with respect to T and T_A, because

$$s_{k+n} \mathbf{1}_A \geq s_k \mathbf{1}_A + (T^k s_n) \mathbf{1}_A \geq s_k \mathbf{1}_A + T^k (s_n \mathbf{1}_A).$$

Positive linear operators on an L_p space (or an Orlicz space) can be uniquely extended to the set L of all measurable functions f from Ω to $(-\infty, \infty]$ such that f^- is integrable (8.4.9). The extension preserves their monotone continuity. This is, in particular, true for the sub-Markovian operator T, and for the conditional expectation $\mathbf{E}^{\mathcal{C}} [\cdot]$. Since the restriction to a set $A \in \mathcal{C}$ of an (extended) superadditive process is (extended) superadditive, choosing $A = C$, we may consider the operator T_C instead of T, and thus assume that $\Omega = C$.

If $A \in \mathcal{C}$, then $T^* \mathbf{1}_A = \mathbf{1}_A$ implies that $\int_A f \, d\mu = \int_A Tf \, d\mu$ and also $\mathbf{E}^{\mathcal{C}} [f] = \mathbf{E}^{\mathcal{C}} [Tf]$. Therefore, applying Lemma 8.6.6, first with $E = \int_A$, then with $E = \mathbf{E}^{\mathcal{C}}$, we obtain, for each $A \in \mathcal{C}$,

(8.6.7b)
$$\lim_n \frac{1}{n} \int_A s'_n \, d\mu = \sup_n \frac{1}{n} \int_A s'_n \, d\mu = \sigma'(A) \leq \infty$$

and also the relations (8.6.7a) hold (recall that $\mu = \rho$ and $r = 1$). Comparing with 8.6.7a and 8.6.4b, we determine that $\sigma = s\mu$ and $\sigma' = s'\mu$ on C. Thus $s = \mathbf{E}^C[\delta]$. Let $F_i = \{0 \le s' < i\}$, $F = \bigcup_{i=1}^{\infty} F_i$, $G = \Omega \setminus F$. Let δ' be the exact dominant of s'_n on F_i with respect to the operator T_{F_i}; then $s' = \mathbf{E}^C[\delta']$ on F_i. Since

$$\frac{s_n}{s'_n} = \frac{s_n}{\sum_{i=0}^{n-1} T^i\delta} \cdot \frac{\sum_{i=0}^{n-1} T^i\delta}{\sum_{i=0}^{n-1} T^i\delta'} \cdot \frac{\sum_{i=0}^{n-1} T^i\delta'}{s'_n},$$

Theorems 8.6.4 shows that s_n/s'_n converges to s/s' on $F_i \cap E$, hence on $F \cap E$.

It remains to consider $G' = G \cap E$. Let $a'_k = s'_k/k$. By Theorem 8.6.4, (8.6.4d), and the primed version of (8.6.4e), we have on G'

$$\lim_n \frac{s_n}{s'_n} \le \lim_n \frac{\sum_{i=0}^{n-1} T^i\delta}{\sum_{i=0}^{n-k} T^i a'_k} = \frac{\mathbf{E}^C[\delta]}{\mathbf{E}^C[a'_k]} \to 0$$

as $k \to \infty$. Hence the limit on G' is 0. The definition of δ' can be extended to G' so that δ' is an extended exact dominant and $s' = \mathbf{E}^C[\delta']$, finite or infinite. ∎

It should be noted that s_n/s'_n need not converge on $D \cap E$ if s_n is not additive on D and T is not Markovian: for example, consider $T = 0$. On the other hand, we prove next that these assumptions can be weakened to $\liminf \int_D \varphi_m \, d\mu < \infty$, where, as before, $\varphi_m = (1/m) \sum_{i=1}^{m}(s_i - Ts_{i-1})$. Indeed, this condition is sufficient for the existence of an exact dominant for s_n, and therefore implies the convergence of s_n/s'_n. In the proof, truncated limits are not used, but the argument is less elementary, since it involves the decomposition of set functions.

(8.6.8)   Theorem. *Let T be a sub-Markovian operator. Let (s_n) be a positive superadditive process. Write $\varphi_m = (1/m) \sum_{i=1}^{m}(s_i - Ts_{i-1})$, and let D be the dissipative part of Ω. Then the limits $\lim_m \int_D \varphi_m \, d\mu$ and $\lim_m \int \varphi_m \, d\mu$ exist, finite or infinite. The following conditions are equivalent:*

   (1) $\lim_m \int_D \varphi_m \, d\mu < \infty$.
   (2) $\lim_m \int \varphi_m \, d\mu < \infty$.
   (3) *There is a dominant.*
   (4) *There is an exact dominant.*

Proof. Recall that T is Markovian on the conservative part C and $\mathbf{1}_C s_n$ is superadditive. Thus

$$\int_C \varphi_m \, d\mu = \frac{1}{m} \int_C s_m \, d\mu \to \sup_m \frac{1}{m} \int_C s_m \, d\mu \le \gamma < \infty.$$

This shows that (1) and (2) are equivalent. (4) $\Longrightarrow$ (3) is trivial.

(2) $\Longrightarrow$ (4). Assume that $\beta = \liminf \int \varphi_m \, d\mu < \infty$. Then there is a subsequence, still denoted φ_m, such that $\lim_m \int \varphi_m \, d\mu = \beta$. Considering the φ_m as elements of L_1^{**}, we may assume that φ_m converges weak* to an element ψ_0 of L_1^{**}, in symbols $\varphi_m \stackrel{*}{\longrightarrow} \psi_0$. Observe that this implies

$$T\varphi_m = T^{**}\varphi_m \stackrel{*}{\longrightarrow} T^{**}\psi_0.$$

Elements of L_1^{**} may be considered to be (signed) charges that vanish on all sets of μ-measure 0. Since all of the φ_m are positive, so is the limit ψ_0. Considering L_1 a subset of L_1^{**} in the usual way, we have by Lemma 8.6.1 $\psi_0 \geq s_1$. Let $\eta_0 = M(\psi_0)$ be the maximal countably additive measure dominated by ψ_0 (see 8.1.4). Then also $\eta_0 \geq s_1\mu$, because the pure charge $\pi_0 = \psi_0 - \eta_0$ does not dominate any measure. The second adjoint operator T^{**} maps L_1 into itself. Let

$$T^{**}\pi_0 = \eta_1 + \pi_1$$

be the decomposition of $T^{**}\pi_0$ into a measure and a pure charge. Then $\eta_0 + T\eta_0 + \eta_1 \geq s_2\mu$. Indeed, applying (8.6.1) with $n = 2$, we have

$$\varphi_m + T\varphi_m \geq \left(1 + \frac{1}{m}\right) s_2,$$

hence $\psi_0 + T^{**}\psi_0 \geq s_2\mu$ and

$$\eta_0 + \pi_0 + T^{**}(\eta_0 + \pi_0) = \eta_0 + \pi_0 + T\eta_0 + \eta_1 + \pi_1 \geq s_2\mu.$$

But π_1 and π_2 are pure charges, and a sum of pure charges is a pure charge, and so does not dominate any measure; so we have

$$\eta_0 + T\eta_0 + \eta_1 \geq s_2\mu.$$

In general, given π_n, let $\eta_{n+1} = M(T^{**}\pi_n)$, and $\pi_{n+1} = T^{**}\pi_n - \eta_{n+1}$. Finally, set $\eta = \sum_{i=0}^{\infty} \eta_i$. Then

$$\sum_{i=0}^{n-1} T^i\eta \geq \sum_{i=0}^{n-1} T^i\eta_0 + \sum_{i=0}^{n-2} T^i\eta_1 + \cdots + \eta_{n-1} \geq s_n\mu.$$

It follows that $\eta(\Omega) = \gamma$, so that $\delta = d\eta/d\mu$ is an exact dominant.

(3) $\Longrightarrow$ (2). Now assume there is a dominant δ. Then

$$0 \leq \int (I - T) \left(\sum_{i=0}^{k-1} T^i\delta - s_k\right) d\mu$$

$$= \int \delta \, d\mu - \int (I - T)s_k \, d\mu - \int T^k\delta \, d\mu.$$

Hence

$$\int \delta \, d\mu \geq \int (I - T) s_k \, d\mu + \int T^k \delta \, d\mu$$
$$= \int \left[(s_k - T s_{k-1}) - T(s_k - s_{k-1}) + T^k \delta \right] d\mu.$$

Take the Cesàro averages on the right.

$$\int \delta \, d\mu \geq \int \left(\varphi_n - \frac{1}{n} T s_n + \frac{1}{n} \sum_{i=1}^n T^i \delta \right) d\mu.$$

Now δ is a dominant and T is positive, so we conclude that $\int \delta \, d\mu \geq \int \varphi_n \, d\mu$. Proceeding now as above, we obtain an *exact* dominant δ' such that $\int \delta' \, d\mu = \liminf \int \varphi_n \, d\mu$. Since δ' is a dominant, $\sup \int \varphi_n \, d\mu \leq \int \delta' \, d\mu$ implies that $\lim \int \varphi_n \, d\mu$ exists. ∎

Ergodic theorems

In the ergodic theorems below, s_n is often of the form

$$\mathbf{1}_C \sum_{i=0}^{n-1} T^i f,$$

the restriction of an additive process to the conservative part C. In such a case, we can obtain an explicit determination of the exact dominant δ appearing in the identification of the limit in terms of the successive contributions of the dissipative part D to the conservative part C, by considering the operator H defined by

$$Hf = \mathbf{1}_C \sum_{i=0}^{\infty} (T I_D)^i f,$$

where the operator I_D is defined by $I_D(f) = \mathbf{1}_D \cdot f$.

The operator H is sub-Markovian since $\gamma = \sup(1/n) \int s_n \, d\mu \leq \int f \, d\mu$, and extends to the set L of measurable functions with integrable positive part in the usual way. Also, $\mathbf{E}_\rho^C \left[H f r^{-1} \right]$ can be defined for $f \in L$.

(8.6.9) Proposition. *Suppose s_n is a positive (extended) superadditive process of the form*

$$\mathbf{1}_C \sum_{i=0}^{n-1} T^i f,$$

with $f \geq 0$. Then Hf is an (extended) exact dominant of s_n.

Proof. It suffices to consider the superadditive case. We have

$$s_n = \sum_{i=0}^{n} T^i(f\,\mathbf{1}_C) + \sum_{i=0}^{n-2} T^i\big(\mathbf{1}_C T(f\,\mathbf{1}_D)\big) + \cdots$$
$$+ (I+T)\big(\mathbf{1}_C(TI_D)^{n-2}f\big) + \mathbf{1}_C(TI_D)^{n-1}f$$
$$\leq \sum_{i=0}^{n-1} T^i Hf.$$

So Hf is a dominant.

If we define operators $H^{(n)}$ by

$$H^{(n)}f = \mathbf{1}_C \sum_{i=0}^{n-1} (TI_D)^i f,$$

then Hf is the limit of $H^{(n)}f$. Now T is Markovian on the conservative part C, so

$$\int s_n\,d\mu = n\int_C f\,d\mu + (n-1)\int_C T(f\,\mathbf{1}_D)\,d\mu + \cdots$$
$$+ 2\int_C (TI_D)^{n-2}f\,d\mu + \int_C (TI_D)^{n-1}f\,d\mu$$
$$= \int_C \sum_{i=0}^{n-1}(TI_D)^i f\,d\mu + \int_C \sum_{i=0}^{n-2}(TI_D)^i f\,d\mu + \cdots + \int_C f\,d\mu$$
$$= \int H^{(n)}f\,d\mu + \int H^{(n-1)}f\,d\mu + \cdots + \int H^{(0)}f\,d\mu.$$

This shows that $(1/n)\int s_n\,d\mu$ is the Cesàro average of $\int H^{(n)}f\,d\mu$, and hence converges to $\lim_n \int H^{(n)}f\,d\mu = \int Hf\,d\mu$. Thus $\int Hf\,d\mu = \gamma$, so the dominant Hf is exact. ∎

We now give some ergodic consequences of the superadditive ratio theorem. The first application is the non-conservative case of the Chacon-Ornstein Theorem.

(8.6.10) Theorem (Chacon-Ornstein). *Let T be a sub-Markovian operator, let $f \in L_1^+$, and let $g \geq 0$. Then*

$$R_n(f,g) = \frac{\sum_{i=0}^{n-1} T^i f}{\sum_{i=0}^{n-1} T^i g}$$

converges to a finite limit h a.e. on the set $E = \{T_P g > 0\}$. Let $\rho = r\mu$ be a probability measure equivalent to μ. The limit h is equal to

$$\frac{\mathbf{E}_\rho^C\left[\frac{Hf}{r}\right]}{\mathbf{E}_\rho^C\left[\frac{Hg}{r}\right]}$$

a.e. on $C \cap E$ and equal to

$$\frac{\sum_{i=0}^{\infty} T^i f}{\sum_{i=0}^{\infty} T^i g}$$

a.e. on $D \cap E$.

Proof. Use 8.6.7 and 8.6.9 with $s_n = \sum_{i=0}^{n-1} T^i f$ and $s_n' = \sum_{i=0}^{n-1} T^i g$. ∎

Next we prove the Dunford-Schwartz Theorem for positive operators. That the supremum of $A_n f$ in the Dunford-Schwartz theorem is finite, follows also from 8.2.6, above.

(8.6.11) Theorem (Dunford-Schwartz). *Let T be a sub-Markovian operator which also satisfies $T\mathbf{1} \leq 1$. If $f \in R_0^+$, then*

$$A_n f = \frac{1}{n} \sum_{i=0}^{n-1} T^i f$$

converges a.e. to a finite limit h. On the conservative part C,

$$h = \frac{\mathbf{E}_\rho^C\left[\frac{Hf}{r}\right]}{\mathbf{E}_\rho^C\left[\frac{1_C}{r}\right]}.$$

On the dissipative part D,

$$h = \frac{\sum_{i=0}^{\infty} T^i f}{\sum_{i=0}^{\infty} T^i \mathbf{1}}.$$

Proof. For each $f \in R_0$, and each constant $\lambda > 0$, there is a function $f^\lambda \in L_1$ such that $|f - f^\lambda| \leq \lambda$. Since $|A_n \lambda| \leq \lambda$ and λ is arbitrarily small, we may assume $f \in L_1$.

Now $s_n = \sum_{i=0}^{n-1} T^i f$ is additive, while $s_n' = n\mathbf{1}$ is extended superadditive, because $k + n \geq k + T(n\mathbf{1})$. We have $s' = \lim_n (1/n)\mathbf{E}_\rho^C\left[1_{Cn} 1r^{-1}\right] = \mathbf{E}_\rho^C\left[1_C r^{-1}\right]$. (See also the identification of the limit in Theorem 8.6.12, below.) ∎

Our last ergodic theorem is Chacon's theorem for positive operators. A sequence (p_i) of positive measurable functions will be called *admissible* iff $Tp_i \leq p_{i+1}$ holds for all $i \geq 0$. For example, (p_i) is admissible if $p_i = T^i g$ for some positive g. Also, if $T\mathbf{1} \leq 1$, then the sequence $p_i = \mathbf{1}$ is admissible. These two examples show that both the Chacon-Ornstein Theorem 8.6.10 and the Dunford-Schwartz Theorem 8.6.11 are consequences of the following.

(8.6.12) Theorem (Chacon). *Let T be a sub-Markovian operator, let $f \in L_1^+$, and let (p_i) be admissible. Then*

$$\frac{\sum_{i=0}^{n-1} T^i f}{\sum_{i=0}^{n-1} p_i}$$

converges to a finite limit h a.e. on the set $E = \{\sum_{i=0}^{\infty} p_i > 0\}$. Let $s' = \lim_n \mathbf{E}_\rho^C \left[\mathbf{1}_C p_n r^{-1} \right]$. The limit h is equal to

$$\frac{\mathbf{E}_\rho^C \left[\frac{Hf}{r} \right]}{s'}$$

a.e. on the set $C \cap E$, and equal to

$$\frac{\sum_{i=0}^{\infty} T^i f}{\sum_{i=0}^{\infty} p_i}$$

a.e. on $D \cap E$.

Proof. Since $Tp_i \leq p_{i+1}$ for each i, we have by induction $T^k p_i \leq p_{i+k}$. Thus

$$\sum_{i=k}^{n+k-1} p_i \geq T^k \sum_{i=0}^{n-1} p_i.$$

Hence $s'_n = \sum_{i=0}^{n-1} p_i$ is extended superadditive. Set $s_n = \sum_{i=0}^{n-1} T^i f$. Now Theorems 8.6.7 and 8.6.9 may be applied. The result follows, except that s' is the limit of the Cesàro averages of $\mathbf{E}^C \left[\mathbf{1}_C p_n \right]$, rather than $\lim \mathbf{E}^C \left[\mathbf{1}_C p_n \right]$ itself. (As before, we assume μ is a probability measure and $r = 1$.) However, $\mathbf{E}^C \left[\mathbf{1}_C p_n \right] = \mathbf{E}^C \left[T_C p_n \right] \leq \mathbf{E}^C \left[\mathbf{1}_C p_{n+1} \right]$, so that the sequence $\mathbf{E}^C \left[\mathbf{1}_C p_n \right]$ is increasing, and the two expressions for s' coincide. ∎

The observation that $\mathbf{E}_\rho^C \left[\mathbf{1}_C p_n r^{-1} \right]$ must be increasing if p_n is admissible indicates that superadditive processes are much more general than processes that are sums of admissible sequences.

Complements

(8.6.13) (Numerical superadditive sequences.) Suppose $x_n \in (-\infty, \infty]$ satisfies $x_{k+n} \geq x_k + x_n$. Then

$$\lim_{n \to \infty} \frac{x_n}{n} = \sup_n \frac{x_n}{n}.$$

To see this, write $\gamma = \sup_n x_n/n$ (finite or infinite). Fix a positive integer d. Each n can be written with quotient k_n and remainder r_n, where $n = k_n d + r_n$ and $1 \leq r_n \leq d$. Then $x_n \geq x_{k_n d} + x_{r_n}$. But $x_{k_n d} \geq k_n x_d$, so

$$\frac{x_n}{n} \geq \frac{(k_n d) x_d/d + x_{r_n}}{n}.$$

Now $k_n d/n \to 1$ and $x_{r_n}/n \to 0$, so we have

$$\liminf_n \frac{x_n}{n} \geq \frac{x_d}{d}.$$

This is true for all d, so $\liminf x_n/n \geq \gamma$. The inequality $\limsup x_n/n \leq \gamma$ is clear.

(8.6.14) The operator H used in 8.6.9 may be generalized as follows. If $A \in \mathcal{F}$, define the operator I_A by $I_A(f) = 1_A \cdot f$. Define H_A by

$$H_A f = 1_A \sum_{i=0}^{\infty} (T I_{\Omega \setminus A})^i f.$$

If $A \in \mathcal{C}$, then Proposition 8.6.9 remains valid with the operator H_A in place of H. However, the operator H_A is sub-Markovian even if A is not absorbing. Hint: Define $H_A^{(n)}$ in a manner analogous to $H^{(n)}$, and show by induction on n that

$$H_A^{(n)} = (I_A + T I_{\Omega \setminus A})^n - (T I_{\Omega \setminus A})^n.$$

Therefore $H_A^{(n)}$ is dominated by powers of the positive contraction

$$I_A + T I_{\Omega \setminus A}.$$

(8.6.15) (Convergence of Cesàro averages $A_n f$.) Suppose there is a positive g such that $\{g > 0\} = C$ and $Tg \leq g$ on C. Then $(1/n) \sum_{i=0}^{n-1} T^i f$ converges a.e. on Ω for each $f \in L_1^+$. Indeed, the process $s_n' = 1_C n g$ is extended superadditive with respect to the operator T_C, so that the ratio $\sum_{i=0}^{n-1} T^i f / n g$ converges a.e. on C. But on D, it is clear that $\sum_{i=0}^{n-1} T^i f / n$ converges to 0.

(8.6.16) For $f \in L_1^+$, set $s_n = \liminf_r \sum_{i=r}^{n+r-1} T^i f$. Then s_n is superadditive, so Theorem 8.6.7 is applicable. This is also true if $\liminf$ is replaced by $\inf$.

(8.6.17) (Subadditive theorem for measure preserving point transformations.) An example of a positive subadditive process to which Theorem 8.6.7 can be applied is

$$s_n = \max_{\varepsilon_i = \pm 1} \left\| \sum_{i=0}^{n-1} \varepsilon_i f \circ \theta^i \right\|,$$

where f is a Bochner integrable function with values in a Banach space, θ is a measure-preserving point transformation on a σ-finite measure space, and the max is taken over all measurable choices $\varepsilon_i(\omega)$ of signs $+1, -1$.

(8.6.18) (More on the superadditive theorem for measure-preserving point transformations.) Let θ be a conservative measure preserving point transformation and p be a number with $0 < p < 1$. Let f be a nonnegative measurable function such that $\int f^p \, d\mu < \infty$. Then, for any positive function g,

$$\lim_n \frac{\left(\sum_{i=0}^{n-1} f \circ \theta^i\right)^p}{\sum_{i=0}^{n-1} g \circ \theta^i} = 0$$

almost everywhere on $\{T_P g > 0\}$. (Akcoglu & Sucheston [1984].) If the measure μ is finite, $g = 1$, and $f \circ \theta^i$ is an independent sequence, then this result is due to Marcinkiewicz. (See also 6.1.15.) The ergodic result remains true if the pth power is replaced by a positive subadditive function φ

(8.6.19) (Markov kernels.) A *Markov transition kernel* is a function $P(\omega, A)$, such that:

    (1) $P(\omega, \cdot)$ is a probability measure defined on $\mathcal{F}$ for each fixed $\omega \in \Omega$.
    (2) $P(\cdot, A)$ is a measurable function on Ω for each fixed $A \in \mathcal{F}$.

The kernel P is *null-preserving* iff $\mu(A) = 0$ implies $P(\omega, A) = 0$ for almost all ω. Such a kernel P defines a Markovian operator T via the Radon-Nikodým isomorphism as follows: If φ is a measure on $\mathcal{F}$, define the measure $T\varphi$ by

$$T\varphi(A) = \int P(\omega, A) \, \varphi(d\omega).$$

Since P is null-preserving, if φ is absolutely continuous, so is $T\varphi$, so T defines an operator from L_1 to L_1. The adjoint operator T^* on L_∞ satisfies

$$T^* h(\omega) = \int h(y) \, P(\omega, dy).$$

An important particular case is when $P(\omega, A) = \mathbf{1}_{\theta^{-1}(A)}(\omega)$, where θ is a measurable invertible point transformation that maps null sets on null sets. Then the Chacon-Ornstein theorem is essentially the ergodic theorem of W. Hurewicz [1944].

    The connections between ergodic theory and Markov processes are studied in Revuz [1974].

(8.6.20) Recall that a sub-Markovian operator is a positive linear contraction on L_1. There is an example, due to Chacon [1964], of a sub-Markovian conservative ($\Omega = C$) operator T such that the Cesàro averages $A_n f = (1/n) \sum_{i=0}^{n-1} T_i f$ diverge a.e. for each non null $f \in L_1^+$. This example also shows that there need not exist a strictly positive function g such that $Tg \leq g$, since the existence of such a g implies convergence of $A_n f$ (see 8.6.15). In fact, in Chacon's example, T is generated by an invertible non-singular (preserving null sets) point transformation θ by

$$T\varphi = \theta^{-1}\varphi, \qquad \varphi = f\mu,$$

which shows that Cesàro convergence may fail also in the setting of the Hurewicz theorem. In the Hurewicz setting, the nonexistence of a g such that $g > 0$ and $Tg \leq g$ (equivalently, $Tg = g$) also provides a negative solution to the famous problem of the existence of a "σ-finite equivalent invariant measure"— this name is given to $\gamma = g\mu$—earlier resolved by D. S. Ornstein [1960].

Even if $A_n f$ converges to a finite limit, as is the case if $T1 \leq 1$, if μ is infinite this limit is often zero (8.6.21), so that the Chacon-Ornstein theorem remains of interest, informing about the relative behavior of sums of iterates of two functions f and g such that $\lim A_n f = \lim A_n g = 0$.

(8.6.21) (The ergodic case.) A sub-Markovian conservative operator T is called *ergodic* iff the σ-algebra $\mathcal{C}$ is trivial, that is, for every element $A \in \mathcal{C}$, either $\mu(A) = 0$ or $\mu(\Omega \setminus A) = 0$. If T is ergodic, $f \in L_1^+$, and $g \geq 0$, then the limit $R(f, g)$ in the Chacon-Ornstein theorem is $\mathbf{E}\left[f\right]/\mathbf{E}\left[g\right]$ a.e. on Ω. Indeed, we observed (following (8.5.3)) that for integrable g the limit in the conservative case is $\mathbf{E}^{\mathcal{C}}\left[f\right]/\mathbf{E}^{\mathcal{C}}\left[g\right]$. The case of general $g \geq 0$ can be reduced to the integrable case by taking integrable positive g_j's with $g_j \uparrow g$ and observing that $\mathbf{E}\left[g_j\right] \uparrow \infty$ and $\mathbf{E}\left[f\right]/\mathbf{E}\left[g_j\right] \geq \mathbf{E}\left[f\right]/\mathbf{E}\left[g\right]$. If T is ergodic, $T1 \leq 1$, $\mu(\Omega) = \infty$ and $f \in L_1^+$, then the limit in the Dunford-Schwartz theorem (8.6.11) is $\lim (1/n) \sum_{i=0}^{n-1} T^i f \leq R(f, 1) = 0$.

(8.6.22) (The ratio theorem if T is not a contraction.) If T is a positive L_1 operator that is power-bounded (i.e., such that $\sup_n \|T^n\|_1 < \infty$), then the space Ω uniquely decomposes into parts Y and Z characterized as follows: if $f \in L_1^+(Y)$, then $\liminf \|T^n f\|_1 > 0$; if $f \in L_1^+(Z)$, then $\|T^n f\|_1 \to 0$. If $f \in L_1^+$ and $g > 0$, then the limit $R(f, g)$ exists a.e. on Y; in general it fails to exist on Z. (This is proved in Sucheston [1967]. Further results about the decomposition $Y + Z$ were obtained by A. Ionescu Tulcea & M. Moretz [1969]; and by Y. Derriennic & M. Lin [1973].)

Remarks

The pioneering paper of E. Hopf [1954] gave the first L_1 operator theorem: Theorem (8.6.11) under the assumption that $T1 = 1$. Theorem (8.6.10) was conjectured by Hopf and proved by Chacon & Ornstein [1960]. The identification of the limit is due independently to Neveu [1961] and Chacon [1962]. The Chacon-Ornstein theorem for measure-preserving transformations is due to E. Hopf [1937]. The identification of the limit is facilitated by the fact that not only C, but also D is absorbing (J. Feldman [1962]). Theorem (8.6.7) and the derivation from it of the general Chacon-Ornstein and Chacon theorems is from Akcoglu & Sucheston [1978]. Theorem (8.6.8) is from Brunel & Sucheston [1979]. The identification (8.6.9) of the operator Hf as an exact dominant is new. The identification of the limit in Chacon's theorem (8.6.12) is due to U. Krengel [1985], p. 130.

The proof via (8.6.7) and (8.6.9) of various ergodic results including Kingman's theorem and Chacon's theorem together with identification of the limit is by far the shortest. However, other approaches may provide additional information. We mention two powerful methods, both presented in detail in Krengel [1985]. The *filling scheme*, the original method of Chacon-Ornstein, was widely applied in ergodic theory and probability. The maximal lemma of Brunel [1963] provides a proof of

Chacon-Ornstein that connects with potential theory; it was applied to identify the limit by P. A. Meyer [1965].

The finite measure case of the superadditive theorem for measure-preserving point transformations is due to the pioneering paper of J. F. C. Kingman [1968]. There exist striking applications, in particular to percolation theory and to the limiting behavior of random matrices (Kingman [1973] and [1976], Kesten [1982]).

9

Multiparameter processes

We present here[*] a unified approach to most of the multiparameter martingale and ergodic theory. In one parameter, the existence of common formulations and proofs is a well known old problem; see e.g. J. L. Doob [1953], p. 342. It has been known that the passage from weak to strong maximal inequalities can be done by a general argument applicable to harmonic analysis, ergodic theory, and martingale theory. In this book a very general such approach is presented in Chapter 3, involving Orlicz spaces and their hearts. There exists also a simple unified (martingales + ergodic theorems) passage from one to many parameters using no multiparameter maximal theorems, based on a general argument valid for order-continuous Banach lattices. This approach gives a unified short proof of many known theorems, namely multiparameter versions of theorems of Doob (Cairoli's theorem [1970] in stronger form, not assuming independence), theorems of Dunford & Schwartz [1956] and Fava [1972], the multiparameter point-transformation case having been earlier proved by Zygmund [1951] and Dunford [1951]. We also obtain multiparameter versions of theorems of Akcoglu [1975], Stein [1961], Rota [1962]. For the Banach lattice argument, the order continuity is needed, which means that the $L \log^k L$ spaces are not acceptable if the measure is infinite: they fail this property and have to be replaced by their hearts, subspaces H_Φ which are closures of simple integrable functions (see Chapter 2).

We will first develop in detail the "multiparameter principle" (Theorem (9.1.3)) that allows the reduction of multiparameter convergence problems to one parameter. There is also a one-sided version of this result, useful to prove "demiconvergence" in many parameters. Also a Banach-valued version of the convergence principle is presented, for operators that have positive dominants (Theorem (9.1.5)); many simple operators are in this class, including point transformations and conditional expectations. As an application, in Section 9.2, we deduce theorems about convergence of multiparameter Cesàro averages of operators. Next, a version of the multiparameter superadditive ratio theorem is obtained (Theorem (9.3.3)); this contains a multiparameter Chacon-Ornstein theorem. Finally, we consider

---

[*]Parts of this chapter are taken from "A principle for almost everywhere convergence of multiparameter processes," by L. Sucheston and L. Szabó, pages 253–273 in *Almost Everywhere Convergence II*, A. Bellow and R. Jones, editors, Copyright © 1991 by Academic Press, Inc. Used by permission of Academic Press.

martingales. It has been known that less than the independence assumptions of Cairoli [1970] suffice for convergence theory; if a martingale is also a "block martingale" then the independence may be dispensed with (Theorem (9.4.5)). Conversely, in the presence of independence (or even properly defined conditional independence—the condition sometimes called F4 or commutation), every martingale is a block martingale (Proposition (9.4.3)). Here we reduce the case of block martingale to successive applications of the conditional expectation operator, which allows the use of Theorem (9.1.3) and simplifies the proofs. Block martingale theorems imply strong laws of large numbers for two parameters (9.4.8).

9.1. A multiparameter convergence principle

Let E be a sigma-complete Banach lattice; that is, a Banach lattice such that every order-bounded sequence has a least upper bound in E (see Lindenstrauss & Tzafriri [1979]). Recall that E is said to have an *order-continuous norm* if for every net (equivalently, sequence) (f_i), $f_i \downarrow 0$ implies $\|f_i\| \downarrow 0$. Let $F \subset E$. A map $T : F \to E$ is *increasing*, if $f \leq g$ implies $Tf \leq Tg$; *positive* if $f \geq 0$ implies $Tf \geq 0$; *linear* if $T(\alpha f + \beta g) = \alpha Tf + \beta Tg$ for any $\alpha, \beta \in \mathbb{R}$; *positively homogeneous* if $|T(\alpha f)| = \alpha |Tf|$ for each $f \in F$, $\alpha \in \mathbb{R}^+$; *subadditive* if $T(f + g) \leq Tf + Tg$. A map that is both positively homogeneous and subadditive is called *sublinear*. Sublinear increasing maps are positive since $0 \leq f$ implies $T0 = 0 \leq Tf$. A map T is *continuous at 0* if for every net (f_i) in F, $\|f_i\| \to 0$ implies $\|Tf_i\| \to 0$; *continuous for order* if $f_i \downarrow f$ implies that $Tf_i \downarrow Tf$; *continuous for order at 0* if $f_i \downarrow 0$ implies that $Tf_i \downarrow 0$.

(9.1.1) Lemma. *Let E be a Banach lattice with order continuous norm, F a Banach sublattice of E, and T a positively homogeneous, increasing map from F^+ to E. Then (i) T is continuous at 0 and continuous at 0 for order. (ii) If in addition T is subadditive (hence sublinear), then T is continuous for order.*

Proof. (i) Assume that T is not continuous at 0. Let I be a directed set, and let $(f_i)_{i \in I}$ be a net of elements of F^+ such that $\lim \|f_i\| = 0$ and $\limsup \|Tf_i\| > \varepsilon > 0$. Choose a sequence (i_n) of indices such that

$$\sum_{n=1}^{\infty} 2^n \|f_{i_n}\| < \infty \text{ and } \inf \|Tf_{i_n}\| \geq \varepsilon.$$

Set

$$g_n = \sum_{k=1}^{n} 2^k f_{i_k}.$$

Since F is closed, $g_n \uparrow g \in F^+$, and for every n,

$$Tg \geq Tg_n \geq T(2^n f_{i_n}) = 2^n T(f_{i_n}),$$

hence

$$\|Tg\| \geq 2^n \varepsilon.$$

This is a contradiction, therefore T must be continuous at 0. If $f_i \downarrow 0$ then $\|f_i\| \downarrow 0$, because E is order continuous. Therefore $\|Tf_i\| \downarrow 0$. Also $Tf_i \downarrow g$ for some $g \in E^+$, and necessarily $\|g\| \leq \lim \|Tf_i\| = 0$. This proves (i).

(ii) Assume T is subadditive. Let (f_i) be a net in F^+ such that $f_i \uparrow f$ or $f_i \downarrow f$. For every index i,

$$Tf_i \leq Tf \leq Tf_i + T(f - f_i) \text{ or } Tf \leq Tf_i \leq Tf + T(f_i - f).$$

In either case, $|Tf - Tf_i| \leq T|f - f_i|$. Now, by the order-continuity of the norm, $\lim \|f_i - f\| = 0$, hence, by part (i), $\lim \||T|f_i - f|\|| = 0$ and $\lim \||Tf - Tf_i\|| = 0$. Therefore $|Tf - Tf_i| \downarrow 0$.   ∎

The following propositions will be useful in deducing many-parameter theorems from one-parameter theorems. The assumption that the directed sets have countable cofinal subsets guarantees that monotone limits are in the lattice if countable suprema are. We state Proposition (9.1.2) (and Theorem (9.1.3)) for increasing sublinear maps, but in all the applications we give, the maps are positive and linear. In that case, by considering separately the positive and negative parts of an element f, we see that the convergence statement in (9.1.2(iii)) [and (9.1.3(iii))] apply to f that need not be positive. The assumption in (9.1.2(i)) [and (9.1.3(i))] about the existence of V_∞ is satisfied if there is *demiconvergence*: the stochastic (or truncated) limit exists and is equal to the lim inf. For applications, see Section 9.4, below. Given an increasing sublinear map $V : F^+ \to E$, let V also denote the extension to F defined by $Vf = V(f^+) - V(f^-)$.

(9.1.2) Proposition. *Let E be a Banach lattice with an order-continuous norm, F a Banach sublattice of E. Let I, J be directed sets with countable cofinal subsets. Let $(V_i)_{i \in I}$ be a net of increasing, sublinear maps $V_i : F^+ \to E$.*

(i) *Assume that there is an increasing, sublinear map $V_\infty : F^+ \to E$ such that for each $f \in F^+$*

$$\liminf_i V_i f \geq V_\infty f.$$

Let $(f_j : j \in J)$ be a net of elements of F^+ such that

$$\liminf_j f_j = f_\infty \in F^+.$$

Then

$$\liminf_{i,j} V_i f_j \geq V_\infty f_\infty.$$

(ii) *Assume that for each $f \in F^+$, $V_\infty f = \limsup_i V_i f \in E$. Let $(f_j)_{j \in J}$ be a net of elements of F^+ such that $\sup f_j \in F^+$. Let $f_\infty = \limsup f_j$. Then*

$$\limsup_{i,j} V_i f_j \leq V_\infty f_\infty.$$

(iii) *Assume that* $\lim V_i f = V_\infty f$ *exists in* E *for each* $f \in F^+$. *Let* $(f_j)_{j \in J}$ *be a net of elements of* F *such that* $\sup |f_j| \in F^+$ *and* $\lim f_j = f_\infty \in F$. *Then*

$$\lim_{i,j} V_i f_j = V_\infty f_\infty.$$

Proof. (i) For each $j \in J$, set $m_j = \inf_{k \geq j} f_k$. By assumption, the net $(m_j)_{j \in J}$ increases to f_∞. Now for each $k \geq j$ we have $f_k \geq m_j$, so $V_u f_k \geq V_u m_j$. Thus

$$\inf_{k \geq j} V_u f_k \geq V_u m_j.$$

Therefore, for each i,

$$\inf_{u \geq i} \inf_{k \geq j} V_u f_k \geq \inf_{u \geq i} V_u m_j.$$

Letting $i \to \infty$ yields

$$\liminf_i \inf_{k \geq j} V_i f_k \geq \liminf_i V_i m_j \geq V_\infty m_j.$$

For each j,

$$\liminf_i \inf_{k \geq j} V_i f_k \leq \liminf_{i,j} V_i f_j.$$

Therefore

$$\liminf_{i,j} V_i f_j \geq V_\infty m_j.$$

Now the net m_j increases to $f_\infty \in F^+$. The operator V_∞ is monotonely continuous by Lemma (9.1.1(ii)). Hence $V_\infty m_j \uparrow V_\infty f_\infty$. It follows that

$$\liminf_{i,j} V_i f_j \geq V_\infty f_\infty.$$

(ii) Necessarily $f_\infty \in F^+$. For each $j \in J$, let $M_j = \sup_{k \geq j} f_k$. Applying Fatou's Lemma as before, we have for each u,

$$\sup_{k \geq j} V_u f_k \leq V_u M_j.$$

Therefore, for each i,

$$\sup_{u \geq i} \sup_{k \geq j} V_u f_k \leq \sup_{u \geq i} V_u M_j.$$

Letting $i \to \infty$ yields

$$\limsup_i \sup_{k \geq j} V_i f_k \leq \limsup_i V_i M_j = V_\infty M_j.$$

For each j,

$$\limsup_i \sup_{k \geq j} V_i f_k \geq \limsup_{i,j} V_i f_j.$$

Therefore

$$\limsup_{i,j} V_i f_j \leq V_\infty M_j.$$

Now the net M_j decreases to $f_\infty \in F^+$. The operator V_∞ is increasing and sublinear being the lim sup of such operators. Therefore, by Lemma (9.1.1), V_∞ is continuous for order. Hence $V_\infty M_j \downarrow V_\infty f_\infty$. It follows that

$$\limsup_{i,j} V_i f_j \leq V_\infty f_\infty.$$

(iii) The operator V_∞ is positive and sublinear being the limit of such operators. We can consider separately the action of V_i and V_∞ on the positive and negative parts of functions, since $\lim f_j^+ = f_\infty^+$, $\lim f_j^- = f_\infty^-$. Now, by parts (i) and (ii),

$$\limsup_{i,j} V_i f_j^+ \leq V_\infty f_\infty^+ \leq \liminf_{i,j} V_i f_j^+,$$

hence $\lim_{i,j} V_i f_j^+ = V_\infty f_\infty^+$, and similarly $\lim_{i,j} V_i f_j^- = V_\infty f_\infty^-$. Hence

$$\lim_{i,j} V_i f_j = V_\infty f_\infty.$$

∎

We now consider more than two parameters. For each i, $1 \leq i \leq d$, let I_i be a directed set with a countable cofinal subset. Let $I = I_1 \times I_2 \times \cdots \times I_d$ be the product set. The partial order on I is defined by $s = (s_1, \ldots, s_d) \leq t = (t_1, \ldots, t_d)$ iff $s_k \leq t_k$ for $k = 1, \ldots, d$. The notation $t \to \infty$ then means that all the indices t_i converge to infinity independently. $L(0)$ has been introduced below to allow for a compact description of the action of the operators $T(i, j)$.

(9.1.3)   Theorem. *Let* $L(0) = L(1) \supset L(2) \supset \cdots \supset L(d)$ *be Banach lattices with order continuous norms and let* I_i, $1 \leq i \leq d$, *be directed sets with countable cofinal subsets. For each* $i = 1, 2, \ldots, d$, *let* $\big(T(i, j)\big)_{j \in I_i}$ *be a net of increasing, sublinear maps* $T(i, j) : L(i)^+ \to L(i-1)^+$.

(i) *Assume that for each* $i = 1, \ldots, d$, *there exists an increasing, sublinear operator* $T(i, \infty) : L(i)^+ \to L(i-1)^+$ *such that for every* $f \in L(i)^+$

$$T(i, \infty)f \leq \liminf_j T(i, j)f \in L(i)^+.$$

Then for every $f \in L(d)^+$ *we have*

$$\liminf_t T(1, t_1)T(2, t_2) \cdots T(d, t_d)f \geq T(1, \infty) \cdots T(d, \infty)f.$$

(ii) *Assume that*
  (a) $\limsup_j T(i,j)f = T(i,\infty)f \in L(i)$ *for* $1 \le i \le d$ *and each*
  $f \in L(1)^+$, *and*
  (b) $\sup_j T(i,j)f \in L(i-1)^+$ *for each* $f \in L(i)^+$, $2 \le i \le d$.
  Then for each $f \in L(d)^+$,

$$\limsup_t T(1,t_1)T(2,t_2)\cdots T(d,t_d)f \le T(1,\infty)\cdots T(d,\infty)f.$$

(iii) *Assume that*
  (a) $\lim_j T(i,j)f = T(i,\infty)f$ *exists and is in* $L(1)$ *for each* $f \in$
  $L(i)^+$ *and*
  (b) $\sup_j T(i,j)f \in L(i-1)$ *for each* $f \in L(i)^+$, $2 \le i \le d$.
  Then for each $f \in L(d)^+$,

$$\lim_t T(1,t_1)T(2,t_2)\cdots T(d,t_d)f = T(1,\infty)\cdots T(d,\infty)f.$$

Proof. (i) By induction on d. For $d = 2$ choose $f \in L(2)^+$ and apply Proposition (9.1.2(i)) with $F = L(1)$, $E = L(0)$, $J = I_2$, $f_j = T(2,j)f$, $f_\infty = \liminf_j f_j$, $I = I_1$, $V_i = T(1,i)$, $V_\infty = T(1,\infty)$. Then $\liminf_i V_i f \ge V_\infty f$ by assumption. It follows that

$$\liminf_{i,j} T(1,i)T(2,j)f = \liminf_{i,j} V_i f_j \ge V_\infty f_\infty$$
$$= T(1,\infty)\liminf_j T(2,j)f$$
$$\ge T(1,\infty)T(2,\infty)f.$$

Now suppose that the inequality holds for any product of d operators and prove it for a product of $d+1$ operators. Let $F = L(d)$, $E = L(0)$, $f \in L(d+1)^+$, $J = I_{d+1}$, $f_j = T(d+1,j)f$, $f_\infty = \liminf_j f_j$, $I = I_1 \times \cdots \times I_d$. For $i = (i_1,\ldots,i_d) \in I$, set $V_i = T(1,i_1)\cdots T(d,i_d)$, $V_\infty = T(1,\infty)\cdots T(d,\infty)$. Since each map $T(i,\infty)$ is increasing and sublinear, the map V_∞ has the same properties. Applying Proposition (9.1.2(i)) and the induction hypothesis we get

$$\liminf_{i,i_{d+1}} T(1,i_1)\cdots T(d,i_d)T(d+1,i_{d+1})f$$
$$= \liminf_{i,j} V_i T(d+1,j)f$$
$$\ge V_\infty f_\infty \ge V_\infty T(d+1,\infty)f$$
$$= T(1,\infty)\cdots T(d,\infty)T(d+1,\infty)f.$$

(ii) By induction on d. For $d = 2$, choose $f \in L(2)^+$ and apply Proposition (9.1.2(ii)) with $F = L(1)$, $E = L(0)$, $J = I_2$, $f_j = T(2,j)f$, $f_\infty = \limsup_j f_j$, $I = I_1$, $V_i = T(1,i)$, $V_\infty = \limsup_i T(1,i)$. Then

$$\limsup_{i,j} T(1,i)T(2,j)f = \limsup_{i,j} V_i f_j \ge V_\infty f_\infty$$
$$= T(1,\infty)\limsup_j T(2,j)f$$
$$= T(1,\infty)T(2,\infty)f.$$

Now suppose that the inequality of (ii) holds for any product of d operators and prove it for a product of $d+1$ operators. Let $F = L(d+1)$, $E = L(d)$, $f \in L(d+1)^+$, $J = I_{d+1}$, $f_j = T(d+1,j)f$, $f_\infty = \limsup_j f_j$, $I = I_1 \times \cdots \times I_d$. For $i = (i_1, \ldots, i_d) \in I$, set $V_i = T(1,i_1) \cdots T(d,i_d)$, $V_\infty = T(1,\infty) \cdots T(d,\infty)$. Since each map $T(i,\infty)$ is increasing and sublinear, so is the map V_∞. By assumption, $\sup_j T(d+1,j)f \in L(d)$. Applying Proposition (9.1.2(ii)) and the induction hypothesis

$$\limsup_{i,i_{d+1}} T(1,i_1) \cdots T(d,i_d)T(d+1,i_{d+1})f$$

$$= \limsup_{i,j} V_i T(d+1,j)f$$

$$\leq \limsup_{i} V_i f_\infty = V_\infty f_\infty$$

$$= T(1,\infty) \cdots T(d,\infty)T(d+1,\infty)f.$$

(iii) By parts (i) and (ii),

$$\limsup_{i} T(1,i_1) \cdots T(d,i_d)f \leq T(1,\infty) \cdots T(d,\infty)f$$

$$\leq \liminf_{i} T(1,i_1) \cdots T(d,i_d)f.$$

Hence the theorem follows.   ∎

Banach-valued processes

Let E be a Banach space with norm $\| \ \|$. We do not look for the greatest generality, restricting discussion to spaces $R_k(E)$. These spaces could be replaced by $L_p(E)$ spaces, $p > 1$ fixed.

We write $L_{max}(E)$ for the largest Orlicz space $L_1(E) + L_\infty(E)$, defined in analogy to the real case (see Proposition (2.2.4)). A linear operator T on $L_{max}(E)$ is said to be *positively dominated* (by $\widehat{T}$) iff there exists a positive linear operator $\widehat{T}$ on $L_{max}(\mathbb{R})$ such that $\|Tf\| \leq \widehat{T}\|f\|$ for all $f \in L_{max}(E)$. Here are some examples of positively dominated operators:

(1) Let θ be a measure preserving point transformation. Then $T(f) = f \circ \theta$, considered as an operator on $L_{max}(E)$ is positively dominated by $\widehat{T}f = f \circ \theta$ considered as an operator on $L_{max}(\mathbb{R})$.

(2) Any linear operator on $L_{max}(\mathbb{R})$ is positively dominated by its *linear modulus* $|T|$ defined by

$$|T|f = \sup_{|g| \leq f} |Tg| \qquad \text{for all } f \geq 0.$$

This follows from Dunford & Schwartz [1958], p. 672; see also Krengel [1985], p. 160.

(3) The operator conditional expectation $\mathbf{E}_\mu^{\mathcal{G}}$ defined on $L_{max}(E)$ (in analogy to the real definition (2.3.9)) is positively dominated by the real conditional expectation $\mathbf{E}_\mu^{\mathcal{G}}$ on $L_{max}(\mathbb{R})$.

We now prove versions of (9.1.2) and (9.1.3) for positively dominated operators in $L_{max}(E)$. The limits below are taken almost everywhere.

(9.1.4) Lemma. *Let k be a positive integer. Let I, J be directed sets with countable cofinal subsets. Let $(V_i)_{i \in I}$ be a net of operators on $L_{\max}(E)$, positively dominated by operators $\widehat{V}_i$ on $L_{\max}(\mathbb{R})$. Suppose that for each $f \in R_0(E)$, the limit $V_\infty f = \lim_i V_i f$ exists and is in $R_0(E)$ and the limit $\lim_i \widehat{V}_i \|f\| = \widehat{V}_\infty \|f\|$ exists and is in $R_0(\mathbb{R})$. Suppose that for each $f \in R_k(E)$ we have $\sup_i \widehat{V}_i \|f\| \in R_{k-1}(\mathbb{R})$. Let $(f_j)_{j \in J}$ be a net of functions in $R_k(E)$ such that $\lim f_j = f_\infty$ exists and $\sup_j \|f_j\| \in R_{k-1}(\mathbb{R})$. Then*

$$\lim_{i,j} V_i f_j = V_\infty f_\infty.$$

Proof. For $m \in J$, define $s_m = \sup_{j \geq m} \|f_j - f_\infty\|$, so $s_m \in R_{k-1}(\mathbb{R}) \subseteq R_0(\mathbb{R})$. Then

$$
\begin{aligned}
\limsup_{i,j} &\|V_i f_j - V_\infty f_\infty\| \\
&\leq \limsup_{i,j} \|V_i(f_j - f_\infty)\| + \limsup_i \|V_i f_\infty - V_\infty f_\infty\| \\
&= \limsup_{i,j} \|V_i(f_j - f_\infty)\| \\
&\leq \limsup_{i,j} \widehat{V}_i \|f_j - f_\infty\| \\
&\leq \limsup_i \widehat{V}_i s_m = \widehat{V}_\infty s_m.
\end{aligned}
$$

But $s_m \to 0$ and $R_0(\mathbb{R})$ has order-continuous norm (2.1.14), so $\widehat{V}_\infty s_m \to 0$. ∎

We now consider several parameters at the same time. As in Theorem (9.1.3), let I_1, $I_2, \cdots, I_d$ be directed sets with countable cofinal subsets, and let $I = I_1 \times I_2 \times \cdots \times I_d$ be the product.

(9.1.5) Theorem. *For $i = 1, 2, \cdots, d$ and $j \in J$, let $T(i,j)$ be an operator on $L_{\max}(E)$, positively dominated by $\widehat{T}(i,j)$. Suppose*

(a) *For each $f \in R_0(E)$, the limits $\lim_j T(i,j)f = T(i,\infty)f \in R_0(E)$ and $\lim_j \widehat{T}(i,j)\|f\| \in R_0(\mathbb{R})$ exist.*

(b) *For each $f \in R_i(E)$, $\sup_j \widehat{T}(i,j)\|f\| \in R_{i-1}(\mathbb{R})$.*

Then for each $f \in R_d(E)$,

$$\lim T(1,t_1)T(2,t_2)\cdots T(d,t_d)\, f = T(1,\infty)T(2,\infty)\cdots T(d,\infty)\, f$$

exists as the indices t_i converge to infinity independently.

Proof. This is proved by induction on d using Lemma (9.1.4). For $d = 2$, choose $f \in R_1(E)$, $f_j = T(2,j)f$ which converges to $T(2,\infty)f = f_\infty$, and $V_i = T(1,i)$. For the general induction step, assume that the theorem holds for d parameters, let $f \in R_{d+1}(E)$, $f_j = T(d+1,j)f$, $f_\infty =$

$T(d+1, \infty)f$, $V_i = T(1, i_1) \cdots T(d, i_d)$. It is easy to see that the operator $\widehat{T}(1, i_1) \cdots \widehat{T}(d, i_d)$ is a positive dominant of V_i ∎

Remarks

The multiparameter convergence principle [Theorem (9.1.3(iii))], with most applications, is from Sucheston [1983]. The one-sided results [(i) and (ii) on demiconvergence] were developed in Millet & Sucheston [1989] and in Sucheston & Szabó [1991]. Theorem (9.1.5) was stated in Frangos & Sucheston [1986]. Demiconvergence in martingale theory was introduced in Edgar & Sucheston [1981], and further studied by Millet & Sucheston [1983].

9.2. Multiparameter Cesàro averages of operators

In 1951, articles of A. Zygmund and N. Dunford appeared under the same title and in the same volume of *Acta Sci. Math. (Szeged)*, proving multiparameter convergence theorems for noncommuting point transformations. Zygmund assumed $f \in L \log^{d-1} L$ of a probability space. Dunford allowed a σ-finite measure but restricted f to L_p, $p > 1$. The first obvious challenge was to find the common generalization of these two settings. The spaces R_k introduced by Fava [1972] fulfilled this role (also in the more general Dunford-Schwartz operator context). However, the theorem of Zygmund, Dunford, and Fava still appeared as a difficult result, depending on a multiparameter maximal theorem: see e.g. Krengel [1985], pp. 196–201. Theorem (9.1.3) reduces the theorem to the one-parameter theory that had been known earlier (cf. N. Wiener [1939]).

We first recall some of the results from Chapter 2 about function spaces and maximal inequalities. A finite Orlicz function (2.1.1) is an increasing convex function $\Phi \colon [0, \infty) \to [0, \infty)$, satisfying $\Phi(0) = 0$ and $\Phi(u) > 0$ for some u. Such a Φ is differentiable a.e.; the derivative φ, defined a.e. by $\Phi(u) = \int_0^u \varphi(x)\, dx$, will be assumed left-continuous. Often we assume in addition that $\Phi(u)/u \to \infty$, which happens if and only if φ is unbounded. Then there is a "left-continuous generalized inverse" ψ of φ, defined by

$$\psi(y) = \inf \{ x \in (0, \infty) : \varphi(x) \geq y \}.$$

The function ψ is a derivative of an Orlicz function Ψ, called the *conjugate* of Φ. We are also interested in the function ξ, defined by $\xi(u) = u\varphi(u) - \Phi(u) = \Psi(\varphi(u))$. It is left-continuous; it may or may not be an Orlicz function.

Let $(\Omega, \mathcal{F}, \mu)$ be a measure space. The *Orlicz modular* is the function $M_\Phi(f) = \int \Phi(|f|)\, d\mu$. The *Luxemburg norm* of a measurable function f is

$$\|f\|_\Phi = \inf \{ a > 0 : M_\Phi(f/a) \leq 1 \}.$$

The space

$$L_\Phi = \{ f : \|f\|_\Phi < \infty \} = \{ f : M_\Phi(f/a) < \infty \text{ for some } a > 0 \}$$

is a Banach space, called an *Orlicz space*. This terminology may also be used when Φ is not an Orlicz function (cf. Φ_0 below). If L_Φ is an order-continuous Banach lattice, then we need not look elsewhere. But in any case, the *heart* H_Φ of L_Φ, defined by

$$H_\Phi = \{\, f : M_\Phi(f/a) < \infty \text{ for all } a > 0 \,\}$$

is an order-continuous Banach sublattice (Theorem (2.1.14)). If Φ satisfies the (Δ_2) condition at 0 and ∞, then $H_\Phi = L_\Phi$ (Theorem (2.1.17)).

Important Orlicz spaces considered here are the spaces $L \log^k L$, defined so as to include $L_1 + L_\infty$ as the case $k = 0$. Set

$$\Phi_k(u) = \begin{cases} u\,(\log u)^k & \text{if } u > 1 \\ 0 & \text{if } 0 \le u \le 1 \end{cases}.$$

Then Φ_k is an Orlicz function for $k \ge 1$; the corresponding ξ is $\xi(u) = k\Phi_{k-1}(u)$, so $L_\xi = L \log^{k-1} L$. The hearts H_{Φ_k} are the Favian spaces R_k. There is no ξ for Φ_0 (so there will be no maximal inequalities for $f \in L_{\Phi_0}$), but nevertheless L_{Φ_0} may be considered as the largest Orlicz space, namely $L_{\Phi_0} = L_1 + L_\infty$; the usual norm of $L_1 + L_\infty$ is equivalent with the norm defined by Φ_0. The space R_0 is the heart of L_{Φ_0}. The spaces R_k can be characterized by

$$R_k = \left\{\, f : \int_{\{|f|>\lambda\}} \Phi_k(|f|)\, d\mu < \infty \text{ for all } \lambda > 0 \,\right\}$$

(see the remarks following (2.2.16)).

The spaces R_k have order-continuous norm by (2.1.14). For each k,

$$\bigcup_{1<p<\infty} L_p \subseteq R_k \subseteq R_{k-1} \subseteq \cdots \subseteq R_0 \subseteq L_1 + L_\infty.$$

If μ is a finite measure, then R_0 coincides with $L_1 = L_1 + L_\infty$. If μ is a σ-finite measure, then R_0 is the appropriate space for the one-parameter martingale theorem and some one-parameter pointwise ergodic theorems: those in which pointwise convergence holds for functions in L_1 and the operator contracts the L_∞ norm. (Conditional expectations and point transformations preserving null sets are in this class.) Indeed, if $f \in R_0$, then $f = g + h$ with $g \in L_1$ and $\|h\|_\infty$ arbitrarily small. Since the operator does not increase the L_∞ norm, h may be disregarded.

In applications of Theorem (9.1.3), the space $L(0) = L(1)$ will be R_0, and the spaces $L(k)$ will be R_{k-1}. The operators $T(i,j)$ will be one-parameter averages defined below. It will be necessary to show that if a function f is in R_k, then the appropriate supremum, called g, is in R_{k-1}. The following lemma follows from (3.1.12(b)), if we note that $\mu\{|f| \ge \lambda\} < \infty$ for all $\lambda > 0$ is satisfied for any $f \in R_k$.

(9.2.1) Lemma. *Let f, g be positive functions such that for some $c > 0$ and every $\lambda > 0$,*

$$\lambda\mu\{g > c\lambda\} \leq \int_{\{f>\lambda\}} f \, d\mu.$$

If $f \in R_k$ then $g \in R_{k-1}$.

This lemma, together with Theorem (8.2.6) shows:

(9.2.2) Lemma. *Let T be a positive linear operator on $L_1 + L_\infty$, and let $A_n(T) = (1/n)(T^0 + T^1 + \cdots + T^{n-1})$. Assume that $\|T\|_1 \leq 1$ and $\sup_n \|A_n(T)\|_\infty = c < \infty$. If $f \in R_k$ then $\sup A_n(T)f = g \in R_{k-1}$.*

Suppose that T is a positive contraction of L_1 and L_∞. The one-parameter averages $A_n(T)f$ converge a.e. for $f \in R_0^+$ by (8.6.11). Thus we have all the elements needed for the application of Theorem (9.1.3), with $L(0) = L(1) = R_0$, $L(i) = R_{i-1}$ for $i \geq 2$, and $T(i, j) = A_j(T_i)$. This yields:

(9.2.3) Theorem. *Let T_i be positive operators defined on $L_1 + L_\infty$ such that $\|T\|_1 \leq 1$ and $\|T\|_\infty \leq 1$, $i = 1, 2, \cdots, d$. For each i, and each $f \in R_0$, denote by $A_\infty(T_i)f$ the a.e. limit of $A_n(T_i)f$. Then for each $f \in R_{d-1}$,*

$$\lim \frac{1}{s_1 s_2 \cdots s_d} \sum_{0 \leq k_1 < s_1} \sum_{0 \leq k_2 < s_2} \cdots \sum_{0 \leq k_d < s_d} T_1^{k_1} T_2^{k_2} \cdots T_d^{k_d} f$$

$$= A_\infty(T_1) A_\infty(T_2) \cdots A_\infty(T_d) f$$

a.e. as the indices s_i go to infinity independently.

Complements

(9.2.4) (Nonpositive multiparameter Dunford-Schwartz.) The statement is the same as (9.2.3) but the T_i's are not assumed positive. The linear modulus operators $|T_i|$ defined by

$$|T_i| f = \sup_{|g| \leq f} |T_i g| \quad \text{for all } f \geq 0$$

are positive contractions on L_1 and L_∞ (see Krengel [1985], p. 160). Hence they have the properties required of positive dominants and the result (9.2.4) follows from Theorem (9.1.5).

(9.2.5) (Multiparameter Mourier theorem.) The one-parameter result, Banach-valued Birkhoff's ergodic theorem, is due to E. Mourier [1953]; for a proof see Krengel [1985], p. 167. Our multiparameter version is as follows:

Theorem. *Let E be a Banach space. Let θ_i be measure-preserving point transformations on a σ-finite measure space $(\Omega, \mathcal{F}, \mu)$. Define $T_i f = f \circ \theta_i$ on $L_{\max}(E) = L_1(E) + L_\infty(E)$. Let $f \in R_{d-1}(E)$. Then the averages*

$$\frac{1}{s_1 s_2 \cdots s_d} \sum_{k_1=0}^{s_1-1} \sum_{k_2=0}^{s_2-1} \cdots \sum_{k_d=0}^{s_d-1} T_1^{k_1} T_2^{k_2} \cdots T_d^{k_d} f$$

converge a.e. as the indices s_i converge to infinity independently.

Proof. The positive dominants are defined as $\widehat{T}_i f = f \circ \theta_i$, $f \in L_{\max}(\mathbb{R})$. They have the required one-parameter convergence and maximal theorems because measure-preserving point transformations on real function spaces do (see Chapter 8); so (9.1.5) is again applicable. ∎

(9.2.6) (Commuting operators.) If the operators T_i in (9.2.3) commute, and we desire only convergence "over squares" where $s_1 = \cdots = s_d$, then it suffices to assume that $f \in L_1$ (or only $f \in R_0$). This result is due to Dunford & Schwartz; for Brunel's proof, see Krengel [1985], p. 215. Our approach via Theorem (9.1.3) does not yield this deep theorem.

(9.2.7) (Multiparameter L_p theorem.) Akcoglu [1975] proved that if T is a positive contraction on an L_p space, $1 < p < \infty$, then for each $f \in L_p^+$, there is a maximal theorem $\|\sup_n A_n(T)f\|_p \le (p/(p-1))\|f\|_p$ and $A_n(f)$ converges a.e. Clearly Theorem (9.1.3) applies with all spaces $L(i) = L_p$. This gives the convergence of averages of the form

$$\frac{1}{s_1 s_2 \cdots s_d} \sum_{k_1=0}^{s_1-1} \sum_{k_2=0}^{s_2-1} \cdots \sum_{k_d=0}^{s_d-1} T_1^{k_1} T_2^{k_2} \cdots T_d^{k_d} f$$

for each $f \in L_p$.

(9.2.8) (Heart of $L_p + L_\infty$.) There is an L_p analog of the theory of R_k spaces. Define the Orlicz space $L_p + L_\infty$ as in (2.2.13), and denote its heart by M_p. Then $f \in M_p$ if and only if, for each $\varepsilon > 0$, there is a decomposition $f = f_1 + f_2$ where $f_1 \in L_1$ and $\|f\|_\infty \le \varepsilon$ (see (2.1.14c)). If T is a positive contraction on L_p that is mean-bounded in L_∞,

$$\sup_n \|A_n(T)\|_\infty = c < \infty,$$

then for $f \in M_p$ we have $A_n(T)f$ converges a.e. and $\sup_n A_n(T)f \in M_p$. Indeed, by (2.1.14c), for each f and each $\varepsilon > 0$, there is a decomposition $f = f_1 + f_2$ with $f_1 \in L_p$ and $\|f_2\|_\infty \le \varepsilon/c$. Hence $\sup_n A_n(T)|f_2| \le \varepsilon$. Now

$$\sup_n A_n(T)|f| \le \sup_n A_n(T)|f_1| + \sup_n A_n(T)|f_2|;$$

the first term on the right is in L_p by Akcoglu's maximal theorem, and the second term on the right is $\le \varepsilon$. The conditions are again satisfied for an application of Theorem (9.1.3) with all the spaces $L(i) = M_p$. It follows that for each $f \in M_p$ the averages

$$\frac{1}{s_1 s_2 \cdots s_d} \sum_{k_1=0}^{s_1-1} \sum_{k_2=0}^{s_2-1} \cdots \sum_{k_d=0}^{s_d-1} T_1^{k_1} T_2^{k_2} \cdots T_d^{k_d} f$$

converge a.e.

Remarks

McGrath [1980], Theorem 3, obtained the convergence result (9.2.7); Yoshimoto [1982] obtained the convergence result (9.2.8) for $f \in L_p$; they used multiparameter maximal theorems. Discussion of the hearts of spaces $L_p + L_\infty$ in the ergodic context is from Edgar & Sucheston [1989].

9.3. Multiparameter ratio ergodic theorems

We recall some results from Chapter 8. Let T be a sub-Markovian operator (positive contraction) on L_1. Then the space Ω decomposes into the conservative part C and the dissipative part D. If $f \in L_1^+$, then $\sum_{i=0}^\infty T^i f = 0$ or ∞ on C, and $\sum_{i=0}^\infty T^i f < \infty$ on D (8.3.1). T is called *Markovian* iff it preserves the integral. A sequence (s_n) of functions in L_1^+ is called an extended superadditive process iff

$$s_{k+n} \geq s_k + T^k s_n$$

for every $k, n \geq 0$, and (s_n) is called a superadditive process iff in addition

$$\gamma = \sup_n \frac{1}{n} \int s_n \, d\mu < \infty.$$

The constant γ is called the *time constant* of the process. The sequence (s_n) is *subadditive* iff $(-s_n)$ is superadditive; (s_n) is *additive* iff it is both superadditive and subadditive.

Note that if (s_n) is superadditive, then since $\gamma < \infty$ we have $s_n \geq \sum_{k=0}^{n-1} T^k s_1$ for all n.

We now restate the main part of Theorem (8.6.7):

(9.3.1) Theorem. *Let T be a sub-Markovian operator on L_1, let (r_n) be a positive superadditive process, and let (s_n) be a positive extended superadditive process. Write $E = \{\sup_n s_n > 0\}$. Then the ratio r_n/s_n converges a.e. to a finite limit on the set $C \cap E$. If either T is Markovian or (r_n) is additive on D, then $\lim (r_n/s_n) = \lim \uparrow r_n / \lim \uparrow s_n < \infty$ exists a.e. on $D \cap E$.*

We are going to use Theorem (9.1.3) to prove a multiparameter variant of this result, in which the sequence in the numerator is additive. A multiparameter version of the Chacon-Ornstein theorem is obtained if the sequence in the denominator is also assumed to be additive.

Assume $s_1 > 0$ and let $f \in L_1^+$. Define

$$h = \sup_n \frac{\sum_{i=0}^{n-1} T^i f}{s_n}$$

$$g = \sup_n \frac{\sum_{i=0}^{n-1} T^i f}{\sum_{i=0}^{n-1} T^i s_1}.$$

Let ν be the finite measure $s_1 \cdot \mu$. Then, by (8.2.3),

$$\nu\{g > \lambda\} \leq \frac{1}{\lambda} \int_{\{g>\lambda\}} \frac{f}{s_1} \, d\nu.$$

Now since $f/s_1 \in L_1(\nu)$, if $f/s_1 \in L\log^k L(\nu)$, then by Corollary (3.1.9b) we have $g \in L\log^{k-1} L$. But $h \leq g$, so we have proved:

(9.3.2) Lemma. *If $f/s_1 \in L\log^k L(\nu)$, then $h \in L\log^{k-1} L(\nu)$.*

We are now ready for the multiparameter ratio ergodic theorem.

(9.3.3) Theorem. *Let $(\Omega, \mathcal{F}, \mu)$ be a probability space. Let T_i, $i = 1, 2, \cdots, d$, be positive contractions on $L_1(\mu)$; for each i let $s_n^{(i)}$ be a positive extended superadditive sequence with respect to T_i, and let ν_i be the measure $s_1^{(i)} \cdot \mu$, $i = 1, 2, \cdots, d$. Suppose that the functions $s_1^{(1)}$, $s_1^{(2)}/s_1^{(1)}, \cdots, s_1^{(d)}/s_1^{(d-1)}$ are bounded away from 0. Then for each f such that $f/s_1^{(d)} \in L\log^{d-1} L(\nu)$,*

$$\frac{\sum_{k_1=0}^{n_1-1} T_1^{k_1}}{s_{n_1}^{(1)}} \frac{\sum_{k_2=0}^{n_2-1} T_2^{k_2}}{s_{n_2}^{(2)}} \cdots \frac{\sum_{k_d=0}^{n_d-1} T_d^{k_d}}{s_{n_d}^{(d)}} f$$

converges a.e. as $n_1, n_2, \cdots, n_d$ go to infinity independently.

To explain the intuitive meaning, consider the case when $d = 2$, $T_1 = T$, $T_2 = U$, $s_m^{(1)} = \sum_{i=0}^{m-1} T^i \mathbf{1}$, $s_n^{(2)} = \sum_{j=0}^{n-1} U^j \mathbf{1}$. Then the theorem asserts that for each $f \in L\log L$,

$$\frac{\sum_{j=0}^{n-1} U^j}{\sum_{j=0}^{n-1} U^j \mathbf{1}} \left(\frac{\sum_{i=0}^{m-1} T^i f}{\sum_{i=0}^{m-1} T^i \mathbf{1}} \right)$$

converges a.e. to a finite limit as $m, n \to \infty$ independently.

Proof of the Theorem. Define the operators

$$T(k, n) = \frac{\sum_{i=0}^{n-1} T_k^i}{s_n^{(k)}}.$$

For $k = 1, 2, \cdots, d$, let $L(k) = \left\{ f : f/s_1^{(k)} \in L\log^{k-1} L(\nu_k) \right\}$, and $L(0) = L(1) = L_1(\nu_1)$. The norm of f in $L(k)$ is defined as the norm of $f/s_1^{(k)}$ in $L\log^{k-1} L(\nu_k)$. Since $L\log^{k-1} L$ is a Banach lattice with order-continuous norm, so is $L(k)$. Also, if $f \in L_1(\mu)$, then $T(n, k)f \in L_1(\nu_k) \subseteq L_1(\nu_{k-1}) \subseteq \cdots \subseteq L_1(\nu_1) \subseteq L_1(\mu)$. Suppose $f \in L(k)$. Then by Lemma (9.3.2),

$$h_k = \sup_n T(k, n)f \in L\log^{k-2} L(\nu_k).$$

Since $s_1^{(k)}/s_1^{(k-1)}$ is bounded away from 0, we have

$$L \log^{k-2} L\left(\nu_k\right) \subseteq L \log^{k-2} L\left(\nu_{k-1}\right).$$

The assumptions also imply that $s_1^{(k-1)}$ is bounded away from 0, so $h_k \in L(k-1)$. The existence of the limits $\lim_{n\to\infty} T(k,n)f$ follows from Theorem (9.3.1); since (r_n) is additive, it is not necessary to assume T Markovian. The proof is completed by Theorem (9.1.3).   ■

Complements

(9.3.4) (A counterexample to a variant of Chacon-Ornstein in two parameters.) A multiparameter version of the Chacon-Ornstein theorem corresponds to the case of (9.3.3) when the processes $(s_n^{(i)})$ are additive for $i = 1, 2, \cdots, d$. It is natural to ask whether a multiparameter version analogous to Theorem (9.2.3) holds; i.e., convergence of ratios of two expressions of the type appearing in (9.2.3) corresponding to two functions. The answer is no: there is a counterexample due to Brunel and Krengel (see Krengel [1985], p. 217). It has $\Omega = \mathbb{N}$ with counting measure, functions in L_1, commuting operators, and convergence to infinity is over "squares." A change of measure with appropriate modification of operators and functions reduces the situation to a probability space. Since in the discrete case $L_1 \subseteq L \log L$, this is also a counterexample to our conjecture where convergence (also over "rectangles") would be claimed for $f \in L \log L$.

Remarks

Theorem (9.3.3) is from Sucheston & Szabó [1991]. The additive version of it is in Frangos & Sucheston [1986].

9.4. Multiparameter martingales

A martingale $(X_t)_{t\in J}$ defined on a probability space $(\Omega, \mathcal{F}, \mathbf{P})$ and indexed by a directed set J need not converge a.e.; we know this from Chapter 4. This is even true for martingales indexed by $\mathbb{N} \times \mathbb{N}$. In fact, the convergence may fail for two reasons:

(1) The integrability is insufficient: In two dimensions, in order for $X_t = \mathbf{E}^{\mathcal{F}_t}[X]$ to converge, the random variable X must be in $L \log L$ even if there is appropriate independence of marginal distributions (see Cairoli [1970]).

(2) The stochastic basis $(\mathcal{F}_t)$ may be such that even L_∞-bounded martingales fail to converge (Dubins & Pitman [1980]).

We will apply a weak maximal inequality for submartingales. Note that $L \log^k L \subseteq L_1$ for a probability space.

(9.4.1) Lemma. *Let (X_n) be a positive submartingale, and let $Y = \sup_n X_n$. Suppose X_n converges in L_1 to $X \in L_1$. Then for each $\lambda > 0$,*

$$\lambda \mathbf{P}\{Y \geq \lambda\} \leq \mathbf{E}\left[\mathbf{1}_{\{Y \geq \lambda\}} X\right].$$

If $X \in L \log^k L$ then $Y \in L \log^{k-1} L$.

Proof. Since X_n converges in L_1, it is uniformly integrable. Therefore we may pass to the limit under the integral sign in (6.1.5), with $\mathbf{P}_1 = \mathbf{P}$. This proves the first assertion. The second assertion follows from (3.1.9(b)). ■

Recall ((1.3.1) and (1.4.3)) that L_1-bounded martingales (and submartingales) indexed by directed sets converge stochastically; and in the presence of uniform integrability converge in L_1 norm.

In this section we will consider martingales in d dimensions. We will see that a "block martingale" converges a.s. if it is properly bounded. A particular case is the convergence under condition (F4). The proof is by reducing block martingales to consecutive applications of conditional expectations and applying Theorem (9.1.3).

Fix a positive integer d. The directed set I will be $\mathbb{N}^d = \mathbb{N} \times \mathbb{N} \cdots \times \mathbb{N}$, with d factors. The ordering on I is defined as usual: if $s = (s_1, s_2, \cdots, s_d)$ and $t = (t_1, t_2, \cdots, t_d)$, then $s \leq t$ iff $s_i \leq t_i$ for all i. Let $(\mathcal{F}_t)_{t \in I}$ be a stochastic basis. For integers i, j with $1 \leq i \leq j \leq d$, write

$$\mathcal{F}_t^{i \cdots j}$$

for the σ-algebra obtained by lumping together the σ algebras on all the axes except for axes numbered from i to j. More precisely, $\mathcal{F}_t^{i \cdots j}$ is the σ-algebra generated by all $\mathcal{F}_s$ with $s_k = t_k$ for $i \leq k \leq j$ and s_k arbitrary for other k. Of course the containment relations remain, so that for fixed i and j, the family $(\mathcal{F}_t^{i \cdots j})_{t \in I}$ is again a stochastic basis. When $i = j$, we will sometimes write $\mathcal{F}_t^i$ for $\mathcal{F}_t^{i \cdots j}$. Denote the conditional expectation $\mathbf{E}^{\mathcal{F}_s^{i \cdots j}}$ by $E_s^{i \cdots j}$, and $\mathbf{E}^{\mathcal{F}_s}$ by E_s.

Since the index set $I = \mathbb{N}^d$ is a directed set, the definition of Chapter 4 applies: a process $(X_t)_{t \in I}$ is a *martingale [submartingale]* iff $X_s = E_s X_t$ $[X_s \leq E_s X_t]$ whenever $s \leq t$. Now I has additional structure, so other variants are possible. If $k \in \mathbb{N}$, $1 \leq k \leq d$, then we will say that the process (X_t) is a *block k-martingale [block k-submartingale]* iff

$$E_s^{1 \cdots k} X_t = X_{(s_1, \cdots, s_k, t_{k+1}, \cdots, t_d)}$$

$$\left[E_s^{1 \cdots k} X_t \geq X_{(s_1, \cdots, s_k, t_{k+1}, \cdots, t_d)}\right].$$

whenever $s \leq t$. An integrable process is a *block martingale [block submartingale]* iff it is a block k-martingale [block k-submartingale] for all $k \leq d$.

Block k-martingales should not be confused with "k-martingales," defined by

$$E_s^k X_t = X_{(t_1, \cdots, t_{k-1}, s_k, t_{k+1}, \cdots, t_d)}.$$

Block martingales may be characterized in terms of "factorization" of conditional expectations:

(9.4.2) Proposition. *(1) Let (X_t) be a uniformly integrable martingale. Write $X = \operatorname{s\,lim} X_t$. Then the following are equivalent:*

(a) $X_t = E_t X$ *is a block martingale.*

(b) *For each $k \leq d$ and each t, $E_t X = E_t^{1\cdots k} E_t^{k+1\cdots d} X$.*

(2) Let (X_t) be a uniformly integrable block submartingale. Write $X = \operatorname{s\,lim} X_t$. Then for each $k \leq d$ and each t, we have $E_t X \leq E_t^{1\cdots k} E_t^{k+1\cdots d} X$.

Proof. First let X_t be a uniformly integrable block submartingale. Let $s = (s_1 \cdots, s_d) \leq t = (t_1, \cdots, t_d)$ with $t_{k+1} = s_{k+1}, \cdots, t_d = s_d$. This is not a loss of generality in what follows, because if $s \leq t$ and $r = (s_1, \cdots, s_k, t_{k+1}, \cdots, t_d)$, then $E_s^{1\cdots k} = E_r^{1\cdots k}$. Since X_t is a block k-submartingale, we have

$$X_s = X_{(s_1, \cdots, s_k, t_{k+1}, \cdots, t_d)} \leq E_s^{1\cdots k} X_t \leq E_s^{1\cdots k} E_t X.$$

Taking the limit in L_1 when $t_1 \to \infty, t_2 \to \infty, \cdots, t_k \to \infty$, we obtain $\lim X_t \leq \lim E_t X = E_s^{k+1\cdots d} X$, hence

$$X_s \leq E_s^{1\cdots k} E_s^{k+1\cdots d} X.$$

This proves (2).

Now let X_t be a uniformly integrable block martingale. Then (a) $\Longrightarrow$ (b) follows from part (2) applied to X_t and $-X_t$. Finally, to prove (b) $\Longrightarrow$ (a), assume that $E_t X = E_t^{1\cdots k} E_t^{k+1\cdots d} X$ for each k. Then

$$E_s^{1\cdots k} X_t = E_s^{1\cdots k} E_t^{1\cdots k} E_t^{k+1\cdots d} X = E_s^{1\cdots k} E_t^{k+1\cdots d} X.$$

Let u be such that $u_i = t_i$ for $k+1 \leq i \leq d$. Then as $u_1 \to \infty, \cdots, u_k \to \infty$, we have $X_u \to E_t^{k+1\cdots d} X$. Since $E_s^{1\cdots k} X_u = X_{(s_1, \cdots, s_k, t_{k+1}, \cdots, t_d)}$, the proof is complete. ∎

(9.4.3) Theorem. *Let (X_t) be a uniformly integrable block submartingale, and write $X = \operatorname{s\,lim} X_t$. Then for all s, we have*

$$X_s \leq E_s^1 E_s^2 \cdots E_s^d X.$$

Let (X_t) be a uniformly integrable block martingale. Then

$$X_s = E_s^1 E_s^2 \cdots E_s^d X.$$

for all s.

Proof. Let (X_t) be a uniformly integrable block submartingale. We claim that for all k, $1 \leq k \leq d$,

(9.4.3a) $$X_s \leq E_s X \leq E_s^1 E_s^2 \cdots E_s^k E_s^{(k+1)\cdots d} X.$$

Only the second inequality has to be proved. The proof is by induction on k.

Case $k = 1$ follows from (9.4.2). Now assume (9.4.3a) holds for some value of k, and consider the next value $k + 1$. Again by (9.4.2),

$$X_s \leq E_s X \leq E_s^{1 - (k+1)} E_s^{k+2-d} X.$$

Taking the limit in L_1 as $s_1 \to \infty, \cdots, s_k \to \infty$,

$$E_s^{(k+1)-d} X \leq E_s^{(k+1)} E_s^{(k+2)-d} X.$$

Now substitute this into (9.4.3a) to complete the induction step.

This completes the proof in the case of a block submartingale. The case of a block martingale follows from this.   ∎

(9.4.4) Theorem. (i) Let (X_t) be a block submartingale that is bounded in the Orlicz space $L \log^{d-1} L$. Then we have upper demiconvergence: $\limsup X_t = \operatorname{s\,lim} X_t$. (ii) If $X \in L \log^{d-1} L$ then $E_t^1 E_t^2 \cdots E_t^d X$ converges a.e. to a finite limit. (iii) Let (X_t) be a block martingale bounded in $L \log^{d-1} L$. Then X_t converges a.e. and in $L \log^{d-1} L$.

Proof. (i) First assume that (X_t) is a *positive* block submartingale bounded in $L \log^{d-1} L$. Apply Theorem (9.1.3(ii)) and Lemma (9.4.1) with: $L(i) = L \log^{i-1} L$ for $1 \leq i \leq d$; $T(i,t) = E_t^i$. Note that as $t_1 \to \infty$, the σ-algebras $\mathcal{F}_t^1$ increase to $\vee_t \mathcal{F}_t^1 = \vee_t \mathcal{F}_t$, so $E_t^1 X$ converges to X as $t_1 \to \infty$. Hence $T(1, \infty) X = X$ a.e. Similarly $T(i, \infty) X = X$ a.e. for all i. Therefore, by (9.4.3) and (9.1.3(ii)), we have

$$\limsup X_t \leq \limsup E_t^1 E_t^2 \cdots E_t^d X \leq T(1, \infty) T(2, \infty) \cdots T(d, \infty) X = X.$$

On the other hand, we always have $X = \operatorname{s\,lim} X_t \leq \limsup X_t$. Hence the statement follows for positive block submartingales; and hence also for block submartingales bounded from below by a constant.

Now let X_t be an arbitrary $L \log^{d-1} L$-bounded block submartingale. Then for any constant a, the process $(X_t \vee a)$ is also a block submartingale, so by the above, we have $\limsup(X_t \vee a) = X \vee a$. By Fatou's lemma, $\limsup X_t > -\infty$ a.e. Thus letting $a \to -\infty$, we obtain $\limsup X_t = X$.

(ii) Apply Theorem (9.1.3(iii)) and Lemma (9.4.1) with $L(i) = L \log^{i-1} L$ for $1 \leq i \leq d$ and $T(i,t) = E_t^i$.

(iii) For a block martingale (X_t), apply part (i) to X_t and $-X_t$, or apply part (ii).   ∎

Complements

(9.4.5) (Reversed processes.) There is also a version of our results with a *dual* directed set I. Fix a positive integer d. Consider

$$I = (-\mathbb{N}) \times (-\mathbb{N}) \times \cdots \times (-\mathbb{N})$$

with d factors. The ordering on I is given by: $s = (s_1, \cdots, s_d) \leq t = (t_1, \cdots, t_d)$ iff $s_i \leq t_i$ for all i. But I is a dual directed set: given any $s, t \in I$, there is $u \in I$ with $u \leq s$ and $u \leq t$. When we discuss limits indexed by I, we are interested in what happens when all the indices go to $-\infty$. Block martingales and block submartingales are defined by the same inequalities.

(9.4.6) (Two-parameter martingales.) Let $I = \mathbb{N} \times \mathbb{N}$. A stochastic basis $(\mathcal{F}_t)_{t \in I}$ is said to satisfy *condition* (F4) iff $\mathcal{F}_t^1$ is conditionally independent of $\mathcal{F}_t^2$ given $\mathcal{F}_t$ (Cairoli & Walsh [1975]). A process (X_t) is a *1-martingale* iff

$$E_s^1 X_t = X_{(s_1, t_2)}$$

whenever $s \leq t$; the process is a *2-martingale* iff

$$E_s^2 X_t = X_{(t_1, s_2)}$$

whenever $s \leq t$. Thus a block martingale is a martingale which is also a 1-martingale.

Since (F4) is symmetric, it can be derived only from symmetric assumptions.

Proposition. *Let $(\mathcal{F}_t)_{t \in I}$ be a stochastic basis indexed by $I = \mathbb{N} \times \mathbb{N}$. The following are equivalent: (a) Every uniformly integrable martingale is a 1-martingale and a 2-martingale. (b) $E_t = E_t^1 E_t^2 = E_t^2 E_t^1$. (c) (F4).*

Proof. A 2-martingale is a 1-martingale after exchanging the coordinates. So the equivalence of (a) and (b) follows from (9.4.2).

(b) $\Longrightarrow$ (c). Assume (b). We must show that $\mathcal{F}_t^1$ is conditionally independent of $\mathcal{F}_t^2$ given $\mathcal{F}_t$. Let $A \in \mathcal{F}_t^1$, $B \in \mathcal{F}_t^2$, $X = \mathbf{1}_A$, $Y = \mathbf{1}_B$. Then we must show that $E_t[XY] = (E_t X)(E_t Y)$. Now

$$E_t[XY] = E_t^2 E_t^1[XY] = \left(E_t^2 X\right)\left(E_t^1 Y\right)$$
$$= \left(E_t^2 E_t^1 X\right)\left(E_t^1 E_t^2 Y\right) = (E_t X)(E_t Y).$$

(c) $\Longrightarrow$ (b). Assume (F4). Let $A \in \mathcal{F}_t^1$, $X \in L_1$. Then

$$\mathbf{E}\left[\mathbf{1}_A E_t^1 E_t^2 X\right] = \mathbf{E}\left[E_t^1(\mathbf{1}_A E_t^2 X)\right] = \mathbf{E}\left[\mathbf{1}_A E_t^2 X\right]$$
$$= \mathbf{E}\left[E_t(\mathbf{1}_A E_t^2 X)\right] = \mathbf{E}\left[E_t(\mathbf{1}_A)E_t(E_t^2 X)\right]$$
$$= \mathbf{E}\left[E_t(\mathbf{1}_A)E_t X\right] = \mathbf{E}\left[\mathbf{1}_A E_t X\right].$$

Since $E_t^1 E_t^2 X$ and $E_t X$ are both measurable with respect to $\mathcal{F}_t^1$, this implies that $E_t^1 E_t^2 X = E_t X$. By symmetry, $E_t^2 E_t^1 X = E_t X$. ∎

(9.4.7) (A two-parameter strong law of large numbers.) Suppose Y_{ij} are integrable independent random variables with expectation zero. Then

$$X_t = X_{m,n} = \sum_{\substack{1 \leq i \leq m \\ 1 \leq j \leq n}} Y_{ij}$$

is a two-parameter martingale, and the stochastic basis $(\mathcal{F}_t)$ generated by it satisfies (F4). On the other hand, there are situations in which a martingale is both a 1-martingale and a 2-martingale, yet (F4) fails. An example is provided by a two-parameter extension of the strong law of large numbers via a theorem proved for one parameter in (6.1.15).

As in (6.1.15), let P_m be the class of permutations on $\mathbb{N}$ with support in $\{1, 2, \cdots, m\}$. A double sequence $(Y_{ij})_{i \in \mathbb{N}, j \in \mathbb{N}}$ of random variables is called *1-exchangeable* iff for each m, and each permutation $\pi \in P_m$, the system $(Y_{i,j})$ has the same distribution as $(Y_{\pi(i),j})$. The double sequence is *2-exchangeable* iff for each m, and each permutation $\pi \in P_m$, the system $(Y_{i,j})$ has the same distribution as $(Y_{i,\pi(j)})$. A double sequence that is both 1-exchangeable and 2-exchangeable is called *row and column exchangeable*.

Theorem. *Let $p > 0$, and let Y_{ij} be positive row and column exchangeable random variables such that $Y_{11}^p \in L_1$. Set*

$$S_{m,n} = \sum_{\substack{i \leq m \\ j \leq n}} Y_{ij}$$

and $X_t = X_{(-m,-n)} = S_{m,n}^p/mn$. If $0 < p \leq 1$, then X_t is a reversed block submartingale; if $1 \leq p < \infty$, then X_t is a reversed block supermartingale. If $0 < p \leq 1$ and $Y_{11}^p \in L \log L$, then X_t converges a.e. The limit is 0 if $p < 1$; the limit is $\mathbf{E}^{\mathcal{E}}[Y_{11}]$ if $p = 1$, where $\mathcal{E}$ is the σ-algebra of row and column exchangeable events.

Proof. Set

$$\mathcal{F}_{(-m,-n)} = \sigma\left\{ X_{(-r,-s)} : r \geq m, s \geq n \right\}.$$

If $\pi \in P_m$, then the two systems

$$\{Y_{i,j}, 1 \leq i \leq m; S_{r,s}, r \geq m\} \text{ and } \{Y_{\pi(i),j}, 1 \leq i \leq m; S_{r,s}, r \geq m\}$$

have the same distribution. Therefore for all $i \leq m$,

$$\mathbf{E}\left[Y_{i,j} \mid \mathcal{F}_{(-m,-1)}\right] = \mathbf{E}\left[Y_{\pi(i),j} \mid \mathcal{F}_{(-m,-1)}\right].$$

Now the lemma proved in (6.1.15) implies that for each $k \leq m$ and each n,

$$\mathbf{E}\left[X_{(-k,-n)} \mid \mathcal{F}_{(-m,-1)}\right] \geq X_{(-m,-n)}$$

if $0 < p \leq 1$. That is, X_t is a reversed 1-submartingale. By symmetry, it is also a reversed 2-submartingale. Applying these two properties one after the other, we see that X_t is a reversed submartingale. Therefore X_t is a reversed block submartingale. So X_t demiconverges by Theorem (9.4.4). If $0 < p < 1$, then $\mathrm{s}\lim X_t = \limsup X_t = 0$ because one-parameter limits are 0 by (6.1.15); hence $\lim X_t = 0$. If $p = 1$, then X_t is a reversed block martingale, so it converges. The limit is identified by integration over row and column exchangeable sets.

The proof that if $p > 1$ then X_t is a reversed block supermartingale is analogous.   ∎

(9.4.8) (Banach-valued martingales in probability spaces.) Let E be a Banach space, and let $\mathcal{F}_t$ be a stochastic basis indexed by $I = \mathbb{N}^d$. What can be said about E-valued martingales X_t ? One-parameter convergence theorems (5.3.20) and maximal theorems (9.4.1) are available for conditional expectations and their positive dominants, which are also conditional expectations applied to real functions. So (9.1.5) is applicable. In the notation of (9.4.3), we see that $E_s^1 E_s^2 \cdots E_s^d X$ converges a.e. if $X \in L \log^{d-1} L(E)$. A more general formulation with the same proof follows. For each fixed k, let $\mathcal{F}_n^k$ be either increasing or decreasing (reversed) stochastic bases indexed by $\mathbb{N}$. Let $E_n^k = \mathbf{E}^{\mathcal{F}_n^k}$. Then

$$X_{(s_1, s_2, \cdots, s_d)} = E_{s_1}^1 E_{s_2}^2 \cdots E_{s_d}^d X$$

converges a.e. if $X \in L \log^{d-1} L(E)$.

The theory of block martingales also extends to the Banach-valued setting. In particular, Theorems (9.4.3) and (9.4.4) are true under the additional assumption that E has the Radon-Nikodým property. It is needed to obtain the conditional expectation representation of uniformly integrable martingales, hence convergence (5.3.30).

(9.4.9) (Infinite measure.) Now consider Banach-valued processes defined on an infinite measure space $(\Omega, \mathcal{F}, \mu)$. The expression

$$E_{s_1}^1 E_{s_2}^2 \cdots E_{s_d}^d X$$

converges a.e. for $X \in R_{d-1}(E)$, under proper assumption of σ-finiteness of measure. Here we prove the one-parameter results for E-valued martingales indexed by $\mathbb{N}$; the extension to several parameters by the methods of this section is easy.

If $\mathcal{F}_n$ are increasing σ-algebras such that μ is σ-finite on $\mathcal{F}_1$, then the problem of a.s. convergence of martingales easily reduces to the previous case, since if $A \in \mathcal{F}_1$ is a set of finite measure and (X_n) is a martingale, then $\mathbf{1}_A X_n$ is a martingale with respect to μ restricted to A; pointwise convergence on A implies pointwise convergence on Ω as $A \uparrow \Omega$. So $\mathbf{E}^{\mathcal{F}_n}[X]$ converges for $X \in L_{\max}(E) = L_1(E) + L_\infty(E)$. Similarly, multiparameter a.e. theorems reduce to those on A. Of more interest are *reversed* martingale theorems, assuming that μ is σ-finite on each $\mathcal{F}_{-n}$, but not on $\bigcap_n \mathcal{F}_{-n}$. In this setting, the one-parameter L_p case was treated by Dellacherie & Meyer [1982], pp. 35–40. Here we limit discussion to martingales given by conditional expectations. Observe that if (X_{-n}) is a reversed martingale, then $X_{-n} = \mathbf{E}^{\mathcal{F}_{-n}}[X_{-1}]$.

(9.4.9 i). *Let E be a Banach space. Let $(\Omega, \mathcal{F}, \mu)$ be a σ-finite measure space, and let $(\mathcal{F}_n)_{-\infty < n < \infty}$ be a stochastic basis. Assume that μ is σ-finite on each $\mathcal{F}_n$.*

(a) *If $X \in L_{\max}(E) = L_1(E) + L_\infty(E)$, then for every $\lambda > 0$,*

$$\mu \left\{ \sup_{-\infty < n < \infty} \left\| \mathbf{E}^{\mathcal{F}_n}[X] \right\| \geq \lambda \right\} \leq \frac{1}{\lambda} \mathbf{E} \left[\|X\| \right].$$

(b) *If $X \in R_k(E)$, $k \geq 1$, then*

$$\sup_{-\infty < n < \infty} \left\| \mathbf{E}^{\mathcal{F}_n}[X] \right\| \in R_{k-1}(E).$$

(c) *If $X \in L_{\max}(E)$, then $\mathbf{E}^{\mathcal{F}_n}[X]$ converges a.e.*
(d) *If $X \in R_0(E)$, then $\mathbf{E}^{\mathcal{F}_{-n}}[X]$ converges a.e.*

Proof. (a) It suffices to consider $\sup_{n \in \mathbb{N}}$ since this is equivalent to taking $\sup_{n \geq m}$ for any fixed m, and the maximal inequality is preserved as $m \to -\infty$. Fix $A \in \mathcal{F}_1$ with $\mu(A) < \infty$. In finite measure spaces, the localization theorem (1.4.2) states that $X_\sigma = \mathbf{E}^{\mathcal{F}_\sigma}[X]$ for simple stopping times σ. The conditional expectation is a contraction on $L_1(E)$ [see the remarks following (5.1.15)]; it follows that $\mathbf{E}[\|X_\sigma\|] \leq \mathbf{E}[\|X\|]$. Hence the maximal inequality (5.2.36) applied with the measure μ restricted to A gives

$$\lambda \mu \left\{ \mathbf{1}_A \sup_{n \in \mathbb{N}} \left\| \mathbf{E}^{\mathcal{F}_n}[X] \right\| \geq \lambda \right\} \leq \mathbf{E} \left[\mathbf{1}_A \|X\| \right].$$

Now let $A \uparrow \Omega$; it follows that

$$\lambda \mu \left\{ \mathbf{1}_A \sup_{n \in \mathbb{N}} \left\| \mathbf{E}^{\mathcal{F}_n}[X] \right\| \geq \lambda \right\} \leq \mathbf{E} \left[\|X\| \right].$$

(b) From (8.2.5) we now obtain

$$\lambda \mu \left\{ \mathbf{1}_A \sup_{n \in \mathbb{N}} \left\| \mathbf{E}^{\mathcal{F}_n}[X] \right\| \geq 2\lambda \right\} \leq \mathbf{E} \left[\mathbf{1}_{\{\|X\| \geq \lambda\}} \|X\| \right].$$

Now apply (9.2.1).

(c) X_n converges on A by the convergence theorem (5.3.20) applied on the set A. Convergence on Ω follows on letting $A \uparrow \Omega$.

(d) Write $X = X_1 + X_2$, with $X_1 \in L_1$ and $\|X_2\|_\infty < \varepsilon$. Since conditional expectations are contractions in L_∞, we may disregard X_2. So we may assume simply that $X \in L_1$. But now the maximal inequality (a) allows the proof of convergence in the same way as in probability space (5.3.36). ∎

(9.4.10) (Multiparameter Rota theorem.) A positive operator is called *bistochastic* iff it preserves the L_1 and L_∞ norms of positive functions. Let T_i be bistochastic operators, and let $U_n = T_1 \cdots T_n T_n^* \cdots T_1^*$. This is sometimes called the "alternating procedure." The one-parameter Rota theorem is: if $f \in R_1$, then $U_n f$ converges a.e. The operator admits a representation

$$U_n f = \mathbf{E}^{\mathcal{G}} \left[\mathbf{E}^{\mathcal{F}_n} [f] \right],$$

where $\mathcal{G}$ is a fixed σ-algebra and $\mathcal{F}_n$ is a decreasing sequence of σ-algebras (Dellacherie & Meyer [1982], p. 56). In order to apply Theorem (9.1.3) to obtain a multiparameter version of the theorem, $f \in R_k$ must imply $\sup U_n f \in R_{k-1}$. This follows from the representation of U_n above, because (by Jensen's inequality) the conditional expectation $\mathbf{E}^{\mathcal{G}}$ respects the classes R_k. So the multiparameter Rota theorem is:

(9.4.10(i)). For $i = 1, \cdots, d$ and $n \in \mathbb{N}$, let T_n^i be a bistochastic operator. Set

$$U_n^i = T_1^i \cdots T_n^i (T_n^i)^* \cdots (T_1^i)^*.$$

If $f \in R_d$, then $\lim U_{s_1}^1 \cdots U_{s_d}^d f$ exists a.e. as the indices s_i converge to infinity independently.

(9.4.11) (Pure L_p alternating procedure.) Let T be a linear operator on L_p, $1 < p < \infty$. Define a (nonlinear) operator $M(T) \colon L_p^+ \to L_p^+$ by

$$M(T)f = \left(T^* \left[(Tf)^{p-1} \right] \right)^{1/(p-1)}.$$

If $(T_n)_{n \in \mathbb{N}}$ is a sequence of positive linear contraction operators on L_p, then for all $f \in L_p^+$, the sequence $M(T_n \cdots T_1) f$ converges a.e. and $\sup_n M(T_n \cdots T_1) f \in L_p^+$ (Akcoglu & Sucheston [1988]).

If $p = 2$, then $M(T) = T^* T$ is linear. In that case, (9.1.3) implies an easy multiparameter version of (9.4.10(i)):

Let $(T_n^{(i)})_{n \in \mathbb{N}}$ be sequences of positive contractions in L_2, for $i = 1$, $2, \cdots, d$. If $f \in L_2^+$, then

$$M(T_{s_1}^{(1)} \cdots T_2^{(1)} T_1^{(1)}) M(T_{s_2}^{(2)} \cdots T_2^{(2)} T_1^{(2)}) \cdots M(T_{s_d}^{(d)} \cdots T_2^{(d)} T_1^{(d)}) f$$

converges a.e. as the indices $s_1, \cdots, s_d$ converge to infinity independently.

In particular, we obtain a multiparameter version of Stein's [1961] theorem, which corresponds to the case when the $T_n^{(i)} = T^{(i)}$ are independent of n and self-adjoint. Then for each $f \in L_2$,

$$\left(T^{(1)} \right)^{2s_1} \left(T^{(2)} \right)^{2s_2} \cdots \left(T^{(d)} \right)^{2s_d} f$$

converges a.e. as the indices $s_1, \cdots, s_d$ converge to infinity independently.

(9.4.12) (Amart approach to two parameters.) Let $J = \mathbb{N} \times \mathbb{N}$. For $t \in J$ we write $t = (t_1, t_2)$ and

$$\mathcal{F}_t^1 = \bigvee_u \mathcal{F}_{(t_1, u)}.$$

A map $\tau \colon \Omega \to J = \mathbb{N} \times \mathbb{N}$ is a 1-*stopping time* iff $\{\tau = t\} \in \mathcal{F}_t^1$ for all $t \in J$. The set of simple 1-stopping times is denoted Σ_1. A 1-*amart* is a process (X_t) such that the net $(\mathbb{E}[X_\tau])_{\tau \in \Sigma_1}$ converges. Assume that the stochastic basis satisfies the conditional independence condition (F4). Let (X_t) be a block martingale; equivalently, let (X_t) be a martingale and a 1-martingale (9.4.6). Then (X_t) is a 1-amart. Since $(\mathcal{F}_t^1)$ is totally ordered, this implies that X_t converges a.s. ((4.2.11) and (4.2.5)). Convergence of X_t also follows from (9.4.4) above, which is a more general argument. But the amart approach was particularly important in the continuous parameter case ($J = \mathbb{R} \times \mathbb{R}$), where it was applied to the problem of regularity of trajectories of martingales bounded in $L \log L$ under condition (F4) (Millet & Sucheston [1981a]). This method proved regularity in the first, second, and fourth quadrants, after Bakry [1979] proved regularity in the first and third quadrants by the method of stochastic integration; see also Meyer [1981]. Again the amart point of view gave the good notions of optional and predictable projections for two-parameter processes; see Bakry [1981].

(9.4.13) (Multiparameter Krengel theorem.) Krengel's stochastic ergodic theorem asserts that if T is a positive linear contraction on L_1 and $f \in L_1^+$, then $(1/n) \sum_{i=0}^{n-1} T^i f$ converges stochastically. Since

$$\liminf \frac{1}{n} \sum_{i=0}^{n-1} T^i f = 0,$$

there is in fact lower demiconvergence to 0 (Krengel [1985], p. 143). Theorem (9.1.3(i)) now gives a multiparameter version of this result (Millet & Sucheston [1989]).

(9.4.14) (Additive amarts.) Multiparameter processes may be studied in another way. A few details are given here.

Let I be a directed set with least element 0, and *locally finite* in the sense that all intervals $[0, t] = \{s \in I : 0 \le s \le t\}$ are finite. A subset $S \subseteq I$ is a (lower) *layer* of I iff from $s \le t$ and $t \in S$ it follows that $s \in S$. We will write $\mathcal{L}(I) = \mathcal{L}$ for the set of all layers. Then $\mathcal{L}$ is a directed set when ordered by inclusion. So we may study processes with index set $\mathcal{L}$, such as martingales or amarts. If $(\mathcal{F}_t)_{t \in I}$ is a stochastic basis indexed by I, there is an *associated stochastic basis* $(\mathcal{H}_S)_{S \in \mathcal{L}}$ indexed by the layers, defined by

$$\mathcal{H}_S = \bigvee_{s \in S} \mathcal{F}_s.$$

A process F_S indexed by $\mathcal{L}$ is called an *additive process* iff

$$F_{S \cup T} + F_{S \cap T} = F_S + F_T \qquad \text{a.e.}$$

for all $S, T \in \mathcal{L}$. Certainly any process (F_S) of the form

$$F_S = \sum_{s \in S} Y_s$$

is additive. An additive process that is also an amart is called an *additive amart*. An additive process that is also a martingale is called an *additive martingale*.

If a process $(X_t)_{t \in I}$ has the form

$$X_t = \sum_{s \leq t} Y_s$$

for some *difference process* (Y_t), then we will say the process (X_t) and the additive process

$$F_S = \sum_{s \in S} Y_s$$

are *associated* processes. In that case:

(i) If we have $\sup_{t \in I} \mathbf{E}\left[|X_t|\right] < \infty$ and $\sup_{\tau \in \Sigma(\mathcal{L})} \left|\mathbf{E}\left[F_\tau\right]\right| < \infty$, then also $\sup_{\tau \in \Sigma(\mathcal{L})} \mathbf{E}\left[|F_\tau|\right] < \infty$.

(ii) If (F_S) is an L_1-bounded amart, then $\left(|F_S|\right)$ is also an L_1-bounded amart.

In some special cases, for example $I = \mathbb{N} \times \mathbb{N}$, there is more. Every process (X_t) can be obtained by adding a difference process (Y_t). In the case $I = \mathbb{N} \times \mathbb{N}$ we also have:

(iii) If (F_S) is an additive amart, and $X_t = F_{[0,t]}$ is L_1-bounded, then X_t converges essentially.

(iv) F_S is an additive martingale if and only if

$$\mathbf{E}\left[X_{i+1,j+1} - X_{i+1,j} - X_{i,j+1} + X_{i,j} \,|\, \mathcal{F}_{i,\infty} \vee \mathcal{F}_{\infty,j}\right] = 0$$

$$\mathbf{E}\left[X_{i+1,0} - X_{i,0} \,|\, \mathcal{F}_{i,\infty}\right] = 0$$

$$\mathbf{E}\left[X_{0,j+1} - X_{0,j} \,|\, \mathcal{F}_{\infty,j}\right] = 0$$

for all $i, j \in \mathbb{N}$.

The conditions in (iv) mean that (X_t) is a *strong martingale* in the sense of Walsh [1979]. So (iii) implies Walsh's result on the convergence of strong martingales. Reference: Edgar [1982].

Remarks

Frangos & Sucheston [1985] introduced block martingales and proved their convergence. Earlier Chatterji [1975] derived convergence of block margingales for $d = 2$ from a Banach-valued martingale convergence theorem. Theorems (9.4.2), (9.4.3), and the proofs of ((9.4.1)(i)) and ((9.4.1)(iii)) are from Sucheston & Szabó [1991]; (9.4.1(ii)) is from Sucheston [1983]. This last result was applied to random fields by Föllmer [1984]; (9.4.7) is from Edgar & Sucheston [1981].

Row and column exchangeable random variables were introduced by D. J. Aldous [1981], and by Edgar & Sucheston [1981] under the name T-exchangeable. There exists now a considerable body of literature on the subject (see Aldous [1985]).

The theory of martingales indexed by $\mathbb{R} \times \mathbb{R}$ was initiated in the memoir of Cairoli & Walsh [1975]. The book Korezlioglu, Mazziotto, & Szpirglas [1981] is a collection of articles devoted to it.

References

M. A. Akcoglu

[1975] A pointwise ergodic theorem for L_p spaces. *Canad. Math. J.* **27**, pp. 1975–1982.

M. A. Akcoglu, L. Sucheston

[1978] A ratio ergodic theorem for superadditive processes. *Z. Wahrscheinlichkeitstheorie und Verw. Gebiete* **44**, pp. 269–278.

[1983] A stochastic ergodic theorem for superadditive processes. *J. Ergodic Theory and Dynamical Systems* **3**, pp. 335–344.

[1984a] On ergodic theory and truncated limits in Banach lattices. In: *Measure Theory Oberwolfach 1983*, Lecture Notes in Mathematics 1089, Springer-Verlag; pp. 241–262.

[1984b] On identification of superadditive ergodic limits. In: *Yale University Symposium Dedicated to S. Kakutani*, Contemporary Mathematics 26, American Mathematical Society; pp. 25–32.

[1985a] On uniform monotonicity of norms and ergodic theorems in function spaces. *Supplemento ai Rend. Circ. Mat. Palermo* **8**, pp. 325–335.

[1985b] An ergodic theorem on Banach Lattices. *Israel J. Math.* **51**, pp. 208–222.

[1988] Pointwise convergence of alternating sequences. *Canad. J. Math.* **40**, pp. 610–632.

[1989] A superadditive ergodic theorem in Banach lattices. *J. Math. Anal. Appl.* **140**, pp. 318–332.

D. J. Aldous

[1981] Representations for partially exchangeable arrays of random variables. *J. Multivariate Anal.* **11**, pp. 581–598.

[1985] Exchangeability and related topics. In: *Ecole d'Eté de Probabilités de Saint-Flour XIII—1983*, Lecture Notes in Mathematics 1117, Springer-Verlag; pp. 1–198.

E. S. Andersen, B. Jessen

[1948] Some limit theorems on set-functions. *Danske Vid. Selsk. Math.-fys. Medd.* **25** no. 5, pp. 1–8.

K. Astbury

[1976] *On Amarts and Other Topics.* Ph. D. Dissertation, Ohio State University.

[1978] Amarts indexed by directed sets. *Ann. Probability* **6**, pp. 267–278.

[1981a] Order convergence of martingales in terms of countably additive and purely finitely additive martingales. *Ann. Probability* **9**, pp. 266–275.

[1981b] The order convergence of martingales indexed by directed sets. *Trans. Amer. Math. Soc.* **265**, pp. 495–510.

D. G. Austin

[1966] A sample function property of martingales. *Ann. Math. Statist.* **37**, pp. 1396–1397.

D. G. Austin, G. A. Edgar, A. Ionescu Tulcea

[1974] Pointwise convergence in terms of expectations. *Z. Wahrscheinlichkeitstheorie und Verw. Gebiete* **30**, pp. 17–26.

S. N. Bagchi

[1983] *On almost sure convergence of classes of multivalued asymptotic martingales.* Ph.D. dissertation, Department of Mathematics, The Ohio State University.

[1985] On a.s. convergence of classes of multivalued asymptotic martingales. *Ann. Inst. H. Poincaré*, Sec. B **21**, pp. 313–321.

D. Bakry

[1979] Sur la régularité des trajectoires des martingales à deux indices. *Z. Wahrscheinlichkeitstheorie und Verw. Gebiete* **50**, pp. 149–157.

[1981] Théorèmes élémentaire des processus à deux indices. *Z. Wahrscheinlichkeitstheorie und Verw. Gebiete* **55**, pp. 55–71.

S. Banach

[1924] Sur un théorème de M. Vitali. *Fundamenta Math.* **5**, pp. 130–136.

H. Bauer

[1981] *Probability Theory and Elements of Measure Theory.* Academic Press.

J. R. Baxter

[1974] Pointwise in terms of weak convergence. *Proc. Amer. Math. Soc.* **46**, pp. 395–398.

[1976] Convergence of stopped random variables. *Advances in Math.* **21**, pp. 112–115.

A. Bellow

[1976a] On vector-valued asymptotic martingales. *Proc. Nat. Acad. Sci. U. S. A.* **73**, pp. 1798–1799.

[1976b] Stability properties of the class of asymptotic martingales. *Bull. Amer. Math. Soc.* **82**, pp. 338–340.

[1977a] Several stability properties of the class of asymptotic martingales. *Z. Wahrscheinlichkeitstheorie und Verw. Gebiete* **37**, pp. 275–290.

[1977b] Les amarts uniformes. *C. R. Acad. Sci. Paris* **284**, pp. A1295–1298.

[1978a] Uniform amarts: A class of asymptotic martingales for which strong almost sure convergence obtains. *Z. Wahrscheinlichkeitstheorie und Verw. Gebiete* **41**, pp. 177–191.

[1978b] Some aspects of the theory of vector-valued amarts. In: *Vector Space Measures and Applications I*, Lecture Notes in Mathematics 644, Springer-Verlag; pp. 57–67.

[1978c] Submartingale characterizations of measurable cluster points. In: *Probability on Banach Spaces*, J. Kuelbs, editor, Advances in Probability and Related Topics 4, Marcel Dekker; pp. 69–80.

[1981] Martingales, amarts and related stopping time processes. In: *Probability in Banach Spaces III*, Lecture Notes in Mathematics 860, Springer-Verlag; pp. 9–24.

[1984] For the historical record. In: *Measure Theory Oberwolfach 1983*, Lecture Notes in Mathematics 1089, Springer-Verlag; p. 271.

A. Bellow, A. Dvoretzky

[1979] A characterization of almost sure convergence. In: *Probability in Banach Spaces II*, Lecture Notes in Mathematics 709, Springer-Verlag; pp. 45–65.

[1980] On martingales in the limit. *Ann. Probability* **8**, pp. 602–606.

A. Bellow, L. Egghe

[1982] Generalized Fatou inequalities. *Ann. Inst. H. Poincaré*, Sec. B **18**, pp. 335–365.

Y. Benyamini, N. Ghoussoub

[1978] Une caractérisation probabiliste de l_1. *C. R. Acad. Sci. Paris* **286**, pp. A795–798.

A. S. Besicovitch

[1945] A general form of the covering principle and relative differentiation of additive functions. *Proc. Cambridge Philos. Soc.* **41**, pp. 103–110.

[1946] A general form of the covering principle and relative differentiation of additive functions II. *Proc. Cambridge Philos. Soc.* **42**, pp. 1–10.

[1947] Corrigenda to the paper "A general form of the covering principle and relative differentiation of additive functions II." *Proc. Cambridge Philos. Soc.* **43**, p. 590.

P. Billingsley

[1979] *Probability and Measure.* Wiley.

G. D. Birkhoff

[1931] Proof of the ergodic theorem. *Proc. Nat. Acad. Sci. U. S. A.* **17**, pp. 656–660.

G. Birkhoff

[1937] Moore-Smith convergence in general topology. *Ann. of Math.* **38**, pp. 39–56.

D. Blackwell, L. E. Dubins

[1963] A converse to the dominated convergence theorem. *Illinois J. Math.* **7**, pp. 508–514.

L. H. Blake

[1970] A generalization of martingales and two subsequent convergence theorems. *Pacific J. Math.* **35**, pp. 279–283.

J. R. Blum, D. L. Hanson, L. H. Koopmans

[1963] On a strong law of large numbers for a class of stationary processes. *Z. Wahrscheinlichkeitstheorie und Verw. Gebiete* **2**, pp. 1–11.

F. A. Boshuizen

[In press] Comparisons of optimal stopping values and prophet inequalities for matrices of random variables.

J. Bourgain

[1977] On dentability and the Bishop-Phelps property. *Israel J. Math.* **28**, pp. 265–271.

R. D. Bourgin

[1983] *Geometric Aspects of Convex Sets with the Radon-Nikodým Property.* Lecture Notes in Mathematics 993, Springer-Verlag.

B. Bru, H. Heinich

[1979] Sur l'espérance des variables aléatoires vectorielles. *C. R. Acad. Sci. Paris* **288**, pp. 65–68.

[1980a] Sur l'espérance des variables aléatoires vectorielles. *Ann. Inst. H. Poincaré*, Sec. B **16**, pp. 177–196.

[1980b] Sur l'espérance des variables aléatoires à valeures dans les espaces de Banach réticulés. *Ann. Inst. H. Poincaré*, Sec. B **16**, pp. 197–210.

B. Bru, H. Heinich, J. C. Lootgieter

[1981] Lois des grands nombres pour les variables échangeables. *C. R. Acad. Sci. Paris* **293**, pp. A485–488.

A. Brunel

[1963] Sur un lemme ergodique voisin du lemme de E. Hopf et sur une de ses applications. *C. R. Acad. Sci. Paris* **256**, pp. A581–584.

A. Brunel, U. Krengel

[1979] Parrier avec un prophéte dans le cas d'un processus sous-additif. *C. R. Acad. Sci. Paris* **288**, pp. A57–60.

A. Brunel, L. Sucheston
[1976a] Sur les amarts faibles à valeurs vectorielles. *C. R. Acad. Sci. Paris* **282**, pp. A1011–1014.
[1976b] Sur les amarts à valeurs vectorielles. *C. R. Acad. Sci. Paris* **283**, pp. A1037–1040.
[1977] Une charactérisation probabiliste de la séparabilité du dual d'un espace de Banach. *C. R. Acad. Sci. Paris* **284**, pp. A1469–1472.
[1979] Sur existence de dominantes exactes pour un processus sur-additif. *C. R. Acad. Sci. Paris* **288**, pp. A153–155.

R. D. Brunk
[1948] The strong law of large numbers. *Duke Math. J.* **15**, pp. 181–195.

Bui Khoi Dam
[1987] On the convergence of amarts in Orlicz spaces. *Ann. Univ. Sci. Budapest. Eötvős Sect. Math.* **30**, pp. 231–239.
[1989] BMO-sequences and amarts. *Acta Math. Hungarica* **53**, pp. 271–279.

D. L. Burkholder
[1964] Maximal inequalities as a necessary condition for almost everywhere convergence. *Z. Wahrscheinlichkeitstheorie und Verw. Gebiete* **33**, pp. 55–59.
[1966] Martingale transforms. *Ann. Math. Statist.* **37**, pp. 1497–1504.
[1973] Distribution function inequalities for martingales. *Ann. Probab.* **1**, pp. 19–42.
[1984] Boundary value problems and sharp inequalities for martingale transforms. *Ann. Probab.* **12**, pp. 647–702.
[1986] An extension of a classical martingale inequality. In: *Probability Theory and Harmonic Analysis*, J.-A. Chao and W. A. Woyczynski, editors, Marcel Dekker, Inc.; pp. 21–30.
[1988] Sharp inequalities for martingales and stochastic integrals. *Astérisque* **157–158**, pp. 75–94.
[1989] On the number of escapes of a martingale and its geometrical significance. In: *Almost Everywhere Convergence*, G. A. Edgar and L. Sucheston, editors, Academic Press; pp. 159–178.
[1991] Explorations in martingale theory and its applications. In: *Ecole d'Eté de Probabilités de Saint-Flour XIX—1989*, Lecture Notes in Mathematics 1464, Springer-Verlag; pp. 1–665.

D. L. Burkholder, B. J. Davis, R. F. Gundy
[1972] Integral inequalities for convex functions of operators on martingales. In: *Proceedings of the Sixth Berkeley Symposium on Mathematical Statistics and Probability*, Volume II, L. M. LeCam, J. Neyman, & E. Scott, editors, University of California Press; pp. 223–240.

D. L. Burkholder, R. F. Gundy
[1970] Extrapolation and interpolation of quasi-linear operators on martingales. *Acta Math.* **124**, pp. 249–304.

H. Busemann, W. Feller
[1934] Zur Differentiation der Lebesgueschen Integrale. *Fund. Math.* **22**, pp. 226–249.

E. Cairoli
[1970] Une inégalité pour martingales à indices multiples et les applications. In: *Séminaire de Probabilités IV*, Lecture Notes in Mathematics 124, Springer-Verlag; pp. 1–28.

E. Cairoli, J. B. Walsh
[1975] Stochastic integrals in the plane. *Acta Math.* **134**, pp. 111–183.

R. V. Chacon
[1962] Identification of the limit of operator averages. *J. Math. Mech.* **11**, pp. 961–968.
[1964] A class of linear transformations. *Proc. Amer. Math. Soc.* **15**, pp. 560–564.
[1974] A "stopped" proof of convergence. *Advances in Math.* **14**, pp. 365–368.

R. V. Chacon, D. S. Ornstein
[1960] A general ergodic theorem. *Illinois J. Math.* **4**, pp. 153–160.

R. V. Chacon, L. Sucheston
[1975] On convergence of vector-valued asymptotic martingales. *Z. Wahrscheinlichkeitstheorie und Verw. Gebiete* **33**, pp. 55–59.

S. D. Chatterji
[1960] Martingales of Banach-valued random variables. *Bull. Amer. Math. Soc.* **66**, pp. 395–398.
[1968] Martingale convergence and the Radon-Nikodým theorem. *Math. Scand.* **22**, pp. 21–41.
[1975] Vector-valued martingales and their applications. In: *Probability in Banach Spaces*, Lecture Notes in Mathematics 526, Springer-Verlag; pp. 33–51.

B. D. Choi, L. Sucheston
[1981] Continuous parameter uniform amarts. In: *Probability in Banach Spaces III*, Lecture Notes in Mathematics 860, Springer-Verlag; pp. 85–98.

G. Choquet
[1956] Existence et unicité des representations intégrales au mouen des points extremaux dans les cônes convexes. *Séminaire Bourbaki* **139**, pp. 1–15.

Y. S. Chow

[1960a] A martingale inequality and the law of large numbers. *Proc. Amer. Math. Soc.* **11**, pp. 107–111.

[1960b] Martingales in a σ-finite measure space indexed by directed sets. *Trans. Amer. Math. Soc.* **97**, pp. 254–285.

[1965] Local convergence of martingales and the law of large numbers. *Ann. Math. Statist.* **36**, pp. 552–558.

[1967a] On the expected value of a stopped submartingale. *Ann. Math. Statist.* **38**, pp. 608–609.

[1967b] On a strong law of large numbers. *Ann. Math. Statist.* **38**, pp. 610–611.

Y. S. Chow, H. Robbins, D. Siegmund

[1971] *Great Expectations: The Theory of Optimal Stopping.* Houghton Mifflin.

Y. S. Chow, H. Teicher

[1988] *Probability Theory, Independence, Interchangeability, Martingales.* Second Edition, Springer-Verlag.

J. P. R. Christensen

[1974] *Topology and Borel Structure.* North-Holland Mathematics Studies 10, North-Holland.

K. L. Chung

[1947] Notes on some strong laws of large numbers. *Amer. J. Math.* **69**, pp. 189–192.

[1951] The strong law of large numbers. In: *Proceedings of the Second Berkeley Symposium on Mathematical Statistics and Probability*, J. Neyman, editor, University of California Press; pp. 341–352.

L. E. Clarke

[1979] Problem Solution 6174. *Amer. Math. Monthly* **86**, p. 313.

D. L. Cohn

[1980] *Measure Theory.* Birkhäuser.

H. H. Corson

[1961] The weak topology of a Banach space. *Trans. Amer. Math. Soc.* **101**, pp. 1–15.

D. Dacunha-Castelle, M. Schreiber

[1974] Techniques probabilistes pour étude de problèmes d'isomorphismes entre espaces de Banach. *Ann. Inst. H. Poincaré*, Sec. B **10**, pp. 229–277.

Bui Khoi Dam listed under Bui

B. J. Davis

[1969] A comparison test for martingale inequalities. *Ann. Math. Statist.* **40**, pp. 505–508.

W. J. Davis, N. Ghoussoub, W. B. Johnson, S. Kwapien, B. Maurey

[1990] Weak convergence of vector valued martingales. In: *Probability in Banach Spaces 6*, U. Haagerup, J. Hoffman-Jørgensen, N. J. Nielsen, editors, Birkhäuser; pp. 41–50.

W. J. Davis, N. Ghoussoub, J. Lindenstrauss

[1981] A lattice renorming theorem and applications to vector-valued processes. *Trans. Amer. Math. Soc.* **263**, pp. 531–540.

W. J. Davis, W. B. Johnson

[1977] Weakly convergent sequences of Banach space valued random variables. In: *Banach Spaces of Analytic Functions*, Lecture Notes in Mathematics 604, Springer-Verlag; pp. 29–31.

W. J. Davis, R. R. Phelps

[1974] The Radon-Nikodým property and dentable sets in Banach spaces. *Proc. Amer. Math. Soc.* **45**, pp. 119–122.

C. Dellacherie, P.-A. Meyer

[1978] *Probabilities and Potential.* North-Holland Mathematics Studies 29, North-Holland.

[1982] *Probabilities and Potential B.* North-Holland Mathematics Studies 72, North-Holland.

A. Denjoy

[1951] Une extension du théorème de Vitali. *Amer. J. Math.* **73**, pp. 314–356.

Y. Derriennic, M. Lin

[1973] On invariant measures and ergodic theorems for positive operators. *J. Funct. Anal.* **13**, pp. 252–267.

J. Diestel, J. J. Uhl, Jr.

[1977] *Vector Measures.* Mathematical Surveys 15, American Mathematical Society.

Dinh Quang Lu'u listed under Lu'u

J. L. Doob

[1953] *Stochastic processes.* Wiley.

[1963] A ratio operator limit theorem. *Z. Wahrscheinlichkeitstheorie und Verw. Gebiete* **1**, pp. 288–294.

[1975] Stochastic process measurablilty conditions. *Ann. Inst. Fourier (Grenoble)* **25**, pp. 163–176.

L. E. Dubins, D. A. Freedman

[1966] On the expected value of a stopped martingale. *Ann. Math. Statist.* **37**, pp. 1505–1509.

L. E. Dubins, J. Pitman

[1980] A divergent, two-parameter, bounded martingale. *Proc. Amer. Math. Soc.* **78**, pp. 414–416.

N. Dunford

[1951] An individual ergodic theorem for non-commutative transformations. *Acta Sci. Math. (Szeged)* **14**, pp. 1–4.

N. Dunford, B. J. Pettis

[1940] Linear operations on summable functions. *Trans. Amer. Math. Soc.* **47**, pp. 323–392.

N. Dunford, J. T. Schwartz
[1956] Convergence almost everywhere of operator averages. *J. Rat. Mech. Anal.* **5**, pp. 129–178.
[1958] *Linear Operators Part I.* Interscience Publishers.

A. Dvoretzky
[1976] On stopping time directed convergence. *Bull. Amer. Math. Soc.* **82**, pp. 347–349.
[1977] Generalizations of martingales. *Advances in Appl. Probability* **9**, pp. 193–194.

G. A. Edgar
[1975] A noncompact Choquet theorem. *Proc. Amer. Math. Soc.* **49**, pp. 354–358.
[1979a] Uniform semiamarts. *Ann. Inst. H. Poincaré*, Sec. B **15**, pp. 197–203.
[1979b] Measurability in a Banach space. *Indiana Univ. Math. J.* **28**, pp. 559–579.
[1980] Asplund operators and a.e. convervence. *J. Multivariate Anal.* **10**, pp. 460–466.
[1982] Additive amarts. *Ann. Probability* **10**, pp. 199–206.
[1983] Two integral representations. In: *Measure Theory and its Applications*, Lecture Notes in Mathematics 1033, Springer-Verlag; pp. 193–198.

G. A. Edgar, A. Millet, L. Sucheston
[1982] On compactness and optimality of stopping times. In: *Martingale Theory in Harmonic Analysis and Banach Spaces*, Lecture Notes in Mathematics 939, Springer-Verlag; pp. 36–61.

G. A. Edgar, L. Sucheston
[1976a] Amarts: a class of asymptotic martingales, A. Discrete parameter. *J. Multivariate Anal.* **6**, pp. 193–221.
[1976b] Amarts: a class of asymptotic martingales, B. Continuous parameter. *J. Multivariate Anal.* **6**, pp. 572–591.
[1976c] The Riesz decomposition for vector-valued amarts. *Z. Wahrscheinlichkeitstheorie und Verw. Gebiete* **36**, pp. 85–92.
[1977a] On vector-valued amarts and dimension of Banach spaces. *Z. Wahrscheinlichkeitstheorie und Verw. Gebiete* **39**, pp. 213–216.
[1977b] Martingales in the limit and amarts. *Proc. Amer. Math. Soc.* **67**, pp. 315–320.
[1981] Démonstrations de lois des grands nombres par les sous-martingales descendents. *C. R. Acad. Sci. Paris* **292**, pp. A967–969.
[1989] On maximal inequalities in Orlicz spaces. In: *Measure and Measurable Dynamics*, edited by R. D. Mauldin, R. M.

Shortt, C. E. Silva, Contemporary Mathematics 94, American Mathematical Society; pp. 113–129.
[1991] A note on weak inequalities in Orlicz and Lorentz spaces. In: *Approximation Theory and Functional Analysis*, edited by C. K. Chui, Academic Press; pp. 73–80.

L. Egghe
[1980a] Some new Chacon-Edgar-type inequalities for stochastic processes, and characterizations of Vitali-conditions. *Ann. Inst. H. Poincaré*, Sec. B **16**, pp. 327–337.
[1980b] Characterizations of nuclearity in Fréchet spaces. *J. Funct. Anal.* **35**, pp. 207–214.
[1981] Strong convergence of pramarts in Banach spaces. *Canad. J. Math.* **33**, pp. 357–361.
[1982a] Weak and strong convergence of amarts in Fréchet spaces. *J. Multivariate Anal.* **12**, pp. 291–305.
[1982b] On sub- and superpramarts with values in a Banach lattice. In: *Measure Theory, Oberwolfach 1981*, Lecture Notes in Mathematics 945, Springer-Verlag; pp. 352–365.
[1984a] *Stopping Time Techniques for Analysts and Probabilists.* London Mathematical Society Lecture Notes Series 100, Cambridge University Press.
[1984b] Convergence of adapted sequences of Pettis-integrable functions. *Pacific J. Math.* **114**, pp. 345–366.

A. Engelbert, H. J. Engelbert
[1979] Optimal stopping and almost sure convergence of random sequences. *Z. Wahrscheinlichkeitstheorie und Verw. Gebiete* **48**, pp. 309–325.
[1980] On a generalization of a theorem of W. Sudderth and some applications. In: *Mathematical Statistics*, Banach Center Publications 6, Polish Scientific Publishers; pp. 111–120.

N. A. Fava
[1972] Weak inequalities for product operators. *Studia Math.* **42**, pp. 271–288.

J. Feldman
[1962] Subinvariant measures for Markov operators. *Duke Math. J.* **29**, pp. 71–98.

D. L. Fisk
[1965] Quasi-martingales. *Trans. Amer. Math. Soc.* **120**, pp. 369–389.

H. Föllmer
[1984] Almost sure convergence of multiparameter martingales for Markov random fields. *Ann. Probability* **12**, pp. 133–140.

J. P. Fouque
[1980a] Régularité des trajectoires des amarts et hyperamarts réels. *C. R. Acad. Sci. Paris* **290**, pp. A107–110.
[1980b] Enveloppe de Snell et théorie générale des processus. *C. R. Acad. Sci. Paris* **290**, pp. A285–288.

J. P. Fouque, A. Millet
[1980] Régularité à gauche des martingales fortes à plusieurs indices. *C. R. Acad. Sci. Paris* **290**, pp. A773–776.

N. E. Frangos
[1985] On regularity of Banach valued processes. *Ann. Probability* **13**, pp. 985–990.

N. E. Frangos, L. Sucheston
[1985] On Convergence and demiconvergence of block martingales and submartingales. In: *Probability in Banach Spaces V*, Lecture Notes in Mathematics 1153, Springer-Verlag; pp. 189–225.
[1986] On multiparameter ergodic and martingale theorems in infinite measure spaces. *Probab. Th. Rel. Fields* **71**, pp. 477–490.

O. Frank
[1966] Generalization of an inequality of Hájek and Rényi. *Scand. Actuarietidskrift* **49**, pp. 85–89.

A. Garsia
[1965] A simple proof of E. Hopf's maximal ergodic theorem. *J. Math. Mech.* **14**, pp. 381–382.
[1970] *Topics in Almost Everywhere Convergence.* Markham.

N. Ghoussoub
[1977] Banach lattices valued amarts. *Ann. Inst. H. Poincaré, Sec. B* **13**, pp. 159–169.
[1979a] Order amarts: A class of asymptotic martingales. *J. Multivariate Anal.* **9**, pp. 165–172.
[1979b] Summability and vector amarts. *J. Multivariate Anal.* **9**, pp. 173–178.
[1982] Riesz spaces valued measures and processes. *Bull. Soc. Math. France* **110**, pp. 146–150.

N. Ghoussoub, L. Sucheston
[1978] A refinement of the Riesz decomposition for amarts and semiamarts. *J. Multivariate Anal.* **8**, pp. 146–150.

N. Ghoussoub, M. Talagrand
[1979] Convergence faible des potentiels de Doob vectoriels. *C. R. Acad. Sci. Paris* **288**, pp. A599–602.

D. Gilat
[1986] The best bound in the $L \log L$ inequality of Hardy and Littlewood and its martingale counterpart. *Proc. Amer. Math. Soc.* **97**, pp. 429–436.

M. Girardi, J. J. Uhl, Jr.
[1990] Slices, RNP, strong regularity, and martingales. *Bull. Austral. Math. Soc.* **41**, pp. 411–415.

C. Goffman, D. Waterman
[1960] On upper and lower limits in measure. *Fundamenta Math.* **48**, pp. 127–133.

R. F. Gundy
[1969] On the class $L \log L$ of martingales and singular integrals. *Studia Math.* **33**, pp. 109–118.

A. Gut
[1982] A contribution to the theory of asymptotic martingales. *Glasgow Math. J.* **23**, pp. 177–186.

A. Gut, K. D. Schmidt
[1983] *Amarts and Set Function Processes.* Lecture Notes in Mathematics 1042, Springer-Verlag.

I. Hájek, A. Rényi
[1956] Generalization of an inequality of Kolmogorov. *Acta Math. Acad. Sci. Hung.* **6**, pp. 281–283.

P. R. Halmos
[1950] *Measure Theory.* Van Nostrand.

G. H. Hardy, J. E. Littlewood, G. Polya
[1952] *Inequalities.* Second edition, Cambridge University Press.

J. A. Hartigan
[1983] *Bayes Theory.* Springer Series in Statistics, Springer-Verlag.

O. Haupt
[1953] Propriété de mesurabilité de bases de dérivation. *Port. Math.* **13**, pp. 37–54.

C. A. Hayes
[1976] Necessary and sufficient conditions for the derivation of integrals of L_Ψ-functions. *Trans. Amer. Math. Soc.* **223**, pp. 385–394.

C. A. Hayes, C. Y. Pauc
[1970] *Derivation and Martingales.* Ergebnisse der Mathematik und ihrer Grenzgebiete 49, Springer-Verlag.

H. Heinich
[1978a] Martingales asymptotiques pour l'ordre. *Ann. Inst. H. Poincaré, Sec. B* **14**, pp. 315–333.
[1978b] Convergence des sous-martingales positives dans un Banach réticulé. *C. R. Acad. Sci. Paris* **286**, pp. 279–280.

F. Hiai
[1985] Convergence of conditional expectations and strong laws of large numbers for multivalued random variables. *Trans. Amer. Math. Soc.* **291**, pp. 613–627.

T. P. Hill

[1983] Prophet inequalities and order selection in optimal stopping problems. *Proc. Amer. Math. Soc.* **88**, pp. 131–137.

[1986] Prophet inequalities for averages of independent non-negative random variables. *Math. Z.* **191**, pp. 427–436.

T. P. Hill, D. P. Kennedy

[1989] Prophet inequalities for parallel processes. *J. Multivariate Anal.* **31**, pp. 236–243.

T. P. Hill, R. P. Kertz

[1981a] Ratio comparisons of supremum and stop rule expectations. *Z. Wahrscheinlichkeitstheorie und Verw. Gebiete* **56**, pp. 283–285.

[1981b] Additive comparisons of stop rule and supremum expectations of uniformly bounded independent random variables. *Proc. Amer. Math. Soc.* **83**, pp. 582–585.

[1982] Comparisons of stop rule and supremum expectations of i.i.d. random variables. *Ann. Probability* **10**, pp. 336–345.

E. Hopf

[1937] *Ergodentheorie.* Ergebnisse der Mathematik 5, Springer-Verlag.

[1954] The general temporally discrete Markov process. *J. Rat. Mech. Anal.* **3**, pp. 13–45.

R. E. Huff

[1974] Dentablilty and the Radon-Nikodým property. *Duke Math. J.* **41**, pp. 111–114.

G. A. Hunt

[1966] *Martingales et Processus de Markov.* Dunod.

W. Hurewicz

[1944] Ergodic theorem without invariant measure. *Ann. Math.* **45**, pp. 192–206.

A. Ionescu Tulcea, C. Ionescu Tulcea

[1961] On the lifting property. *J. Math. Anal. Appl.* **3**, pp. 537–546.

[1963] Abstract ergodic theorems. *Trans. Amer. Math. Soc.* **107**, pp. 107–124.

[1969] *Topics in the Theory of Lifting.* Ergebnisse der Mathematik und ihrer Grenzgebiete 48, Springer-Verlag.

A. Ionescu Tulcea, M. Moretz

[1969] Ergodic properties of semi-Markovian operators on the Z^1 part. *Z. Wahrscheinlichkeitstheorie und Verw. Gebiete* **13**, pp. 119–122.

W. B. Johnson, G. Schechtman

[1988] Martingale inequalities in rearrangement invariant function spaces. *Israel J. Math.* **64**, pp. 267–275.

J. L. Kelley

[1950] Convergence in topology. *Duke Math. J.* **17**, pp. 277–283.

[1955] *General Topology.* D. Van Nostrand Company.

D. Kennedy

[1985] Optimal stopping of independent random variables and maximizing prophets. *Ann. Probabab.* **10**, pp. 566–571.

H. Kesten

[1982] *Percolation Theory for Mathematicians.* Progress in Probability and Statistics 2, Birkhäuser.

J. F. C. Kingman

[1968] The ergodic theory of subadditive stochastic processes. *J. Royal Stat. Soc. B* **30**, pp. 499–510.

[1973] Subadditive ergodic theory. *Ann. Probab.* **1**, pp. 883–909.

[1975] Subadditive processes. In: *Ecole d'Eté de Probabilités de Saint-Flour V*, Lecture Notes in Mathematics 539, Springer-Verlag; pp. 167–223.

H. Korezlioglu, G. Mazziotto, J. Szpirglas

[1981] *Processus aléatoires a deux indices.* Springer-Verlag.

A. Korzeniowski

[1978] Martingales in Banach spaces for which the convergence with probability one, in probability and in law coincide. *Colloq. Math.* **39**, pp. 153–159.

M. A. Krasnosel'skii, Ya. B. Rutickii

[1961] *Convex Functions and Orlicz Spaces.* Gordon and Breach Science Publishers.

U. Krengel

[1985] *Ergodic Theorems.* Walter de Gruyter.

U. Krengel, L. Sucheston

[1977] Semi-amarts and finite values. *Bull. Amer. Math. Soc.* **83**, pp. 745–747.

[1978] On semiamarts, amarts and processes with finite value. In: *Probability on Banach Spaces*, J. Kuelbs, editor, Advances in Probability and Related Topics 4, Marcel Dekker; pp. 197–266.

[1980] Temps d'arrêt et tactiques pour des processus indexés par un ensemble ordonné. *C. R. Acad. Sci. Paris* **290**, pp. A192–196.

[1981] Stopping rules and tactics for processes indexed by a directed set. *J. Multivariate Anal.* **11**, pp. 199–229.

[1987] Prophet compared to gambler: An inequality for transforms of processes. *Ann. Probab.* **15**, pp. 1593–1599.

K. Krickeberg

[1956] Convergence of martingales with a directed index set. *Trans. Amer. Math. Soc.* **83**, pp. 313–337.

[1957] Stochastische Konvergenz von Semimartingalen. *Math. Z.* **66**, pp. 470–486.

[1959] Notwendige Konvergenzbedingungen bei Martingalen und verwandted Processen. In: *Transactions of the Second Prague Conference on Information Theory*, Czech. Acad. Sci.; pp. 279–305.

K. Krickeberg, C. Pauc

[1963] Martingales et dérivation. *Bull. Soc. Math. France* **91**, pp. 455–554.

K. Kunen, H. Rosenthal

[1982] Martingale proofs of some geometrical results in Banach space theory. *Pacific J. Math.* **100**, pp. 153–175.

C. W. Lamb

[1973] A ratio limit theorem for approximate martingales. *Canad. J. Math.* **25**, pp. 772–779.

P. Lévy

[1937] *Théorie de l'addition des variables aléatoires.* Gauthier-Villars, Paris.

D. R. Lewis, C. Stegall

[1973] Banach spaces whose duals are isomorphic to $l_1(\Gamma)$. *J. Functional Anal.* **12**, pp. 177–187.

M. Lin and R. Wittmann

[1991] Pointwise ergodic theorems for certain order preserving mappings in L_1. In: *Almost Everywhere Convergence II*, A. Bellow and R. Jones, editors, Academic Press; pp. 253–273.

W. Linde

[1976] An operator ideal in connection with the Radon-Nikodým property of Banach spaces. *Math. Nachr.* **71**, pp. 65–73.

J. Lindenstrauss, L. Tzafriri

[1977] *Classical Banach Spaces I.* Ergebnisse der Mathematik und ihrer Grenzgebiete 92, Springer-Verlag.

[1979] *Classical Banach Spaces II.* Ergebnisse der Mathematik und ihrer Grenzgebiete 97, Springer-Verlag.

L. H. Loomis

[1975] Dilations and extremal measures. *Advances in Math.* **17**, pp. 1–13.

Dinh Quang Lu'u

[1981] On convegence of vector-valued amarts of finite order. *Math. Nachr.* **113**, pp. 39–45.

[1985a] Amarts of finite order and Pettis Cauchy sequences of Bochner integrable functions in locally convex spaces. *Ann. Sci. Univ. Clermont-Ferrand II. Probab. Appl.* No. 3, pp. 91–106.

[1985b] Stability and convergence of amarts in Fréchet spaces. *Acta Math. Hungarica* **45**, pp. 99–108.

W. A. J. Luxemburg

[1955] *Banach function spaces.* Thesis, Delft University.

D. Maharam

[1958] On a theorem of von Neumann. *Proc. Amer. Math. Soc.* **9**, pp. 987–994.

V. Marraffa

[1988] On almost sure convergence of amarts and martingales without the Radon-Nikodým property. *J. Theoretical Probab.* **1**, pp. 255–261.

H. B. Maynard

[1973] A geometrical characterization of Banach spaces having the Radon-Nikodým property. *Trans. Amer. Math. Soc.* **185**, pp. 493–500.

S. M. McGrath

[1980] Some ergodic theorems for commuting L_1 contractions. *Studia Math.* **70**, pp. 153–160.

J. F. Mertens

[1972] Théorie des processus stochastiques généraux; applications aux surmartingales. *Z. Wahrscheinlichkeitstheorie und Verw. Gebiete* **22**, pp. 45–68.

P.-A. Meyer

[1965] Théorie ergodique et potentiels. *Ann. Inst. Fourier (Grenoble)* **15**, pp. 89–96.

[1966] *Probability and Potentials.* Blaisdell Publishing Company.

[1971] Le retournement du temps, d'après Chung et Walsh. In: *Séminaire de Probabilités V*, Lecture Notes in Mathematics 191, Springer-Verlag; pp. 213–236.

[1981] Théorie élémentaire des processus à deux indices. In: *Processus Aleatoires a Deux Indices*, Lecture Notes in Mathematics 863, Springer-Verlag; pp. 1–39.

A. Millet

[1978] Sur la caractérisation des conditions de Vitali par la convergence essentielle des martingales. *C. R. Acad. Sci. Paris* **287**, pp. A887–890.

[1981] Convergence and regularity of strong submartingales. In: *Processus Aleatoires a Deux Indices*, Lecture Notes in Mathematics 863, Springer-Verlag; pp. 50–58.

A. Millet, L. Sucheston

[1978] Classes d'amarts filtrants et conditions de Vitali. *C. R. Acad. Sci. Paris* **286**, pp. A835–837.

[1979a] Characterizations of Vitali conditions with overlap in terms of convergence of classes of amarts. *Canad. J. Math.* **31**, pp. 1033–1046.

[1979b] La convergence essentielle des martingales bornées dans L_1 n'implique pas

la condition de Vitali V. *C. R. Acad. Sci. Paris* **288**, pp. A595–598.

[1980a] Convergence et régularité des martingales à indices multiples. *C. R. Acad. Sci. Paris* **291**, pp. A147–150.

[1980b] Convergence of classes of amarts indexed by directed sets. *Canad. J. Math.* **32**, pp. 86–125.

[1980c] On covering conditions and convergence. In: *Measure Theory, Oberwolfach 1979*, Lecture Notes in Mathematics 794, Springer-Verlag; pp. 431–454.

[1980d] A characterization of Vitali conditions in terms of maximal inequalities. *Ann. Probabability* **8**, pp. 339–349.

[1980e] On convergence of L_1-bounded martingales indexed by directed sets. *Probab. Math. Statist.* **1**, pp. 151–169.

[1981a] On regularity of multiparameter amarts and martingales. *Z. Wahrscheinlichkeitstheorie und Verw. Gebiete* **56**, pp. 21–45.

[1981b] Demi-convergence des processus à deux indices. *C. R. Acad. Sci. Paris* **293**, pp. 435–438.

[1983] Demiconvergence of processes indexed by two indices. *Ann. Inst. H. Poincaré*, Sec. B **19**, pp. 175–187.

[1989] On fixed points and multiparameter ergodic theorems in Banach lattices. *Canad. Math. J.* **40**, pp. 429–458.

J. Mogyoródi

[1978] Remark on a theorem of J. Neveu. *Ann. Univ. Sci. Budapest. Eőtvős Sect. Math.* **21**, pp. 77–81.

E. H. Moore

[1915] Definition of limit in general integral analysis. *Proc. Nat. Acad. Sci. U. S. A.* **1**, pp. 628.

E. H. Moore, H. L. Smith

[1922] A general theory of limits. *Amer. J. Math.* **44**, pp. 102–121.

A. P. Morse, W. Transue

[1950] Functionals F bilinear over the product $A \times B$ of two pseudo-normed vector spaces II. *Ann. of Math.* **51**, pp. 576–614.

E. Mourier

[1953] Éléments aléatoires dans un espace de Banach. *Ann. Inst. H. Poincaré*, Sec. B **13**, pp. 161–244.

A. G. Mucci

[1973] Limits for martingale-like sequences. *Pacific J. Math.* **48**, pp. 197–202.

[1976] Another martingale convergence theorem. *Pacific J. Math.* **64**, pp. 539–541.

I. Namioka, E. Asplund

[1967] A geometric proof of Ryll-Nardzewski's fixed point theorem. *Bull. Amer. Math. Soc.* **73**, pp. 443–445.

I. Namioka, R. R. Phelps

[1975] Banach spaces which are Asplund spaces. *Duke Math. J.* **42**, pp. 735–750.

P. A. Nelson

[1970] A class of orthgonal series related to martingales. *Ann. Math. Statist.* **41**, pp. 1684–1694.

J. von Neumann

[1931] Algebraische Repräsentanten der Funktionen bis auf eine Menge von Masse Null. *J. Crelle* **165**, pp. 109–115.

[1949] On rings of operators, reduction theory. *Ann. of Math.* **30**, pp. 401–485.

J. Neveu

[1961] Sur le théorème ergodique ponctuel. *C. R. Acad. Sci. Paris* **252**, pp. 1554–1556.

[1964] Potentiels markoviens discret. *Ann. Univ. Clermont* **24**, pp. 37–89.

[1965a] *Mathematical Foundations of the Calculus of Probability.* Holden-Day.

[1965b] Relations entre la théorie des martingales et la théorie ergodique. *Ann. Inst. Fourier (Grenoble)* **15**, pp. 31–42.

[1972] Convergence presque sur des martingales multivoques. *Ann. Inst. H. Poincaré*, Sec. B **8**, pp. 1–7.

[1975] *Discrete Parameter Martingales.* North-Holland Mathematical Library 10, North-Holland.

S. Orey

[1967] F-processes. In: *Proceedings of the Fifth Berkeley Symposium on Mathematical Statistics and Probability*, Volume II, Part 1, L. M. LeCam & J. Neyman, editors, University of California Press; pp. 301–303.

W. Orlicz

[1932] Über eine gewisse Klasse von Räumen vom typus B. *Bull. Acad. Sci. Polonaise A*, pp. 207–220.

[1936] Über Räume (L^M). *Bull. Acad. Sci. Polonaise A*, pp. 93–107.

D. S. Ornstein

[1960] On invariant measures. *Bull. Amer. Math. Soc.* **66**, pp. 297–300.

R. E. A. C. Paley

[1932] A remarkable series of orthogonal functions. *Proc. London Math. Soc.* **34**, pp. 241–264.

C. Y. Pauc

[1953] Ableitungsbasen, Prätopologie und starker Vitalischer Satz. *J. Reine Angew. Math.* **191**, pp. 69–91.

V. C. Pestien

[1982] An extended Fatou equation and continuous-time gambling. *Advances in Appl. Probability* **14**, pp. 309–323.

R. R. Phelps
[1966] *Lectures on Choquet's Theorem*. Van Nostrand Mathematical Studies 7, D. Van Nostrand Company.
[1974] Dentability and extreme points in Banach spaces. *J. Functional Anal.* **17**, pp. 78–90.

G. Pisier
[1986] *Factorization of Linear Operators and Geometry of Banach Spaces*. CBMS Regional Conference Series in Mathematics 60, American Mathematical Society.

R. Pol
[1980] On a question of H. H. Corson and some related problems. *Fundamenta Math.* **109**, pp. 143–154.

R. de Possel
[1936] Dérivation abstraite des fonctions d'ensemble. *J. Math. Pures et Appl.* **15**, pp. 391–409.

K. M. Rao
[1969] Quasi-martingales. *Math. Scand.* **24**, pp. 79–92.

O. J. Reinov
[1975] Operators of type RN in Banach spaces (Russian). *Dokl. Akad. Nauk SSSR* **220**, pp. 528–531; English translation: *Soviet Math. Dokl.* **16** (1975) 119–123.

P. Révész
[1968] *The Laws of Large Numbers*. Academic Press.

D. Revuz
[1974] *Markov Chains*. North-Holland.

M. A. Rieffel
[1968] The Radon-Nikodým theorem for the Bochner integral. *Trans. Amer. Math. Soc.* **131**, pp. 466–487.

H. Robbins, D. Siegmund
[1971] A convergence theorem for non negative almost supermartingales and some applications. In: *Optimizing Methods in Statistics*, Academic Press; pp. 233–257.

U. Rønnow
[1967] On integral representations of vector-valued measures. *Math. Scand.* **21**, pp. 45–53.

G. C. Rota
[1962] An "alternierende Verfahren" for general positive operators. *Bull. Amer. Math. Soc.* **68**, pp. 95–102.

H. L. Royden
[1968] *Real Analysis*. Second edition, Macmillan.

M. Rubinstein
[1974] Properties of uniform integrability and convergence for families of random variables. *Rend. Accad. Naz. dei Lincei* **57**, pp. 95–99.

W. Rudin
[1973] *Functional Analysis*. McGraw-Hill.

C. Ryll-Nardzewski
[1967] On fixed points of semigroups of endomorphisms of linear spaces. In: *Proceedings of the Fifth Berkeley Symposium on Mathematical Statistics and Probability*, Volume II, Part 1, L. M. LeCam & J. Neyman, editors, University of California Press; pp. 55–61.

E. Samuel-Cahn
[1984] Comparison of threshold stopping rules and maximum for independent nonnegative random variables. *Ann. Probab.* **12**, pp. 1213–1216.

R. Sato
[1986] Individual ergodic theorem for superadditive processes. *Acta Math. Hungarica* **47**, pp. 153–155.

F. S. Scalora
[1961] Abstract martingale convergence theorems. *Pacific J. Math.* **11**, pp. 347–374.

M. Slaby
[1982] Convergence of submartingales and amarts in Banach lattices. *Bull. Acad. Polon. Sci. Sér. Sci. Math.* **30**, pp. 291–299.

R. Solovay
[1970] A model of set theory in which every set of reals is Lebesgue measurable. *Ann. of Math.* **92**, pp. 1–56.

C. Stegall
[1975] The Radon-Nikodým property in conjugate Banach spaces. *Trans. Amer. Math. Soc.* **206**, pp. 213–223.
[1981] The Radon-Nikodým property in conjugate Banach spaces, II. *Trans. Amer. Math. Soc.* **264**, pp. 507–519.

E. Stein
[1961] On the maximal ergodic theorem. *Proc. Nat. Acad. Sci. U. S. A.* **47**, pp. 1894–1897.

R. Subramanian
[1973] On a generalization of martingales due to Blake. *Pacific J. Math.* **48**, pp. 275–278.

L. Sucheston
[1964] On the existence of finite invariant measures. *Math. Z.* **34**, pp. 327–336.
[1967] On the ergodic theorem for positive operators. *Z. Wahrscheinlichkeitstheorie und Verw. Gebiete* **8**, pp. 1–11, 353–356.
[1983] On one-parameter proofs of almost sure convergence of multiparameter processes. *Z. Wahrscheinlichkeitstheorie und Verw. Gebiete* **63**, pp. 43–49.

L. Sucheston, L. Szabó

[1991] A principle for almost everywhere convergence of multiparameter processes. In: *Almost Everywhere Convergence II*, A. Bellow and R. Jones, editors, Academic Press; pp. 253–273.

L. Sucheston, Z. Yan

[In press] On the prophet problem for transforms.

W. D. Sudderth

[1971] A "Fatou's equation" for randomly stopped variables. *Ann. Math. Statist.* **42**, pp. 2143–2146.

L. I. Szabó

[1991] The converse of the dominated ergodic theorem in Hurewicz setting. *Canad. Math. Bull.* **34**, pp. 405–411.

J. Szulga

[1978a] On the submartingale characterization of Banach lattices isomorphic to l^1. *Bull. Acad. Polon. Sci. Sér. Sci.* **26**, pp. 65–68.

[1978b] Boundedness and convergence of Banach lattice valued submartingales. In: *Probability Theory on Vector Spaces*, Lecture Notes in Mathematics 656, Springer-Verlag; pp. 251–256.

[1979] Regularity of Banach lattice valued martingales. *Colloq. Math.* **41**, pp. 303–312.

J. Szulga, W. A. Woyczynski

[1976] Convergence of submartingales in Banach lattices. *Ann. Probability* **4**, pp. 464–469.

M. Talagrand

[1984] *Pettis Integral and Measure Theory.* Memoirs of the Amer. Math. Soc. 307, American Mathematical Society.

[1985] Some structure results for martingales in the limit and pramarts. *Ann. Probability* **13**, pp. 1192–1203.

[1986] Derivation, L^Ψ-bounded martingales and covering conditions. *Trans. Amer. Math. Soc.* **293**, pp. 257–289.

R. L. Taylor, P. Z. Daffer, R. F. Patterson

[1985] *Limit Theorems for Sums of Exchangeable Random Variables.* Rowman & Allanheld Publishers.

J. J. Uhl, Jr.

[1977] Pettis mean convergence of vector-valued asymptotic martingales. *Z. Wahrscheinlichkeitstheorie und Verw. Gebiete* **37**, pp. 291–295.

J. B. Walsh

[1979] Convergence and regularity of multiparameter strong martingales. *Z. Wahrscheinlichkeitstheorie und Verw. Gebiete* **46**, pp. 177–192.

N. Wiener

[1939] The ergodic theorem. *Duke Math. J.* **5**, pp. 1–18.

R. Wittmann

[In press] On the prophet inequality for subadditive processes. *Stochatic Analysis and Applications.*

V. Yankov

[1941] On the uniformization of A-sets (Russian). *Dokl. Akad. Nauk SSSR* **30**, pp. 491–592.

K. Yoshida, E. Hewitt

[1953] Finitely aditive measures. *Trans. Amer. Math. Soc.* **72**, pp. 44–66.

T. Yoshimoto

[1982] Pointwise ergodic theorem and function spaces M_p^α. *Studia Math.* **72**, pp. 253–271.

W. H. Young

[1912] On classes of summable functions and their Fourier series. *Proc. Royal Soc. London* **87**, pp. 225–229.

A. C. Zaanen

[1949] Note on a certain class of Banach spaces. *Proc. Nederl. Acad. Sci. A* **52**, pp. 488–498.

[1983] *Riesz Spaces II.* North-Holland Mathematical Library 30, North-Holland.

A. Zygmund

[1934] On the differentiability of multiple integrals. *Fundamenta Math.* **23**, pp. 143–149.

[1951] An individual ergodic theorem for noncommutative transformations. *Acta Sci. Math. (Szeged)* **14**, pp. 103–110.

[1967] A note on differentiability of integrals. *Colloquium Mathematicum* **16**, pp. 199–204.

Index of names

Index of terms